D1715015

Progress in Mathematics
Volume 161

Mathematical Essays in honor of Gian-Carlo Rota

Bruce E. Sagan
Richard P. Stanley
Editors

Birkhäuser
Boston • Basel • Berlin

Bruce E. Sagan
Department of Mathematics
Michigan State University
East Lansing, MI 48824

Richard P. Stanley
Department of Mathematics
MIT
Cambridge, MA 02139

Library of Congress Cataloging-in-Publication Data

Mathematical essays in honor of Gian-Carlo Rota / Bruce E. Sagan,
Richard P. Stanley, editors.
p. cm. -- (Progress in mathematics ; v. 161)
Includes bibliographical references.
ISBN 0-8176-3872-5 (hardcover : alk. paper).
1. Mathematics I. Rota, Gian-Carlo, 1932- II. Sagan, Bruce
Eli. III. Stanley, Richard P., 1944- . IV. Series: Progress in
mathematics (Boston, Mass.) ; vol. 161.
QA7.M34445 1998 98-15395
510--dc21 CIP

Printed on acid-free paper

Birkhäuser 

ISBN 0-8176-3872-5
ISBN 3-7643-3872-5

Reformatted from authors' disks by Texniques, Boston, MA
Printed and bound by Quinn-Woodbine, Woodbine, NJ.

9 8 7 6 5 4 3 2 1

Contents

Preface

In April of 1996 an array of mathematicians converged on Cambridge, Massachusetts, for the Rotafest and Umbral Calculus Workshop, two conferences celebrating Gian-Carlo Rota's 64th birthday. It seemed appropriate when fêting one of the world's great combinatorialists to have the anniversary be a power of 2 rather than the more mundane 65. The over seventy-five participants included Rota's doctoral students, coauthors, and other colleagues from more than a dozen countries. As a further testament to the breadth and depth of his influence, the lectures ranged over a wide variety of topics from invariant theory to algebraic topology.

This volume is a collection of articles written in Rota's honor. Some of them were presented at the Rotafest and Umbral Workshop while others were written especially for this Festschrift. We will say a little about each paper and point out how they are connected with the mathematical contributions of Rota himself.

One of Rota's earliest loves in combinatorics was the theory of partially ordered sets (posets). It is no accident that his seminal series of papers "On the Foundations of Combinatorial Theory" begins with an article on Möbius inversion [*Z. Wahrscheinlichkeitstheorie* **2** (1964), 340–368], which is most naturally done in the poset setting. In later work Rota and Metropolis [*Siam J. Appl. Math.* **35** (1978) 689–694] characterized the lattice of faces of the n-cube, calling the structure a *cubical algebra*. In the present volume, Oliveira and Bailey show that one can give a list of universal axioms for these algebras inside the variety of implication algebras. In a different direction, Crapo investigates the representation of lattices in general by unities.

Crapo and Rota wrote the next Foundations installment, this one on combinatorial geometries [*Studies in Appl. Math.* **49** (1970), 109–133]. These objects, also known as matroids, provide a setting which simultaneously generalizes ideas from linear algebra, matching theory, and the theory of graphs. Kung and Bonin in their contribution to this Festschrift provide a bound for the number of points in a combinatorial geometry with no minor isomorphic to the 8-point line. Whiteley is concerned with matroid applications to static rigidity and splines. Diaconis, Eisenbud, and Sturmfels develop new connections between graph theory and commutative algebra by showing how primary decomposition of an ideal describes the components of a graph arising in problems from combinatorics, statistics, and operations research.

The idea of bringing techniques from algebra and other branches of mathematics to bear on combinatorial problems is one that runs throughout Rota's work. In particular, the concept of generating function has turned out to be a fundamental tool which was the subject of Foundations VI, coauthored with Doubilet and Stanley [in "Proceedings of the Sixth Berkeley Symposium on Mathematical Statistics and Probability," Vol. II: Probability Theory, University of California Press, Berkeley, CA, 1971, 267-318]. Stanley's paper in this collection deals with the generating function for distance to a chamber from a fixed base chamber in the extended Shi hyperplane arrangement, generalizing the well-known enumeration of trees by number of inversions. Also in a geometric vein, Billera, Ehrenborg, and Readdy prove that the flag f-vectors of zonotopes satisfy no additional affine relations over and above those for all polytopes. It follows that the **cd**-index, essentially the generating function for the flag f-vector, is the most efficient way of encoding the affine information. Freeman presents a method for determining when a sequence of polynomials is orthogonal from its generating function. There will be more to say about orthogonal polynomials in connection with the umbral calculus.

Some of Rota's most substantial contributions have been to invariant theory. In particular, the ninth Foundations article, written with Doubilet and Stein [*Studies in Appl. Math.* **53** (1974), 185–216], gives a characteristic-free approach to this topic. Its crowning achievement is a new proof of the First Fundamental Theorem of Invariant Theory using a straightening algorithm in the letter-place algebra. In this anthology, Buchsbaum demonstrates how letter-place methods can be used to construct homotopies for resolutions of certain Weyl modules, while Chan classifies the invariants and covariants of skew symmetric tensors of step three and dimension six in a characteristic-free way. One of the important applications of invariants is to representation theory, especially of the symmetric group S_n, and to the theory of symmetric functions. Ram's paper contains an elementary proof of Roichman's formula for the irreducible characters of the type A Iwahori-Hecke algebra. Garsia and Remmel consider the Kostka-Foulkes polynomials $K_{\lambda\mu}(q,t)$, which are the change of basis coefficients for two symmetric function bases, and prove Macdonald's conjecture that they have nonnegative integer coefficients for the special case when λ is an augmented hook and μ is arbitrary. Partitions not only index the irreducible representations of S_n, but also have many interesting properties of their own. Andrews uses MacMahon's partition analysis technique to prove a theorem of Bousquet-Mélou and Eriksson which generalizes Euler's classic result that the number of partitions of n into odd parts equals the number into distinct parts. Krattenthaler evaluates the Andrews-Burge determinant for a certain symmetry class of plane partitions using a new method involving linearly independent linear combinations of the rows or columns of the determinant.

Another area in which Rota's work has had widespread influence is the umbral calculus. In Foundations III [with Mullin in "Graph Theory and its

Applications," Academic Press, New York, NY, 1970, 168–213] and VIII [with Kahaner and Odlyzko, *J. Math. Anal. Appl.* **42** (1973), 685–760], Rota put substitution techniques from the 19th century on a rigorous footing via linear operators. Previous extensions of the theory to many variables had not been done in a basis-free way. But in their contribution to this Festschrift, Loeb, Di Bucchianico, and Rota show how one can use Hilbert space to accomplish this end. Taylor is interested in applications of the classical calculus to difference equations using Bernoulli umbrae, while D'Antona seeks to provide algebraic motivation for certain symbolic substitutions using his notion of a target ring.

It is implicit in Rota's article with Roman [*Adv. in Math.* **27(2)** (1978), 95–188] and explicit in his paper with Joni [*Studies in Appl. Math.* **61** (1979), 93–138] that the umbral calculus is connected with a Hopf algebra structure on the algebra of polynomials and the corresponding algebra structure on the dual. Ray's article herein explores what happens to umbral calculations done in the category of coassociative coalgebras over a commutative ring rather than a field, while Mendez investigates the relationship with the Hopf algebra of symmetric functions. One of the main motivations for this calculus is to generalize and unify results about orthogonal polynomials and hypergeometric series. Four of the articles in this volume apply umbral methods in this setting: Ismail and Stanton obtain orthogonal polynomials as moments for other orthogonal polynomials, Di Bucchianico and Loeb study the relationship between natural exponential families of probability measures and Sheffer polynomials, Wimp addresses the problem of finding closed forms for hypergeometric series, and Chen derives hypergeometric identities by parameter augmentation.

Of course, one-sentence summaries can not do justice to the papers in this Festschrift, just as this preface only modestly surveys a small portion of the large body of Rota's important work. We encourage the reader to delve more deeply!

Bruce E. Sagan — East Lansing

Richard P. Stanley — Cambridge

August, 1997

Gian-Carlo Rota, M.I.T., Cambridge, MA

ROTAFEST PROGRAM

Tuesday, April 16

7:00–9:30 p.m.	reception at Charles Hotel

Wednesday, April 17

9:00–12:00	morning session (André Joyal, chair)
9:00–9:30	Adriano Garsia, The $n!$-conjecture and the q, t Kostka polynomials
9:45–10:15	Stephen Grossberg, Nonlinear dynamics of neural networks
10:45–11:15	Mark Haiman, $N!$ is all you need
11:30–12:00	Lawrence Harper, The peaks of partition numbers
2:00–5:00	afternoon session (Erwin Lutwak, chair)
2:00–2:30	Jay Goldman, Combinatorics and knot theory
2:45–3:15	Daniel Klain, Invariant valuations on convex bodies
3:45–4:15	Joseph Kung, Line sizes and the number of points in a matroid
4:30–5:00	Andrew Odlyzko, Increasing subsequences in random permutations

Thursday, April 18

9:00–12:00	morning session (Steve Tanny, chair)
9:00–9:30	Willaim Schmitt
9:45–10:15	Bruce Sagan, Beyond semimodular lattices
10:45–11:15	Pat O'Neil: CANCELLED
11:30–12:00	David Sharp, Raleigh-Taylor instability, chaotic mixing layer and stochastic PDE's
2:00–5:00	afternoon session (Peter Doubilet, chair)
2:00–2:30	Richard Stanley, Hyperplane arrangements, inversions, and trees
2:45–3:15	Joel Stein, The future of invariant theory
3:45–4:15	Bernd Sturmfels, Lattice walks and primary decomposition

4:30–5:00	Neil White, Coxeter matroids
6:00–10:00	banquet, Hyatt Regency Hotel

Friday, April 19

9:00–12:00	morning session (Joseph Oliviera, chair)
9:00–9:30	Walter Whiteley, Two matroids from geometric homology: an analogy with digressions
9:45–10:15	Kenneth Baclawski, Politically correct ordered sets: Socially responsible combinatorics
10:45–11:15	Wendy Chan, Classification of trivectors in 6-D space
11:30–12:00	David Buchsbaum, Letter-place methods and homotopy
	afternoon free
6:30	dinner at Salamander
around 6:30	alternate dinner at Royal East

Saturday, April 20

9:00–12:00	morning session (Curtis Greene, chair)
9:00–9:30	Henry Crapo, Unities and negation: On the representation of lattices
9:45–10:15	**Gian-Carlo Rota**, Ten lessons I should have been taught
10:45–11:15	Peter Duren, Recent progress on Bergman spaces
11:30–12:00	Richard Ehrenborg, Coproducts and the *cd*-index
2:00–5:00	afternoon session (Michael Hawrylycz, chair)
2:00–2:30	Steven Fisk, Q-analogs of simplicial complexes
2:45–3:15	Jack Freeman

Mathematical Essays in honor of Gian-Carlo Rota

MacMahon's Partition Analysis: I. The Lecture Hall Partition Theorem

*George E. Andrews**

Dedicated to my friend, Gian-Carlo Rota

Abstract

In this paper, we analyze the beautiful theorem of M. Bousquet-Mélou and K. Eriksson, the Lecture Hall Partition Theorem, via MacMahon's Partition Analysis. Their theorem asserts that the number of partitions of n of the form $b_j + b_{j-1} + \cdots + b_1$ wherein

$$\frac{b_j}{j} \geqq \frac{b_{j-1}}{j-1} \geqq \cdots \geqq \frac{b_1}{1} \geqq 0$$

equals the number of partitions of n into odd parts each $\leqq 2j-1$. As they have noted the theorem reduces to Euler's classical partition theorem when $j \to \infty$.

The central object of the paper is to demonstrate the power of the method of Partition Analysis, developed by P. A. MacMahon over 80 years ago. In the early 1970's, Richard Stanley successfully utilized Partition Analysis in his monumental treatment of magic labelings of graphs. Apart from this one shining moment, Partition Analysis has lain dormant. In this paper we hope to point to its further utility by proving the deep theorem of Bousquet-Mélou and Eriksson.

*Partially supported by National Science Foundation Grant DMS-8702695-04 and by the Australian Research Council. Concerning the latter, the gracious support and interest of Omar Foda made possible my visit to Melbourne University and my introduction to Mireille Bousquet-Mélou and Lecture Hall Partitions.

1. Introduction

Recently M. Bousquet-Mélou and K. Eriksson proved the following result [2].

Theorem 1. (The Lecture Hall Partition Theorem). *The number of partitions of n of the form $b_j + b_{j-1} + \cdots + b_1$ wherein*

$$\frac{b_j}{j} \geqq \frac{b_{j-1}}{j-1} \geqq \cdots \geqq \frac{b_1}{1} \geqq 0$$

equals the number of partitions of n into odd parts each $\leqq 2j-1$.

For example, if $n = 13$ and $j = 3$, the ten Lecture Hall partitions are $13+0+0$, $12+1+0$, $11+2+0$, $10+3+0$, $10+2+1$, $9+4+0$, $9+3+1$, $9+5+0$, $8+4+1$, $7+4+2$ while the ten relevant partition with odd parts each $\leqq 5$ are $5+5+3$, $5+5+1+1+1$, $5+3+3+1+1$, $5+3+1+1+1+1+1$, $5+1+\cdots+1$, $3+3+3+3+1$, $3+3+3+1+1+1+1$, $3+3+1+\cdots+1$, $3+1+\cdots+1$, $1+1\cdots+1$.

Bousquet-Mélou and Eriksson actually refine the Lecture Hall Partition Theorem by proving [2].

Theorem 2. *The number of partitions of n of the form $b_j + b_{j-1} + \cdots + b_1$ wherein*

$$\frac{b_j}{j} \geqq \frac{b_{j-1}}{j-1} \geqq \cdots \geqq \frac{b_1}{1} \geqq 0$$

and

$$b_j - b_{j-1} + b_{j-2} - \cdots + (-1)^{j-1} b_1 = m$$

equals the number of partitions of n into exactly m odd parts each $\leqq 2j-1$.

Thus in the above example if we choose $m = 5$ in addition to $n = 13$, $j = 3$, we get three Lecture Hall partitions $9+4+0$, $8+4+1$, $7+4+2$, and three relevant partitions with odd parts $5+5+1+1+1$, $5+3+3+1+1$, and $3+3+3+3+1$.

Of course, we can let $j \to \infty$ to obtain the following generalization of Euler's classic result [1; p. 5] as did Bousquet-Mélou and Eriksson.

Corollary. *The number of partitions of n into m odd parts equals the number of partitions of n into distinct parts written in decreasing size, $b_1 + b_2 + \cdots + b_k$, where $(b_1 + b_3 + b_5 + \cdots) - (b_2 + b_4 + b_6 \cdots) = m$.*

We will have more to say about the work of Bousuet-Mélou and Eriksson in Section 9. For now we observe that they have extended their discoveries in [2] still further and that they have found two methods of proof, one a bijective proof and one that relies on Weyl group theory.

Our method of proof is the method of Partition Analysis first suggested by Cayley [3] but primarily developed by P. A. MacMahon in [4], [5], [6], surveyed by him again in [8; Vol. 2, Section VIII, 91–170] and recapped in his Collected Papers [9; Ch. 10, pp. 1119–1314].

A careful examination of MacMahon's work shows that MacMahon developed a magnificent machine which according to MacMahon's own admission [7; Vol. 2, p. 187] failed to prove the major theorems on plane partitions. Very much to his credit, MacMahon developed alternative tools necessary to prove his plane partition discoveries. However that admission of the failure of Partition Analysis to prove general results on plane partitions led to near total neglect of the method. The single major exception to this comment is the beautiful solution to the Anand-Dumir-Gupta Conjecture by Richard Stanley [10]. However, since 1973 there has been (to my knowledge) no further work on Partition Analysis.

In Section 2, we provide a review of Partition Analysis. We try to make clear in Section 3, Partition Analysis is an obvious method to apply to Lecture Hall Partitions. Section 4 is devoted to the direct application of Partition Analysis to the generation of Lecture Hall Partitions. In Section 5, we factor the numerator of the relevant generating function. In Section 6 we consider an example of the arguments in Section 5. In Section 7, we prove the theorems of Bousquet-Mélou and Eriksson. We conclude with a look at the future for MacMahon's Partition Analysis.

It is appropriate to point out that my current interest in MacMahon's Partition Analysis is due entirely to the influence of Gian-Carlo Rota. In 1971, he suggested that I edit MacMahon's Collected Papers for MIT Press, a project that took many years. The first volume [8] was published in 1978 and the second [9] in 1986. I had decided, since MacMahon had died in 1929, to try to connect his work with the research of the intervening half century. Some topics (Partition Analysis being a perfect example) appeared to me to be more important than subsequent work suggested; so I tried to tag them in my memory. When Bousquet-Mélou told me of the Lecture Hall Partition Theorem, a bell went off in my head. Here at long last was a really hard theorem that was obviously tailor-made for Partition Analysis.

Rota's great respect for MacMahon is abundantly revealed in the two marvelous introductions he wrote for the Collected Papers [8; pp. xiii–xiv], [9; p. xi]. I am indebted to Gian-Carlo for many many things among them his encouragement of a project which 25 years later led to this paper.

2. A review of MacMahon's partition analysis

Since MacMahon's method has not been used for some time, we provide a general review here. Perhaps we best begin with his own words [7; vol. 2 pp. 91–92]

"Consider i numbers in descending order of magnitude

$$\alpha_1, \alpha_2, \alpha_3, \ldots, \alpha_i.$$

The order is defined by the Diophantine relations

$$\begin{gathered} \alpha_1 \geqslant \alpha_2, \\ \alpha_2 \geqslant \alpha_3, \\ \vdots \\ \alpha_{i-1} \geqslant \alpha_i, \end{gathered}$$

and subject to them we consider the sum

$$\Sigma x^{\alpha_1+\alpha_2+\alpha_3+\cdots+\alpha_i}.$$

Now observe that the algebraic fraction

$$\frac{1}{(1-\lambda_1 x)\left(1-\frac{\lambda_2}{\lambda_1}x\right)\left(1-\frac{\lambda_3}{\lambda_2}x\right)\cdots\left(1-\frac{\lambda_i}{\lambda_{i-1}}x\right)},$$

when expanded in ascending powers of x, has the general term

$$(\lambda_1)^{\alpha_1}\left(\frac{\lambda_2}{\lambda_1}\right)^{\alpha_2}\left(\frac{\lambda_3}{\lambda_2}\right)^{\alpha_3}\cdots\left(\frac{\lambda_i}{\lambda_{i-1}}\right)^{\alpha_i} x^{\alpha_1+\alpha_2+\alpha_3+\cdots+\alpha_i},$$

or

$$\lambda_1^{\alpha_1-\alpha_2}\lambda_2^{\alpha_2-\alpha_3}\cdots\lambda_{i-1}^{\alpha_{i-1}-\alpha_i}\lambda_i^{\alpha_i}x^{\alpha_1+\alpha_2+\alpha_3+\cdots+\alpha_i},$$

and that if the Diophantine relations are to be satisfied this must be free from negative powers of $\lambda_1, \lambda_2, \lambda_3, \ldots$.

It is thus evident that we have only to expand the fraction

$$\frac{1}{(1-\lambda_1 x)\left(1-\frac{\lambda_2}{\lambda_1}x\right)\left(1-\frac{\lambda_3}{\lambda_2}x\right)\cdots\left(1-\frac{\lambda_i}{\lambda_{i-1}}x\right)},$$

and reject all the terms involving negative powers of $\lambda_1, \lambda_2, \lambda_3, \ldots,$ and subsequently put

$$\lambda_1 = \lambda_2 = \lambda_3 = \cdots = \lambda_i = 1$$

to obtain the desired sum

$$\Sigma\, x^{\alpha_1+\alpha_2+\alpha_3+\cdots+\alpha_i} .$$

The performance of these operations upon the fraction we shall denote by prefixing the symbol $\underset{\geqslant}{\Omega}$, so that

$$\Sigma\, x^{\alpha_1+\alpha_2+\alpha_3+\cdots+\alpha_i} = \underset{\geqslant}{\Omega} \frac{1}{(1-\lambda_1 x)\left(1-\frac{\lambda_2}{\lambda_1}x\right)\left(1-\frac{\lambda_3}{\lambda_2}x\right)\cdots\left(1-\frac{\lambda_i}{\lambda_{i-1}}x\right)} ."$$

More generally

$$\underset{\geqslant}{\Omega} \sum_{s_1=-\infty}^{\infty} \cdots \sum_{s_r=-\infty}^{\infty} A_{s_1\cdots s_r} \lambda_1^{s_1} \cdots \lambda_r^{s_r} = \sum_{s_1=0}^{\infty} \cdots \sum_{s_r=0}^{\infty} A_{s_1\cdots s_r} , \tag{2.1}$$

where the $A_{s_1\cdots s_r}$ may be functions of several complex variables.

Let us now return to MacMahon. He notes that to make effective use of this method, one requires a number of simple identities for the Ω symbol. As he remarks [7; Vol. 2 p. 102].

"We may add conveniently the easily verifiable results

$$\underset{\geqslant}{\Omega} \frac{1}{(1-\lambda x)\left(1-\frac{y}{\lambda}\right)\left(1-\frac{z}{\lambda}\right)} = \frac{1}{(1-x)(1-xy)(1-xz)},$$

$$\underset{\geqslant}{\Omega} \frac{1}{(1-\lambda x)(1-\lambda y)\left(1-\frac{z}{\lambda}\right)} = \frac{1-xyz}{(1-x)(1-y)(1-xz)(1-yz)},$$

$$\underset{\geqslant}{\Omega} \frac{1}{(1-\lambda x)\left(1-\frac{y}{\lambda^2}\right)} = \frac{1}{(1-x)(1-x^2y)},$$

$$\underset{\geqslant}{\Omega} \frac{1}{(1-\lambda^2 x)\left(1-\frac{y}{\lambda}\right)} = \frac{1+xy}{(1-x)(1-xy^2)},$$

$$\underset{\geqslant}{\Omega} \frac{1}{(1-\lambda x)(1-\lambda y)\left(1-\frac{z}{\lambda^2}\right)} = \frac{1+xyz-x^2yz-xy^2z}{(1-x)(1-y)(1-x^2z)(1-y^2z)},$$

$$\underset{\geqslant}{\Omega} \frac{1}{(1-\lambda x)\left(1-\frac{y}{\lambda^s}\right)} = \frac{1}{(1-x)(1-x^sy)},$$

$$\underset{\geqslant}{\Omega} \frac{1}{(1-\lambda^s x)\left(1-\frac{y}{\lambda}\right)} = \frac{1+xy\frac{1-y^{s-1}}{1-y}}{(1-x)(1-xy^s)}."$$

In order to gain a feel for this method, the above may be viewed as elementary exercises for the reader.

For the actual applications required in this paper we require the following results which are easily established generalizations of those of MacMahon. For integers $0 \leqq \alpha < s$, $b \geqq 0$,

$$\underset{\geqslant}{\Omega} \frac{\lambda^\alpha}{(1-\lambda^s A)\left(1-\frac{B}{\lambda^{s+b}}\right)} = \underset{\geqslant}{\Omega} \frac{\lambda^\alpha}{(1-\lambda^s A)\left(1-\frac{AB}{\lambda^b}\right)} \tag{2.2}$$

$$\begin{aligned}\underset{\geqslant}{\Omega} \frac{\lambda^\alpha}{(1-\lambda^s x)\left(1-\frac{y}{\lambda}\right)} &= \frac{\frac{1-y^{\alpha+1}}{1-y} + xy^{\alpha+1}\frac{1-y^{s-\alpha-1}}{1-y}}{(1-x)(1-xy^s)} \\ &= \frac{1+y+y^2+\cdots+y^\alpha+xy^{\alpha+1}+xy^{\alpha+2}+\cdots+xy^{s-1}}{(1-x)(1-xy^s)}\end{aligned} \tag{2.3}$$

We also need the rather obvious contraction principle: let $C \geqq 0$, $M > 0$ and $0 \leqq \nu < M$ all integers,

$$\begin{aligned}&\underset{\geqslant}{\Omega} H(\alpha_1, \alpha_2, \ldots, \alpha_n; \lambda_1^M, \lambda_2, \ldots, \lambda_n)\lambda_1^{CM+\nu} \\ &= \underset{\geqslant}{\Omega} H(\alpha_1, \alpha_2, \ldots, \alpha_n; \lambda_1, \lambda_2, \ldots, \lambda_n)\lambda_1^C .\end{aligned} \tag{2.4}$$

3. Why use partition analysis?

The main point to be made is this: The application of Partition Analysis to Lecture Hall partitions:

(1) Directly reduces the problem to the combinatorial interpretation of a recursively defined polynomial;
(2) Allows consideration of the problem in full generality so that possible refinements of the problem present themselves automatically.

The full argument that we begin in the next section obscures the initial insights that come from the consideration of a special case. To see this let us consider $j = 4$.

In the notation of Partition Analysis, we see that the generating function in the case $j = 4$ is given by

$$\underset{\geqq}{\Omega} \sum_{n_1,n_2,n_3,n_4 \geqq 0} a_1^{n_1} a_2^{n_2} a_3^{n_3} a_4^{n_4} \lambda_1^{3n_1-4n_2} \lambda_2^{2n_2-3n_3} \lambda_3^{n_3-2n_4} \tag{3.1}$$

$$= \underset{\geqq}{\Omega} \frac{1}{(1-\lambda_1^3 a_1)\left(1-\frac{\lambda_2^2 a_2}{\lambda_1^4}\right)\left(1-\frac{\lambda_3 a_3}{\lambda_2^3}\right)\left(1-\frac{a_4}{\lambda_3^2}\right)}$$

$$= \underset{\geqq}{\Omega} \frac{1}{(1-\lambda_1^3 a_1)\left(1-\frac{\lambda_2^2 a_2 a_1}{\lambda_1}\right)\left(1-\frac{\lambda_3 a_3}{\lambda_2^3}\right)\left(1-\frac{a_4}{\lambda_3^2}\right)}$$

(by (2.2) applied to λ_1 with $s = 3$)

$$= \underset{\geqq}{\Omega} \frac{1 + a_1^2 a_2 \lambda_2^2 + a_1^3 a_2^2 \lambda_2^4}{(1-a_1)(1-\lambda_2^6 a_2^3 a_1^4)\left(1-\frac{\lambda_3 a_3}{\lambda_2^3}\right)\left(1-\frac{a_4}{\lambda_3^2}\right)}$$

(by (2.3) applied to λ_1)

$$= \underset{\geqq}{\Omega} \frac{1 + a_1^2 a_2 + a_1^3 a_2^2 \lambda_2}{(1-a_1)(1-\lambda_2^2 a_2^3 a_1^4)\left(1-\frac{\lambda_3 a_3}{\lambda_2}\right)\left(1-\frac{a_4}{\lambda_3^2}\right)}$$

(by the contraction principle (2.4) applied to λ_2)

$$= \underset{\geqq}{\Omega} \frac{(1 + a_1^2 a_2)(1 + a_2^3 a_1^4 a_3 \lambda_3) + a_1^3 a_2^2 (1 + \lambda_3 a_3)}{(1-a_1)(1-a_1^4 a_2^3)(1-a_1^4 a_2^3 a_3^2 \lambda_3^2)\left(1-\frac{a_4}{\lambda_3^2}\right)}$$

(by (2.3) applied to λ_2)

$$= \frac{(1 + a_1^2 a_2)(1 + a_2^3 a_1^4 a_3) + a_1^3 a_2^2 (1 + a_3)}{(1-a_1)(1-a_1^4 a_2^3)(1-a_1^4 a_2^3 a_3^2)(1-a_1^4 a_2^3 a_3^2 a_4)}$$

(by the first contracting using (2.4) and then applying (2.3))

$$= \frac{1 + a_1^2 a_2 + a_1^4 a_2^3 a_3 + a_1^6 a_2^4 a_3 + a_1^3 a_2^2 + a_3 a_1^3 a_2^2}{(1 - a_1)(1 - a_1^4 a_2^3)(1 - a_1^4 a_2^3 a_3^2)(1 - a_1^4 a_2^3 a_3^2 a_4)} .$$

The above involved two successive applications of the sequence (2.2), (2.3) and (2.4) (although in the final sequence (2.2) was actually not required because its application with $b = 0$ is trivial). The general case will reduce to a rational function of the a_i precisely by continuation of this sequence.

If we now set $a_1 = a_2 = a_3 = a_4 = q$, then our generating function becomes

$$\begin{aligned} &\frac{1 + q^3 + q^8 + q^{11} + q^5 + q^6}{(1-q)(1-q^7)(1-q^9)(1-q^{10})} \\ &= \frac{(1 + q^3 + q^6)(1 + q^5)}{(1-q)(1-q^7)(1-q^9)(1-q^{10})} \\ &= \frac{1}{(1-q)(1-q^3)(1-q^5)(1-q^7)} , \end{aligned} \tag{3.2}$$

thus confirming Theorem 1 in the case $j = 4$.

However, it is natural to ask (without knowing Theorem 2 in advance) if any weaker substitutions will effect a dramatic factorization of the numerator. Indeed, if we merely set $a_3 = a_1$ and $a_4 = a_2$, we find

$$\begin{aligned} &\frac{1 + a_1^2 a_2 + a_1^5 a_2^3 + a_1^7 a_2^4 + a_1^3 a_2^2 + a_1^4 a_2^2}{(1 - a_1)(1 - a_1^4 a_2^3)(1 - a_1^6 a_2^3)(1 - a_1^6 a_2^4)} \\ &= \frac{(1 + a_1^2 a_2 + a_1^4 a_2^2)(1 + a_1^3 a_2^2)}{(1 - a_1)(1 - a_1^4 a_2^3)(1 - a_1^6 a_2^3)(1 - a_1^6 a_2^4)} \\ &= \frac{1}{(1 - a_1)(1 - a_1^4 a_2^3)(1 - a_1^2 a_2)(1 - a_1^3 a_2^2)} . \end{aligned} \tag{3.3}$$

Upon setting $a_1 = qa$ and $a_2 = q/a$ we derive the instance $j = 4$ of Theorem 2.

In the following sections, we shall treat the general case. Our main point of difficulty will be the analysis of the numerator polynomials such as (in the case $j = 4$) the numerator in the last line of (3.1).

The point to stress here is that we have carried off the case $j = 4$ with no effective combinatorial argument or knowledge. In other words, the entire problem is reduced by Partition Analysis to the factorization of an explicitly given polynomial.

4. The fundamental generating functions

Given the Partition Analysis methods developed in Section 2, we define for $0 \leqq j < n$

$$F_{n,j}(a_1, a_2, \ldots, a_{n+1}) \\ = \underset{\geqslant}{\Omega} \frac{\lambda_1^j}{(1-\lambda_1^n a_1)\left(1-\frac{\lambda_2^{n-1}}{\lambda_1^{n+1}}a_2\right)\left(1-\frac{\lambda_3^{n-2}}{\lambda_2^n}a_3\right)\cdots\left(1-\frac{a_{n+1}}{\lambda_n^2}\right)}, \quad (4.1)$$

and

$$G_{n,j}(a_1, a_2, \ldots, a_{n+1}) \quad (4.2) \\ = \underset{\geqslant}{\Omega} \frac{\lambda_1^j}{(1-\lambda_1^n a_1)\left(1-\frac{\lambda_2^{n-1}}{\lambda_1}a_2\right)\left(1-\frac{\lambda_3^{n-2}}{\lambda_2^n}a_3\right)\left(1-\frac{\lambda_4^{n-3}}{\lambda_3^{n-1}}a_4\right)\cdots\left(1-\frac{a_{n+1}}{\lambda_n^2}\right)}.$$

We note that the only difference in the two expressions is the power of λ_1 in the second denominator factor. The $F_{n,j}$ is the expression of interest here (see (4.4)); however the $G_{n,j}$ allow simpler computations and are thus introduced for convenience.

When comparing the work in this and subsequent sections with that of Section 3, it is important to keep in mind that in Section 3, $G_{3,0}(a_1, a_1a_2, a_3, a_4)$ was computed.

Furthermore, by (2.2)

$$F_{n,j}(a_1, a_2, \ldots, a_{n+1}) \\ = \underset{\geqslant}{\Omega} \frac{\lambda_1^j}{(1-\lambda_1^n a_1)\left(1-\frac{\lambda_2^{n-1}a_1a_2}{\lambda_1}\right)\left(1-\frac{\lambda_3^{n-2}a_3}{\lambda_2^n}\right)\cdots\left(1-\frac{a_{n+1}}{\lambda_n^2}\right)} \\ = G_{n,j}(a_1, a_1a_2, a_3, \ldots, a_{n+1}). \quad (4.3)$$

Most importantly following from the basic definitions of Section 2, we see that

$$F_{n,0}(a_1, a_2, \ldots, a_{n+1}) = \sum_{\frac{b_{n+1}}{n+1} \geqq \frac{b_n}{n} \geqq \frac{b_{n-1}}{n-1} \geqq \cdots \geqq \frac{b_1}{1} \geqq 0} a_1^{b_{n+1}} a_2^{b_n} \cdots a_{n+1}^{b_1}; \quad (4.4)$$

i.e., we have the $(n+1)$-variable generating function for Lecture Hall partitions.

Next, for $0 \leqq j < n$, by (2.3), we see that

$$
\begin{aligned}
&G_{n,j}(a_1, a_2, \ldots, a_{n+1}) \\
&\quad = \underset{\geqslant}{\Omega} \frac{\lambda_1^j}{(1-\lambda_1^n a_1)\left(1-\frac{\lambda_2^{n-1}}{\lambda_1}a_2\right)\left(1-\frac{\lambda_3^{n-2}}{\lambda_2^n}a_3\right)\cdots\left(1-\frac{a_{n+1}}{\lambda_n^2}\right)} \\
&= \underset{\geqslant}{\Omega} \frac{1+\lambda_2^{n-1}a_2+\cdots+(\lambda_2^{n-1}a_2)^j+a_1(\lambda_2^{n-1}a_2)^{j+1}+\cdots+a_1(\lambda_2^{n-1}a_2)^{n-1}}{(1-a_1)(1-a_1(\lambda_2^{n-1}a_2)^n)\left(1-\frac{\lambda_3^{n-2}}{\lambda_2^n}a_3\right)\cdots\left(1-\frac{a_{n+1}}{\lambda_n^2}\right)} \\
&\quad = \underset{\geqslant}{\Omega} \frac{1+\sum_{h=0}^{j-1} a_2^{h+1}\lambda_2^h + a_1\sum_{h=j}^{n-2} a_2^{h+1}\lambda_2^h}{(1-a_1)(1-a_1a_2^n\lambda_2^{n-1})\left(1-\frac{\lambda_3^{n-2}}{\lambda_2}a_3\right)\cdots\left(1-\frac{a_{n+1}}{\lambda_n^2}\right)}
\end{aligned}
$$

(by (2.4)) and the fact that $\lambda_2^{i(n-1)} = \lambda_2^{(i-1)n+n-i}$

$$
\begin{aligned}
= \frac{1}{(1-a_1)} \Bigg(& G_{n-1,0}(a_1a_2^n, a_3, \ldots, a_{n+1}) \\
& + \sum_{h=0}^{j-1} a_2^{h+1} G_{n-1,h}(a_1a_2^n, a_3, \ldots, a_{n+1}) \\
& + a_1 \sum_{h=j}^{n-2} a_2^{h+1} G_{n-1,h}(a_1a_2^n, a_3, \ldots, a_{n+1}) \Bigg)
\end{aligned}
\tag{4.5}
$$

Furthermore, if we define

$$
\begin{aligned}
&G_{n,j}(a_1, a_2, \ldots, a_{n+1}) \\
&\quad = \frac{g_{n,j}(a_1, a_2, \ldots, a_{n+1})}{(1-a_1)(1-a_1a_2^n)(1-a_1a_2^na_3^{n-1})\cdots(1-a_1a_2^n\cdots a_{n+1})},
\end{aligned}
\tag{4.6}
$$

then we deduce directly from (4.5) that

$$
\begin{aligned}
& g_{n,j}(a_1, a_2, \ldots, a_{n+1}) \\
& \quad = g_{n-1,0}(a_1a_2^n, a_3, \ldots, a_{n+1}) \\
& \quad + \sum_{h=0}^{j-1} a_1^{h+1} g_{n-1,h}(a_1a_2^n, a_3, \ldots, a_{n+1}) \\
& \quad + a_1 \sum_{h=j}^{n-2} a_1^{h+1} g_{n-1,h}(a_1a_2^n, a_3, \ldots, a_{n+1})
\end{aligned}
\tag{4.7}
$$

which together with the obvious fact that

$$g_{1,0}(a_1, a_2) = 1 \tag{4.8}$$

completely defines the polynomials $g_{n,j}(a_1, a_2, \ldots, a_{n+1})$.

In the next sections we shall study with special care the following specializations of these functions.

Namely

$$\begin{aligned}\Phi_{n+1}(a;q) &= F_{n,0}\left(a, \frac{q}{a}, a, \frac{q}{a}, a, \frac{q}{a}, \cdots\right)\\ &= G_{n,0}\left(a, q, a, \frac{q}{a}, a, \frac{q}{a}, \ldots\right)\\ &= \frac{g_{n,0}\left(a, q, a, \frac{q}{a}, a, \frac{q}{a}, \cdots\right)}{(1-a)(1-aq^n)(1-a^nq^n)(1-a^2q^{2(n-1)})(1-a^{n-1}q^{2(n-1)})\cdots}\\ &= \frac{g_{n,0}\left(a, q, a, \frac{q}{a}, a, \frac{q}{a}, \cdots\right)}{(1-a)\prod_{j=1}^{n}\left(1-a^jq^{j(n-j+1)}\right)}.\end{aligned} \tag{4.9}$$

Additionally, we let

$$\gamma(n) = g_{n,0}\left(a, q, a, \frac{q}{a}, a, \frac{q}{a}, \cdots\right). \tag{4.10}$$

5. The factorization of $\gamma(\mathbf{n})$, Part I

The entire object of this section and the next is the establishment of

Lemma 1.

$$\begin{aligned}\gamma(n) &= \prod_{j=1}^{n-1}\left(1 + a\,q^j + a^2q^{2j} + \cdots + a^{n-j}q^{j(n-j)}\right)\\ &= \prod_{j=1}^{n-1}\left(\frac{1-a^{n-j+1}q^{j(n-j+1)}}{1-a\,q^j}\right).\end{aligned} \tag{5.1}$$

We shall lead up to the proof gradually. First we note that iteration of

(4.7) gives us every term of the polynomial $\gamma(n)$. We rewrite (4.7) as

$$g_{n,j}(a_1, a_2, \ldots, a_{n+1}) = \sum_{i=0}^{n-1} a_1^{\chi(i>j)} a_2^i g_{n-1,i-\chi(i>0)}(a_1 a_2^n, a_3, \ldots, a_{n+1}), \tag{5.2}$$

where $\chi(S)$ is 1 if S is true and zero if S is false. For ease of notation we define for nonnegative i and j

$$x(i,j) = \chi(i > j - \chi(j,0)) = \begin{cases} 1 & \text{if } i \geqq j \text{ and } i \neq 0 \\ 0 & \text{otherwise.} \end{cases} \tag{5.3}$$

Consequently iterating (5.2), we find

$$\begin{aligned} g_{n,j}(a_1, a_2, \ldots, a_{n+1}) &= \sum_{\substack{0 \leqq i_h \leqq n-h \\ 1 \leqq h \leqq n}} a_1^{\chi(i_1>j)} a_2^{i_1} (a_1 a_2^n)^{x(i_2,i_1)} a_3^{i_2} \\ &\quad \times (a_1 a_2^n a_3^{n-1})^{x(i_3,i_2)} a_4^{i_3} \\ &\quad \times (a_1 a_2^n a_3^{n-1} a_4^{n-2})^{x(i_4,i_3)} a_5^{i_4} \\ &\quad \vdots \end{aligned} \tag{5.4}$$

and this, of course, implies

$$\gamma(n) = \sum_{\substack{0 \leqq i_h \leqq n-h \\ 1 \leqq h \leqq n}} E(i_1, i_2, \ldots, i_n) \tag{5.5}$$

where

$$\begin{aligned} E(i_1, i_2, \ldots, i_n) &= a^{\chi(i_1>0)} q^{i_1} (a\, q^n)^{x(i_2,i_1)} a^{i_2} \\ &\quad \times (a^n q^n)^{x(i_3,i_2)} \left(\frac{q}{a}\right)^{i_3} (a^2 q^{2(n-1)})^{x(i_4,i_3)} a^{i_4} \\ &\quad \times (a^{n-1} q^{2(n-1)})^{x(i_5,i_4)} \left(\frac{q}{a}\right)^{i_5} (a^3 q^{3(n-2)})^{x(i_6,i_5)} a^{i_6} \\ &\quad \vdots \qquad\qquad \vdots \qquad\qquad \vdots \end{aligned} \tag{5.6}$$

Our general procedure is to group the terms of this sum so that the factorization becomes obvious. The following grouping (whose application follows in the next section) yields the required result when $n = 4$.

$$\begin{aligned}
\gamma(4) =&(E(0,0,0,0) + E(1,0,0,0) + E(2,1,0,0) + E(3,2,0,0)) \\
&+ (E(2,0,0,0) + E(3,1,0,0) + E(0,2,0,0) + E(1,2,0,0)) \\
&+ (E(3,0,0,0) + E(0,1,0,0) + E(1,1,0,0) + E(2,2,0,0)) \\
&+ (E(3,2,1,0) + E(0,0,1,0) + E(1,0,1,0) + E(2,1,1,0)) \\
&+ (E(0,2,1,0) + E(1,2,1,0) + E(2,0,1,0) + E(3,1,1,0)) \\
&+ (E(2,2,1,0) + E(3,0,1,0) + E(0,1,1,0) + E(1,1,1,0)) \\
=&(E(0,0,0,0)(1 + aq + a^2q^2 + a^3q^3) \\
&+ E(2,0,0,0)(1 + aq + a^2q^2 + a^3q^3) \\
&+ E(3,0,0,0)(1 + aq + a^2q^2 + a^3q^3) \\
&+ E(3,2,1,0)(1 + aq + a^2q^2 + a^3q^3) \\
&+ E(0,2,1,0)(1 + aq + a^2q^2 + a^3q^3) \\
&+ E(2,2,1,0)(1 + aq + a^2q^2 + a^3q^3) \\
=&(1 + aq + a^2q^2 + a^3q^3) \\
&\times \{(E(0,0,0,0) + E(2,0,0,0) + E(3,2,1,0)) \\
&+ (E(3,0,0,0) + E(0,2,1,0) + E(2,2,1,0))\} \\
=&(1 + aq + a^2q^2 + a^3q^3) \\
&\times (E(0,0,0,0)(1 + aq^2 + a^2q^4) \\
&+ E(3,0,0,0)(1 + aq^2 + a^2q^4)) \\
=&(1 + aq + a^2q^2 + a^3q^3)(1 + aq^2 + a^2q^4) \\
&\times (E(0,0,0,0) + E(3,0,0,0)) \\
=&(1 + aq + a^2q^2 + a^3q^3)(1 + aq^2 + a^2q^4)(1 + aq^3) \,.
\end{aligned} \tag{5.7}$$

The next section will explain the grouping that is illustrated above. The basic object, however, may be observed above in the $n = 4$ case; namely, we peel off the factors one at a time.

6. The factorization of $\gamma(n)$, Part II

We now consider a sequence of subsets of the set of indices appearing in (5.5). Namely we first set

$$I_{m,j} = \begin{cases} \{(0,0),(j+1,j),(j+2,j+1),\ldots,(m,m-1)\}, & \text{if } j < m \\ \{(0,0)\} & \text{if } j \geqq m. \end{cases} \tag{6.1}$$

Then we set

$$\begin{aligned} T_n^{(r)} = \{(i_1,i_2,\ldots,i_n)\big| i_1 = 0 \text{ or } i_1 \in [r+1,n-1], \\ \text{and } (i_{2h},i_{2h+1}) \in I_{n-2h,r-h+1} \\ \text{for } 1 \leqq h \leqq r, \quad 0 \leqq i_j \leqq n-j \text{ for } j > 2r+1\}\,. \end{aligned} \tag{6.2}$$

Thus the sum in (5.5) is over $T_n^{(0)}$, while, for example,

$$\begin{aligned} T_n^{(1)} = \{(i_1,\ldots,i_n)\big|\,, \qquad & 0 \leqq i_1 \leqq n-1, i_1 \neq 1\,, \\ & 0 \leq i_2 \leqq n-2, i_2 \neq 1\,, \\ & i_3 = 0 \text{ if } i_2 = 0 \text{ and} \\ & i_3 = i_2 - 1 \text{ if } i_2 > 0\}. \end{aligned} \tag{6.3}$$

Clearly

$$T_n^{(0)} \supset T_n^{(1)} \supset T_n^{(2)} \supset \cdots \supset T_n^{(n-1)} = \{(0,0,\cdots,0)\}\,. \tag{6.4}$$

We note in fact that there are $(n-j)!$ elements of $T_n^{(j)}$. Furthermore to each element of $T_n^{(j)}$ we associate $n-j$ elements of $T_n^{(j-1)}$ as follows:

Let

$$\tau = (i_1,i_2,i_2^*,i_4,i_4^*,\ldots,i_{2j},i_{2j}^*,i_{2j+2},\ldots,i_n) \tag{6.5}$$

be a given element of $T_n^{(j)}$ where $i_{2h}^* = i_{2h} - 1$ if $i_{2h} > 0$ and otherwise

$i_{2h}^* = i_{2h} = 0$. We observe that

$$\begin{aligned} i_1 &\in \{0, j+1, \ldots, n-1\} \\ i_2 &\in \{0, j+1, \ldots, n-2\} \\ i_4 &\in \{0, j, \ldots, n-4\} \\ &\vdots \\ i_{2h} &\in \{0, j-h+2, \ldots, n-2h\} \\ &\vdots \\ i_{2j} &\in \{0, 2, \ldots, n-2j\} . \end{aligned}$$

We now define an augmentation operation, A_j, on τ which, by successive application, produces exactly $n-j+1$ distinct elements of $T_n^{(j-1)}$. First, A_j alters only coordinates among the initial $2j$ coordinates and alters each coordinate of τ up to and including the first 0 entry (if there is a 0 entry among the first $2j$) and leaves the remaining entries fixed. The effect of A_j on the entries it alters is to "increase each entry by 1." This last expression is in quotes for two reasons. First it might be that this is the unique 0 entry affected by A; so if i_{2h} is 0, then A replaces $i_{2h} = 0$ by $i_{2h} = j-h$. Similarly if $i_1 = 0$, then A_j replaces $i_1 = 0$ by $i_1 = j$. Furthermore if either i_1 or i_h is maximal relative to the conditions defining $T_n^{(j-1)}$, then the replacement coordinate is 0. Finally the entries i_{2h+1} $(0 < h < j)$ are replaced so that the image of τ under A_j is still in $T_n^{(j-1)}$.

Proof. Proof of Lemma 1

We construct a sequence of polynomials

$$\gamma_j(n) = \sum_{(i_1,\ldots,i_n)\in T_n^{(j)}} E(i_1, i_2, \ldots, i_n) . \tag{6.6}$$

The proof relies on two assertions:

$$T_n^{(j-1)} = \cup_{h=0}^{n-j} A_j^h \, T_n^{(j)} . \tag{6.7}$$

$$E(A_j^h(i_1, i_2, \ldots, i_n)) = a^h q^{hj} E(i_1, i_2, \ldots, i_n) , \tag{6.8}$$

for each $(i_1, i_2, \ldots, i_n) \in T_n^{(j)}$, $h = 0, 1, \ldots, n-j$.

Let us verify that these results are all that is required. The union in (6.5) must be a disjoint union because the cardinality of the left-hand side is $(n-j+1)!$ and each of the $n-j+1$ sets on the right hand side has at most $(n-j)!$ entries (and consequently must have exactly $(n-j)!$ entries). So

$$\begin{aligned} \gamma_{j-1}(n) &= \sum_{(i_1,\ldots,i_n)\in T_n^{(j-1)}} E(i_1,i_2,\ldots,i_n) \\ &= \sum_{(i_1,i_2,\ldots,i_n)\in \bigcup_{h=0}^{n-j} A_j^h T_n^{(j)}} E(i_1,i_2,\ldots,i_n) \\ &= \sum_{(i_1,i_2,\ldots,i_n)\in T_n^{(j)}} E(A_j^h(i_1,i_2,\ldots,i_n)) \qquad (6.9) \\ &= \sum_{(i_1,i_2,\ldots,i_n)\in T_n^{(j)}} \sum_{h=0}^{n-j} a^h\, q^{hj}\, E(i_1,i_2,\ldots,i_n) \\ &= (1+aq^j+a^2q^{2j}+\cdots+a^{n-j}q^{(n-j)j})\gamma_j(n)\,. \end{aligned}$$

Equation (5.1) now follows by iteration of (6.9) once we note that $\gamma(n) = \gamma_0(n)$, and $\gamma_{n-1}(n) = 1$ (by (6.4)).

First we prove (6.7). We begin by observing that the $n-j+1$ n-tuples $\tau, A_j\tau, A_j^2\tau, \ldots, A_j^{n-j}\tau$ (where $(i_1,\ldots,i_n) \in T_n^{(j)}$) are all distinct because the first entries of each n-tuple are distinct and are, in fact, the numbers $0, j, \ldots, n-1$ in some order.

Next we note that $A_j^{n-j+1}\tau' = \tau'$ for each $\tau' \in T_n^{(j-1)}$. This works as follows. The successive applications of A_j to the first coordinate i_1 of τ' advances i_1 by 1 cyclicly through $0, j, j+1, \ldots, n-1$, and so wherever you start you wind up where you began on the $n-j+1$ step. For columns 2 and 3 (i.e., i_2 and i_3) we have $n-j$ possible pairs $(0,0)$, $(j, j-1)$, $(j+1, j), \ldots, (n-2, n-3)$; however as we successively apply A_j we find that when 0 arises in the first column the entries i_2 and i_3 are unchanged, and this is the only time that happens. Hence again after $n-j+1$ steps we have the same second and third coordinates. In general for i_{2h} and i_{2h+1}, we find that they must be among the $n-h-j$ pairs $(0,0)$, $(j-h+2, j-h+1), \ldots (n-2h, n-2h-1)$ but there will be h times when the application of A_j does not change the pair. Hence again after $n-j+1$ steps we are back to where we started from. Thus A_j^{n-j+1} is the identity on $T_n^{(j-1)}$.

Also we note that if $\tau' \in T_n^{(j-1)}$, then at least one of $\tau', A_j\tau', \ldots, A_j^{n-j}\tau'$ must be in $T_n^{(j)}$. The difference between elements of $T_n^{(j-1)}$ and $T_n^{(j)}$ is that for each coordinate i_{2h}, $j-h$ might appear for an n-tuple of $T_n^{(j-1)}$ but not for $T_n^{(j)}$ (or i_1 might be j in $T_n^{(j-1)}$ but not in $T_n^{(j)}$). Among τ', $A_j\tau', \ldots, A_j^{n-j}\tau'$ there must be an entry wherein either $i_{2j} = i_{2j+1} = 0$ or $i_{2j} = i_{2j+1} + 1$, in other words, there is at least one entry which is a candidate for being in $T_n^{(j)}$. This is because either i_{2j} and i_{2j+1} must always be zero or else i_{2j} cycles through $\{0, 2, \ldots, n-2j\}$. Suppose such an entry is $A_j^{\lambda}\tau'$. If $A_j^{\lambda}\tau' \in T_n^{(j)}$ we are done. If not, it means that there is some minimal pair (or minimal first element) relative to $T_n^{(j-1)}$ to the left of (i_{2j}, i_{2j+1}); otherwise $A_j^{\lambda}\tau'$ would necessarily be in $T_n^{(j)}$. Hence $A_j^{-1}A_j^{\lambda}$ moves the minimal element farthest to the left back to a zero element. If $A_j^{\lambda-1}\tau' \notin T_n^{(j)}$ repeat this argument. Within at most j steps all minimal elements (relative to $T_n^{(j)}$) will be removed while (i_{2j}, i_{2j+1}) will be unaltered. Hence we will have found an element of $T_n^{(j)}$.

The preceding two observations show that we may take the $(n-j)!$ elements of $T_n^{(j)}$ and associate with each of them a set of $n-j+1$ elements of $T_n^{(j-1)}$. Furthermore these sets are disjoint because if two had a common element, the successive applications of A_j would prove they were identical sets. This establishes (6.7).

Now we turn to (6.8). We successively apply $A_j, A_j^2, \ldots, A_j^{n-j}$ to an element $(i_1, \ldots, i_n) \in T_n^{(j)}$. By the very definition of $T_n^{(j)}$ we see that each of $x(i_3, i_2)$, $x(i_5, i_4), \ldots, x(i_{2j-1}, i_{2j-2})$ will obviously be zero and since we apply A_j at most $n-j$ times we see that in these instances $x(i_{2j+1}, i_{2j})$ will be zero as well. Hence examining equation (5.6), we see that in passing from $A_j^h\tau$ to $A_j^{h+1}\tau$ $(0 \leqq h \leqq n-j-1)$, we change (in a non-exceptional case) from

$$\begin{aligned} &a^{\chi(i_1>0)} q^{i_1} (aq^n)^{x(i_2,i_1)} a^{i_2} \\ &\times (a^n q^n)^0 \left(\frac{q}{a}\right)^{i_2-1} (a^2 q^{2(n-1)})^{x(i_4,i_3)} a^{i_4} \\ &\times (a^{n-1} q^{2(n-1)})^0 \left(\frac{q}{a}\right)^{i_4-1} (a^3 q^{3(n-1)})^{x(i_6,i_5)} a^{i_6} \\ &\vdots \end{aligned} \tag{6.10}$$

to

$$\begin{aligned}
&a^{\chi(i_1>0)}q^{i_1+1}(aq^n)^{x(i_2,i_1)}a^{i_2+1}\\
&\times (a^nq^n)^0\left(\frac{q}{a}\right)^{i_2}(a^2q^{2(n-1)})^{x(i_4,i_3)}a^{i_4+1}\\
&\times (a^{n-1}q^{2(n-1)})^0\left(\frac{q}{a}\right)^{i_4}(a^3q^{3(n-2)})^{x(i_6,i_5)}a^{i_6+1}\\
&\vdots
\end{aligned} \tag{6.11}$$

which (since there are 1's added to each of $i_1, i_2, \ldots, i_{2j}$) introduces the factor q^j arising from each increase of $i_1, i_2, i_4, \ldots, i_{2j-2}$ and the simple factor a arising only from the increase of i_{2j}.

Of course, it will often happen that we are facing an exceptional case where either a maximal pair $(i_{2h,}, i_{2h+1}) = (n-2h, n-2h-1)$ is mapped to $(0,0)$ or a $(0,0)$ is mapped to $(j-h+1, j-h)$. At first glance this would appear to dramatically effect the relevant portion of (6.10). Namely in the first instance

$$\begin{gathered}
(a^hq^{h(n-h+1)})^{x(i_{2h},i_{2h-1})}a^{i_{2h}}\\
(a^{n-h+1}q^{h(n-h+1)})^{x(i_{2h+1},i_{2h})}\left(\frac{q}{a}\right)^{i_{2h+1}}\\
(a^{h+1}q^{(h+1)(n-h)})^{x(i_{2h+2},i_{2h+1})}a^{i_{2h+2}}
\end{gathered}$$

changes to

$$\begin{gathered}
(a^hq^{h(n-h+1)})^{x(i_{2h},i_{2h-1})-1}a^{i_{2h}-(n-2h)}\\
(a^{n-h+1}q^{h(n-h+1)})^{x(i_{2h+1},i_{2h})}\left(\frac{q}{a}\right)^{i_{2h+1}-(n-2h-1)}\\
(a^{h+1}q^{(h+1)(n-h)})^{x(i_{2h+2},i_{2h+1})+1}a^{i_{2h+2}+1}
\end{gathered}$$

for an alteration due to these particular terms of

$$a^{-h-(n-2h)+(n-2h-1)+h+1+1}q^{-h(n-h+1)-(n-2h+1)+(h+1)(n-h)} = aq\,,$$

which is precisely what the effect would have been if each of $i_{2h-1}, i_{2h}, i_{2h+1}$ and i_{2h+2} had increased by 1. Similarly the same precise effect occurs if successive maximal pairs occur.

Finally if (i_{2h}, i_{2h+1}) is the first $(0,0)$ pair then we recall that nothing to the right of this pair is altered by A; while this pair is replaced by

$(j-h, j-h-1)$, and in this instance

$$\begin{gathered}(a^h q^{h(n-h+1)})^{x(i_{2h},i_{2h-1})} a^{i_{2h}} \\ (a^{n-h+1} q^{h(n-h+1)})^{x(i_{2h+1},i_{2h})} \left(\frac{q}{a}\right)^{i_{2h+1}} \\ (a^{h+1} q^{(h+1)(n-h)})^{x(i_{2h+2},i_{2h+1})} a^{i_{2h+2}}\end{gathered}$$

changes to

$$\begin{gathered}(a^h q^{h(n-h+1)})^{x(i_{2h},i_{2h-1})} a^{i_{2h}+j-h} \\ (a^{n-h+1} q^{h(n-h+1)})^{x(i_{2h+1},i_{2h})} \left(\frac{q}{a}\right)^{i_{2h+1}+j-h-1} \\ (a^{h+1} q^{(h+1)(n-h)})^{x(i_{2h+2},i_{2h+1})} a^{i_{2h+2}}\end{gathered}$$

for an alteration due to these particular terms of

$$a^{j-h-(j-h-1)} q^{j-h-1} = a\, q^{j-h-1}$$

which is precisely the factor we need to obtain in order to have the same effect as the nonexceptional case. To be more specific we have no further alterations to the right so we must pick up the a factor here while there are now the unaltered pairs $(i_{2h+4}, i_{2h+5}), \dots (i_{2j-2}, i_{2j-1})$, where an additional factor q^{j-h-2} would have been produced in the exceptional case. Hence the $q^{j-h-1} = q \cdot q^{j-h-2}$ yields the same effect as in the non-exceptional case.

Thus in every instance the application of A_j to $A_j^h \tau$ (where $\tau \in T_n^{(j)}$) produces a factor aq^j times the corresponding term in (5.6).

Therefore (6.8) is valid and consequently Lemma 1 is proved.

7. An example

Before we make the very easy derivation of Theorem 2 from Lemma 1, we wish to illustrate precisely what happened at the end of Section 5 in the case $n = 4$. In that case $T_4^{(0)}$ has 24 elements which we list in the six sets below with an element of $T_4^{(1)}$ topping each set.

0000	2000	3000	3210	0210	2210
1000	3100	0100	0010	1210	3010
2100	0200	1100	1010	2010	0110
3200	1200	2200	2110	3110	1110

Each line above is obtained by applying A_1 (relative to $T_4^{(0)}$) to the preceding line. Furthermore, the structure of the first line of (6.6) is now clear.

In turn $T_4^{(1)}$ may be divided into two sets with an element of $T_4^{(2)}$ topping each set.

0000	3000
2000	0210
3210	2210

Each line above is obtained by applying A_2 (relative to $T_4^{(1)}$) to the preceding line, and the second line of (6.6) is revealed.

8. The lecture hall partition theorem

We prove Theorem 2 which is a refinement of Theorem 1, and we remind the reader that this theorem was proved originally in [2].

The generating function for the partitions given in Theorem 2 is by (3.4) and (3.7)

$$\begin{aligned}
\Phi_{n+1}(aq, q^2) &= F_{n,0}\left(aq, \frac{q}{a}, aq, \frac{q}{a}, aq, \frac{q}{a}, \dots\right) \\
&= G_{n,0}(aq, q^2, aq, \frac{q}{a}, aq, \frac{q}{a}, \dots) \\
&= \frac{g_{n,0}\left(aq, q^2, aq, \frac{q}{a}, aq, \frac{q}{a}, \dots\right)}{(1-aq)\prod_{j=1}^{n}(1-a^j q^{2j(n-j+1)+j})} \\
&= \frac{\prod_{j=2}^{n}(1-(aq)^j q^{2j(n+1-j)}) \cdot \prod_{j=1}^{n-1}\frac{1}{(1-aq^{2j+1})}}{(1-aq)\prod_{j=1}^{n}(1-(aq)^j q^{2j(n-j+1)})} \\
&= \frac{1}{(1-aq)(1-aq^{2n+1})\prod_{j=1}^{n-1}(1-aq^{2j+1})} \\
&= \prod_{j=0}^{n}\frac{1}{(1-aq^{2j+1})},
\end{aligned}$$

and Theorem 2 follows immediately.

9. Conclusion

There are several items to be noted in our concluding remarks.

First, there is some way in which our approach (apart from being analytic-combinatorial as opposed to bijective-combinatorial) is different from the outcome of the Bousquet-Mélou and Eriksson approach. This can be most easily illustrated by the case $n = 1$, where they obtain the generating function (after change of variable)

$$\frac{1+aq}{(1-a^2q^2)(1-aq^3)} = \frac{1}{(1-aq)(1-aq^3)} .$$

On the other hand, the proof presented here yields

$$\frac{g_{1,0}(aq,q^2)}{(1-aq)(1-aq^3)} = \frac{1}{(1-aq)(1-aq^3)} .$$

More generally it appears that Bousquet-Mélou and Eriksson have an additional factor $(1+aq+\cdots+a^nq^n)$ in both numerator and denominator.

Finally there is the matter of the future of applications of MacMahon's Partition Analysis. It would seem to me to be quite bright. Its difficulty lies in understanding a variety of numerators of rational functions, in this paper the

$$g_{n,j}(a_1,a_2,\ldots,a_{n+1}) .$$

Furthermore one's combinatorial knowledge of the problem in question can be quite minimal. The natural process of Partition Analysis will reveal relevant refinements as we showed in Section 3.

Finally I would like to suggest that Partition Analysis is potentially a very useful tool in the age of computer algebra.

The next project is a return to MacMahon's Partition Analysis "Waterloo": plane partitions. MacMahon [7: Vol. 2, p. 187] concluded his attempt to study plane partitions via Partition Analysis with the comment: "Our knowledge of the Ω operation is not sufficient to enable us to establish the final form of the result."

I hope to use AXIOM to reexamine this problem with the fond hope that MacMahon's Magnificent Machine may be fully refurbished.

References

[1] G. E. Andrews, *The Theory of Partitions, Encyclopedia of Mathematics and Its Applications*, Vol. 2, G.-C. Rota ed., Addison-Wesley, Reading, 1976. [Reissued: Cambridge University Press, Cambridge, 1985].

[2] M. Bousquet-Mélou and K. Eriksson, Lecture hall partitions I and II, *The Ramanujan Journal*, **1** (1997), 101–111 and **1** (1997), 165–185.

[3] A. Cayley, On an algebraical operation, *Quart. J. of Pure and Appl. Math.*, **13** (1875), 369–375. [also Coll. Math Papers of A. Cayley, Vol. 9, pp. 537–542].

[4] P. A. MacMahon, Memoir on the theory of the partition of numbers — Part I, *Phil. Trans.* **187** (1897), 619–673 (cf., Reference 8, pp. 1026–1080).

[5] P. A. MacMahon, Memoir on the theory of the partition of numbers — Part II, *Phil. Trans.* **192** (1899), 351–401 (cf., Reference 8, pp. 1138–1188).

[6] P. A. MacMahon, Memoir on the theory of the partition of numbers — Part III, *Phil. Trans.* **205** (1906), 35–58 (cf. Reference 8, pp. 1255–1277).

[7] P. A. MacMahon, *Combinatory Analysis*, 2 vols., Cambridge University Press, Cambridge, 1915–1916 (Reprinted: Chelsea, New York, 1960).

[8] P. A. MacMahon, *Collected Papers, Vol. 1, Combinatorics*, G. E. Andrews, ed., MIT Press, Cambridge, 1978.

[9] P. A. MacMahon, *Collected Papers, Vol. 2, Number Theory, Invariants, and Applications*, G. E. Andrews ed., MIT Press, Cambridge, 1986.

[10] R. P. Stanley, Linear homogeneous Diophantine equations and magic labelings of graphs, *Duke Math. J.*, **40** (1973), 607–632.

The Department of Mathematics
Melbourne University
Parkville, Victoria, 3052
Australia
and
The Department of Mathematics
The Pennsylvania State University
University Park, PA 16802
e-mail: andrews@math.psu.edu

The cd-Index of Zonotopes and Arrangements

Louis J. Billera,* *Richard Ehrenborg, and Margaret Readdy*

To Gian-Carlo Rota, for years of inspiration.

Abstract

We investigate a special class of polytopes, the zonotopes, and show that their flag f-vectors satisfy only the affine relations fulfilled by flag f-vectors of *all* polytopes. In addition, we determine the lattice spanned by flag f-vectors of zonotopes. By duality, these results apply as well to the flag f-vectors of central arrangements of hyperplanes.

1. Introduction

The flag f-vector of a convex polytope is an enumerative invariant of its lattice of faces, containing more information than the usual f-vector. While the latter counts the numbers of faces in each dimension, the former counts the numbers of chains (flags) having any possible set of dimensions.

The Euler relation is the only affine relation satisfied by f-vectors of all polytopes. For simplicial (or simple) d-polytopes, there are $\lfloor \frac{d-1}{2} \rfloor$ additional relations, called the Dehn–Sommerville equations, which provide a complete description of the affine space generated by all such f-vectors [10]. The information contained in the f-vector of a simplicial polytope is nicely summarized in the form of the h-vector [18].

In the case of the flag f-vector, there is a large set of equations that is satisfied for all polytopes. The corresponding affine space has dimension given by the Fibonacci sequence [1]. The **cd**-index provides an efficient way to summarize this information [2].

In the case of simplicial, simple and cubical polytopes, the flag f-vector reduces directly to the f-vector. In this paper we investigate another special class of polytopes, the zonotopes, and show for these that there is no reduction whatsoever; that is, we show that the flag f-vectors of zonotopes satisfy only the affine relations satisfied by flag f-vectors of *all* polytopes.

*Supported in part by NSF grant DMS-9500581.

This strengthens a result of Liu [14, Theorem 4.7.1]. Zonotopes are of particular interest in the study of hyperplane arrangements (see [20]), to which they are dual. A direct consequence of our result is that the **cd**-index of a central hyperplane arrangement is the most efficient encoding of the affine information of its flag f-vector.

We define the basic terminology used throughout this paper. For a convex d-dimensional polytope Q, and for a subset $S \subseteq \{0, \ldots, d-1\}$, we denote by f_S the number of chains of faces (*flags*) in Q, $F_1 \subset \cdots \subset F_k$, with $S = \{\dim F_1, \ldots, \dim F_k\}$. The vector consisting of all the numbers f_S, $S \subseteq \{0, \ldots, d-1\}$, is called the *flag f-vector* of Q. The affine span of the flag f-vectors of all polytopes (more generally, of all Eulerian posets) is described by a system of linear equations, known as the generalized Dehn–Sommerville equations [1].

For any $S \subseteq \{0, \ldots, d-1\}$, we set $h_S = \sum_{T \subseteq S} (-1)^{|S-T|} f_T$. Define a polynomial in the non-commuting variables $\mathbf{a}$ and $\mathbf{b}$, called the **ab**-*index*, by

$$\Psi(Q) = \sum_{S} h_S \cdot u_S,$$

where $u_S = z_0 \cdots z_{d-1}$, $z_i = \mathbf{b}$ if $i \in S$ and $z_i = \mathbf{a}$ if $i \notin S$. An implicit encoding of the generalized Dehn–Sommerville equations is given by the fact that $\Psi(Q)$ is always a polynomial in the variables $\mathbf{c} = \mathbf{a}+\mathbf{b}$ and $\mathbf{d} = \mathbf{a}\cdot\mathbf{b}+\mathbf{b}\cdot\mathbf{a}$. We call the polynomial the **cd**-*index* of Q.

As an example, the **cd**-index of a polygon Q is given by

$$\Psi(Q) = \mathbf{c}^2 + (f_0 - 2) \cdot \mathbf{d} \tag{1.1}$$

and the **cd**-index of a 3-dimensional polytope Q is given by

$$\Psi(Q) = \mathbf{c}^3 + (f_0 - 2) \cdot \mathbf{dc} + (f_2 - 2) \cdot \mathbf{cd}. \tag{1.2}$$

In Section 2, we discuss the operations of taking pyramids and prisms, and we use them to give a direct proof that the flag f-vectors of all polytopes span the linear space determined by the generalized Dehn–Sommerville equations. We next discuss zonotopes and three operations on them – projection, Minkowski sum with a line segment and prism. We use the coalgebra techniques of [7] to determine their effect on the **cd**-index. In Section 4, we show the **cd**-index of an n-fold iterated Minkowski sum is a polynomial function of n, and we use this in Section 5 to show that the flag f-vectors of zonotopes also span the space of all flag f-vectors. This result is extended in Section 6 by determining the lattice spanned by the **cd**-indices of all zonotopes. It is the ring of all integral polynomials in $\mathbf{c}$ and $2\mathbf{d}$. In terms of flag f-vectors, this is equivalent to saying that f_S is divisible by $2^{|S|}$. Some observations and concluding remarks are indicated in the final section.

The authors thank Gábor Hetyei and the two referees for making useful comments on an earlier version of this paper.

2. Polytopes span

For a field $\mathbf{k}$ of characteristic 0, let $\mathcal{F}$ be the polynomial algebra in non-commuting variables $\mathbf{c}$ and $\mathbf{d}$ over the field $\mathbf{k}$, that is, $\mathcal{F} = \mathbf{k}\langle \mathbf{c}, \mathbf{d} \rangle$. (In fact, everything we do here works in any characteristic other than 2.) If we set the degree of $\mathbf{c}$ to 1 and the degree of $\mathbf{d}$ to 2, we define $\mathcal{F}_d$ to be all polynomials in $\mathcal{F}$ that are homogeneous of degree d.

Recall that a derivation f on an algebra A is a linear map satisfying the product rule $f(x \cdot y) = f(x) \cdot y + x \cdot f(y)$, and that $f(1) = 0$. Observe that it is enough to determine how the derivation acts on a set of generators, and hence we may describe a derivation on $\mathcal{F}$ by giving its value on the elements $\mathbf{c}$ and $\mathbf{d}$. Define two derivations D and G on $\mathcal{F}$ by $D(\mathbf{c}) = 2 \cdot \mathbf{d}$, $D(\mathbf{d}) = \mathbf{c} \cdot \mathbf{d} + \mathbf{d} \cdot \mathbf{c}$, $G(\mathbf{c}) = \mathbf{d}$, and $G(\mathbf{d}) = \mathbf{c} \cdot \mathbf{d}$. Observe that both these derivations increase the degree by 1, that is, they are maps from $\mathcal{F}_d$ to $\mathcal{F}_{d+1}$.

For a polytope Q we denote the pyramid over Q by $\mathrm{Pyr}(Q)$. Likewise, denote the prism over Q by $\mathrm{Pri}(Q)$. We similarly denote two linear maps $\mathrm{Pyr}, \mathrm{Pri} : \mathcal{F} \longrightarrow \mathcal{F}$, by

$$\mathrm{Pyr}(w) = w \cdot \mathbf{c} + G(w)$$

and

$$\mathrm{Pri}(w) = w \cdot \mathbf{c} + D(w).$$

The following results are proved in [7] using coalgebra techniques (see [7, Theorems 4.4 and 5.2]).

Proposition 2.1. *For a polytope Q we have that*

$$\begin{aligned} \Psi(\mathrm{Pyr}(Q)) &= \mathrm{Pyr}(\Psi(Q)), \\ \Psi(\mathrm{Pri}(Q)) &= \mathrm{Pri}(\Psi(Q)). \end{aligned}$$

Lemma 2.2. *The linear span of the two sets* $\mathrm{Pyr}(\mathcal{F}_d)$ *and* $\mathrm{Pri}(\mathcal{F}_d)$ *is the linear space* $\mathcal{F}_{d+1}$.

Proof. Define a third derivation G' on the algebra $\mathcal{F}$ by $G'(\mathbf{c}) = \mathbf{d}$ and $G'(\mathbf{d}) = \mathbf{d} \cdot \mathbf{c}$. It follows that $w \cdot \mathbf{c} + G(w) = \mathbf{c} \cdot w + G'(w)$ for all $w \in \mathcal{F}$ (see [7, Lemma 5.1]).

Observe that $\mathrm{Pri}(w) - \mathrm{Pyr}(w) = D(w) - G(w) = G'(w)$. Thus the statement of the lemma is equivalent to that $\mathrm{Pyr}(\mathcal{F}_d)$ and $G'(\mathcal{F}_d)$ span the space $\mathcal{F}_{d+1}$. Let $V = \mathrm{Pyr}(\mathcal{F}_d) + G'(\mathcal{F}_d)$.

Let w be an element in $\mathcal{F}_d$. Then we have that $\mathbf{c} \cdot w = w \cdot \mathbf{c} + G(w) - G'(w) = \mathrm{Pyr}(w) - G'(w)$. Hence $\mathbf{c} \cdot w$ belongs to V.

Let v be in $\mathcal{F}_{d-1}$. Then we have that $G'(\mathbf{c} \cdot v) = \mathbf{d} \cdot v + \mathbf{c} \cdot G'(v)$. Since $\mathbf{c} \cdot G'(v)$ belongs to V by the previous paragraph and $G'(\mathbf{c} \cdot v)$ also belongs to V, we have $\mathbf{d} \cdot v \in V$.

Since a monomial in $\mathcal{F}_{d+1}$ begins either with a $\mathbf{c}$ or a $\mathbf{d}$, we conclude $V = \mathcal{F}_{d+1}$. ■

From Lemma 2.2, we conclude directly the basic result that the linear span of all flag f-vectors has dimension given by the Fibonacci numbers [1].

Theorem 2.3. *Beginning with a point one can produce, by repeated use of the operations* Pyr *and* Pri, *a set of polytopes whose* **cd**-*indices span* $\mathcal{F}$.

We note that this approach does not identify a specific basis, as was done in [1] and [12]. We end this section with a few useful facts.

Lemma 2.4. *The two linear maps* Pri *and* Pyr *are injective. The linear map G has kernel generated by 1, and the linear map D has kernel generated by the elements of the form* $(\mathbf{c}^2 - 2 \cdot \mathbf{d})^j$, $j \geq 0$.

Proof. Let $\mathcal{F}_d^{(i)}$ be the linear span of all monomials of degree d containing i $\mathbf{d}$'s. Define two derivations D_0 and D_1 by: $D_0(\mathbf{c}) = 0$, $D_0(\mathbf{d}) = \mathbf{cd} + \mathbf{dc}$, $D_1(\mathbf{c}) = 2 \cdot \mathbf{d}$, and $D_1(\mathbf{d}) = 0$. Define two linear maps Pri_0 and Pri_1 by: $\mathrm{Pri}_0(v) = D_0(v) + v \cdot \mathbf{c}$ and $\mathrm{Pri}_1(v) = D_1(v)$. We have that $\mathrm{Pri} = \mathrm{Pri}_0 + \mathrm{Pri}_1$, and Pri_j is a linear map from $\mathcal{F}_d^{(i)}$ to $\mathcal{F}_{d+1}^{(i+j)}$.

Define a linear map $\phi: \ \mathcal{F}_d^{(i)} \to \mathbf{k}[x_0, \ldots, x_i]$ by

$$\phi(\mathbf{c}^{n_0}\mathbf{d}\mathbf{c}^{n_1}\mathbf{d} \cdots \mathbf{d}\mathbf{c}^{n_i}) = x_0^{n_0} x_1^{n_1} \cdots x_i^{n_i}.$$

This map takes the linear space $\mathcal{F}_d^{(i)}$ isomorphically onto the linear space of homogeneous polynomials of degree $d - 2 \cdot i$ in the variables $x_0, \ldots, x_i$. Moreover, we have

$$\phi(\mathrm{Pri}_0(w)) = (x_0 + 2 \cdot x_1 + \cdots + 2 \cdot x_i) \cdot \phi(w).$$

Since the ring of polynomials is an integral domain, we have $(x_0 + 2 \cdot x_1 + \cdots + 2 \cdot x_i)$ is not a zero divisor. Hence $\mathrm{Pri}_0 : \mathcal{F}_d^{(i)} \longrightarrow \mathcal{F}_{d+1}^{(i)}$ is an injective map. The linear map Pri corresponds to a block matrix B, where the row labels of the block $B_{i,j}$ correspond to monomials having i $\mathbf{d}$'s and the column labels correspond to monomials having j $\mathbf{d}$'s. The blocks on the diagonal of B are described by Pri_0 and the blocks below the diagonal are equal to zero. We thus conclude Pri is also injective.

The proof is similar for the linear map Pyr. The only difference is that we obtain another polynomial of degree 1, namely the polynomial $x_0 + \cdots + x_i$.

For the two linear maps D and G we need to modify the argument. For D we get the polynomial $x_0+2x_1+\cdots+2x_{i-1}+x_i$ and for G we get $x_0+\cdots+x_{i-1}$. We can now obtain that the two linear maps $D_0, G_0 : \mathcal{F}_d^{(i)} \longrightarrow \mathcal{F}_{d+1}^{(i)}$ are injective when $i \geq 1$, while when $i = 0$ they are the zero maps. Since $\dim(\mathcal{F}_d^{(0)}) = 1$, the dimensions of the kernels of the linear maps $D, G : \mathcal{F}_d \longrightarrow \mathcal{F}_{d+1}$ are each at most 1.

For $d \geq 1$ it is easy to see that G restricted to $\mathcal{F}_d^{(0)} \oplus \mathcal{F}_d^{(1)}$ is an injective map. Thus G can still be divided into blocks so that the blocks on the main diagonal are injective. Hence we conclude that for $d \geq 1$ the map $G : \mathcal{F}_d \longrightarrow \mathcal{F}_{d+1}$ is injective, that is, the kernel of G is generated by the polynomial 1.

When d is odd, one can similarly obtain that D restricted to $\mathcal{F}_d^{(0)} \oplus \mathcal{F}_d^{(1)}$ is an injective map. Hence $D : \mathcal{F}_d \longrightarrow \mathcal{F}_{d+1}$ is an injective map for d odd. Finally, when d is even it is easy to see that $D\left((\mathbf{c}^2 - 2 \cdot \mathbf{d})^{d/2}\right) = 0$. Hence the kernel of D is generated by elements of the form $(\mathbf{c}^2 - 2 \cdot \mathbf{d})^j$. ∎

Corollary 2.5. *For all non-negative integers k we have that $D^k(\mathbf{c})$ is non-zero.*

Proof. The proof is by induction on k. It follows directly for $k = 0$ and $k = 1$. Assume for $k \geq 1$ that $D^k(\mathbf{c})$ is non-zero. Observe that the coefficient of $\mathbf{c}^{k+1}$ in $D^k(\mathbf{c})$ is zero. Hence $D^k(\mathbf{c})$ is not a scalar multiple of $(\mathbf{c}^2 - 2 \cdot \mathbf{d})^j$ and so $D^k(\mathbf{c})$ does not belong to the kernel of D. We conclude that $D^{k+1}(\mathbf{c})$ is non-zero. ∎

3. Zonotopes

The *Minkowski sum* of two subsets X and Y of $\mathbb{R}^d$ is defined as

$$X + Y = \{\mathbf{x} + \mathbf{y} \in \mathbb{R}^d \ : \ \mathbf{x} \in X, \mathbf{y} \in Y\}.$$

Notably, the Minkowski sum of two convex polytopes is another convex polytope. For a vector $\mathbf{x}$ we denote the set $\{\lambda \cdot \mathbf{x} \ : \ 0 \leq \lambda \leq 1\}$ by $[\mathbf{0}, \mathbf{x}]$. We denote by $\mathrm{aff}(X)$ the *affine span* of X, that is, the intersection of all affine subspaces containing the set X.

We say that the nonzero vector $\mathbf{x} \in \mathrm{aff}(Q)$ lies in *general position* with respect to the convex polytope Q if the line $\{\lambda \cdot \mathbf{x} + \mathbf{u} \in \mathbb{R}^d \ : \ \lambda \in \mathbb{R}\}$ intersects the boundary of the polytope Q in at most two points for all $\mathbf{u} \in \mathbb{R}^d$. Alternatively, $\mathbf{x} \in \mathrm{aff}(Q)$ is in general position if $\mathbf{x}$ is parallel to no proper face of Q.

From [7, Prop. 6.5] we have the following result. Let Q be a d-dimensional convex polytope and $\mathbf{x}$ a nonzero vector that lies in general position with respect to the polytope Q. Let H be a hyperplane orthogonal to the vector $\mathbf{x}$, and let $\mathrm{Proj}(Q)$ be the orthogonal projection of Q onto the hyperplane H. Observe that $\mathrm{Proj}(Q)$ is a $(d-1)$-dimensional convex polytope.

Proposition 3.1. *The* **cd**-*index of the Minkowski sum of* Q *and* $[\mathbf{0}, \mathbf{x}]$ *is given by*

$$\Psi(Q + [\mathbf{0}, \mathbf{x}]) = \Psi(Q) + D(\Psi(\mathrm{Proj}(Q))).$$

A *zonotope* is the Minkowski sum of line segments. That is, if $\mathbf{x}_1, \ldots, \mathbf{x}_n \in \mathbb{R}^d$, then the zonotope they generate is the Minkowski sum

$$Z = [\mathbf{0}, \mathbf{x}_1] + \cdots + [\mathbf{0}, \mathbf{x}_n].$$

A (central) *hyperplane arrangement* is a finite collection $\mathcal{H}$ of linear hyperplanes in $\mathbb{R}^d$. An arrangement is called *essential* if the intersection of all its hyperplanes is the origin. An arrangement $\mathcal{H}$ induces a subdivision of $\mathbb{R}^d$ into relatively open cells whose closures are ordered by inclusion. The resulting poset is a lattice, called the *face lattice* of $\mathcal{H}$. An arrangement $\mathcal{H} \subset \mathbb{R}^d$ has a natural flag f-vector with components $f_S(\mathcal{H})$, where $S \subseteq \{1, \ldots, d\}$. The face lattice of Z is anti-isomorphic to that of the central arrangement $\mathcal{H}$ of the n hyperplanes with normals $\mathbf{x}_1, \ldots, \mathbf{x}_n$ [5, Prop. 2.2.2]. If Z is d-dimensional, then its flag f-vector and that of its dual hyperplane arrangement are related by $f_S(Z) = f_{d-S}(\mathcal{H})$, where $S = \{i_1, \ldots, i_k\} \subseteq \{0, \ldots, d-1\}$ and $d - S = \{d - i_k, \ldots, d - i_1\}$.

Two important and useful facts about the combinatorial behavior of zonotopes are the following:

1. The face lattice of Z is determined by the *oriented matroid* of the point configuration $\{\mathbf{x}_1, \ldots, \mathbf{x}_n\}$ [5, Prop. 2.2.2], and

2. The flag f-vector of Z is determined by the *matroid* of the configuration $\{\mathbf{x}_1, \ldots, \mathbf{x}_n\}$ [5, Cor. 4.6.3].

Since we are only interested here in invariants derivable from the flag f-vector, we will consider two zonotopes to be equal if they have the same underlying matroid. In this case it will be important, when defining operations on zonotopes, to show that they only depend on their matroids.

For a zonotope Z we note that the combinatorial type of the prism over Z can be realized as the zonotope $\mathrm{Pri}(Z) = Z + [\mathbf{0}, \mathbf{x}]$ for any $\mathbf{x} \notin \mathrm{aff}(Z)$. At the level of matroids, this involves adding a new element independent of all the original ones, that is, forming a free extension of one higher rank.

We define a zonotope $M(Z)$ by

$$M(Z) = Z + [\mathbf{0}, \mathbf{x}],$$

where $\mathbf{x}$ lies in general position with respect to Z. While the combinatorial type of $M(Z)$ depends on the choice of $\mathbf{x}$, its matroid is well-defined. This follows since the underlying matroid of $M(Z)$ is always a free extension (of the same rank) of the matroid of Z, that is, an extension such that $\mathbf{x}$ lies on no proper subspace spanned by the generators $\mathbf{x}_1, \ldots, \mathbf{x}_n$.

Finally, we define the zonotope $\pi(Z)$ to be the projection of $M(Z)$ along the direction $\mathbf{x}$, that is, onto the hyperplane orthogonal to $\mathbf{x}$. Observe that $\pi(Z)$ is the projection $\mathrm{Proj}(Z)$ in a general direction. The matroid of the zonotope $\pi(Z)$ is well-defined, since it is obtained by contracting $\mathbf{x}$ in the matroid of $M(Z)$.

Directly as a corollary of Proposition 3.1 we have

Corollary 3.2. *For a zonotope Z we have*

$$\Psi(M(Z)) - \Psi(Z) = D(\Psi(\pi(Z))).$$

The operations Pri, M, and π were used by Liu [14] to give a lower bound on the dimension of the span of the flag f-vectors of zonotopes. The relationship between these operations is given by the following lemma. The second relation was first observed by Liu in [14, Theorem 4.2.7].

Lemma 3.3. *For a zonotope Z we have, up to matroid,*

$$\pi(M(Z)) = M(\pi(Z))$$

and

$$\pi(\mathrm{Pri}(Z)) = M(Z).$$

Proof. In each pair we check that the underlying matroids are the same.

For $\pi(M(Z))$ one makes a free extension of Z by $\mathbf{x}$ and again by $\mathbf{y}$ (both in $\mathrm{aff}(Z)$), then contracting $\mathbf{y}$. The image of $\mathbf{x}$ under this contraction is still free with respect to the images of $\mathbf{x}_1, \ldots, \mathbf{x}_n$, so the resulting matroid is the same as that of $M(\pi(Z))$.

For $\pi(\text{Pri}(Z))$ the description is the same, except now neither $\mathbf{x}$ nor $\mathbf{y}$ is in $\text{aff}(Z)$. In this case, the images of $\mathbf{x}_1, \ldots, \mathbf{x}_n$ will have the same matroid as $M(Z)$. ∎

4. Polynomial functions

In this section we define polynomial functions and derive some of their properties. These functions play a role in the proof of our main theorem. Let V and W be vector spaces over the field $\mathbf{k}$.

Definition 4.1. *A function $f : \mathbb{N} \longrightarrow V$ is called a* polynomial function of degree d *if it can be written in the form*

$$f(n) = \mathbf{v}_d \cdot \binom{n}{d} + \mathbf{v}_{d-1} \cdot \binom{n}{d-1} + \cdots + \mathbf{v}_0 \cdot \binom{n}{0},$$

where $\mathbf{v}_0, \ldots, \mathbf{v}_d \in V$ and $\mathbf{v}_d \neq \mathbf{0}$. We call $\mathbf{v}_d$ the leading coefficient.

Observe that $\binom{n}{d}$ is defined by the Pascal relation in any characteristic. We define the *difference operator* Δ by $\Delta f(n) = f(n+1) - f(n)$. The following proposition contains the essential results we will need about polynomial functions.

Proposition 4.2. *Let $f : \mathbb{N} \longrightarrow V$ be a polynomial function of degree d.*

(i) If $\phi : V \longrightarrow W$ is a linear map then the composition $\phi \circ f : \mathbb{N} \longrightarrow W$ is a polynomial function of degree at most d. If ϕ applied to the leading coefficient is non-zero then the degree is d.

(ii) The function $\Delta f(n)$ is a polynomial function of degree $d-1$.

(iii) If g is a function from $\mathbb{N}$ to V such that $\Delta g(n) = f(n)$ then g is a polynomial function of degree $d+1$ with the same leading coefficient as f.

(iv) The vector $f(0)$ is in the linear span of $f(1), \ldots, f(d+1)$.

Proof. Let $f(n)$ be the polynomial function of degree d

$$f(n) = \mathbf{v}_d \cdot \binom{n}{d} + \mathbf{v}_{d-1} \cdot \binom{n}{d-1} + \cdots + \mathbf{v}_0 \cdot \binom{n}{0}.$$

(i) Observe that

$$(\phi \circ f)(n) = \phi(\mathbf{v}_d) \cdot \binom{n}{d} + \phi(\mathbf{v}_{d-1}) \cdot \binom{n}{d-1} + \cdots + \phi(\mathbf{v}_0) \cdot \binom{n}{0},$$

which is a polynomial function of degree at most d. When $\phi(\mathbf{v}_d) \neq \mathbf{0}$ we have that $\phi \circ f$ is of degree d.

(ii) It is straightforward to obtain

$$\Delta f(n) = \mathbf{v}_d \cdot \binom{n}{d-1} + \mathbf{v}_{d-1} \cdot \binom{n}{d-2} + \cdots + \mathbf{v}_1 \cdot \binom{n}{0},$$

which proves (ii).

(iii) By induction on n we have

$$g(n) = \mathbf{v}_d \cdot \binom{n}{d+1} + \mathbf{v}_{d-1} \cdot \binom{n}{d} + \cdots + \mathbf{v}_0 \cdot \binom{n}{1} + g(0),$$

which is a polynomial function of degree $d+1$. The leading coefficient is $\mathbf{v}_d$, which is the leading coefficient of f.

(iv) By property (ii) we know that $\Delta^d f(n)$ is a polynomial function of degree 0, hence it is a constant. Thus $\Delta^d f(0) = \Delta^d f(1)$. But $\Delta^d f(0)$ is a linear combination of $f(0), \ldots, f(d)$ and $\Delta^d f(1)$ is a linear combination of $f(1), \ldots, f(d+1)$. The coefficient of $f(0)$ in $\Delta^d f(0)$ is $(-1)^d$, which is nonzero, and hence the relation $\Delta^d f(0) = \Delta^d f(1)$ gives the desired result. ∎

Observe that Proposition 4.2 and its proof are valid in any characteristic for the field $\mathbf{k}$ since $(-1)^d$ is never zero. Moreover, it applies to Abelian groups ($\mathbb{Z}$-modules) as well. This last fact will be used in Section 6.

The main result of this section shows that the **cd**-index of iterates of the operation M is a polynomial function.

Theorem 4.3. *Let Z be a d-dimensional zonotope. Then the mapping $n \mapsto \Psi(M^n(Z))$ is a polynomial function of degree $d-1$ into $\mathcal{F}_d$ with leading coefficient $D^{d-1}(\mathbf{c})$.*

Proof. The proof is by induction on d. The base case is $d = 2$. Assume that Z is a 2-dimensional zonotope, that is, Z is a $2k$-gon. Then $M(Z)$ is a $(2k+2)$-gon, and $M^n(Z)$ is a $(2k+2n)$-gon. By equation (1.1) we have the **cd**-index of $M^n(Z)$ is given by $\Psi(M^n(Z)) = \mathbf{c}^2 + (2k+2n-2) \cdot \mathbf{d} = 2 \cdot n \cdot \mathbf{d} + \mathbf{c}^2 + (2k-2) \cdot \mathbf{d}$. This is a polynomial function of degree 1 in n with leading coefficient $2 \cdot \mathbf{d} = D(\mathbf{c})$.

Assume that $d \geq 3$ and let $W = \pi(Z)$. Observe that W is a $(d-1)$-dimensional zonotope. Now by Corollary 3.2 and Lemma 3.3 we have

$$\begin{aligned} \Delta(\Psi(M^n(Z))) &= \Psi(M^{n+1}(Z)) - \Psi(M^n(Z)) \\ &= D(\Psi(M^n(\pi(Z)))) \\ &= D(\Psi(M^n(W))). \end{aligned}$$

By the induction hypothesis we know that $n \mapsto \Psi(M^n(W))$ is a polynomial function of degree $d-2$ with leading coefficient $D^{d-2}(\mathbf{c})$. By Corollary 2.5 and by property (i) in Proposition 4.2, we have $n \mapsto D(\Psi(M^n(W)))$ is a polynomial function of degree $d-2$ with non-zero leading term $D^{d-1}(\mathbf{c})$. Finally, by property (iii) in the same proposition we complete the induction. ■

5. Zonotopes span

Let $\mathcal{G}_d$ be the linear span of the **cd**-indices of zonotopes of dimension d. Liu proved that $\dim \mathcal{G}_d \geq \dim \mathcal{G}_{d-1} + \dim \mathcal{G}_{d-3}$ [14, Theorem 4.7.1]. In this section we prove that $\dim \mathcal{G}_d = \dim \mathcal{G}_{d-1} + \dim \mathcal{G}_{d-2}$, that is, $\mathcal{G}_d$ equals $\mathcal{F}_d$.

Since zonotopes are polytopes, we know that $\mathcal{G}_d \subseteq \mathcal{F}_d$. We first prove a variation of Lemma 2.2 that substitutes D for Pyr in order to be able to operate solely with zonotopes.

Lemma 5.1. *The linear span of the two sets $D(\mathcal{F}_d)$ and* $\mathrm{Pri}(\mathcal{F}_d)$ *is the whole space $\mathcal{F}_{d+1}$.*

Proof. Let V be the subspace of $\mathcal{F}_d$ which is spanned by $D(\mathcal{F}_d)$ and $\mathrm{Pri}(\mathcal{F}_d)$, that is, $V = D(\mathcal{F}_d) + \mathrm{Pri}(\mathcal{F}_d)$.

Let $w \in \mathcal{F}_d$. Since $w \cdot \mathbf{c} = \mathrm{Pri}(w) - D(w) \in V$, we know that every **cd**-monomial which ends with a $\mathbf{c}$ belongs to $\mathrm{Pri}(\mathcal{F}_d) + D(\mathcal{F}_d)$.

Consider $v \in \mathcal{F}_{d-1}$. We have that $D(v \cdot \mathbf{c}) = D(v) \cdot \mathbf{c} + 2 \cdot v \cdot \mathbf{d}$, and hence $v \cdot \mathbf{d} = \frac{1}{2} \cdot (D(v \cdot \mathbf{c}) - D(v) \cdot \mathbf{c})$. We have $D(v \cdot \mathbf{c}) \in V$. Moreover $D(v) \cdot \mathbf{c} \in V$ by the previous paragraph. Hence $v \cdot \mathbf{d} \in V$, and we conclude that every **cd**-monomial belongs to V. ■

The following result shows that the flag f-vectors of zonotopes made by the successive application of the operators Pri and M, beginning with $Z = \mathbf{0}$, span the space of all flag f-vectors of polytopes.

Theorem 5.2. *The* **cd**-*indices of d-dimensional zonotopes linearly span the space of* **cd**-*polynomials of degree d, that is, $\mathcal{G}_d = \mathcal{F}_d$.*

Proof. The proof is by induction on the dimension d; the case $d \leq 2$ is clear. We assume that the theorem holds for $d \geq 2$, hence $\mathcal{G}_d = \mathcal{F}_d$, and prove it for $d+1$. Assume that $\{Z_1, \ldots, Z_N\}$ form a spanning set of zonotopes of dimension d. Since $\Psi(\text{Pri}(Z_i)) = \text{Pri}(\Psi(Z_i))$ we have that $\text{Pri}(\mathcal{F}_d) \subseteq \mathcal{G}_{d+1}$.

By combining Theorem 4.3 and property (iv) in Proposition 4.2, we know that $\Psi(Z_i)$ lies in the linear span of $\Psi(M(Z_i)), \ldots, \Psi(M^d(Z_i))$. Hence, we know that $\{M^j(Z_i) \mid 1 \leq i \leq N, 1 \leq j \leq d\}$ is a spanning set of zonotopes. Observe that every zonotope in this spanning set is the Minkowski sum of a line segment with a d-dimensional zonotope. Hence we can describe this spanning set as $\{M(W_1), \ldots, M(W_{N \cdot d})\}$.

By Lemma 3.3 and Corollary 3.2 we have

$$\begin{aligned} \Psi(M(\text{Pri}(W_i))) - \Psi(\text{Pri}(W_i)) &= D(\Psi(\pi(\text{Pri}(W_i)))) \\ &= D(\Psi(M(W_i))). \end{aligned}$$

Since both $M(\text{Pri}(W_i))$ and $\text{Pri}(W_i)$ are $(d+1)$-dimensional zonotopes, we have $D(\Psi(M(W_i))) \in \mathcal{G}_{d+1}$. But since $\{M(W_i)\}$ forms a spanning set for $\mathcal{F}_d$, we obtain that $D(\mathcal{F}_d) \subseteq \mathcal{G}_{d+1}$. By Lemma 5.1 we obtain that $\mathcal{G}_{d+1} = \mathcal{F}_{d+1}$, which completes the induction. ■

Since the face lattice of a central hyperplane arrangement is an Eulerian poset, it has a **cd**-index, obtainable from that of its dual zonotope by reversing each **cd**-monomial.

Corollary 5.3. *The* **cd**-*indices of essential hyperplane arrangements in* $\mathbb{R}^d$ *linearly span the space of* **cd**-*polynomials of degree* d.

6. The integral span

We turn now to the problem of finding the *integral* span of flag f-vectors of zonotopes. This leads to an integral **c**-2**d**–index for zonotopes and central arrangements.

Let $\mathcal{R}$ be the ring in the non-commuting variables **c** and **d** over the integers $\mathbb{Z}$, that is, $\mathcal{R} = \mathbb{Z}\langle \mathbf{c}, \mathbf{d} \rangle$. As before let the degree of **c** be 1 and the degree of **d** be 2. Let $\mathcal{R}_d$ be all polynomials in $\mathcal{R}$ that are homogeneous of degree d. We view $\mathcal{R}_d$ as an Abelian group. Similarly, let $\mathcal{T} = \mathbb{Z}\langle \mathbf{c}, 2\mathbf{d} \rangle$ and let $\mathcal{T}_d = \mathcal{T} \cap \mathcal{R}_d$. For a **cd**-monomial w, let $p(w)$ be the number of **d**'s that occur in w. A generating set of $\mathcal{T}_d$ is $2^{p(w)} \cdot w$, where w ranges over all **cd**-monomials of degree d.

Observe that Lemma 2.2 and Theorem 2.3 have the following integer analogues.

Lemma 6.1. *The Abelian group $\mathcal{R}_{d+1}$ is generated by* $\mathrm{Pyr}(\mathcal{R}_d)$ *and* $\mathrm{Pri}(\mathcal{R}_d)$.

Theorem 6.2. *The Abelian group $\mathcal{R}_d$ is generated by the* **cd**-*index of d-dimensional polytopes.*

The goal of this section is to prove the analogous result of Theorem 6.2 for zonotopes. Let $\mathcal{S}_d$ be the subgroup of $\mathcal{R}_d$ generated by the elements $\Psi(Z)$, where Z ranges over all d-dimensional zonotopes. We begin by showing that $\mathcal{T}_d \subseteq \mathcal{S}_d$. This proof is essentially the same as the proof of Theorem 5.2. We need the following lemma.

Lemma 6.3. *The Abelian group $\mathcal{T}_{d+1}$ is generated by* $\mathrm{Pri}(\mathcal{T}_d)$ *and* $D(\mathcal{T}_d)$.

The proof differs from the proof of Lemma 5.1 in only one point. We do not divide by 2; we instead use the relation $2 \cdot v \cdot \mathbf{d} = D(v \cdot \mathbf{c}) - D(v) \cdot \mathbf{c}$ and the fact that the monomial $v \cdot \mathbf{d}$ contains one more $\mathbf{d}$ than v, that is, $p(v \cdot \mathbf{d}) = p(v) + 1$. We thus have that the generating set of $\mathcal{T}_{d+1}$ lies in the integral span of $\mathrm{Pri}(\mathcal{T}_d)$ and $D(\mathcal{T}_d)$.

The results in Section 4 also apply to Abelian groups as well as vector spaces. Hence the proof of Theorem 5.2 generalizes to a proof of the following result.

Proposition 6.4. *The Abelian group $\mathcal{T}_d$ is contained in the group $\mathcal{S}_d$.*

It remains to show the inclusion in the other direction, that is, $\mathcal{S}_d \subseteq \mathcal{T}_d$. For S a subset of $\{0, 1, \ldots, d-1\}$, we call S *sparse* if for all i, $\{i, i+1\} \not\subseteq S$ and $d-1 \notin S$. Suppose that S has cardinality p. Let w be a **cd**-monomial of degree d containing p **d**'s. We say that w *covers* the sparse set S if u_S appears in the expansion of $w = w(\mathbf{c}, \mathbf{d})$ as an **ab**-polynomial $w = w(\mathbf{a} + \mathbf{b}, \mathbf{ab} + \mathbf{ba})$. More explicitly, we can write $w = \mathbf{c}^{i_0} \cdot \mathbf{d} \cdot \mathbf{c}^{i_1} \cdot \mathbf{d} \cdots \mathbf{d} \cdot \mathbf{c}^{i_p}$, where $i_k \geq 0$. Define $j_0, \ldots, j_{p-1}$ by $j_0 = i_0$ and $j_{h+1} = j_h + 2 + i_{h+1}$. Observe that the hth $\mathbf{d}$ in w covers the positions j_h and j_{h+1}. Then w covers the sparse set S if and only if S is contained in the set $\{j_0, j_0+1, j_1, j_1+1, \ldots, j_{p-1}, j_{p-1}+1\}$. (Compare this notion with Stanley's definition of $\mathcal{W}_S$ [16].)

For a **cd**-monomial w and a **cd**-polynomial $F(\mathbf{c}, \mathbf{d})$, we denote the coefficient of w in $F(\mathbf{c}, \mathbf{d})$ by $[w]F(\mathbf{c}, \mathbf{d})$.

Definition 6.5. *For a d-dimensional polytope Q and a sparse subset S of $\{0, 1, \ldots, d-1\}$, define k_S by*

$$k_S = \sum_w [w]\Psi(Q),$$

where the sum ranges over all **cd**-*monomials w of degree d that cover S and contain exactly $|S|$* **d***'s.*

We call the vector (k_S), where S ranges over all sparse subsets, the *flag k-vector*. As an example, let $d = 8$ and $S = \{0, 3, 5\}$. Then we have

$$k_{\{0,3,5\}} = [\mathbf{d}^3\mathbf{c}^2]\Psi(Q) + [\mathbf{d}^2\mathbf{cdc}]\Psi(Q) + [\mathbf{dcd}^2\mathbf{c}]\Psi(Q).$$

As a refinement of Proposition 1.3 in [16] we obtain the following relation.

Proposition 6.6. *The coefficients of the* **cd**-*monomials containing* p **d**'*s can be expressed as an integer linear combination of* k_S'*s where* S *has cardinality* p. *That is, for* w *containing* p **d**'*s we have*

$$[w]\Psi(Q) = \sum_{|S|=p} q_{w,S} \cdot k_S,$$

where the sum ranges over sparse sets S *and* $q_{w,S}$ *are integers.*

The proof follows by ordering the sets and the monomials by lexicographic order. It is then easy to see that the defining relation of k_S corresponds to a lower triangular matrix with 1's on the main diagonal. Thus this linear relation is invertible over the integers.

Lemma 6.7. *For* T *a sparse subset of* $\{0, 1, \ldots, d-1\}$ *we have that*

$$h_T = \sum_{U \subseteq T} k_U.$$

The proof is by expanding the **cd**-index in terms of **a**'s and **b**'s and collecting terms.

Combining Lemma 6.7 with the relation $f_S = \sum_{T \subseteq S} h_T$, we obtain

$$f_S = \sum_{U \subseteq S} 2^{|S-U|} \cdot k_U. \tag{6.1}$$

By the Principle of Inclusion-Exclusion, the inverse of this relation is

$$k_S = \sum_{U \subseteq S} (-1)^{|S-U|} \cdot 2^{|S-U|} \cdot f_U. \tag{6.2}$$

Lemma 6.8. *For a zonotope* Z *we have that* $2^{|S|}$ *divides* f_S.

Proof. Observe that a zonotope is centrally symmetric and every face of a zonotope is a zonotope. Hence, every face of the zonotope Z is centrally symmetric. (Zonotopes are characterized by the central symmetry of all their faces, in fact, of their 2-dimensional faces. See [5, Proposition 2.2.14].)

We may count f_S, where $S = \{i_1 < \cdots < i_k\}$, by first choosing a face F_{i_k} of dimension i_k, then choosing an i_{k-1}-dimensional face of F_{i_k}, and so on. But since at each selection the face F_{i_j} is centrally symmetric (including Z), we know that there is an even number of choices of $F_{i_{j-1}}$. By multiplying together all these factors of 2, we obtain $2^{|S|}$. ■

Lemma 6.9. *For a zonotope Z we have that $k_S \equiv 0 \bmod 2^{|S|}$.*

Proof. It is enough to observe that $2^{|S|}$ divides $2^{|S-U|} \cdot f_U$. ∎

By combining Proposition 6.6 and Lemma 6.9 we obtain

Proposition 6.10. *The* **cd**-*index of a zonotope Z of dimension d belongs to $\mathcal{T}_d$. That is, $\mathcal{S}_d \subseteq \mathcal{T}_d$.*

Proof. It is enough to prove for a zonotope Z and a **cd**-monomial w that the coefficient of w in $\Psi(Z)$ is divisible by $2^{p(w)}$ where $p(w) = p$ is the number of **d**'s occurring in w. That is, $[w]\Psi(Z) \equiv 0 \bmod 2^p$.

Indeed, by Proposition 6.6 and Lemma 6.9 we have

$$[w]\Psi(Z) = \sum_{|S|=p} q_{w,S} \cdot k_S \equiv 0 \bmod 2^p,$$

where S ranges over all sparse subsets of $\{0, 1, \ldots, d-1\}$ having cardinality p. ∎

Combining Propositions 6.4 and 6.10 gives us the main result of this section.

Theorem 6.11. *The Abelian group generated by the* **cd**-*indices of zonotopes of dimension d is precisely $\mathcal{T}_d$, that is, all integral polynomials of degree d in the variables* **c** *and* 2**d**.

As a direct consequence of this theorem, Proposition 6.6, and equation (6.2), we get the following:

Corollary 6.12. *The lattice spanned by flag f-vectors of all d-zonotopes is the set of all integral vectors $f = (f_S)$ where f_S is divisible by $2^{|S|}$.*

Since in the relation $f_S(Z) = f_{d-S}(\mathcal{H})$ between a d-zonotope Z and its dual (essential) hyperplane arrangement $\mathcal{H}$ the sets S and $d - S$ have the same cardinality, we obtain the following.

Corollary 6.13. *The lattice spanned by flag f-vectors of all essential hyperplane arrangements in $\mathbb{R}^d$ is the set of all integral vectors $f = (f_S)$ where f_S is divisible by $2^{|S|}$.*

7. Concluding remarks

Our method proves that zonotopes span, but is there a nice basis? We describe one possible basis, suggested in [14]. To do so, we define two operations P and B on a zonotope Z, where $PZ := \mathrm{Pri}(Z)$ and $BZ := M(\mathrm{Pri}(Z))$. Note that both result in a zonotope of one higher dimension. Now if we write a BP-word of length d, that is, a word of length d made with the letters B and P, we may view this as a sequence of d operations performed on the 0-dimensional zonotope $\mathbf{0}$, and so as a d-dimensional zonotope. Liu [14] conjectured that a basis for the flag f-vectors (and hence, the **cd**-indices) of all d-dimensional zonotopes could be constructed by forming all BP-words of length d ending in P and having no two consecutive B's. This should be compared to the basis for all polytopes given in [1] which was made up of similar combinations of pyramid and bipyramid operations.

We have described the lattice spanned by all **cd**-indices of zonotopes. The next natural problem is to determine all linear inequalities that they must satisfy. It is known that the **cd**-index of any polytope must be non-negative [16]. What more can be said about zonotopes? There is a family of linear inequalities known to be satisfied by flag f-vectors of zonotopes.

Theorem 7.1 (Varchenko/Liu). *If Z is a d-dimensional zonotope and $S = \{i_1, \ldots, i_k\}$, then*

$$\frac{f_S(Z)}{f_{i_1}(Z)} < \binom{d - i_1}{i_2 - i_1, \ldots, i_k - i_{k-1}, d - i_k} \cdot 2^{i_k - i_1}.$$

For the case $k = 2$ this was proved in [19] (see also [5, §4.6]) and stated in [9]. For this generality a proof is given in [14]. Theorem 7.1 bounds the average number of $S - \{i_1\}$ chains in links of i_1-faces of a d-dimensional zonotope by the number of $S - \{i_1\}$ chains in a $(d - i_1)$-dimensional crosspolytope (all with the dimensions shifted appropriately). It is easy to find polytopes for which the inequalities in Theorem 7.1 fail. For example, the cyclic polytope $C_d(n)$ does not satisfy the inequality for $S = \{0, 1\}$ when $n \geq 2d + 1$.

For 3-zonotopes it is enough to consider the pairs (f_0, f_2). In this case the convex hull taken over all 3-zonotopes can be completely described as the cone with apex $(8, 6)$ (corresponding to the 3-cube), and extreme rays $(1, 1)$ and $(2, 1)$. All zonotopes of the form $M^n(\mathrm{Pri}(\text{square}))$ can be found along the ray $(1, 1)$, while on the other ray one finds all those of the form $\mathrm{Pri}(M^n(\text{square}))$ (prisms over even polygons). Other than the fact that only even points appear, the problem of determining which lattice points in this cone are actually realized by zonotopes (or by oriented matroids) appears to be a difficult one. See, for example, [10, Chap. 18].

The **cd**-index of a zonotope does depend only on the underlying matroid, and not on the oriented matroid. This suggests that there is a **cd**–index for matroids, in fact, a **c**-2**d**–index, independent of whether they are orientable or not. The authors are currently investigating the **cd**-index without reference to orientation.

For polytopes and certain classes of Eulerian posets the flag h-vector has been given a combinatorial interpretation. We wonder if the flag k-vector can also be given a combinatorial interpretation. Observe that we only define k_S for sparse sets S. We may extend the flag k-vector to all sets by inverting the relation given in Lemma 6.7. It is known that the flag h-vector of a polytope, being the "fine" h-vector of a balanced Cohen–Macaulay complex, must be the fine f-vector of another balanced simplicial complex Δ [18, Thm 4.6]. Thus, in this case, the flag k-vector can be interpreted as the fine h-vector of this Δ.

By the definition of the flag k-vector, $k_S \geq 0$ for all sparse S whenever the **cd**-index is non-negative. Is there a larger interesting class of posets for which the sparse flag k-vector is always non-negative? For example, in the case just described this will occur if the complex Δ is itself Cohen–Macaulay. Here the full flag k-vector will be non-negative. The full flag k-vector is not always non-negative as can be seen by examining the flag k-vector of a tetrahedron, for which $k_{\{0,1\}} = -4$.

References

[1] M. Bayer and L. Billera, Generalized Dehn–Sommerville relations for polytopes, spheres and Eulerian partially ordered sets, *Invent. Math.* **79** (1985), 143–157.

[2] M. Bayer and A. Klapper, A new index for polytopes, *Discrete Comput. Geom.* **6** (1991), 33–47.

[3] M. Bayer and B. Sturmfels, Lawrence polytopes, *Canad. J. Math.* **42** (1990), 62–79.

[4] L. Billera and N. Liu, Noncommutative enumeration in graded posets, in preparation.

[5] A. Björner, M. Las Vergnas, B. Sturmfels, N. White, G. Ziegler, "Oriented Matroids," Cambridge University Press, Cambridge, 1993.

[6] R. Ehrenborg, On posets and Hopf algebras, *Adv. Math.* **119** (1996), 1–25.

[7] R. Ehrenborg and M. Readdy, Coproducts and the **cd**-index, to appear in *J. Algebraic Combin.*

[8] K. Fukuda, S. Saito, and A. Tamura, Combinatorial face enumeration in arrangements of hyperplanes and oriented matroids, *Discrete Appl. Math.* **31** (1991), 141–149.

[9] K. Fukuda, S. Saito, A. Tamura, and T. Tokuyama, Bounding the number of k-faces in arrangements of hyperplanes, *Discrete Appl. Math.* **31** (1991), 151–165.

[10] B. Grünbaum, "Convex Polytopes," John Wiley and Sons, London, 1967.

[11] S.A. Joni and G.-C. Rota, Coalgebras and Bialgebras in Combinatorics, *Stud. Appl. Math.* **61** (1979), 93–139.

[12] G. Kalai, A new basis of polytopes, *J. Combin. Theory Ser. A* **49** (1988), 191–209.

[13] Gian-Carlo Rota on Combinatorics, Introductory Papers and Commentaries, J.P.S. Kung ed., Birkhäuser, Boston, 1995.

[14] N. Liu, "Algebraic and combinatorial methods for face enumeration in polytopes," Doctoral dissertation, Cornell University, Ithaca, New York, 1995.

[15] R. P. Stanley, "Enumerative Combinatorics, Vol. I," Wadsworth and Brooks/Cole, Pacific Grove, 1986.

[16] R.P. Stanley, Flag f-vectors and the *cd*-index, *Math. Z.* **216** (1994), 483–499.

[17] R.P. Stanley, A survey of Eulerian posets, in: "Polytopes: Abstract, Convex, and Computational," T. Bisztriczky, P. McMullen, R. Schneider, A.I. Weiss, eds., NATO ASI Series C, vol. 440, Kluwer Academic Publishers, 1994.

[18] R. P. Stanley, "Combinatorics and Commutative Algebra," 2nd Edition, Birkhäuser, Boston, 1996.

[19] A. Varchenko, On the numbers of faces of a configuration of hyperplanes, *Soviet Math. Dokl.* **38** (1989), 291–295.

[20] G.M. Ziegler, "Lectures on Polytopes," Graduate Texts in Mathematics 152, Springer-Verlag, New York, 1995.

Louis J. Billera
Richard Ehrenborg
Margaret Readdy
Department of Mathematics
Cornell University
White Hall
Ithaca, NY 14853-7901.

Letter-Place Methods and Homotopy

David A. Buchsbaum

Dedicated to Gian-Carlo Rota

1. Introduction

In this paper I want to illustrate the use of letter-place methods in the construction of homotopies for resolutions of certain Weyl modules. I'll look at two cases: the resolutions of two-rowed skew-shapes (already in the literature [B-R 1]), and the resolutions of hooks (work in progress; only the first three stages of the homotopy appear here). The desire to construct such homotopies arose from the attempt to answer Rota's question: What do the syzygies of these modules look like? Since, in representation theory and combinatorics, the 'correct' form of an answer to such a question is in terms of a basis consisting of easily described tableaux, the double standard tableaux that appear in letter-place algebras lend themselves admirably to this topic.

Throughout this paper, we will be considering a fixed free module, F, over a commutative ring, and the representations of the general linear group of F. In general, we will omit the letter F. When we speak of projective resolutions of Weyl modules, the ring over which we are taking these resolutions is the Schur algebra of appropriate degree. Since the resolutions we are describing are universal and not necessarily minimal, we may as well assume that we are always working over the ring of integers, and hence the Schur algebra in question is the integral Schur algebra.

2. The two-rowed case

In this section we look at the general two-rowed skew-shape:

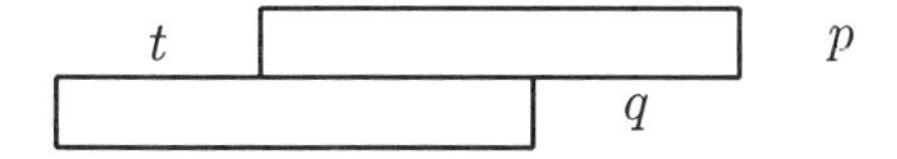

where the top row has p boxes, the second row has q boxes, and the protuberance of the second row to the left of the first is t. Expressed in terms of skew partitions this is simply $(p+t,q)/(t,0)$.

The Weyl module associated to this shape has the resolution:

$$0 \to M_{q-t} \xrightarrow{d_{q-t}} M_{q-t-1} \xrightarrow{d_{q-t-1}} \cdots \xrightarrow{d_3} M_2 \xrightarrow{d_2} M_1 \xrightarrow{d_1} M_0 \qquad (*)$$

where

$$M_0 = D_p \otimes D_q;$$
$$M_1 = \sum_{l>t} Z_{21}^{(l)} x D_{p+l} \otimes D_{q-l};$$
$$\vdots$$
$$M_k = \sum_{\substack{l_1>t \\ l_1>0}} Z_{21}^{(l_1)} x Z_{21}^{(l_2)} x \cdots x Z_{21}^{(l_k)} x D_{p+|l|} \otimes D_{q-|l|}$$
$$\vdots$$
$$M_{q-1} = \underbrace{Z_{21}^{(t+1)} x Z_{21}^{(1)} x \cdots x Z_{21}^{(1)}}_{q-1} x D_{p+q} \otimes D_0.$$

By D_r we mean the divided power of degree r of the underlying free module F. The symbol $Z_{21}^{(l)}$ stands for the divided power of degre l of the free generator Z_{21}, and the action of $Z_{21}^{(l)}$ on any term $D_u \otimes D_v$ is a place polarization of degree l from place one to place two. By $|l|$ we mean the sum $l_1 + \cdots + l_k$, and the symbol x stands for a separator variable (in the sense of [B-R 2]). The boundary maps are those of the Bar Complex as described in the same article. For example, using letter-place notation for the basis elements of $D_u \otimes D_v$, we can describe the boundary map $d_2 : M_2 \to M_1$ as follows:

$$d_2 \left(Z_{21}^{(l_1)} x Z_{21}^{(l_2)} x \left\langle \begin{array}{c|cc} w & 1^{(p+l_1+l_2)} & 2^{(q_1)} \\ w' & 2^{(q_2)} & \end{array} \right\rangle \right)$$
$$= Z_{21}^{(l_1)} x \binom{l_2+q_1}{q_1} \left\langle \begin{array}{c|cc} w & 1^{(p+l_1)} & 2^{(l_2+q_1)} \\ w' & 2^{(q_2)} & \end{array} \right\rangle$$
$$- \binom{l_1+l_2}{l_2} Z_{21}^{(l_1+l_2)} x \left\langle \begin{array}{c|cc} w & 1^{(p+l_1+l_2)} & 2^{(q_1)} \\ w' & 2^{(q_2)} & \end{array} \right\rangle.$$

As the letter-place notation implies, we are assuming that

$$q_1 + q_2 = q - l_1 - l_2;$$
$$q_2 \leq p + l_1 + l_2.$$

(Of course, given that $p + t \geq q$, the second condition is always satisfied.)

We describe the homotopy maps $s_k : M_k \to M_{k+1}$, $k=0,\ldots,q-t-1$ in [B-R 2] as follows:

$$s_0\left(\left\langle \begin{array}{c|cc} w & 1^{(p)} & 2^{(q_1)} \\ w' & 2^{(q_2)} & \end{array} \right\rangle\right) = \begin{cases} Z_{21}^{(q_1)} \quad x \left\langle \begin{array}{c|c} w & 1^{(p+q_1)} \\ w' & 2^{(q_2)} \end{array} \right\rangle & \text{if } q_1 > t\,, \\ 0 \qquad\qquad \text{otherwise} & \end{cases}$$

$$s_k\left(Z_{21}^{(l_1)} x \cdots x Z_{21}^{(l_k)} x \left\langle \begin{array}{c|cc} w & 1^{(p+|l|)} & 2^{(q_1)} \\ w' & 2^{(q_2)} & \end{array} \right\rangle\right),$$

$$= \begin{cases} Z_{21}^{(l_1)} x \cdots x Z_{21}^{(l_k)} x Z_{21}^{(q_1)} \quad x \left\langle \begin{array}{c|c} w & 1^{(p+|l|+q_1)} \\ w' & 2^{(q_2)} \end{array} \right\rangle & \text{if } q_1 > 0\,. \\ 0 \qquad\qquad \text{otherwise} & \end{cases}$$

It is easy to see that this is a splitting contracting homotopy for the resolution. It's well-known that, if we have any free complex $X = \{X_i; d_i\}$ with a splitting contracting homotopy $\{s_i\}$, if $\{x_{i\alpha}\}$ is the subset of the basis of X_i such that $s_i(x_{i\alpha}) \neq 0$, and the set $\{s_i(x_{i\alpha})\}$ is linearly independent, then the set $\{d_{i+1}s_i(x_{i\alpha})\}$ is a basis for the cycles in dimension i. Applying this to our case, we see that in dimension 0 the basis for the syzygies can be taken to be the set:

$$\left\{ Z_{21}^{(t+r)} x \left\langle \begin{array}{c|c} w & 1^{(p+t+r)} \\ w' & 2^{(q-t-r)} \end{array} \right\rangle;\ r > 0 \right\},$$

while in positive dimension k, the basis can be taken to be the set:

$$\left\{ Z_{21}^{(t+r_1)} x Z_{21}^{(r_2)} x \cdots x Z_{21}^{(r_{k+1})} x \left\langle \begin{array}{c|c} w & 1^{(p+t+|r|)} \\ w' & 2^{(q-t-|r|)} \end{array} \right\rangle;\ r_i > 0 \right\}.$$

3. The hooks

In this section we will write down a resolution of the Weyl modules corresponding to the partitions known as 'hooks'. This kind of partition is of the form $(p, 1^{(q)})$, and corresponds to the picture on the following page.

If one simply were looking for a resolution of this Weyl module, and were not interested in describing a basis for the syzgies, one could proceed as follows.

We first write down the projective resolution of Λ^k for each k. Our contention is that it's the following:

$$X^{(k)} : D_k \xrightarrow{\partial} \cdots \xrightarrow{\partial} \sum_{\substack{i_j>0 \\ i_1+\cdots+i_{k-p}=k}} D_{i_1} \otimes \cdots \otimes D_{i_{k-p}} \xrightarrow{\partial}$$

$$\sum_{\substack{i_j>0 \\ i_1+\cdots+i_{k-p+1}=k}} D_{i_1}\otimes\cdots\otimes D_{i_{k-p+1}} \xrightarrow{\partial} \cdots \xrightarrow{\partial} \underbrace{D_1\otimes\cdots\otimes D_1}_{k}$$

where

$$\partial(a_1\otimes\cdots\otimes a_{k-p}) = \sum_{j=1}^{k-p}(-1)^j a_1\otimes\cdots\otimes\Delta'(a_j)\otimes a_{j+1}\otimes\cdots\otimes a_{k-p}.$$

By $\Delta'(a_j)$ we mean:

$$\Delta'(a_j) = \sum_{0<l<i_j} a_j(l)\otimes a_j(i_j-l).$$

This is a complex that can be seen from the fact that Δ is coassociative.

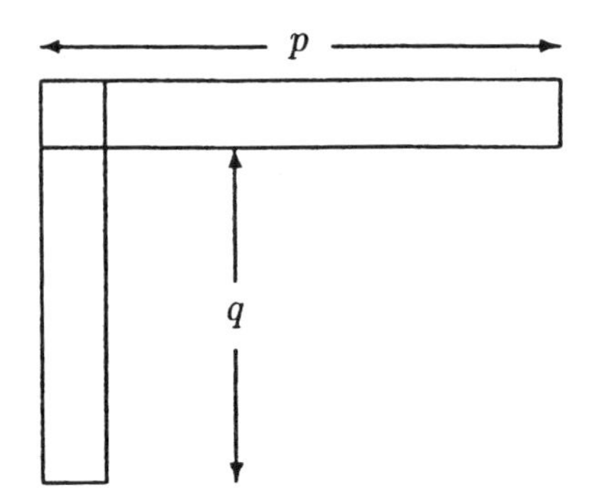

To prove acyclicity, we define a filtration on this complex, and assume by induction that $X^{(q)}$ is acyclic (i.e., has vanishing homology in positive dimensions, and has homology equal to Λ^q in dimension 0).

Define $F_p(X^{(k)})$ to be the subcomplex of $X^{(k)}$ consisting of those terms with $i_1 \leq p$. We have

$$0 \subset F_1(X^{(k)}) \subset F_2(X^{(k)}) \subset \cdots \subset F_k(X^{(k)}) = X^{(k)}.$$

Then

$$F_p(X^{(k)})/F_{p-1}(X^{(k)}) \approx D_p \otimes X^{(k-p)}$$

with a dimension shift of $p-1$, i.e.,

$$F_p(X^{(k)})/F_{p-1}(X^{(k)}) \approx 0$$

in dimension $0,\ldots,p-2$, while

$$F_p(X^{(k)})/F_{p-1}(X^{(k)}) \approx D_p \otimes X_l^{(k-p)}$$

in dimensions $p-1+l$, for $l \geq 0$.

Therefore, the E^1 term of the spectral sequence (i.e., the homology of the E^0 term described above) is $D_p \otimes \Lambda^{k-p}$. Thus the E^2 term is just Λ^k, and standard arguments finish the proof.

Now we can write down the resolution of the hook, $K_{(p,1^q)}$. In terms of the preceding notation, it's simply

$$X^{(p+q)}/F_{p-1}(X^{(p+q)})$$

with a dimension shift of $p-1$.

To see this easily, we use induction on q, and the exact sequence:

$$0 \to K_{(p+1,1^{q-1})} \to D_p \otimes \Lambda^q \to K_{(p,1^q)} \to 0.$$

Without going into too much detail, we just state that the map from the resolution of $K_{(p+1,1^{q-1})}$, which is $X^{(p+1)}/F_p(X^{(p+q)}$, into $D_p \otimes \Lambda^{k-p}$ is induced by:

$$a_1 \otimes \cdots \otimes a_{k-p} \mapsto \sum a_1(p) \otimes a_1(i_1 - p) \otimes a_2 \otimes \cdots \otimes a_{k-p}.$$

(The notation $\sum a_1(p) \otimes a_1(i_1 - p)$ denotes the diagonalization of a_1 into its degree p and degree $i_1 - p$ components.)

With this, it's easy to see that the mapping cone is the indicated resolution of our hook, and we have succeeded in our first effort. The problem now is to set these resolutions up in letter-place terminology so that we will be in a position to start defining a splitting contracting homotopy. (I know of no way of describing such a homotopy without resorting to the letter-place basis. That no equivariant homotopy exists is, I believe, fairly obvious.)

In old-fashioned terms, we know that the modules that comprise the n-dimensional chains of the resolution of the hook are direct sums of modules of the form

$$D_{p-1+i_0} \otimes D_{i_1} \otimes \cdots \otimes D_{i_{q-n}} \tag{$*$}$$

with all $i_j > 0$ and $i_0 + \cdots + i_{q-n} = q+1$.

What we want to do is prefix these terms with the appropriate operators and separators, and say which of the $q+1$ places the identified terms occupy. The boundary operator will be the usual sum of the polarizations of the separators to 1, but to make sense of this boundary, we have to make explicit certain Capelli-like identities which will be described soon. First, we specify which places the factors in the term $(*)$ occupy:

D_{p-1+i_0} is in place number 1;

D_{i_1} is in place number $i_0 + 1$;

D_{i_2} is in place number $i_0 + i_1 + 1$;

$\vdots$

D_{i_j} is in place number $i_0 + i_1 + \cdots + i_{j-1} + 1$;

$\vdots$

$D_{i_{q-n}}$ is in place number $i_0 + i_1 + \cdots + i_{q-n-1} + 1 \quad (\leq q+1)$.

Of course this means that in all the other places we have D_0.

Next, we define some operators.

For each integer $i \geq 1$ and each $k \geq 1$, define the element in the appropriate free-product algebra (see [B-R 2]):

$$Z(i;k) \\ = Z^{(1)}_{k+i-1,k+i-2}x_{k+i-2}Z^{(2)}_{k+i-2,k+i-3}x_{k+i-3}\cdots x_{k+2}Z^{(i-2)}_{k+2,k+1}x_{k+1}Z^{(i-1)}_{k+1,k}x_k\,.$$

We agree that if $i = 1$, then this element is just equal to the identity.

I claim that the term $(*)$ above is, in the resolution of the hook, preceded by:

$$Z(i_{q-n};k_{q-n})Z(i_{q-n-1};k_{q-n-1})\cdots Z(i_1;k_1)Z(i_0;k_0)\,,$$

where $k_j = i_0 + \cdots + i_{j-1} + 1$ for $j = 1,\cdots,q-n$, and $k_0 = 1$. (It is easy to see that this is indeed an operator of length n.)

To make sense of the boundary map, we introduce the Capelli-like identities mentioned above:

$$\begin{aligned} Z^{(a)}_{k+\beta,k}Z^{(b)}_{k,k-a}x_{k-\alpha} &= \sum_{b>u\geq 0} Z^{(b-u)}_{k,k-a}x_{k-a}Z^{(a-u)}_{k+\beta,k}Z^{(u)}_{k+\beta,k-a}\,; \\ Z^{(a)}_{k+\beta_1+\beta_2,k+\beta_2}Z^{(b)}_{k,k-a}x_{k-a} &= Z^{(b)}_{k,k-a}x_{k-a}Z^{(a)}_{k+\beta_1+\beta_2,k+\beta_2}\,. \end{aligned} \tag{C}$$

With these identities, it is easy to see that we have reproduced the resolution in letter-place terms, although we haven't yet put the divided power terms into letter-place notation. To make the conversion to letter-place complete, and to make calculation with the boundary map easier to handle, we'll proceed as follows.

Let's suppose we have words, w, in letters, and places $1, 2, \ldots, q+1$. Then the terms we're looking at in dimension n correspond to sequences of places

$$k_0, k_1.\cdots, k_{q-n}$$

satisfying the following conditions

$$\begin{aligned} &k_0 = 1, \\ &k_j - k_{j-1} = i_{j-1} > 0. \end{aligned}$$

We set $i_{q-n} = q + 2 - k_{q-n} > 0$. Then the typical term in dimension n is

$$Z(i_{q-n}; k_{q-n}) Z(i_{q-n-1}; k_{q-n-1}) \cdots$$
$$Z(i_1; k_1) Z(i_0; k_0)(w_0|k_0^{(p-1+i_0)})(w_1|k_1^{(i_1)}) \cdots (w_{q-n}|k_{q-n}^{(i_{q-n})}).$$

Since this notation is a little obscure until one becomes used to it, it's probably worthwhile to look at a simple, but not altogether trivial, example. Suppose we start out with a term in $D_{p+3} \otimes \underbrace{D_1 \otimes \cdots \otimes D_1}_{q-3}$. In this case, we have the sequence $\{i_0, i_1, \cdots, i_{q-3}\}$, where $i_0 = 4, i_1 = 1, \cdots, i_{q-3} = 1$. This then gives rise to the sequence of 'places': $\{k_0, k_1, \cdots, k_{q-3}\}$, where $k_0 = 1, k_1 = 5, k_2 = 6, \cdots, k_{q-3} = q + 2$ and we see that a typical generator of $D_{p+3} \otimes \underbrace{D_1 \otimes \cdots \otimes D_1}_{q-3}$ is of the form

$$(w_0|1^{(p+3)})(w_1|5^{(1)}) \cdots (w_{q-3}|(q+1)^{(1)})$$

where we are using the integers to stand for the positive places. The operator prefix for this term is simply $Z(4;1)$ since all the i_j are equal to 1 except for $j = 0$. Thus we have to compute the boundary of

$$Z_{43}^{(1)} x_3 Z_{32}^{(2)} x_2 Z_{21}^{(3)} x_1 (w_0|1^{(p+3)})(w_1|5^{(1)}) \cdots (w_{q-3}|(q+1)^{(1)}).$$

Consider:

$$\begin{aligned}
& Z_{43}^{(1)} x_3 Z_{32}^{(2)} x_2 Z_{21}^{(3)} x_1 (w_0|1^{(p+3)})(w_1|5^{(1)}) \cdots (w_{q-3}|(q+1)^{(1)}) \\
& \to \; Z_{43}^{(1)} x_3 Z_{32}^{(2)} x_2 (w_0|1^{(p)} 2^{(3)})(w_1|5^{(1)}) \cdots (w_{q-3}|(q+1)^{(1)}) \\
& - Z_{43}^{(1)} x_3 Z_{32}^{(2)} Z_{21}^{(3)} x_1 (w_0|1^{(p+3)})(w_1|5^{(1)}) \cdots (w_{q-3}|(q+1)^{(1)}) \\
& + Z_{43}^{(1)} Z_{32}^{(2)} x_2 Z_{21}^{(3)} x_1 (w_0|1^{(p+3)})(w_1|5^{(1)}) \cdots (w_{q-3}|(q+1)^{(1)}).
\end{aligned}$$

The first summand on the right is one of our familiar terms. What we have to do is make sense of the remaining two summands. We do this by invoking the Capelli identities (C). These give us:

$$Z_{32}^{(2)} Z_{21}^{(3)} x_1 = Z_{21}^{(1)} x_1 Z_{31}^{(2)} + Z_{21}^{(2)} x_1 Z_{32}^{(1)} Z_{31}^{(1)} + Z_{21}^{(3)} x_1 Z_{32}^{(2)}; \tag{1}$$

$$Z_{43}^{(1)} Z_{32}^{(2)} x_2 = Z_{32}^{(1)} x_2 Z_{42}^{(1)} + Z_{32}^{(2)} x_2 Z_{43}^{(1)}. \tag{2}$$

Applying (1) to the term

$$Z_{43}^{(1)} x_3 Z_{32}^{(2)} Z_{21}^{(3)} x_1 (w_0|1^{(p+3)})(w_1|5^{(1)}) \cdots (w_{q-3}|(q+1)^{(1)}),$$

and observing that the second place doesn't occur, we're left with

$$Z_{43}^{(1)}x_3Z_{21}^{(1)}x_1(w_0|1^{(p+1)}3^{(2)})(w_1|5^{(1)})\cdots(w_{q-3}|(q+1)^{(1)})\,.$$

Applying (2) to the term

$$Z_{43}^{(1)}Z_{32}^{(2)}x_2Z_{21}^{(3)}x_1(w_0|1^{(p+3)})(w_1|5^{(1)})\cdots(w_{q-3}|(q+1)^{(1)})\,,$$

observing that the second and third places don't occur, and that

$$Z_{43}^{(1)}Z_{21}^{(3)}x_1 = Z_{21}^{(3)}x_1Z_{43}^{(1)}\,,$$

we get

$$\begin{aligned} &Z_{43}^{(1)}Z_{32}^{(2)}x_2Z_{21}^{(3)}x_1(w_0|1^{(p+3)})(w_1|5^{(1)})\cdots(w_{q-3}|(q+1)^{(1)}) \\ &= Z_{32}^{(1)}x_2Z_{21}^{(2)}x_1Z_{41}^{(1)}(w_0|1^{(p+3)})(w_1|5^{(1)})\cdots(w_{q-3}|(q+1)^{(1)}) \\ &= Z_{32}^{(1)}x_2Z_{21}^{(2)}x_1(w_0|1^{(p+2)}4^{(1)})(w_1|5^{(1)})\cdots(w_{q-3}|(q+1)^{(1)}). \end{aligned}$$

Thus we see that the boundary of

$$Z_{43}^{(1)}x_3Z_{32}^{(2)}x_2Z_{21}^{(3)}x_1(w_0|1^{(p+3)})(w_1|5^{(1)}\cdots(w_{q-3}|(q+1)^{(1)})$$

is

$$\begin{aligned} &Z_{43}^{(1)}x_3Z_{32}^{(2)}x_2(w_0|1^{(p)}2^{(3)})(w_1|5^{(1)})\cdots(w_{q-3}|(q+1)^{(1)}) \\ &- Z_{43}^{(1)}x_3Z_{21}^{(1)}x_1(w_0|1^{(p+1)}e^{(2)})(w_1|5^{(1)})\cdots(w_{q-3}|(q+1)^{(1)}) \\ &+ Z_{32}^{(1)}x_2Z_{21}^{(2)}x_1(w_0|1^{(p+2)}4^{(1)})(w_1|5^{(1)})\cdots(w_{q-3}|(q+1)^{(1)}). \end{aligned}$$

This is precisely the result (up to sign) that we get in the non-letter-place formulation. (Perhaps a word should be added about the signs in these two versions of these resolutions. Given the sequence $k_0, k_1, \cdots, k_{q-n}$ satisfying the conditions

$$\begin{aligned} &k_0 = 1, \\ &k_j - k_{j-1} > 0, \\ &\text{and } q+2-k_{q-n} > 0, \end{aligned}$$

we can of course define the sequence $\{i_0, i_1, \cdots, i_{q-3}\}$, where $i_{j-1} = k_j - k_{j-1}$ for $i = 1, \cdots, q-n$, and $i_{q-n} = q+2-k_{q-n}$. We then map the term

$$Z(i_{q-n};k_{q-n})Z(i_{q-n-1};k_{q-n-1})\ \cdots Z(i_1;k_1)Z(i_0;k_0)(w_0|k_0^{(p-1+i_0)})(w_1|k_1^{(i_1)})\cdots(w_{q-n}|k_{q-n}^{(i_{q-n})})$$

to the term $(-1)^{\sum k_j} w_0 \otimes w_1 \otimes \cdots \otimes w_{q-n} \in D_{p-1+i_0} \otimes D_{i1} \otimes \cdots \otimes D_{i_{q-n}}$, which provides us with an isomorphism of complexes.)

We now proceed to the definition of the first three steps of a splitting contracting homotopy for this resolution of the hook Weyl module, i.e., of $K_{(p,1^q)}$. The standard basis of this module consists of standard tableaux:

$$\left\langle \begin{matrix} x_{i_1} & x_{i_2} & \cdots & x_{i_p} \\ x_{j_1} & & & \\ x_{j_2} & & & \\ \vdots & & & \\ x_{j_q} & & & \end{matrix} \right\rangle,$$

where the subscripted letters are basis elements of our underlying free module (or, in more orthodox terminology, are words in the letter alphabet)), $i_1 \leq i_2 \leq \cdots \leq i_p$ and $i_1 < j_1 < \cdots < j_q$. If we denote by $s_{-1} : K_{(p,1^q)} \to D_p \otimes \underbrace{D_1 \otimes \cdots \otimes D_1}_{q}$ the first step of the homotopy, we define

$$s_{-1}\left(\left\langle \begin{matrix} x_{i_1} & x_{i_2} & \cdots & x_{i_p} \\ x_{j_1} & & & \\ x_{j_2} & & & \\ \vdots & & & \\ x_{j_q} & & & \end{matrix} \right\rangle\right) = \left\langle \begin{matrix} x_{i_1}x_{i_2}\cdots x_{i_p} \\ x_{j_1} \\ x_{j_2} \\ \vdots \\ x_{j_q} \end{matrix} \left| \begin{matrix} 1^{(p)} \\ 2 \\ 3 \\ \\ q+1 \end{matrix} \right. \right\rangle.$$

Double standard tableaux of the type

$$\left\langle \begin{matrix} w_1 \\ w_2 \\ w_3 \\ \vdots \\ w_{q+1} \end{matrix} \left| \begin{matrix} q^{(p)} \\ 2 \\ 3 \\ \\ q+1 \end{matrix} \right. \right\rangle \in D_p \otimes \underbrace{D_1 \otimes \cdots \otimes D_1}_{q},$$

we will call VS (for very standard). On such elements, the next stage of the homotopy

$$\begin{aligned} s_0 : D_p \otimes \underbrace{D_1 \otimes \cdots \otimes D_1}_{q} \to & \\ & \left\{Z^{(1)}_{2,1} x_1 (w_1|1^{(p+1)})(w_3|3)(w_4|4)\cdots(w_{q+1}|q+1)\right\} \\ \oplus \quad & \left\{Z^{(1)}_{3,2} x_2 (w_1|1^{(p)})(w_2|2^{(2)})(w_4|4)\cdots(w_{q+1}|q+1)\right\} \\ \vdots \quad & \qquad \vdots \\ \oplus \quad & \left\{Z^{(1)}_{q,q-1} x_{q-1} (w_1|1^{(p)})(w_2|2)(w_3 13)\cdots(w_{q-1}|(q-1)^{(2)})(w_{q+1}|q+1)\right\} \\ \oplus \quad & \left\{Z^{(1)}_{q+1,q} x_q (w_1|1^{(p)})(w_2|2)(w_3|3)\cdots(w_{q-1}|q-1)(w_q|(q)^{(2)})\right\} \end{aligned}$$

will take the value zero. (The notation is designed to indicate the terms we have in dimension 1 of our resolution. For instance, the first term is $D_{p+1}\otimes\underbrace{D_1\otimes\cdots\otimes D_1}_{q-1}$ prefixed by $Z_{2,1}^{(1)}x_1$; the second is $D_p\otimes D_2\otimes\underbrace{D_1\otimes\cdots\otimes D_1}_{q-2}$ prefixed by $Z_{3,2}^{(1)}x_2$, etc. The place notation indicates which place is left out in this tensor product and also which place has degree equal to two [or $p+1$ in the first instance]. In general, the terms prefixed by $Z_{i+1,i}^{(1)}x_i$ have the $(i+1)^{th}$ place omitted and the ith place increased in degree.)

Since we have already defined the value of s_0 to be zero on the VS tableaux of $D_p\otimes\underbrace{D_1\otimes\cdots\otimes D_1}_{q}$, we have to define s_0 on the double standard tableaux that are not VS. Such a double standard tableau must either be of the form (again we leave out the words and focus on the places):

$$(I)\quad \left\langle\left|\begin{array}{llll} 1^{(p)} & & j & \cdots \\ 2 & & \cdots & \cdots \\ \vdots & & \vdots & \vdots \\ & j-1 & \cdots & \cdots \\ \vdots & \vdots & \vdots & \vdots \end{array}\right.\right\rangle \quad \text{or} \quad (II)\quad \left\langle\left|\begin{array}{llll} 1^{(p)} & & & \\ 2 & & j & \cdots \\ \vdots & & \vdots & \vdots \\ & j-1 & \cdots & \cdots \\ \vdots & \vdots & \vdots & \vdots \end{array}\right.\right\rangle.$$

In the case $p=1$, of course, the second type of tableau would not appear. We define

$$s_0\left(\left\langle\left|\begin{array}{llll} 1^{(p)} & & 2 & \cdots \\ x & & \cdots & \cdots \\ \vdots & & \vdots & \vdots \\ & y & \cdots & \cdots \\ \vdots & \vdots & \vdots & \end{array}\right.\right\rangle\right) = Z_{2,1}^{(1)}x_1\left\langle\left|\begin{array}{llll} 1^{(p+1)} & & & \cdots \\ x & & \cdots & \cdots \\ \vdots & & \vdots & \vdots \\ & y & \cdots & \cdots \\ \vdots & \vdots & \vdots & \vdots \end{array}\right.\right\rangle;$$

$$s_0\left(\left\langle\left|\begin{array}{llll} 1^{(p)} & & j & \cdots \\ 2 & & \cdots & \cdots \\ \vdots & & \vdots & \vdots \\ & j-1 & \cdots & \cdots \\ \vdots & \vdots & \vdots & \vdots \end{array}\right.\right\rangle\right) = Z_{j,j-1}^{(1)}x_{j-1}\left\langle\left|\begin{array}{llll} I^{(}p) & & j-1 & \cdots \\ 2 & & \cdots & \cdots \\ \vdots & & \vdots & \vdots \\ & j-1 & \cdots & \cdots \\ \vdots & \vdots & \vdots & \vdots \end{array}\right.\right\rangle$$

$$-\,s_0\left(\left\langle\left|\begin{array}{llll} 1^{(p)} & & j-1 & \cdots \\ 2 & & \cdots & \cdots \\ \vdots & & \vdots & \vdots \\ & j & \cdots & \cdots \\ \vdots & \vdots & \vdots & \vdots \end{array}\right.\right\rangle\right),\ j>2;$$

while

$$s_0\left(\left\langle\left|\begin{array}{cccc} 1^{(p)} & & & \\ 2 & & 3 & \cdots \\ \vdots & & \vdots & \vdots \\ & & \cdots & \cdots \\ \vdots & \vdots & \vdots & \vdots \end{array}\right.\right\rangle\right) = Z^{(1)}_{3,2}x_2\left\langle\left|\begin{array}{cccc} 1^{(p)} & & & \\ 2^{(2)} & & \cdots & \\ \vdots & & \vdots & \vdots \\ & & \cdots & \cdots \\ \vdots & \vdots & \vdots & \vdots \end{array}\right.\right\rangle;$$

$$s_0\left(\left\langle\left|\begin{array}{cccc} 1^{(p)} & & & \\ 2 & & j\cdots & \\ \vdots & & \vdots & \vdots \\ & j-1 & \cdots & \cdots \\ \vdots & \vdots & \vdots & \vdots \end{array}\right.\right\rangle\right) = Z^{(1)}_{j,j-1}x_{j-1}\left\langle\left|\begin{array}{cccc} 1^{(p)} & & & \\ 2 & & j-1 & \cdots \\ \vdots & & \vdots & \vdots \\ & j-1 & \cdots & \cdots \\ \vdots & \vdots & \vdots & \vdots \end{array}\right.\right\rangle$$

$$- s_0\left(\left\langle\left|\begin{array}{cccc} 1^{(p)} & & j-1 & \cdots \\ 2 & & \cdots & \cdots \\ \vdots & & \vdots & \vdots \\ & j & \cdots & \cdots \\ \vdots & \vdots & \vdots & \vdots \end{array}\right.\right\rangle\right), j > 3\,.$$

With this inductive definition, it is trivial to check that we have the first two steps of a splitting contracting homotopy. It is also clear that a basis for the first syzygies is parametrized by the linearly independent elements

$$\left\{ Z^{(1)}_{2,1}x_1\left\langle\left|\begin{array}{cccc} 1^{(p+1)} & & & \cdots \\ x & & \cdots & \cdots \\ \vdots & & \vdots & \vdots \\ & y & \cdots & \cdots \\ \vdots & \vdots & \vdots & \vdots \end{array}\right.\right\rangle, Z^{(1)}_{j,j-1}x_{j-1}\left\langle\left|\begin{array}{cccc} 1^{(p)} & & j-1 & \cdots \\ 2 & & \cdots & \cdots \\ \vdots & & \vdots & \vdots \\ & j-1 & \cdots & \cdots \\ \vdots & \vdots & \vdots & \vdots \end{array}\right.\right\rangle,\right.$$

$$\left. Z^{(1)}_{3,2}x_2\left\langle\left|\begin{array}{cccc} 1^{(p)} & & & \\ 2^{(2)} & & \cdots & \\ \vdots & & \vdots & \vdots \\ & & \cdots & \cdots \\ \vdots & \vdots & \vdots & \vdots \end{array}\right.\right\rangle, Z^{(1)}_{j,j-1}x_{j-1}\left\langle\left|\begin{array}{cccc} 1^{(p)} & & & \\ 2 & & j-1 & \cdots \\ \vdots & & \vdots & \vdots \\ & j-1 & \cdots & \cdots \\ \vdots & \vdots & \vdots & \vdots \end{array}\right.\right\rangle\right\}.$$

To define the next step of the homotopy, we have to consider the various double standard tableaux that occur in dimension 1 and indicate where they are to be sent. In order to facilitate the description of the map s_1, it's useful to observe that the map s_0, as defined on the tableaux above, could be regarded as a formal operator on those tableaux in the sense that the dots may be filled in any way we like (even if the terms wouldn't make sense in our particular context). We will see this principle used in what follows.

It is clear that we may define the map s_1 on all the elements in the brackets above to be zero. Thus all the terms of the form $Z_{2,1}^{(1)}x_1\langle|\rangle$ are sent to zero. We now consider elements of the form $Z_{j,j-1}^{(1)}x_{j-1}T$, where $j > 2$ and T is a double standard tableau. There are a number of different cases that have to be considered.

$$(1) \qquad T = \left\langle \left| \begin{array}{cccccc} 1^{(p)} & i & \cdots & (j-1)^{(\varepsilon)} & \cdots & \\ \vdots & \vdots & & & & \vdots \end{array} \right. \right\rangle, \; 1 < i < j-1; \; \varepsilon = 1, 2$$

$$(2) \qquad T = \left\langle \left| \begin{array}{cccccc} 1^{(p)} & & & & & \\ 2 & i & \cdots & (j-1)^{(\varepsilon)} & \cdots & \\ \vdots & \vdots & & & & \vdots \end{array} \right. \right\rangle, \; 2 < i < j-1; \; \varepsilon = 1, 2$$

$$(3) \qquad T = \left\langle \left| \begin{array}{cccccc} 1^{(p)} & & & i & \cdots & \\ \vdots & & & & & \\ & (j-1)^{(\varepsilon)} & & \cdots & & \\ \vdots & \vdots & & \vdots & \vdots & \end{array} \right. \right\rangle, \; i < j-1; \; \varepsilon = 1, 2$$

$$(4) \qquad T = \left\langle \left| \begin{array}{ccccc} 1^{(p)} & & & & \\ 2 & i & & \cdots & \\ \vdots & & & & \\ & & (j-1)^{(\varepsilon)} & \cdots & \\ & & \vdots & & \vdots \end{array} \right. \right\rangle, \; i < j-1; \; \varepsilon = 1, 2.$$

In all of these cases, we set $s_1(Z_{j,j-1}^{(1)}x_{j-1}T) = Z_{j,j-1}^{(1)}x_{j-1}s_0(T)$.

(Here is where we are treating s_0 as the formal operator alluded to above; we may simply disregard the fact that the map is not legitimately applicable to the tableau T, since the formula for $s_0(T)$ used before just depends on the indices less than or equal to i):

$$(5) \qquad T = \left\langle \left| \begin{array}{ccc} 1^{(p)} & (j-1)^{(2)} & \cdots \\ \vdots & & \vdots \end{array} \right. \right\rangle .$$

For this situation, we extend the definition of the operator s_0 as follows:

$$s_0\left(\left\langle \left| \begin{array}{ccc} 1^{(p)} & 2^{(2)} & \cdots \\ \vdots & & \vdots \end{array} \right. \right\rangle \right) = Z_{2,1}^{(2)}x_1 \left\langle \left| \begin{array}{cc} 1^{(p+2)} & \cdots \\ \vdots & \vdots \end{array} \right. \right\rangle ;$$

$$s_0\left(\left\langle\begin{array}{|llll} 1^{(p)} & & (j-1)^{(2)} & \vdots \\ 2 & & \cdots & \vdots \\ \vdots & & & \vdots \\ & j-2 & \cdots & \vdots \\ \vdots & \vdots & & \vdots \end{array}\right\rangle\right) = Z^{(2)}_{j-1,j-2}x_{j-2}\left\langle\begin{array}{|llll} 1^{(p)} & & (j-2)^{(2)} & \cdots \\ 2 & & \cdots & \vdots \\ \vdots & & & \vdots \\ & j-2 & \cdots & \\ \vdots & \vdots & & \vdots \end{array}\right\rangle$$

$$-\, s_0\left(\left\langle\begin{array}{|lllll} 1^{(p)} & & j-2 & j-1 & \cdots \\ 2 & & & & \\ \vdots & & & & \\ & j-1 & \cdots & & \\ \vdots & \vdots & & & \vdots \end{array}\right\rangle\right)$$

Here we set

$$s_1(Z^{(1)}_{j,j-1}x_{j-1}T) = Z^{(1)}_{j,j-1}x_{j-1}s_0(T) + s_1\left(Z^{(1)}_{j-1,j-2}x_{j-2}\left\langle\begin{array}{|llll} 1^{(p)} & & (j-2)^{(2)} & \cdots \\ 2 & & \cdots & \vdots \\ \vdots & & & \vdots \\ & j & \cdots & \\ \vdots & \vdots & & \vdots \end{array}\right\rangle\right),$$

where, when $j = 3$, there is no additional summand, and the explicitly exhibited tableau shows that the index j is in the place originally occupied by $j-2$ in the tableau T.

$$T = \left\langle\begin{array}{|lll} 1^{(p)} & & \\ 2 & (j-1)^{(2)} & \cdots \\ \vdots & \vdots & \vdots \\ j-2 & \cdots & \\ \vdots & & \vdots \end{array}\right\rangle \quad j > 3. \tag{6}$$

Again we extend the definition of s_0:

$$s_0\left(\left\langle\begin{array}{|lll} 1^{(p)} & & \\ 2 & 3^{(2)} & \cdots \\ \vdots & \vdots & \vdots \end{array}\right\rangle\right) = Z^{(2)}_{3,2}x_2\left\langle\begin{array}{|lll} 1^{(p)} & & \\ 2^{(3)} & \cdots & \\ \vdots & \vdots & \vdots \end{array}\right\rangle;$$

$$s_0\left(\left\langle\begin{array}{|llll} 1^{(p)} & & & \\ 2 & (j-1)^{(2)}\cdots & & \\ \vdots & \vdots & & \vdots \\ j-2 & & \cdots & \\ \vdots & \vdots & & \vdots \end{array}\right\rangle\right) = Z^{(2)}_{j-1,j-2}x_{j-2}\left\langle\begin{array}{|llll} 1^{(p)} & & & \\ 2 & (j-2)^{(2)}\cdots & & \\ \vdots & & & \vdots \\ j-2 & \cdots & \cdots & \\ \vdots & \vdots & & \vdots \end{array}\right\rangle$$

$$- s_0\left(\left\langle\begin{array}{|cccc} 1^{(p)} & & & \\ 2 & j-2 & j-2 & \cdots \\ \vdots & \vdots & \vdots & \vdots \\ j-1 & \cdots & & \\ \vdots & \vdots & \vdots & \vdots \end{array}\right\rangle\right).$$

With this operation, we can define

$$s_1(Z^{(1)}_{j,j-1}x_{j-1}T) = Z^{(1)}_{j,j-1}x_{j-1}s_0(T) + s_1\left(Z^{(1)}_{j-1,j-2}x_{j-2}\left\langle\begin{array}{|ccc} 1^{(p)} & & \\ 2 & (j-2)^{(2)} & \cdots \\ \vdots & \vdots & \vdots \\ j-3 & \cdots & \\ j & \cdots & \\ \vdots & \vdots & \vdots \end{array}\right\rangle\right),$$

where we recall that $j > 3$, and when $j = 4$ there is no additional summand.

$$\text{(7}a\text{)}\qquad T = \left\langle\begin{array}{|cccc} 1^{(p)} & & & j+k \quad \cdots \\ \vdots & & & \vdots \\ & & j-1 & \\ & j-1 & & \\ \vdots & & & \vdots \end{array}\right\rangle, \ k \geq 1;$$

$$\text{(7}b\text{)}\qquad T = \left\langle\begin{array}{|ccc} 1^{(p)} & & j+k \quad \cdots \\ \vdots & & \vdots \\ & (j-1)^{(2)} & \\ \vdots & & \vdots \end{array}\right\rangle. \ k \geq 1;$$

In these situations we set
(a)

$$s_1\left(Z^{(1)}_{j,j-1}x_{j-1}\left\langle\begin{array}{|ccccc} 1^{(p)} & & & j+1 & \cdots \\ \vdots & & & \vdots & \vdots \\ & & j-1 & \cdots & \\ & j-1 & \cdots & & \\ \vdots & \vdots & \vdots & \vdots & \end{array}\right\rangle\right)$$

$$= Z^{(1)}_{j+1,j}x_jZ^{(2)}_{j,j-1}x_{j-1}\left\langle\begin{array}{|ccccc} 1^{(p)} & & & j-1 & \cdots \\ \vdots & & & \vdots & \vdots \\ & & j-1 & \cdots & \\ & j-1 & \cdots & & \\ \vdots & \vdots & \vdots & \vdots & \end{array}\right\rangle$$

$$+\, Z^{(1)}_{j+1,j}x_j s_0\left(\left\langle\left|\begin{array}{ccccc} 1^{(p)} & & & j-1 & \cdots \\ \vdots & & & \vdots & \vdots \\ & & j & \cdots & \\ & j & \cdots & & \\ \vdots & \vdots & \vdots & \vdots & \end{array}\right.\right\rangle\right);$$

$$s_1\left(Z^{(1)}_{j,j-1}x_{j-1}\left\langle\left|\begin{array}{ccccc} 1^{(p)} & & & j+k & \cdots \\ \vdots & & & \vdots & \vdots \\ & & j-1 & \cdots & \\ & j-1 & \cdots & & \\ \vdots & \vdots & \vdots & \vdots & \end{array}\right.\right\rangle\right)$$

$$= Z^{(1)}_{j+k,j+k-1}x_{j+k-1}Z^{(1)}_{j,j-1}x_{j-1}\left\langle\left|\begin{array}{cccccc} 1^{(p)} & & & & j+k-1 & \cdots \\ \vdots & & & \vdots & \vdots & \vdots \\ & & & j+k-1 & & \\ \vdots & & j-1 & & & \\ & j-1 & & & & \\ \vdots & \vdots & & & & \vdots \end{array}\right.\right\rangle$$

$$+\, s_1\left(Z^{(1)}_{j,j-1}x_{j-1}\left\langle\left|\begin{array}{cccccc} 1^{(p)} & & & & j+k-1 & \cdots \\ \vdots & & & \vdots & \vdots & \vdots \\ & & & j+k & & \\ \vdots & & j-1 & & & \\ & j-1 & & & & \\ \vdots & \vdots & & & & \vdots \end{array}\right.\right\rangle\right);$$

(b)

$$s_1\left(Z^{(1)}_{j,j-1}x_{j-1}\left\langle\left|\begin{array}{ccccc} 1^{(p)} & & & j+1 & \cdots \\ \vdots & & & \vdots & \vdots \\ & & (j-1)^{(2)} & \cdots & \\ & & \cdots & & \\ \vdots & \vdots & \vdots & \vdots & \end{array}\right.\right\rangle\right)$$

$$= Z^{(1)}_{j+1,j}x_j Z^{(2)}_{j,j-1}x_{j-1}\left\langle\left|\begin{array}{ccccc} 1^{(p)} & & & j-1 & \cdots \\ \vdots & & & \vdots & \vdots \\ & & (j-1)^{(2)} & \cdots & \\ & & \cdots & & \\ \vdots & \vdots & \vdots & \vdots & \end{array}\right.\right\rangle$$

$$+\, Z^{(1)}_{j+1,j}x_j s_0\left(\left\langle\left|\begin{array}{ccccc} 1^{(p)} & & & j-1 & \cdots \\ \vdots & & & \vdots & \vdots \\ & & (j)^{(2)} & \cdots & \\ & & \cdots & & \\ \vdots & \vdots & \vdots & \vdots & \end{array}\right.\right\rangle\right);$$

$$s_1\left(Z^{(1)}_{j,j-1}x_{j-1}\left\langle \left| \begin{array}{llllll} 1^{(p)} & & & j+k & \cdots \\ \vdots & & & \vdots & \vdots \\ & & (j-1)^{(2)} & \cdots & \\ & & \cdots & & \\ \vdots & \vdots & \vdots & \vdots & \end{array} \right. \right\rangle \right)$$

$$= Z^{(1)}_{j+k,j+k-1}x_{j+k-1}Z^{(1)}_{j,j-1}x_{j-1}\left\langle \left| \begin{array}{llllll} 1^{(p)} & & & & j+k-1 & \cdots \\ \vdots & & & \vdots & \vdots & \vdots \\ & & & j+k-1 & & \\ \vdots & & (j-1)^{(2)} & & & \\ \vdots & \vdots & & & & \vdots \end{array} \right. \right\rangle$$

$$+\, s_1\left(Z^{(1)}_{j,j-1}x_{j-1}\left\langle \left| \begin{array}{llllll} 1^{(p)} & & & & j+k-1 & \cdots \\ \vdots & & & \vdots & \vdots & \vdots \\ & & & j+k & & \\ \vdots & & (j-1)^{(2)} & & & \\ \vdots & \vdots & & & & \vdots \end{array} \right. \right\rangle \right).$$

Up to now we have seen the definition of our maps s_{-1}, s_0, and s_1, but have had no demonstration that these do, indeed, define a homotopy. I'll now give some examples of how one shows this so that we can see how trivial it is to verify that these formulae do the job they're supposed to do. It will become clear that the trick lies in finding the formulae. It can only be hoped that, as more proficiency is gained with this letter-place technique, the degree of complication will decrease.

Let's take a double standard tableau of type (I), i.e., we have the tableau:

$$T = \left\langle \left| \begin{array}{llll} 1^{(p)} & & j & \cdots \\ 2 & & \cdots & \cdots \\ \vdots & & \vdots & \vdots \\ & j-1 & \cdots & \cdots \\ \vdots & \vdots & \vdots & \vdots \end{array} \right. \right\rangle,$$

and we are assuming that $j > 2$ (the case $j = 2$ is easy to handle). This goes to zero under the boundary map, so what we must show is that

$$\partial s_0(T) = T.$$

By the definition of s_0, we have

$$s_0(T) = Z^{(1)}_{j,j-1} x_{j-1} \left\langle \begin{array}{|llll} 1^{(p)} & & j-1 & \cdots \\ 2 & & \cdots & \cdots \\ \vdots & & \vdots & \vdots \\ & j-1 & \cdots & \cdots \\ \vdots & \vdots & \vdots & \vdots \end{array} \right\rangle - s_0 \left\langle \begin{array}{|llll} 1^{(p)} & & j-1 & \cdots \\ 2 & & \cdots & \cdots \\ \vdots & & \vdots & \vdots \\ & j & \cdots & \cdots \\ \vdots & \vdots & \vdots & \vdots \end{array} \right\rangle.$$

(Don't forget that $j > 2$.) If we apply the boundary map to $s_0(T)$, we get

$$\left\langle \begin{array}{|llll} 1^{(p)} & & j & \cdots \\ 2 & & \cdots & \cdots \\ \vdots & & \vdots & \vdots \\ & j-1 & \cdots & \cdots \\ \vdots & \vdots & \vdots & \vdots \end{array} \right\rangle + \left\langle \begin{array}{|llll} 1^{(p)} & & j-1 & \cdots \\ 2 & & \cdots & \cdots \\ \vdots & & \vdots & \vdots \\ & j & \cdots & \cdots \\ \vdots & \vdots & \vdots & \vdots \end{array} \right\rangle - \partial s_0 \left\langle \begin{array}{|llll} 1^{(p)} & & j-1 & \cdots \\ 2 & & \cdots & \cdots \\ \vdots & & \vdots & \vdots \\ & j & \cdots & \cdots \\ \vdots & \vdots & \vdots & \vdots \end{array} \right\rangle.$$

By a suitable induction hypothesis (starting, of course, with $j = 2$), we may assume that

$$\partial s_0 \left\langle \begin{array}{|llll} 1^{(p)} & & j-1 & \cdots \\ 2 & & \cdots & \cdots \\ \vdots & & \vdots & \vdots \\ & j & \cdots & \cdots \\ \vdots & \vdots & \vdots & \vdots \end{array} \right\rangle = \left\langle \begin{array}{|llll} 1^{(p)} & & j-1 & \cdots \\ 2 & & \cdots & \cdots \\ \vdots & & \vdots & \vdots \\ & j & \cdots & \cdots \\ \vdots & \vdots & \vdots & \vdots \end{array} \right\rangle,$$

so that we end up with our desired result. (The fact that $\partial s_0(T) = T$ if

$$T = \left\langle \begin{array}{|lccc} 1^{(p)} & & 2 & \cdots \\ x & & \cdots & \cdots \\ \vdots & & \vdots & \vdots \\ & y & \cdots & \cdots \\ \vdots & \vdots & \vdots & \vdots \end{array} \right\rangle, \text{ i.e., if } j = 2, \text{ comes from the fact that}$$

$$s_0 \left\langle \begin{array}{|lccc} 1^{(p)} & & 2 & \cdots \\ x & & \cdots & \cdots \\ \vdots & & \vdots & \vdots \\ & y & \cdots & \cdots \\ \vdots & \vdots & \vdots & \vdots \end{array} \right\rangle = Z_{2,1}^{(1)} x_1 \left\langle \begin{array}{|lccc} 1^{(p+1)} & & & \cdots \\ x & & \cdots & \cdots \\ \vdots & & \vdots & \vdots \\ & y & \cdots & \cdots \\ \vdots & \vdots & \vdots & \vdots \end{array} \right\rangle,$$

so that, with only this one term, we immediately get our result.) The same type of calculation verifies our homotopy for tableaux of type (II).

The next step is a little more complicated; we will demonstrate that for tableaux, T, of type (1) and ($7a$), the formula

$$s_0\partial(Z_{j,j-1}^{(1)}x_{j-1}T) + \partial s_1(Z_{j,j-1}^{(1)}x_{j-1}T) = Z_{j,j-1}^{(1)}x_{j-1}T,$$

is valid. First we consider the tableau of type 1),with $\varepsilon = 2:$

$$T = \left\langle \begin{array}{|lcccc} 1^{(p)} & i & \cdots & (j-1)^{(2)} & \cdots \\ \vdots & \vdots & & & \vdots \end{array} \right\rangle, \quad 1 < i < j-1.$$

We then have

$$s_0\partial(Z_{j,j-1}^{(1)}x_{j-1}T) + \partial s_1(Z_{j,j-1}^{(1)}x_{j-1}T) =$$

$$s_0 \left\langle \begin{array}{|lcccc} 1^{(p)} & i & \cdots & (j-1)^{(1)}j & \cdots \\ \vdots & \vdots & & & \vdots \end{array} \right\rangle + \partial(Z_{j,j-1}^{(1)}x_{j-1}s_0(T)) =$$

$$s_0 \left\langle \begin{array}{|lcccc} 1^{(p)} & i & \cdots & (j-1)^{(1)}j & \cdots \\ \vdots & \vdots & & & \vdots \end{array} \right\rangle$$
$$+ \partial \left(Z_{j,j-1}^{(1)}x_{j-1}s_0 \left\langle \begin{array}{|lcccc} 1^{(p)} & i & \cdots & (j-1)^{(2)} & \cdots \\ \vdots & \vdots & & & \vdots \end{array} \right\rangle \right).$$

But

$$\partial \left(Z_{j,j-1}^{(1)}x_{j-1}s_0 \left\langle \begin{array}{|lcccc} 1^{(p)} & i & \cdots & (j-1)^{(2)} & \cdots \\ \vdots & \vdots & & & \vdots \end{array} \right\rangle \right)$$

$$
\begin{aligned}
&= Z^{(1)}_{j,j-1} x_{j-1} \partial \left(s_0 \left\langle \begin{array}{|lllll} 1^{(p)} & i & \cdots & (j-1)^{(2)} & \cdots \\ \vdots & \vdots & & & \vdots \end{array} \right\rangle \right) \\
&\quad - Z^{(1)}_{j,j-1} s_0 \left\langle \begin{array}{|lllll} 1^{(p)} & i & \cdots & (j-1)^{(2)} & \cdots \\ \vdots & \vdots & & & \vdots \end{array} \right\rangle \\
&= Z^{(1)}_{j,j-1} x_{j-1} \left\langle \begin{array}{|lllll} 1^{(p)} & i & \cdots & (j-1)^{(2)} & \cdots \\ \vdots & \vdots & & & \vdots \end{array} \right\rangle \\
&\quad - s_0 \left\langle \begin{array}{|lllll} 1^{(p)} & i & \cdots & (j-1)^{(1)} j & \cdots \\ \vdots & \vdots & & & \vdots \end{array} \right\rangle \\
&= Z^{(1)}_{j,j-1} x_{j-1} T - s_0 \left\langle \begin{array}{|lllll} 1^{(p)} & i & \cdots & (j-1)^{(1)} j & \cdots \\ \vdots & \vdots & & & \vdots \end{array} \right\rangle
\end{aligned}
$$

This gives us our result for tableaux of type 1) with $\varepsilon = 2$. The case $\varepsilon = 1$ proceeds in the same way; the main point is that the operator s_0 is sensitive to the position of the index, i, and doesn't pay any attention to the indices j and $j-1$. The cases 2), 3) and 4) proceed very similarly.

We'll now look at the situation when our tableau, T, is of the form $7a$), i.e.,

$$
T = \left\langle \begin{array}{|lllll} 1^{(p)} & & & j+k & \cdots \\ \vdots & & & \vdots & \\ & & j-1 & & \\ & j-1 & & & \\ \vdots & & & & \vdots \end{array} \right\rangle, \quad k \geq 1.
$$

Here, too, we must verify that

$$
s_0 \partial (Z^{(1)}_{j,j-1} x_{j-1} T) + \partial s_1 (Z^{(1)}_{j,j-1} x_{j-1} T) = Z^{(1)}_{j,j-1} x_{j-1} T.
$$

As before, we calculate $s_0 \partial (Z^{(1)}_{j,j-1} x_{j-1} T)$, and get

$$
s_0 \left(\left\langle \begin{array}{|lllll} 1^{(p)} & & & j+k & \cdots \\ \vdots & & & \vdots & \\ & & j & & \\ & j-1 & & & \\ \vdots & & & & \vdots \end{array} \right\rangle + \left\langle \begin{array}{|lllll} 1^{(p)} & & & j+k & \cdots \\ \vdots & & & \vdots & \\ & & j-1 & & \\ & j & & & \\ \vdots & & & & \vdots \end{array} \right\rangle \right).
$$

If $k = 1$, the above equals

$$Z^{(1)}_{j+1,j} x_j \partial_{j,j-1} \left(\left\langle \begin{array}{|llll} 1^{(p)} & & & j \ \cdots \\ \vdots & & & \vdots \\ & & j-1 & \\ & j-1 & & \\ \vdots & & & \vdots \end{array} \right\rangle \right) -$$

$$s_0 \partial_{j+1,j-1} \left(\left\langle \begin{array}{|llll} 1^{(p)} & & & j \ \cdots \\ \vdots & & & \vdots \\ & & j-1 & \\ & j-1 & & \\ \vdots & & & \vdots \end{array} \right\rangle \right).$$

(This is seen directly from the definition of the operator s_0.) Working at these terms a little more, we see that the above is

$$Z^{(1)}_{j+1,j} x_j \left\{ \left\langle \begin{array}{|llll} 1^{(p)} & & & j \ \cdots \\ \vdots & & & \vdots \\ & & j & \\ & j-1 & & \\ \vdots & & & \vdots \end{array} \right\rangle + \left\langle \begin{array}{|llll} 1^{(p)} & & & j \ \cdots \\ \vdots & & & \vdots \\ & & j-1 & \\ & j & & \\ \vdots & & & \vdots \end{array} \right\rangle \right\} -$$

$$Z^{(1)}_{j,j-1} x_{j-1} \left\{ \left\langle \begin{array}{|llll} 1^{(p)} & & & j-1 \ \cdots \\ \vdots & & & \vdots \\ & & j+1 & \\ & j-1 & & \\ \vdots & & & \vdots \end{array} \right\rangle \right.$$

$$\left. + \left\langle \begin{array}{|llll} 1^{(p)} & & & j-1 \ \cdots \\ \vdots & & & \vdots \\ & & j-1 & \\ & j+1 & & \\ \vdots & & & \vdots \end{array} \right\rangle \right\}$$

$$+ s_0 \left\{ \left\langle \begin{array}{|llll} 1^{(p)} & & & j-1 \ \cdots \\ \vdots & & & \vdots \\ & & j+1 & \\ & j & & \\ \vdots & & & \vdots \end{array} \right\rangle + \left\langle \begin{array}{|llll} 1^{(p)} & & & j-1 \ \cdots \\ \vdots & & & \vdots \\ & & j & \\ & j+1 & & \\ \vdots & & & \vdots \end{array} \right\rangle \right\}.$$

We next have, when $k = 1$,

$$s_1(Z^{(1)}_{j,j-1} x_{j-1} T) \quad = \quad Z^{(1)}_{j+1,j} x_j Z^{(2)}_{j,j-1} x_{j-1} \left\langle \begin{array}{|llll} 1^{(p)} & & & j-1 \ \cdots \\ \vdots & & & \vdots \\ & & j-1 & \\ & j-1 & & \\ \vdots & & & \vdots \end{array} \right\rangle +$$

$$Z^{(1)}_{j+1,j}x_j s_0\left(\left\langle\begin{array}{|lllll} 1^{(p)} & & & j-1 & \cdots \\ \vdots & & & \vdots & \\ & & j & & \\ & j & & & \\ \vdots & & & & \vdots \end{array}\right\rangle\right),$$

so that $\partial s_1(Z^{(1)}_{j,j-1}x_{j-1}T)$ becomes

$$Z^{(1)}_{j+1,j}x_j\left\{\left\langle\begin{array}{|lllll} 1^{(p)} & & & j & \cdots \\ \vdots & & & \vdots & \\ & & j & & \\ & j-1 & & & \\ \vdots & & & & \vdots \end{array}\right\rangle+\left\langle\begin{array}{|lllll} 1^{(p)} & & & j & \cdots \\ \vdots & & & \vdots & \\ & & j-1 & & \\ & j & & & \\ \vdots & & & & \vdots \end{array}\right\rangle\right.$$

$$\left.+\left\langle\begin{array}{|lllll} 1^{(p)} & & & j-1 & \cdots \\ \vdots & & & \vdots & \\ & & j & & \\ & j & & & \\ \vdots & & & & \vdots \end{array}\right\rangle\right\}$$

$$-Z^{(1)}_{j,j-1}x_{j-1}\left\{\left\langle\begin{array}{|lllll} 1^{(p)} & & & j+1 & \cdots \\ \vdots & & & \vdots & \\ & & j-1 & & \\ & j-1 & & & \\ \vdots & & & & \vdots \end{array}\right\rangle\right.$$

$$+\left\langle\begin{array}{|lllll} 1^{(p)} & & & j-1 & \cdots \\ \vdots & & & \vdots & \\ & & j+1 & & \\ & j-1 & & & \\ \vdots & & & & \vdots \end{array}\right\rangle$$

$$\left.+\left\langle\begin{array}{|lllll} 1^{(p)} & & & j-1 & \cdots \\ \vdots & & & \vdots & \\ & & j-1 & & \\ & j+1 & & & \\ \vdots & & & & \vdots \end{array}\right\rangle\right\}$$

$$
-s_0\left\{\left\langle\;\left|\begin{array}{lllll}
1^{(p)} & & & j-1 & \cdots \\
\vdots & & & \vdots & \\
 & & j+1 & & \\
 & j & & & \\
\vdots & & & & \vdots
\end{array}\right.\right\rangle\right.
$$

$$
\left.+\left\langle\;\left|\begin{array}{lllll}
1^{(p)} & & & j-1 & \cdots \\
\vdots & & & \vdots & \\
 & & j & & \\
 & j+1 & & & \\
\vdots & & & & \vdots
\end{array}\right.\right\rangle\right\}.
$$

Taking care of signs, one sees that all the terms that should cancel do cancel, and we're left with the one term that we want. Once the case $k = 1$ is settled, the proof for arbitrary k proceeds by induction (the inductive definition of the homotopy is rigged so that this is possible).

References

[B-R1] Projective Resolutions of Weyl Modules, (with G.-C. Rota), *Proc. Natl. Acad. Sci. USA*, **90** (March 1993), 2448–2450.

[B-R2] A New Construction in Homological Algebra, (with G.-C. Rota). *Proc. Natl. Acad. Sci. USA*, **91** (May, 1994), 4115–4119.

Brandeis University
Department of Mathematics
Waltham, MA 02254

Classification of Trivectors in 6-D Space

Wendy Chan

Abstract

We give a characteristic-free classification of the invariants and covariants of skew symmetric tensors of step three in six-dimensional space (valid in all infinite fields of characteristics other than two or three). In particular, we prove that there is only one invariant. Our work leads to a notable new conjecture on the covariants of supersymmetric tensors.

1 Introduction

The purpose of this introduction is to broadly survey the topics developed in this paper. The paper in general falls into two parts. Part I includes Sections 2, 3 and 4. In these sections, we give the first complete, self contained exposition of the background material on supersymmetric invariant theory. In theory, these results could be found in previous work by Grosshans, Rota, Stein and others; however, the work of these authors is often incomplete, and is certainly difficult to locate in various papers.

Part II of this paper is Section 5. We present the invariant theory of a general skew-symmetric tensor of step three in a space of dimension 6. The novelties are the following. First, we have completely adopted the positive variable notation for the description of invariants and covariants of skew-symmetric tensors, as introduced by Grosshans *et al* [7]. We show how ordinary computations relating to such tensors can be substantially simplified by use of positive notation. For example, such classical notions as the span of a tensor and the Grassmann conditions for decomposability, become all but trivial in positive notation. We further pursue the development of positive notation to the point of bcing able to express in positive notation all canonical forms for such tensors (Theorem 3). As a byproduct of our use of positive notation, the relationship between the various possible canonical forms of a skew symmetric tensor and the covariants whose vanishing corresponds to each of the canonical forms becomes transparent, and one sees almost by inspection why it is that a given invariant corresponds to a certain canonical form.

Lastly we prove that such tensors have a unique invariant (Theorem 2). Our proof is not only purely combinatorial, it has the advantage of being valid over all infinite fields of characteristic other than 2 or 3 (unlike all previous work on this subject, which was exclusively confined to characteristic zero). Our classification has an unexpected consequence. On staring at the symbolic forms of the covariants in positive notation, one observes that the covariants are obtained by taking sub-tableau of the unique invariant, and one is thus led to conjecture that this is a general phenomenon. In other words, we conjecture that for all skew symmetric tensors, covariants are invariably obtained as subtableaux of invariants. This conjecture would be immediately proved, were it possible to complete a partial tableau expressing a covariant into an invariant (that is, a product of brackets) by using only elements from the span of the tensor. In general, this is not possible for arithmetical reasons; nevertheless, we surmise that partial tableau can always be successfully completed with "symbols" equivalent to the given tensor.

2 The supersymmetric algebra

Most of the material presented in current and next sections can be found in [7]. We begin by defining a generalization of an ordinary polynomial algebra on a *signed set* A with integer coefficients. The variables in A are of three types: positively signed, neutral, and negatively signed, and we write $A = A^+ \cup A^0 \cup A^-$.

2.1 Construction of Super[A]

The **symmetric algebra** Symm(A^0) is the familiar commutative algebra of polynomials in the variables A^0. The coefficients of these polynomials will be integers, although (here and everywhere below) an arbitrary commutative ring with identity could be taken as the ring of coefficients. We shall denote the set of all integers by Z and the set of all positive integers by N^+.

The **exterior algebra** Ext(A^-) is the algebra generated by the variables A^- modulo the relations $ab = -ba$ and $a^2 = 0$ for $a, b \in A^-$. Thus, Ext(A^-) is the algebra of "polynomials in anticommutative variables A^-." A nonzero monomial in Ext(A^-) is a product of a finite sequence of variables $a_1 a_2 \cdots a_n$, $a_i \in A^-$, where no two a_i coincide. In particular,

$$a_{\sigma(1)} a_{\sigma(2)} \cdots a_{\sigma(n)} = sgn(\sigma) a_1 a_2 \cdots a_n$$

for any permutation σ of the set $\{1, 2, \ldots, n\}$.

The **divided powers algebra** $\mathrm{Divp}(A^+)$ is the commutative algebra generated by the variables $a^{(i)}$ subject to the following identities:

$$a^{(i)}a^{(j)} = \binom{i+j}{i} a^{(i+j)},$$

as a ranges over A^+ and as $i = 0, 1, 2, \ldots$. We set $a^{(0)} = 1$ and $a^{(1)} = a$. (other identities usually imposed in the definition of the divided power algebra will not be needed, and need not be recalled here.)

The **extended signed set** A_{exd} of A is defined as

$$A_{\text{exd}} = A^- \cup A^0 \cup \{a^{(i)} : a \in A^+, i \in N^+\}.$$

We denote by $\mathrm{Mon}(A_{\text{exd}})$ the **free monoid** generated by A_{exd}. In other words, $\mathrm{Mon}(A_{\text{exd}})$ is the set of all finite sequences of elements of A_{exd}. (Including the empty sequence.) We construct a Z-algebra, $\mathrm{Tens}[A]$, as the free associative algebra generated by the set A_{exd} with identity. Thus, if $p \in \mathrm{Tens}[A]$, then p can be written uniquely as a finite sum, $p = \sum c_i w_i$, where $c_i \in N^+$ and $w_i \in \mathrm{Mon}(A_{\text{exd}})$. Multiplication in $\mathrm{Tens}[A]$ is defined as

$$\left(\sum_i c_i w_i\right)\left(\sum_j d_j w_j\right) = \sum_{i,j} c_i d_j w_i w_j,$$

where the product $w_i w_j$ is taken simply to be juxtaposition.

For a monomial w in $\mathrm{Tens}[A]$, we define the **parity** of w, denoted by $|w|$, to be 0 or 1 if the number of all negative variables in w is even or odd, respectively. The **length** of w, denoted by $\mathrm{length}(w)$, is the total number of variables in w, where each $a^{(i)}$ is counted i times. The **content** of a in w, denoted by $\mathrm{cont}(w; a)$, is the number of occurrences of variable a in w, where each $a^{(i)}$ is also counted i times.

In $\mathrm{Tens}[A]$, let I_A be the ideal generated by all expressions of the following forms:

1. $uv - (-1)^{|u||v|}vu$ for $u, v \in \mathrm{Mon}(A_{\text{exd}})$,

2. aa where $a \in A^-$,

3. $a^{(i)}a^{(j)} - \binom{i+j}{i} a^{(i+j)}$ where $a \in A^+$.

Definition 1 *The supersymmetric algebra* $\mathrm{Super}[A]$ *is the quotient algebra*

$$\mathrm{Tens}[A]/I_A.$$

It is clear that if $a_1, a_2, \ldots, a_r$ are distinct elements in A^+, $b_1, b_2, \ldots, b_s$ are distinct elements in A^-, and $c_1, c_2, \ldots, c_u$ are distinct elements in A^0, then all elements of the form $a_1^{(e_1)} a_2^{(e_2)} \cdots a_r^{(e_r)} b_1 b_2 \cdots b_s c_1^{m_1} c_2^{m_2} \cdots c_u^{m_u}$ constitute a basis for Super[A]. This basis is called the **trivial basis**.

2.2 Super[A] is a Hopf Algebra

The **Tensor product** Super[A] $\otimes$ Super[A] is defined to be the algebra with the following multiplication rule:

$$w_1 \otimes w_2 \cdot w_1' \otimes w_2' = (-1)^{|w_2||w_1'|} w_1 w_1' \otimes w_2 w_2',$$

where w_1, w_2, w_1', w_2' are monomials in Super[A]. The module structure on Super[A] $\otimes$ Super[A] agrees with the usual tensor product module structure. We may define $|w_1 \otimes w_2| = |w_1| + |w_2|$. Similarly, Super$[A]^{\otimes 3}$ can be defined by setting

$$w_1 \otimes w_2 \otimes w_3 \cdot w_1' \otimes w_2' \otimes w_3' = \epsilon \cdot w_1 w_1' \otimes w_2 w_2' \otimes w_3 w_3',$$

where $\epsilon = (-1)^{(|w_2||w_1'| + |w_3||w_1'| + |w_3||w_2'|)}$. Also, $|w_1 \otimes w_2 \otimes w_3| \stackrel{def}{=} |w_1| + |w_2| + |w_3|$.

Proposition 1 *Let* π: Tens[A] $\to$ Super[A] *be the natural mapping. Then the mapping* $\pi \otimes \pi$: Tens[A] $\otimes$ Tens[A] $\to$ Super[A] $\otimes$ Super[A] *has kernel* $J_A = I_A \otimes$ Tens[A] + Tens[A] $\otimes$ I_A.

Definition 2 *The coproduct* $\triangle$ *of this Hopf algebra is an algebra homomorphism from* Super[A] *to* Super[A] $\otimes$ Super[A] *defined by*

- $\triangle(1) = 1 \otimes 1$,
- $\triangle(a) = 1 \otimes a + a \otimes 1 \quad$ *if* $a \in A^-$,
- $\triangle(a^{(n)}) = a^{(n)} \otimes 1 + a^{(n-1)} \otimes a + \cdots + 1 \otimes a^{(n)} \quad$ *if* $a \in A^+$.

If u is a monomial in Super[A], then $\triangle u$ is a sum of terms of the form $r \otimes s$, $\triangle u = r_1 \otimes s_1 + r_2 \otimes s_2 + \cdots + r_p \otimes s_p$. We shall denote this series by the **Sweedler notation**:

$$\triangle u = \sum_u u_{(1)} \otimes u_{(2)}.$$

For example, let $a \in A^+$ and c, $d \in A^-$, then

$$\begin{aligned} & \triangle(a^{(2)}cd) \\ = \; & \triangle(a^{(2)}) \cdot \triangle(c) \cdot \triangle(d) \\ = \; & (1 \otimes a^{(2)} + a \otimes a + a^{(2)} \otimes 1)(1 \otimes c + c \otimes 1)(1 \otimes d + d \otimes 1) \\ = \; & 1 \otimes a^{(2)}cd + a \otimes acd + c \otimes a^{(2)}d - d \otimes a^{(2)}c + a^{(2)} \otimes cd + ac \otimes ad \\ & -ad \otimes ac + cd \otimes a^{(2)} + a^{(2)}c \otimes d - a^{(2)}d \otimes c + acd \otimes a + a^{(2)}cd \otimes 1. \end{aligned}$$

Proposition 2 (COASSOCIATIVE LAW) *Let A be a signed set. The homomorphisms* $(1 \otimes \triangle) \cdot \triangle$ *and* $(\triangle \otimes 1) \cdot \triangle$ *from* Super$[A]$ *to* Super$[A] \otimes$ Super$[A] \otimes$ Super$[A]$ *coincide.*

Therefore we can define a sequence of algebra homomorphisms

$$\triangle^{(k)} : \text{Super}[A] \to \text{Super}[A]^{\otimes k}$$

recursively as $\triangle^{(k)} = (\overbrace{1 \otimes \cdots \otimes 1}^{k-1} \otimes \triangle) \cdot \triangle^{(k-1)}$. In general, if w is a monomial in Super$[A]$, then $\triangle^{(k)} w$ is a sum of terms having the form $r \otimes s \otimes \cdots \otimes t$, each term having $k+1$ factors. We shall denote this by **Sweedler notation**,

$$\triangle^{(k)} w = \sum_{w} w_{(1)} \otimes w_{(2)} \otimes \cdots \otimes w_{(k+1)}.$$

Next, define the **augmentation map** ϵ: Super$[A] \to Z$ by setting

$$\epsilon(w) = \begin{cases} w & \text{if length}(w) = 0 \\ 0 & \text{if length}(w) > 0 \end{cases}$$

and extending to an algebra homomorphism.

The **antipode** S of the Hopf algebra Super$[A]$ is defined as the linear extension of the map

$$S(w) = (-1)^{\text{length}(w)} w,$$

where w is a monomial in Super$[A]$.

Proposition 3 *Let w be a monomial in* Super$[A]$. *Then*

$$\sum_{w} w_{(1)} S(w_{(2)}) = \epsilon(w),$$

where $w_{(1)}$ and $w_{(2)}$ are such that $\triangle w = \sum_w w_{(1)} \otimes w_{(2)}$.

2.3 Polarizations

Let A be a proper signed alphabet. The polarization operators $D(b,a)$ for a, $b \in A$ are signed derivations in the supersymmetric algebra Super$[A]$. The **parity** of $D(b,a)$, denoted by $|D(b,a)|$, is defined to be 0 when a,b have the same sign and to be 1 when a, b have the different sign. $D(b,a)$ is called a **positive polarization** if its parity is zero and a **negative polarization** if its parity is 1. The formal definition of $D(b,a)$ is as follows.

Let w, w' be two monomials in Super$[A]$. The **polarization operator** $D(b,a)$ for a, $b \in A$ is a linear map from Super$[A]$ to Super$[A]$ such that

- $D(b,a)1 = 0$,
- $D(b,a)c = 0 \quad \text{if } c \neq a$,
- $D(b,a)a^{(k)} = ba^{(k-1)} \quad \text{if } a \in A^+ \text{ and } k > 0, \text{ where } a^{(0)} = 1$,
- $D(b,a)a = b \quad \text{if } a \in A^-$,
- $D(b,a)(ww') = D(b,a)w \cdot w' + (-1)^{|D(b,a)|\cdot|w|} w \cdot D(b,a)w'$.

If $D(b,a)$ is a positive polarization, we can define its divided powers by setting

- $D^{(0)}(b,a)w = w$,
- $D^{(1)}(b,a) = D(b,a)$,
- $D^{(k)}(b,a)a^{(r)} = 0 \quad \text{if } r < k$,
- $D^{(k)}(b,a)a^{(r)} = b^{(k)}a^{(r-k)} \quad \text{if } r \geq k$,
- $D^{(k)}(b,a)(ww') = \sum_{i=0}^{k} D^{(i)}(b,a)w \cdot D^{(k-i)}(b,a)w'$.

If $D(b,a)$ is a negative polarization, it is easy to see $D(b,a)D(b,a) = 0$ and thus we set

- $D^{(0)}(b,a)w = w$,
- $D^{(1)}(b,a) = D(b,a)$,
- $D^{(k)}(b,a) = 0 \quad \text{if } k > 1$.

Proposition 4 *The polarization operators $D(b,a)$ in* Super$[A]$ *are well defined.*

Proposition 5 *Suppose $b \neq c$ and $a \neq d$, then*

$$D(d,c)D(b,a) = \begin{cases} -D(b,a)D(d,c) & \text{if } |D(b,a)| = |D(d,c)| = 1, \\ D(b,a)D(d,c) & \text{otherwise.} \end{cases}$$

2.4 Young Diagrams

Let A be a set. A **Young diagram** on A, $D = (w_1, w_2, \ldots, w_n)$, is a sequence of elements $w_i \in \mathrm{Mon}(A)$ such that $\mathrm{length}(w_1) \geq \mathrm{length}(w_2) \geq \cdots \geq \mathrm{length}(w_n)$. The integer n is called the number of rows of the Young diagram D and the vector $\lambda = (\lambda_1, \lambda_2, \ldots, \lambda_n)$ is called the **shape** of D; here $\lambda_i = \mathrm{length}(w_i)$. If $D = (w_1, w_2, \ldots, w_n)$ where $w_1 = x_1 x_2 \cdots x_{\lambda_1}$, $w_2 = y_1 y_2 \cdots y_{\lambda_2}$, etc., then we write D as follows:

$$D = \begin{array}{lllll} x_1 & x_2 & x_3 & \cdots & x_{\lambda_1} \\ y_1 & y_2 & \cdots & y_{\lambda_2} & \\ & & \vdots & & \end{array} .$$

The **content** of D, denoted by $\mathrm{cont}(D)$, is the multiset consisting of all entries in D. That is, $\mathrm{cont}(D)=\{x_1, \ldots, x_{\lambda_1}, y_1, \ldots, y_{\lambda_2}, \ldots\}$.

The shapes of Young diagrams may be partially ordered. Indeed, if $\lambda = (\lambda_1, \lambda_2, \ldots, \lambda_n)$ and $\lambda' = (\lambda'_1, \lambda'_2, \ldots, \lambda'_m)$, then we write $\lambda \leq \lambda'$ when $\sum_i \lambda_i = \sum_i \lambda'_i$ and $\lambda_1 \leq \lambda'_1$, $\lambda_1 + \lambda_2 \leq \lambda'_1 + \lambda'_2$, and so on.

Let $D = (w_1, w_2, \ldots, w_n)$ be a Young diagram with $w_1 = x_{11} x_{12} \cdots x_{1\lambda_1}$, $w_2 = x_{21} x_{22} \cdots x_{2\lambda_2}$, ... We define a Young diagram $\tilde{D} = (\tilde{w}_1, \tilde{w}_2, \ldots, \tilde{w}_k)$, called the **dual** of D, by setting $\tilde{w}_1 = x_{11} x_{21} \cdots x_{n1}$, $\tilde{w}_2 = x_{12} x_{22} \cdots$, and so on. The shape of $\tilde{D}$ is $\tilde{\lambda} = (\tilde{\lambda}_1, \tilde{\lambda}_2, \ldots, \tilde{\lambda}_k)$ where $\tilde{\lambda}_i$ is the number of λ_j which is greater than or equal to i. We call $k = \lambda_1$ the number of columns of D. Furthermore, we say that two entries of D are in the same row (resp. column) if they are in the same w_i (resp. $\tilde{w}_i$). Also, in a similar way, we may speak of "adjacent entries" in D.

From now on, we assume that A is a proper signed alphabet.

Definition 3 *Let $D = (w_1, w_2, \ldots, w_n)$ be a Young diagram on A and let $\tilde{D} = (\tilde{w}_1, \tilde{w}_2, \ldots, \tilde{w}_k)$ be its dual. The Young diagram D is said to be standard when, for each of the words*

$$w_i = x_1 x_2 \cdots x_r,$$

$$\tilde{w}_i = y_1 y_2 \cdots y_s,$$

the following conditions are satisfied:

1. *if $x_j \in A^+$, then $x_j \leq x_{j+1}$,*
2. *if $x_j \in A^-$, then $x_j < x_{j+1}$,*
3. *if $y_j \in A^+$, then $y_j < y_{j+1}$,*
4. *if $y_j \in A^-$, then $y_j \leq y_{j+1}$.*

3 Letter-place pairings

The supersymmetric algebra Super$[A]$ is to be extended to a more complex structure, called the *fourfold algebra.* We shall see that the fourfold algebra is the suitable machinery to develop the characteristic-free invariant theory of skew-symmetric tensors.

3.1 The Letter-Place Pairings

Let A be any signed set. If $k \in N^+$, we denote by $\text{Tens}_k[A]$ the submodule of Tens$[A]$ spanned by 0 and all $w \in \text{Mon}(A_{\text{exd}})$ such that $\text{length}(w) = k$, and similarly we define $\text{Super}_k[A]$.

Let L and P be disjoint proper signed sets. We shall call the elements of L **letters** and the elements of P **places**. Define a new signed set, which is not proper, by $[L|P] = \{(x|\alpha) : x \in L, \alpha \in P\}$. The sign of an element in $[L|P]$ is defined by the following rules:

1. if $x \in L^+$ and $\alpha \in P^+$, then $(x|\alpha) \in [L|P]^+$;
2. if $x \in L^-$ and $\alpha \in P^-$, then $(x|\alpha) \in [L|P]^0$;
3. otherwise, $(x|\alpha) \in [L|P]^-$.

The Z-algebra Super$[L|P]$ will be called the **fourfold algebra**. An element of $[L|P]_{\text{exd}}$ will be called a **letterplace**.

Definition 4 *Let w be a monomial in* Super$[L]$ *of length k and u be a monomial in* Super$[P]$ *of length l. Define a bilinear mapping Ω from* Super$[L] \times$ Super$[P]$ *to* Super$[L|P]$ *by the following rules :*

R. 1. *if $k \neq l$, $\Omega = 0$. For $k = l = n$, we define Ω inductively and will denote $\Omega(w, u)$ by $(w|u)$, which is called the biproduct of w and u;*

R. 2. $(1|1) = 1$;

R. 3. *for $x \in L$ and $\alpha \in P$, set $\Omega(x, \alpha) = (x|\alpha) \in [L|P]$;*

R. 4. $(x^{(n)}|\alpha^{(n)}) = (x|\alpha)^{(n)}$;

R. 5. *Laplace expansion by rows*

$$(w|u'u'') = \sum_w (-1)^{(|w_{(2)}||u'|)} (w_{(1)}|u')(w_{(2)}|u''),$$

where $\triangle w = \sum_w w_{(1)} \otimes w_{(2)}$.

R. 6. *Laplace expansion by columns*

$$(w'w''|u) = \sum_u (-1)^{(|w''||u_{(1)}|)}(w'|u_{(1)})(w''|u_{(2)}),$$

where $\Delta u = \sum_u u_{(1)} \otimes u_{(2)}$.

Proposition 6 *The bilinear mapping* $\Omega : \mathrm{Super}[L] \times \mathrm{Super}[P] \to \mathrm{Super}[L|P]$, *is well defined.*

We shall denote $(-1)^{\sum_{i>j}|w_i||u_j|}(w_1|u_1)(w_2|u_2)\cdots(w_n|u_n)$ by

$$\left(\begin{array}{c|c} w_1 & u_1 \\ w_2 & u_2 \\ \vdots & \vdots \\ w_n & u_n \end{array}\right). \tag{1}$$

Thus, Rule 5 becomes

$$(w|u'u'') = \sum_w \left(\begin{array}{c|c} w_{(1)} & u' \\ w_{(2)} & u'' \end{array}\right),$$

and Rule 6 may be written as

$$(w'w''|u) = \sum_w \left(\begin{array}{c|c} w' & u_{(1)} \\ w'' & u_{(2)} \end{array}\right).$$

3.2 Polarizations

We may now define a polarization $\mathbf{D(b,a)}$ in $\mathrm{Super}[L|P]$ for $a, b \in L$. Our development parallels the definition of the polarization $D(b,a)$ in $\mathrm{Super}[L]$.

- $\mathbf{D(b,a)}1 = 0$;
- $\mathbf{D(b,a)}(c|x) = 0 \quad \text{if } c \neq a$;
- $\mathbf{D(b,a)}(a|x)^{(k)} = (b|x)(a|x)^{(k-1)} \quad$ if $a \in L^+$, $x \in P^+$ and $k > 0$;
- $\mathbf{D(b,a)}(a|x) = (b|x)$;
- Let w, w' be monomials in $\mathrm{Super}[L|P]$, then

  $$\mathbf{D(b,a)}(ww') = \mathbf{D(b,a)}w \cdot w' + (-1)^{|\mathbf{D(b,a)}|\cdot|w|} w \cdot \mathbf{D(b,a)}w',$$

  where $|\mathbf{D(b,a)}|$ denotes the parity of $\mathbf{D(b,a)}$ which is the same as $|D(b,a)|$, and $|w|$ denotes the parity of w which is equal to $(-1)^k$ here; k is the number of negative letter-place pairings in w.

If $\mathbf{D}(\mathbf{b},\mathbf{a})$ is a positive polarization, we can define its divided powers by setting

- $\mathbf{D^{(0)}}(\mathbf{b},\mathbf{a})w = w$,
- $\mathbf{D^{(1)}}(\mathbf{b},\mathbf{a}) = \mathbf{D}(\mathbf{b},\mathbf{a})$,
- $\mathbf{D^{(k)}}(\mathbf{b},\mathbf{a})(a|x)^{(r)} = 0 \quad \text{if } r < k$,
- $\mathbf{D^{(k)}}(\mathbf{b},\mathbf{a})(a|x)^{(r)} = (b|x)^{(k)}(a|x)^{(r-k)} \quad \text{if } r \geq k$,
- $\mathbf{D^{(k)}}(\mathbf{b},\mathbf{a})(ww') = \sum_{i=0}^{k} \mathbf{D^{(i)}}(\mathbf{b},\mathbf{a})w \cdot \mathbf{D^{(k-i)}}(\mathbf{b},\mathbf{a})w'$.

If $\mathbf{D}(\mathbf{b},\mathbf{a})$ is a negative polarization, then $\mathbf{D}(\mathbf{b},\mathbf{a})\mathbf{D}(\mathbf{b},\mathbf{a}) = 0$ and therefore we set

- $\mathbf{D^{(0)}}(\mathbf{b},\mathbf{a})w = w$,
- $\mathbf{D^{(1)}}(\mathbf{b},\mathbf{a}) = \mathbf{D}(\mathbf{b},\mathbf{a})$,
- $\mathbf{D^{(k)}}(\mathbf{b},\mathbf{a}) = 0 \quad \text{if } k > 1$.

Proposition 7 *The polarization operators* $\mathbf{D}(\mathbf{b},\mathbf{a})$ *in* $\mathrm{Super}[L|P]$ *are well-defined.*

Proposition 8 *Let* w *be a monomial in* $\mathrm{Super}[L]$, ϖ *be a monomial in* $\mathrm{Super}[P]$. *Then* $\mathbf{D}(\mathbf{b},\mathbf{a})$ *is an induced map of* $D(b,a)$, *i.e.,*

$$\mathbf{D^{(k)}}(\mathbf{b},\mathbf{a})(w|\varpi) = \left(D^{(k)}(b,a)w\middle|\,\varpi\right).$$

Similarly, each polarization operator $(a,b)T$ in $\mathrm{Super}[P]$ induces a place polarization operator $(\mathbf{a},\mathbf{b})\mathbf{T}$ in $\mathrm{Super}[L|P]$ such that

$$(w \mid \varpi)(\mathbf{a},\mathbf{b})\mathbf{T} = (w \mid \varpi(a,b)T),$$

and

$$(ww')(\mathbf{a},\mathbf{b})\mathbf{T} = w \cdot (w'(\mathbf{a},\mathbf{b})\mathbf{T}) + (-1)^{|(\mathbf{a},\mathbf{b})\mathbf{T}|\cdot|w'|}\,(w(\mathbf{a},\mathbf{b})\mathbf{T}) \cdot w'.$$

If we use tableau notation, it is easy to see that the polarization operators in $\mathrm{Super}[L|P]$ can be computed as

$$\mathbf{D}\left(\begin{array}{c|c} w_1 & \varpi_1 \\ w_2 & \varpi_2 \\ \vdots & \vdots \\ w_n & \varpi_n \end{array}\right) = \sum_j (-1)^{\sum_{i<j}|\mathbf{D}||w_i|} \left(\begin{array}{c|c} w_1 & \varpi_1 \\ \vdots & \vdots \\ D(w_j) & \varpi_j \\ \vdots & \vdots \\ w_n & \varpi_n \end{array}\right) \tag{2}$$

or

$$\begin{pmatrix} w_1 & \varpi_1 \\ w_2 & \varpi_2 \\ \vdots & \vdots \\ w_n & \varpi_n \end{pmatrix} T = \sum_j (-1)^{\sum_{i<j} |T||\varpi_i|} \begin{pmatrix} w_1 & \varpi_1 \\ \vdots & \vdots \\ w_j & (\varpi_j)T \\ \vdots & \vdots \\ w_n & \varpi_n \end{pmatrix} \tag{3}$$

From now on, we shall use the simplified notation $D(b, a)$ and $(a, b)T$ instead of $\mathbf{D}(\mathbf{b}, \mathbf{a})$ and $(\mathbf{a}, \mathbf{b})\mathbf{T}$, respectively, to denote the polarizations in $\mathrm{Super}[L|P]$.

Proposition 9 *Let D represent any letter polarization and T represent any place polarization. Then for any element $h \in \mathrm{Super}[L|P]$,*

$$(D(h))T = D((h)T).$$

Examples:

Let Λ be a proper signed set disjoint from L and P. We shall consider the supersymmetric algebra $\mathrm{Super}[L'|P']$, where $L' = L \cup \Lambda$ and $P' = P \cup \Lambda$. The variables in Λ will be called **virtual symbols**. For $w = u_1 u_2 \cdots u_k \in \mathrm{Mon}(L_{\mathrm{exd}})$ and $\theta \in \Lambda$, we use $D(w, \theta)$ to denote $D(u_1, \theta) \cdots D(u_i, \theta) \cdots D(u_k, \theta)$, where $D(u_i, \theta) = D^{(m)}(a, \theta)$ if $u_i = a^{(m)} \in L_{\mathrm{exd}}$. Similarly, we define $(\theta, w)T$ to be $(\theta, u_1)T(\theta, u_2)T \cdots (\theta, u_k)T$ if $w \in \mathrm{Mon}(P_{\mathrm{exd}})$.

1. Let $v_i \in \mathrm{Mon}(L_{\mathrm{exd}})$ and $u_i \in \mathrm{Mon}(P_{\mathrm{exd}})$ for $1 \leq i \leq p$, having length λ_i. The vector $\lambda = (\lambda_1 \geq \lambda_2 \geq \cdots \geq \lambda_p)$ is a Young shape. Let $\{1^+, 2^+, \ldots, p^+\}$ be a set of positive virtual symbols. Then

$$\begin{pmatrix} v_1 & u_1 \\ v_2 & u_2 \\ \vdots & \vdots \\ v_p & u_p \end{pmatrix} = \begin{pmatrix} v_1 & (1^+)^{(\lambda_1)} \\ v_2 & (2^+)^{(\lambda_2)} \\ \vdots & \vdots \\ v_p & (p^+)^{(\lambda_p)} \end{pmatrix} (1^+, u_1)T(2^+, u_2)T \cdots (p^+, u_p)T.$$

2. Let v_i, u_i be defined as above, then

$$\begin{pmatrix} v_1 & u_1 \\ v_2 & u_2 \\ \vdots & \vdots \\ v_p & u_p \end{pmatrix} = D(v_1, 1^+)D(v_2, 2^+) \cdots D(v_p, p^+) \begin{pmatrix} (1^+)^{(\lambda_1)} & u_1 \\ (2^+)^{(\lambda_2)} & u_2 \\ \vdots & \vdots \\ (p^+)^{(\lambda_p)} & u_p \end{pmatrix}.$$

3.3 The Standard Basis Theorem

The standard basis theorem is the most fundamental result in the supersymmetric algebra. Its proof requires the following exchange identity.

Proposition 10 (THE SUPERSYMMETRIC EXCHANGE IDENTITY) *Let a, b, c be monomials in* $\mathrm{Super}[L]$ *with* $\Delta a = \sum_a a_{(1)} \otimes a_{(2)}$ *and* $\Delta b = \sum_b b_{(1)} \otimes b_{(2)}$. *Let x and y be monomials in* $\mathrm{Super}[P]$ *with* $\Delta y = \sum_y y_{(1)} \otimes y_{(2)}$. *Then*

$$\sum_b \left(\begin{array}{c|c} ab_{(1)} & x \\ b_{(2)}c & y \end{array} \right) = (-1)^{|a||b|} \sum_{a,y} \left(\begin{array}{c|c} ba_{(1)} & xy_{(1)} \\ S(a_{(2)})c & y_{(2)} \end{array} \right),$$

where $S(a_{(2)})$ is the antipode.

Let A be a signed set. The homomorphism Disp: $\mathrm{Mon}(A_{\text{exd}}) \to \mathrm{Mon}(A)$ is defined by the following rule: if $a \in A^-$ or $a \in A^0$, set Disp(a)=a; if $a \in A^+$, set $\mathrm{Disp}(a^{(n)}) = aa\cdots a$ (n-times). For example, if $w = bc^{(3)}a^{(2)}a^{(3)}cd$, then

$$\mathrm{Disp}(w) = bcccaaaaacd.$$

One can also define an inverse mapping from the set $\mathrm{Mon}(A)$ to the set $\mathrm{Mon}(A_{\text{exd}})$, denoted by $w \to \mathrm{stand}(w)$, which is however not a homomorphism of monoids. For $w \in \mathrm{Mon}(A)$, define stand(w) to be the unique element in $\mathrm{Mon}(A_{\text{exd}})$ such that

- $\mathrm{Disp}(\mathrm{stand}(w)) = w$,
- if $\mathrm{stand}(w) = y_1 y_2 \cdots y_k$ where $y_i \in A_{\text{exd}}$ and if $y_j = a^{(i)}$ for some $a \in A^+$, $i \in N^+$, then $y_{j-1} \neq a^{(r)}$ and $y_{j+1} \neq a^{(s)}$ for all integers r and s.

For example, $\mathrm{stand}(\mathrm{Disp}(bc^{(3)}a^{(2)}a^{(3)}cd)) = bc^{(3)}a^{(5)}cd$.

Returning to $\mathrm{Super}[L|P]$, let $D = (w_1, w_2, \ldots, w_n)$ and $E = (w'_1, w'_2, \ldots, w'_n)$ be Young diagrams of the same shape on L and P, respectively. The **Young tableau** of the diagram pair (D, E) is defined by

$$\mathrm{Tab}(D|E) = (\mathrm{stand}(w_1) \mid \mathrm{stand}(w'_1)) \cdots (\mathrm{stand}(w_n) \mid \mathrm{stand}(w'_n)).$$

With this terminology in place, we now state the standard basis theorem. The proof of this theorem can be found in [7].

Theorem 1 *Let L and P be disjoint proper signed alphabets. The set of* $\mathrm{Tab}(D_i|E_i)$, *where D_i and E_i are standard and have the same shape, form a* *basis of* Super$[L|P]$ *over* Z. *In particular, let* D *and* E *be Young diagrams, having the same shape, on* L *and* P, *respectively. Then*

$$\mathrm{Tab}(D|E) = \sum_i c_i \mathrm{Tab}(D_i|E_i)$$

in Super$[L|P]$, *where*

(i) D_i *and* E_i *are standard on* L *and* P, *respectively,*

(ii) $\mathrm{cont}(D_i)=\mathrm{cont}(D)$ *and* $\mathrm{cont}(E_i)=\mathrm{cont}(E)$,

(iii) $\mathrm{shape}(D) \leq \mathrm{shape}(D_i)$,

(iv) *the integer coefficients* c_i *are unique.*

4 The umbral module

Most of the material in this section can be found in [3]. We begin with the theory of the White module, which generalizes to the supersymmetric case the bracket module introduced by White in his thesis.

4.1 The White Module

Recall that a **Peano space** of step n, denoted by $\mathrm{P_n}(V)$, is a vector space V of dimension n over a field F with a non-degenerate antisymmetric n-linear form $\{v_1 v_2 \cdots v_n\}$. We shall construct a "universal vector space" over a "universal" field, in which all facts of projective geometry that are true irrespective of the choice of field will hold, and conversely. All we are given is an infinite set L^- (letters) whose elements will be the "generic points" of the universal vector space, and a step n. The role of the universal field will be fulfilled by the abstract *bracket ring* of step n.

We start with the supersymmetric bracket algebra. Let $L = L^+ \cup L^-$ be a proper signed alphabet and let $P = P^- = \{y_1, y_2, \dots, y_n\}$ be a disjoint negatively signed alphabet with n distinct variables. For a monomial $w \in$ Super$[L]$, we define the **bracket** $[w] \in$ Super$[L|P]$ to be

$$(w|y_1 y_2 \cdots y_n).$$

Therefore if length(w) is not equal to n, $[w] = 0$. Furthermore for w_1, $w_2 \in$ Super$[L]$ with length n, we get

$$[w_1][w_2] = (-1)^{(|w_1|+\mathrm{length}(w_1))(|w_2|+\mathrm{length}(w_2))}[w_2][w_1].$$

An element $p \in \mathrm{Super}[L|P]$ is called a **bracket monomial** if $p = [w_1][w_2]\cdots[w_k]$, where $w_1, w_2, \ldots, w_k$ are monomials in $\mathrm{Super}[L]$, and a **bracket polynomial** if it is a linear combination of bracket monomials. The **bracket algebra** of step n, denoted by $\mathrm{B_n}[L]$, is the subalgebra of $\mathrm{Super}[L|P]$ generated by the brackets $[w] \in \mathrm{Super}[L|P]$. We use

$$\begin{bmatrix} w_1 \\ w_2 \\ \vdots \\ w_k \end{bmatrix} \tag{4}$$

to denote

$$(-1)^{\sum_{i=2}^{k} |w_i|\cdot n^{i-1}}[w_1][w_2]\cdots[w_k]. \tag{5}$$

The fundamental relation in $\mathrm{B_n}[L]$ is the following **bracket exchange identity**, which is a specialized form of the exchange identity in the fourfold algebra.

Proposition 11 (The BRACKET EXCHANGE IDENTITY) *Let a, b, c be monomials in* $\mathrm{Super}[L]$ *with* $\Delta a = \sum_a a_{(1)} \otimes a_{(2)}$ *and* $\Delta b = \sum_b b_{(1)} \otimes b_{(2)}$. *Then in* $\mathrm{B_n}[L]$,

$$\sum_b (-1)^{|b_{(2)}|\cdot n}[ab_{(1)}][b_{(2)}c] = (-1)^{|a||b|}\sum_a (-1)^{|a_{(2)}|\cdot n + \mathrm{length}(a_{(2)})}[ba_{(1)}][a_{(2)}c].$$

As the exchange identity leads to the standard basis theorem in the fourfold algebra, the bracket exchange identity leads to a corresponding standard basis theorem, which is easy to state (and is omitted here) in the bracket algebra $\mathrm{B_n}[L]$. Thus, any bracket monomial $[w_1][w_2]\cdots[w_k]$, in which a positive letter $a \in L$ appears more than n times (i.e., $\mathrm{cont}(w_1;a)+\mathrm{cont}(w_2;a)+\cdots+\mathrm{cont}(w_k;a) > n$), is equal to zero. This is because such a monomial can not be straightened to standard forms defined in Definition 3 (check the 3rd condition).

Until the end of this section, we shall restrict L to be a negatively signed alphabet, denoted by L^-. The bracket algebra $\mathrm{B_n}[L^-]$ is then isomorphic to the commutative and associative algebra generated by the set of symbols $\{[a_1 a_2 \cdots a_n] : a_1, \ldots, a_n \in L^-\}$ subject to:

- for any permutation $\sigma \in S_n$,

$$[a_{\sigma(1)} a_{\sigma(2)} \cdots a_{\sigma(n)}] = sgn(\sigma)[a_1 a_2 \cdots a_n];$$

- whenever $a_i = a_j$, $[a_1 \cdots a_i \cdots a_j \cdots a_n] = 0$;

- for any $k \in \{1, 2, \ldots, n\}$,

$$\sum sgn(\sigma)[a_{\sigma(1)} \cdots a_{\sigma(n)}][a_{\sigma(n+1)} \cdots a_{\sigma(n+k)} b_1 \cdots b_{n-k}] = 0,$$

where the sum ranges over all $\sigma \in S_{n+k}$ with $\sigma(1) < \cdots < \sigma(n)$ and $\sigma(n+1) < \cdots < \sigma(n+k)$.

The antisymmetric n-linear form $\{\cdot\}$ can be completely characterized by the bracket form $[\cdot]$ in $\mathrm{B_n}[L^-]$ as the following proposition shows.

Proposition 12 *Let* $\mathrm{B_n}[L^-]$ *be a bracket algebra over* Z *and* $\mathrm{P_n}(V)$ *be a Peano space over a field* F*, where* $\{(x|i) : x \in L^-, i \in \{1, 2, \ldots n\}\}$ *is a set of generators of all transcendentals of* F*. Let* $\{e_1, e_2, \ldots, e_n\}$ *be a basis of* V *with* $[e_1 e_2 \ldots e_n] = 1$*. Let* ϕ*:* $L^- \to V$ *be a map such that* $\phi(x) = \sum_{i=1}^{n} (x|i) e_i$*. Then the corresponding algebra homomorphism* ϕ'*:* $\mathrm{B_n}[L^-] \to F$ *such that*

$$\phi'([x_1 x_2 \cdots x_n]) = \{\phi(x_1)\phi(x_2) \cdots \phi(x_n)\} = det((x_i|j))$$

is well defined. Furthermore, for any $p \in \mathrm{B_n}[L^-]$*,* $p = 0$ *if and only if* $\phi'(p) = 0$*.*

Proposition 13 $\mathrm{B_n}[L^-]$ *is an integral domain.*

Now, we are ready to define a universal vector space with a universal field. The **White module** of step n over Z, denoted by $\mathrm{W_n}(L^-)$, is the module spanned by all elements of the form

$$\alpha x \qquad \text{where } \alpha \in \mathrm{B_n}[L^-] \text{ and } x \in L^-,$$

subject to the following condition: given $\alpha_i, \beta_j \in \mathrm{B_n}[L^-]$,

$$\sum_i \alpha_i x_i = \sum_j \beta_j y_j$$

if and only if for every $z_1, z_2, \ldots, z_{n-1} \in L^-$ we get the following identity in $\mathrm{B_n}[L^-]$.

$$\sum_i \alpha_i [x_i z_1 z_2 \cdots z_{n-1}] = \sum_j \beta_j [y_j z_1 z_2 \cdots z_{n-1}].$$

Proposition 14 *The White module* $\mathrm{W_n}(L^-)$ *is a free module with dimension* n*. Any* n *distinct elements* $x_1, x_2, \ldots, x_n \in L^-$ *are independent.*

Proposition 15 *Let* $\mathrm{W_n}(L^-)$ *be the White module over* Z *and* $\mathrm{P_n}(V)$ *be the Peano space over a field* F*. Then for any map* ϕ*:* $L^- \to V$*, there exists*

1. *a unique algebra homomorphism* $\hat{\phi}$: $\mathrm{B_n}[L^-] \to F$ *such that*

$$\hat{\phi}([x_1 x_2 \cdots x_n]) = \{\phi(x_1)\phi(x_2)\cdots\phi(x_n)\}$$

for $x_1, x_2, \ldots, x_n \in L^-$,

2. *a unique linear map* $\tilde{\phi}$: $\mathrm{W_n}(L^-) \to V$ *such that*

$$\tilde{\phi}\left(\sum_i \alpha_i x_i\right) = \sum_i \hat{\phi}(\alpha_i)\phi(x_i)$$

for $\alpha_i \in \mathrm{B_n}[L^-]$ *and* $x_i \in L^-$.

We would like to extend the idea of the White module to include skew-symmetric tensors as well as vectors. Fix the step k of the skew-symmetric tensors one wants to represent. Then consider the algebra generated by all elements of the form

$$\sum_i c_i w_i,$$

where $c_i \in \mathrm{B_n}[L^-]$, and w_i are monomials in $\mathrm{Super}[L^-]$ having the same length k, subject to the following condition:

$$\sum_i c_i w_i = 0$$

if and only if

$$\sum_i c_i [w_i w] = 0$$

for every monomial $w \in \mathrm{Super}[L^-]$ of length $n-k$. This algebra has a universal property that can be inferred from the universal property of the White module, and is called the **exterior algebra of the White module**, denoted by $\bigwedge(\mathrm{W_n}(L^-))$. We have not defined here the multiplication rule in the exterior algebra of the White module; the rule can be found in Brini's paper [2].

An element of $\bigwedge(\mathrm{W_n}(L^-))$ is called a **skew-symmetric tensor**. From now on we shall say "tensors" instead of "skew-symmetric tensors." A tensor of step 1 is called a **vector**. If t is a tensor of step k represented by $x_1, x_2, \ldots, x_n \in L^-$, i.e.,

$$t = \sum_{i_1 < i_2 < \cdots < i_k} c_{i_1, i_2, \ldots, i_k} x_{i_1} x_{i_2} \cdots x_{i_k}, \tag{6}$$

then the scalars $c_{i_1, i_2, \ldots, i_k}$ will be called the **Plücker** coordinates of the tensor t. The **Plücker** coordinates are well defined since any n distinct elements of L^- are independent in the White module $\mathrm{W_n}(L^-)$.

4.2 The Umbral Module

In view of our definition for the supersymmetric bracket algebra, it is natural to ask next whether one can define an analogous generalization of the White module, that will play for the exterior algebra of a Peano space the role played by the White module for Peano space. This can be done and we shall first define the White module analog of the component of degree k of the exterior algebra of a Peano space. We call such a module the *umbral module* of step k.

Let $L = L^+ \cup L^-$ be a proper signed alphabet and T be a subset of $\bigwedge(\mathrm{W_n}(L^-))$. $\mathcal{F}$ is an arbitrary map from L^+ to T, called the **umbral map**. If $\mathcal{F}(a) = t$ (a tensor of step k), we say a is an **umbra** for t, having **arity** k. We shall use $a^{(k)}$ as the symbolic representation of the tensor t. If t has the form (6), then the polarization of a to t, denoted by $D(t, a)$, is defined as follows:

$$D(t,a) = \sum_{i_1<\cdots<i_k} c_{i_1,i_2,\ldots,i_k} D(x_{i_1},a)D(x_{i_2},a)\cdots D(x_{i_k},a).$$

If $\mathcal{F}(a) = \mathcal{F}(b)$, we write

$$a \propto b,$$

which is read as a is **equivalent** to b.

For the umbral map $\mathcal{F}$: $L^+ \to T$, we define an **equivalence relation** $\sim$, which is compatible with the linear structure, in the supersymmetric bracket algebra $\mathrm{B_n}[L]$ as follows:

- if $a \in L^+$ and $0 < \mathrm{cont}(m; a) \neq$ the arity of a, then

  $$m \sim 0,$$

  where m is a supersymmetric bracket monomial;

- if $\{a < b < \cdots < c\}$ is the set of all positive letters appearing in m with arities k_a, k_b, ... and k_c, respectively, and $cont(m; a) = k_a$, $cont(m; b) = k_b$, ..., $cont(m; c) = k_c$, then

  $$m \sim D(t_c, c)\cdots D(t_b, b)D(t_a, a)m,$$

  where $\mathcal{F}(a) = t_a$, $\mathcal{F}(b) = t_b$, ..., $\mathcal{F}(c) = t_c$.

Since the equivalence relation $\sim$ is defined through polarization, it is not difficult to see $\sim$ is well defined in the supersymmetric bracket algebra $\mathrm{B_n}[L]$. Furthermore, the quotient algebra

$$\mathrm{B_n}[L]/\sim$$

is isomorphic to

$$\mathrm{B_n}[L^-].$$

We shall call the quotient algebra the **umbral bracket algebra** of step n, denoted by $\mathrm{UB_n}[L]$.

The **umbral module** of step n over $\mathrm{UB_n}[L]$, denoted by $\mathrm{U_n}(L)$, is defined as the quotient module of Super$[L]$ by imposing the equivalence relation $\propto$ defined as follows: for c_i, $d_j \in \mathrm{UB_n}[L]$ and w_i, $w'_j \in$ Super$[L]$,

$$\sum_i c_i w_i \propto \sum_j d_j w'_j$$

in $\mathrm{U_n}(L)$ if and only if

$$\sum_i c_i[w_i w] = \sum_j d_j[w'_j w]$$

in $\mathrm{UB_n}[L]$, for every monomial $w \in$ Super$[L]$ containing no positive letters appearing in either $\sum_i c_i w_i$ or $\sum_j d_j w_j$.

The direct sum of the umbral modules $\mathrm{U_k}(L)$ for all k up to n can be given the structure of an algebra (the "umbral algebra" for skew-symmetric tensors), where the coefficients are elements of the bracket ring of the supersymmetric algebra Super$[L]$. This algebra now plays the role of the exterior algebra of the White module, and thus it is a universal model for all computations in the exterior algebra of any vector space. We are led to expect that every notion of exterior algebra, in particular of linear algebra, has a corresponding notion in the umbral algebra. Such a corresponding notion can be obtained by reasoning with the exterior algebra of the White module, but eventually one learns to work with the umbral algebra directly. One thus discovers a new way of doing linear algebra. We shall not give here a precise description of linear algebra done with positive letters; instead, we give a few examples.

(i) Cramer's rule, expressing a vector b as a linear combination of vectors belonging to the basis $a^{(n)}$, is expressed by the elegant identity

$$[a^{(n)}]b = -[a^{(n-1)}b]a.$$

The classical version of Cramer's rule is, of course, obtained by polarizing $a^{(n)}$ to $x_1 x_2 \cdots x_n$, where the x_i are negative letters. One then obtains

$$[x_1 x_2 \cdots x_n]b = \sum_{i=1}^{n} [x_1 x_2 \cdots \hat{x}_i \cdots x_n b] x_i,$$

which is the usual version of Cramer's rule. More generally, a tensor $c^{(k)}$ of step k, where c is a positive letter, can be expressed as the sum of decomposable tensors made up in terms of the vectors of the basis $a^{(n)}$ by the identity

$$[a^{(n)}]b^{(k)} = [a^{(n-k)}b^{(k)}]a^{(k)}.$$

(ii) Let $a^{(2)}$ and $b^{(2)}$ be two lines in the plane. Their intersection can be given by either of the two equal expressions

$$[a^{(2)}b]b = [b^{(2)}a]a.$$

If we polarize $a^{(2)}$ to xy and $b^{(2)}$ to uv, where x,y,u, v are negative letters, we obtain the classical expression for the intersection point of two lines spanned by x and y and by u and v, respectively:

$$[xyu]v - [xyv]u = [uvx]y - [uvy]x.$$

The brackets here are, of course, determinants of vectors. One can also obtain expressions for general intersections of subspaces in projective space. For example,

$$[a^{(p)}b^{(n-p)}]b^{(i)}$$

represents the intersection of 2 subspaces of dimension p and $n-p+i$ in a projective space of dimension n, where a, b are positive letters having arities p and $n-p+i$ respectively.

Now we proceed to define the covariant module. The **covariant module** of step n over $\mathrm{UB_n}[L]$, denoted by $\mathrm{Cov_n}(L)$, is the module $\mathrm{Super}[L] \oplus \mathrm{Super}[L]^{\otimes 2} \oplus \mathrm{Super}[L]^{\otimes 3} \oplus \cdots$ by imposing the following equivalence relation $\simeq$, which is compatible with the linear structure:

$$w \otimes w' \otimes w'' \otimes \cdots \simeq u \otimes u' \otimes u'' \otimes \cdots$$

in $\mathrm{Cov_n}(L)$ if and only if

$$[ww_1][w'w_2][w''w_3]\cdots = [uw_1][u'w_2][u''w_3]\cdots$$

in $\mathrm{UB_n}[L]$ for every choice of $\{w_i : w_i \in \mathrm{Super}[L]\}$ subject to the condition that no positive letters in any w_i of $\{w_i\}$ occur in either $w \otimes w' \otimes w'' \otimes \cdots$ or in $u \otimes u' \otimes u''$, where w, w', w'', $\ldots$ and u, u', u'', $\ldots$ are monomials in $\mathrm{Super}[L]$ with $\mathrm{length}(w) = \mathrm{length}(u), \ldots, \mathrm{length}(w'') = \mathrm{length}(u''), \ldots$.

An element of $\mathrm{Cov_n}(L)$ is called a **covariant**.

4.3 Skew-Symmetric Tensors

The notation sketched above, using $a^{(k)}$ to denote a skew-symmetric tensor in the umbral module, can be used to describe facts about skew-symmetric tensors.

In the umbral module $\mathrm{U_n}(L)$, a tensor $a^{(k)}$ is said to be divisible by a vector b if there exists a tensor $c^{(k-1)}$ with $c \neq b$ such that $a^{(k)} \propto bc^{(k-1)}$. The tensor $a^{(k)}$ is called **decomposable** if there are vectors $b_1, b_2, \ldots, b_k \in L^+$ such that $a^{(k)} \propto b_1 \cdots b_k$.

Proposition 16 *1. A vector v divides a tensor $a^{(k)}$ iff $a^{(k)}v \propto 0$.*

2. If $a^{(k)}$ is not decomposable, but is divisible by a vector, then it must be divisible by the vector

$$[a^{(k)}b^{(k-1)}u^{(n-2k+1)}]b,$$

where a and b are equivalent and $u^{(n-2k+1)}$ is some tensor of step $n-2k+1$.

3. Let $a^{(k)}$ be a tensor of step k. Vectors $v_1, v_2, \ldots, v_r$ are linearly independent vectors, each of which divides $a^{(k)}$ ($r \leq k$). Then there exists a tensor $c^{(k-r)}$ such that $a^{(k)} \propto v_1v_2 \cdots v_rc^{(k-r)}$.

4. Every tensor $a^{(n-1)}$ of step $n-1$ is decomposable.

Let $a^{(k)}$ be a tensor of step k. We define the **span** of $a^{(k)}$, denoted by $\mathrm{span}(a^{(k)})$, to be the submodule of $\mathrm{U_n}(L)$ spanned by all vectors having the form

$$[a^{(k-1)}u^{(n-k+1)}]a$$

as $u^{(n-k+1)}$ ranges over all decomposable tensors of step $n-k+1$. The dimension of the $\mathrm{span}(a^{(k)})$ is called the **rank** of $a^{(k)}$. If $a^{(k)}$ is decomposable, say $a^{(k)} \propto v_1v_2 \cdots v_k$ where v_i are of step one, then $\mathrm{span}(a^{(k)})$ has a basis $\{v_1, v_2, \ldots v_k\}$. This can be seen as follows.

Let $v^{(n-k)}$ be a decomposable tensor of step $n-k$ such that

$$[a^{(k)}v^{(n-k)}] \neq 0$$

in $\mathrm{UB_n}[L]$. Then by the bracket exchange identity, we obtain

$$[a^{(k-1)}v^{(n-k)}v_i]a \propto -[a^{(k)}v^{(n-k)}]v_i - [a^{(k)}v^{(n-k-1)}v_i]v.$$

The last term above vanishes since $a^{(k)} \propto v_1v_2 \cdots v_k$. Therefore, the vector $[a^{(k)}v^{(n-k)}]v_i$ is in the span of $a^{(k)}$ for $i = 1, 2, \ldots, k$, as does every vector of the form $[a^{(k-1)}u^{(n-k+1)}]a$. So, $\mathrm{span}(a^{(k)})$ has the basis $\{v_1, v_2, \ldots, v_k\}$.

Proposition 17 *Let $a^{(k)}$ be a tensor of step k. Then*

1. $\operatorname{rank}(a^{(k)}) \geq k$.
2. $\operatorname{span}(a^{(k)})$ *is the smallest subspace of* $\mathrm{U_n}(L)$ *such that $a^{(k)}$ can be expressed by using only vectors in* $\operatorname{span}(a^{(k)})$.
3. $\operatorname{rank}(a^{(k)}) = k$ *if and only if $a^{(k)}$ is decomposable.*
4. *if* $\operatorname{rank}(a^{(k)}) > k$, *then* $\operatorname{rank}(a^{(k)}) \geq k+2$.

Proposition 18 *$a^{(k)}$ has rank r if and only if the covariant*

$$\xi_1\colon a_1 a_2 \cdots a_{r+1} \otimes a_1^{(k-1)} \otimes a_2^{(k-1)} \otimes \cdots \otimes a_{r+1}^{(k-1)}$$

vanishes , but not the covariant $\xi_2\colon a_1 a_2 \cdots a_r \otimes a_1^{(k-1)} \otimes a_2^{(k-1)} \otimes \cdots \otimes a_r^{(k-1)}$, where all symbols a_i are equivalent to a.

Proposition 19 *Let $b^{(k)}$ be the umbral representation of a skew symmetric tensor of step k, with the equivalent umbral representations*

$$a_1^{(k)} \propto a_2^{(k)} \propto \cdots \propto a_{p+1}^{(k)} \propto b^{(k)}.$$

Then $b^{(k)}$ is divisible by at least $k-p$ $(0 \leq p \leq k-1)$ linearly independent vectors if and only if the covariant

$$a_1^{(k)} b \otimes a_2^{(k)} b \otimes \cdots \otimes a_{p+1}^{(k)} b \otimes b^{(k-p-1)}$$

is zero.

5 Classification of trivectors in 6-D space

The notation sketched in last section, using the divided powers algebra of an alphabet to describe facts about skew-symmetric tensors, leads to a new way of describing the orbits of the action of the general linear group on the exterior algebra. References on the subject of classification include [14], [7]. Unlike the old methods, our approach is purely algebraic and combinatorial. We work directly with the umbral module of step 6 over a set of equivalent letters representing a tensor of step 3. All the techniques we require for the classification are the bracket exchange identity (Prop. 11), the standard basis theorem, and Lemma 1 described below. We prove that the invariant ring (under the action of the general linear group) of trivectors of a 6-D space has a unique generator, whose symbolic representation is

$$\begin{bmatrix} a^{(3)} b^{(2)} c \\ b c^{(2)} d^{(3)} \end{bmatrix},$$

where a, b, c, d are equivalent letters representing the same trivector. Our result is valid over all infinite fields of characteristic other than 2 or 3. In addition, we observed that the covariants are obtained by taking sub-tableau of the unique invariant, and we are thus led to conjecture that for all skew-symmetric tensors, covariants are invariably obtained as subtableaux of invariants.

In the current section we assume that the umbral module $\mathrm{U}_6(L)$ is over a field K of characteristic zero. However, this assumption is not essential. It is for the simplification of description. In fact, we only need to assume that Char $K \neq 2$ and 3.

5.1 Covariants

We shall state and prove various facts about the covariants of trivectors of a 6-D space. The notation (4) will be used to denote the bracket polynomial (5). In particular, if w_1, w_2, ...,w_k are monomials having only positive letters, then

$$\begin{bmatrix} w_1 \\ w_2 \\ \vdots \\ w_k \end{bmatrix} = [w_1][w_2]\cdots[w_k].$$

We start with a key lemma which deals with the changes of equivalent letters in the umbral module.

Lemma 1 *Let a, b, ..., c be equivalent letters in L^+, representing a tensor t of step 3 in $\mathrm{U_n}(L)$, and let $f(a,b,\ldots,c)$ be a bracket polynomial. Then*

$$f(a,b,\ldots,c) = sgn(\sigma)f(\sigma(a),\sigma(b),\ldots,\sigma(c))$$

for every permutation σ of the set $\{a,b,\ldots,c\}$. Furthermore, if σ is an odd permutation with

$$f(a,b,\ldots,c) = f(\sigma(a),\sigma(b),\ldots,\sigma(c))$$

or if σ is an even permutation with

$$f(a,b,\ldots,c) = -f(\sigma(a),\sigma(b),\ldots,\sigma(c)),$$

then

$$f(a,b,\ldots,c) = 0$$

in $\mathrm{U_n}(L)$.

For example, $[a^{(3)}b^{(3)}] = 0$ in $\mathrm{U}_6(L)$.

Proof. Assume $a < b < c < \cdots$. By definition, the equality

$$f(a,b,c,\ldots,d) = D(t,d)\cdots D(t,c)D(t,b)D(t,a)f(a,b,c,\ldots,d)$$

holds in $\mathrm{U_n}(L)$. Furthermore, by Proposition 5 we obtain

$$\begin{aligned} & D(t,d)\cdots D(t,b)D(t,a)f(a,b,c,\ldots,d) \\ = \; & D(t,\sigma(d))\cdots D(t,\sigma(b))D(t,\sigma(a))f(\sigma(a),\sigma(b),\sigma(c),\ldots,\sigma(d)) \\ = \; & sgn(\sigma)D(t,d)\cdots D(t,b)D(t,a)f(\sigma(a),\sigma(b),\sigma(c),\ldots,\sigma(d)), \end{aligned}$$

which implies

$$f(a,b,c,\ldots,d) = sgn(\sigma)f(\sigma(a),\sigma(b),\sigma(c),\ldots,\sigma(d)).$$

If σ is an odd permutation, then

$$f(a,b,c,\ldots,d) = -f(\sigma(a),\sigma(b),\sigma(c),\ldots,\sigma(d)).$$

Given the condition that

$$f(a,b,c,\ldots,d) = f(\sigma(a),\sigma(b),\ldots,\sigma(d)),$$

we get

$$f(a,b,c,\ldots,d) = 0.$$

The proof of the other case is similar and is omitted. ∎

Note: In the following context, bold face letters in a bracket monomial indicate how the exchange identity Prop. 11 is applied. More specifically, the bold face letters indicate $b_{(1)}$ and $b_{(2)}$ at the left hand side of the exchange identity. If length$(b_{(1)})$+length$(b_{(2)})$ is greater than n, then the right hand side of the exchange identity will be equal to 0.

Lemma 2 *Let Roman letters a, b be equivalent and represent a tensor of step 3. Then in $\mathrm{U}_6(L)$ the following three conditions are equivalent:*

(i) $\begin{matrix} a^{(3)}b^{(2)} \\ b \end{matrix} \propto 0,$

(ii) *The Grassmann Condition*
$\begin{matrix} a^{(3)}b \\ b^{(2)} \end{matrix} \propto 0,$

(iii) $\begin{matrix} a^{(3)}b \\ b \\ b \end{matrix} \propto 0.$

Proof. (i) $\Leftrightarrow$ (ii). Suppose that $\begin{matrix} a^{(3)}b^{(2)} \\ b \end{matrix} \propto 0$, then for any tensor $\alpha^{(2)}$ of step 2 and any tensor $\beta^{(4)}$ of step 4,

$$\begin{aligned} 2\begin{bmatrix} a^{(3)}b\alpha^{(2)} \\ b^{(2)}\beta^{(4)} \end{bmatrix} &= \begin{bmatrix} \mathbf{a}^{(3)}\mathbf{b}\boldsymbol{\alpha}^{(2)} \\ \mathbf{b}b\beta^{(4)} \end{bmatrix} \\ &= -\begin{bmatrix} a^{(2)}b^{(2)}\alpha^{(2)} \\ ab\beta^{(4)} \end{bmatrix} - \begin{bmatrix} a^{(3)}b^{(2)}\alpha \\ \alpha b\beta^{(4)} \end{bmatrix}. \end{aligned} \tag{7}$$

By Lemma 1, the first term on the right hand side vanishes, and the second term equals zero by the assumption. So, we obtain that

$$\begin{matrix} a^{(3)}b \\ b^{(2)} \end{matrix} \propto 0.$$

Secondly, assume that $\begin{matrix} a^{(3)}b \\ b^{(2)} \end{matrix} \propto 0$, then for the above $\beta^{(5)}$ and any vector v, the equality

$$2\begin{bmatrix} a^{(3)}b^{(2)}v \\ b\beta^{(5)} \end{bmatrix} = \begin{bmatrix} a^{(3)}b\mathbf{b}v \\ \mathbf{b}\boldsymbol{\beta}^{(5)} \end{bmatrix} = -\begin{bmatrix} a^{(3)}b\beta v \\ b^{(2)}\beta^{(4)} \end{bmatrix} = 0$$

implies

$$\begin{matrix} a^{(3)}b^{(2)} \\ b \end{matrix} \propto 0.$$

Next, we show that (ii) $\Leftrightarrow$ (iii). First, if $\begin{matrix} a^{(3)}b \\ b^{(2)} \end{matrix} \propto 0$, then for any tensor $\beta^{(5)}$ of step 5, $\alpha^{(2)}$ of step 2 and $\gamma^{(5)}$ of step 5, the equality

$$\begin{aligned} \begin{bmatrix} a^{(3)}b\alpha^{(2)} \\ \mathbf{b}\boldsymbol{\beta}^{(5)} \\ \mathbf{b}\gamma^{(5)} \end{bmatrix} &= -\begin{bmatrix} a^{(3)}b\alpha^{(2)} \\ b^{(2)}\beta^{(4)} \\ \beta\gamma^{(5)} \end{bmatrix} \\ &= 0 \end{aligned} \tag{8}$$

implies

$$\begin{matrix} a^{(3)}b \\ b \\ b \end{matrix} \propto 0.$$

Secondly, if $\begin{array}{l} a^{(3)}b \\ b \\ b \end{array} \propto 0$, then for any decomposable tensor $\beta^{(4)}$ of step 4 and any tensor $\alpha^{(2)}$ of step 2, let $\gamma^{(4)} \propto \beta^{(4)}$ and choose vectors x and y satisfying $[xy\beta^{(4)}] \not\propto 0$. The following identity (the first term disappears by the assumption; the 2nd term disappears since $\gamma^{(4)}$ is equivalent to $\beta^{(4)}$ which is decomposable):

$$\begin{aligned} \left[\begin{array}{l} a^{(3)}b\alpha^{(2)} \\ \mathbf{b}^{(2)}\boldsymbol{\beta}^{(4)} \\ \mathbf{x}y\gamma^{(4)} \end{array}\right] &= -\left[\begin{array}{l} a^{(3)}b\alpha^{(2)} \\ bx\beta^{(4)} \\ by\gamma^{(4)} \end{array}\right] - \left[\begin{array}{l} a^{(3)}b\alpha^{(2)} \\ b^{(2)}x\beta^{(3)} \\ \beta y\gamma^{(4)} \end{array}\right] \\ &= 0 \end{aligned}$$

implies

$$\begin{array}{l} a^{(3)}b \\ b^{(2)} \end{array} \propto 0$$

since the umbral bracket algebra $\mathrm{UB}_{\mathrm{n}}[L]$ (p. 80) is an integral domain. ∎

Lemma 3 *Let a, b, c be equivalent letters, representing a tensor of step 3. Then in $\mathrm{U}_6(L)$ the following two conditions are equivalent:*

(i) $\begin{array}{l} a^{(3)}b^{(2)}c \\ bc^{(2)} \end{array} \propto 0;$

(ii) $\begin{array}{l} a^{(3)}b^{(2)} \\ bc^{(3)} \end{array} \propto 0.$

Furthermore, these two conditions imply

$$\begin{array}{l} a^{(3)}b^{(2)}c \\ bc \\ c \end{array} \propto 0.$$

Proof. Suppose that $\begin{array}{l} a^{(3)}b^{(2)}c \\ bc^{(2)} \end{array} \propto 0$. Then for any vector α and any tensor $\beta^{(2)}$ of step 2, we get by the bracket exchange identity

$$\begin{aligned} 3\left[\begin{array}{l} a^{(3)}b^{(2)}\alpha \\ bc^{(3)}\beta^{(2)} \end{array}\right] &= \left[\begin{array}{l} \mathbf{a}^{(3)}\mathbf{b}^{(2)}\boldsymbol{\alpha} \\ bc^{(2)}\mathbf{c}\beta^{(2)} \end{array}\right] \\ &= -\left[\begin{array}{l} a^{(2)}b^{(2)}c\alpha \\ abc^{(2)}\beta^{(2)} \end{array}\right] - 2\left[\begin{array}{l} a^{(3)}bc\alpha \\ b^{(2)}c^{(2)}\beta^{(2)} \end{array}\right] - \left[\begin{array}{l} a^{(3)}b^{(2)}c \\ bc^{(2)}\alpha\beta^{(2)} \end{array}\right]. \end{aligned}$$

The first two terms on the right hand side vanish by Lemma 1. The third term is equal to zero by the assumption. Thereby,

$$\begin{array}{l} a^{(3)}\mathfrak{b}^{(2)} \\ bc^{(3)} \end{array} \propto 0.$$

On the other hand, if $\begin{array}{l} a^{(3)}\mathfrak{b}^{(2)} \\ bc^{(3)} \end{array} \propto 0$, then for any tensor $\alpha^{(3)}$ of step 3, the identity

$$\left[\begin{array}{l} a^{(3)}\mathfrak{b}^{(2)}\mathbf{c} \\ \mathbf{bc^{(2)}\boldsymbol{\alpha}^{(3)}} \end{array}\right] = -3\left[\begin{array}{l} a^{(3)}\mathfrak{b}^{(3)} \\ c^{(3)}\alpha^{(3)} \end{array}\right] - \left[\begin{array}{l} a^{(3)}\mathfrak{b}^{(2)}\alpha \\ bc^{(3)}\alpha^{(2)} \end{array}\right] = 0$$

implies

$$\begin{array}{l} a^{(3)}\mathfrak{b}^{(2)}c \\ bc^{(2)} \end{array} \propto 0.$$

Thirdly, for any tensor $\alpha^{(4)}$ of step 4 and any tensor $\beta^{(5)}$ of step 5, we obtain

$$\begin{aligned} \left[\begin{array}{l} a^{(3)}\mathfrak{b}^{(2)}c \\ \mathbf{bc}\boldsymbol{\alpha}^{(4)} \\ \mathbf{c}\beta^{(5)} \end{array}\right] &= -\frac{1}{2}\left[\begin{array}{l} \mathbf{a^{(3)}b^{(2)}c} \\ \mathbf{c}c\alpha^{(4)} \\ \mathfrak{b}\beta^{(5)} \end{array}\right] - \left[\begin{array}{l} a^{(3)}\mathfrak{b}^{(2)}c \\ bc^{(2)}\alpha^{(3)} \\ \alpha\beta^{(5)} \end{array}\right] \\ &= \frac{1}{2}\left[\begin{array}{l} a^{(3)}bc^{(2)} \\ bc\alpha^{(4)} \\ \mathfrak{b}\beta^{(5)} \end{array}\right] + \frac{1}{2}\left[\begin{array}{l} a^{(2)}\mathfrak{b}^{(2)}c^{(2)} \\ ac\alpha^{(4)} \\ \mathfrak{b}\beta^{(5)} \end{array}\right] - \left[\begin{array}{l} a^{(3)}\mathfrak{b}^{(2)}c \\ bc^{(2)}\alpha^{(3)} \\ \alpha\beta^{(5)} \end{array}\right]. \end{aligned}$$

By Lemma 1, the first term on the right hand side equals

$$-\frac{1}{2}\left[\begin{array}{l} a^{(3)}\mathfrak{b}^{(2)}c \\ bc\alpha^{(4)} \\ c\beta^{(5)} \end{array}\right]$$

and the second term vanishes. Therefore, we get the following equality:

$$\frac{3}{2}\left[\begin{array}{l} a^{(3)}\mathfrak{b}^{(2)}c \\ bc\alpha^{(4)} \\ c\beta^{(5)} \end{array}\right] = -\left[\begin{array}{l} a^{(3)}\mathfrak{b}^{(2)}c \\ bc^{(2)}\alpha^{(3)} \\ \alpha\beta^{(5)} \end{array}\right]. \tag{9}$$

If $\begin{array}{l} a^{(3)}\mathfrak{b}^{(2)}c \\ bc^{(2)} \end{array} \propto 0$, then

$$\begin{array}{l} a^{(3)}\mathfrak{b}^{(2)}c \\ bc \\ c \end{array} \propto 0.$$

∎

Lemma 4 *Let a, b, c, d be equivalent letters, representing a tensor of step 3 in* $\mathrm{U}_6(L)$. *If*

$$\begin{bmatrix} a^{(3)}b^{(2)}c \\ bc^{(2)}d^{(3)} \end{bmatrix} = 0,$$

then $\begin{matrix} a^{(3)}b^{(2)}c \\ bc^{(2)}d^{(2)} \\ d \end{matrix} \propto 0$ *and* $[a^{(3)}b^{(2)}c]bc^{(2)}$ *is decomposable.*

Proof. We prove the equivalence relation first. Let α be any vector and $\beta^{(5)}$ be any tensor of step 5, then by lemma 1 and the bracket exchange identity we get

$$\begin{aligned}
\begin{bmatrix} a^{(3)}b^{(2)}c \\ \mathbf{bc^{(2)}d^{(2)}\alpha} \\ \mathbf{d}\beta^{(5)} \end{bmatrix} &= -\begin{bmatrix} a^{(3)}b^{(2)}c \\ bcd^{(3)}\alpha \\ c\beta^{(5)} \end{bmatrix} - \begin{bmatrix} a^{(3)}b^{(2)}c \\ c^{(2)}d^{(3)}\alpha \\ b\beta^{(5)} \end{bmatrix} - \begin{bmatrix} a^{(3)}b^{(2)}c \\ bc^{(2)}d^{(3)} \\ \alpha\beta^{(5)} \end{bmatrix} \\
&= \begin{bmatrix} \mathbf{a^{(3)}bc^{(2)}} \\ \mathbf{b}cd^{(3)}\alpha \\ b\beta^{(5)} \end{bmatrix} - \begin{bmatrix} a^{(3)}b^{(2)}c \\ c^{(2)}d^{(3)}\alpha \\ b\beta^{(5)} \end{bmatrix} - \begin{bmatrix} a^{(3)}b^{(2)}c \\ bc^{(2)}d^{(3)} \\ \alpha\beta^{(5)} \end{bmatrix} \\
&= -2\begin{bmatrix} a^{(3)}b^{(2)}c \\ c^{(2)}d^{(3)}\alpha \\ b\beta^{(5)} \end{bmatrix} - \begin{bmatrix} a^{(3)}b^{(2)}c \\ c^{(2)}d^{(3)}\alpha \\ b\beta^{(5)} \end{bmatrix} - \begin{bmatrix} a^{(3)}b^{(2)}c \\ bc^{(2)}d^{(3)} \\ \alpha\beta^{(5)} \end{bmatrix} \\
&= -3\begin{bmatrix} a^{(3)}b^{(2)}c \\ c^{(2)}d^{(3)}\alpha \\ b\beta^{(5)} \end{bmatrix} - \begin{bmatrix} a^{(3)}b^{(2)}c \\ bc^{(2)}d^{(3)} \\ \alpha\beta^{(5)} \end{bmatrix}. \qquad (10)
\end{aligned}$$

On the other hand, we can apply the exchange identity (Prop. 11) and lemma 1 in a slightly different way as follows:

$$\begin{aligned}
\begin{bmatrix} a^{(3)}b^{(2)}c \\ \mathbf{bc^{(2)}d^{(2)}\alpha} \\ \mathbf{d}\beta^{(5)} \end{bmatrix} &= -\frac{1}{3}\begin{bmatrix} \mathbf{a^{(3)}b^{(2)}c} \\ bcd^{(2)}\mathbf{d}\alpha \\ c\beta^{(5)} \end{bmatrix} - \begin{bmatrix} a^{(3)}b^{(2)}c \\ c^{(2)}d^{(3)}\alpha \\ b\beta^{(5)} \end{bmatrix} - \begin{bmatrix} a^{(3)}b^{(2)}c \\ bc^{(2)}d^{(3)} \\ \alpha\beta^{(5)} \end{bmatrix} \\
&= \frac{2}{3}\begin{bmatrix} a^{(3)}b^{(2)}d \\ bc^{(2)}d^{(2)}\alpha \\ c\beta^{(5)} \end{bmatrix} - \begin{bmatrix} a^{(3)}b^{(2)}c \\ c^{(2)}d^{(3)}\alpha \\ b\beta^{(5)} \end{bmatrix} - \begin{bmatrix} a^{(3)}b^{(2)}c \\ bc^{(2)}d^{(3)} \\ \alpha\beta^{(5)} \end{bmatrix} \\
&= -\frac{2}{3}\begin{bmatrix} a^{(3)}b^{(2)}c \\ bc^{(2)}d^{(2)}\alpha \\ d\beta^{(5)} \end{bmatrix} - \begin{bmatrix} a^{(3)}b^{(2)}c \\ c^{(2)}d^{(3)}\alpha \\ b\beta^{(5)} \end{bmatrix} - \begin{bmatrix} a^{(3)}b^{(2)}c \\ bc^{(2)}d^{(3)} \\ \alpha\beta^{(5)} \end{bmatrix}.
\end{aligned}$$

This implies

$$\frac{5}{3}\left[\begin{array}{l} a^{(3)}b^{(2)}c \\ bc^{(2)}d^{(2)}\alpha \\ d\beta^{(5)} \end{array}\right] = -\left[\begin{array}{l} a^{(3)}b^{(2)}c \\ c^{(2)}d^{(3)}\alpha \\ b\beta^{(5)} \end{array}\right] - \left[\begin{array}{l} a^{(3)}b^{(2)}c \\ bc^{(2)}d^{(3)} \\ \alpha\beta^{(5)} \end{array}\right]. \tag{11}$$

Combining (10) and (11), we get

$$4\left[\begin{array}{l} a^{(3)}b^{(2)}c \\ bc^{(2)}d^{(2)}\alpha \\ d\beta^{(5)} \end{array}\right] = -2\left[\begin{array}{l} a^{(3)}b^{(2)}c \\ bc^{(2)}d^{(3)} \\ \alpha\beta^{(5)} \end{array}\right], \tag{12}$$

i.e.,

$$\begin{array}{l} a^{(3)}b^{(2)}c \\ bc^{(2)}d^{(2)} \\ d \end{array} \propto 0.$$

Next we prove the decomposability of $[a^{(3)}b^{(2)}c]bc^{(2)}$. Let $\alpha^{(3)}$ denote $[a^{(3)}b^{(2)}c]bc^{(2)}$. To show that $\alpha^{(3)}$ is decomposable, we only need to check that

$$\begin{array}{l} \alpha^{(3)}\beta^{(2)} \\ \beta \end{array} \propto 0,$$

where $\beta^{(3)} \propto \alpha^{(3)}$. Let $\beta^{(3)} = [a_1^{(3)}b_1^{(2)}c_1]b_1c_1^{(2)}$, where a_1, b_1, c_1 are equivalent symbols of a, b, c. Then we get

$$\begin{array}{l} \alpha^{(3)}\beta^{(2)} \\ \beta \end{array} \propto \left[\begin{array}{l} a^{(3)}b^{(2)}c \\ a_1^{(3)}b_1^{(2)}c_1 \end{array}\right] \begin{array}{l} bc^{(2)}c_1^{(2)} \\ b_1 \end{array} + \left[\begin{array}{l} a^{(3)}b^{(2)}c \\ a_1^{(3)}b_1^{(2)}c_1 \end{array}\right] \begin{array}{l} bc^{(2)}b_1c_1 \\ c_1 \end{array}.$$

However, for any vector v and any tensor $\pi^{(5)}$ of step 5, the identity

$$\left[\begin{array}{l} a^{(3)}b^{(2)}c \\ \mathbf{a_1^{(3)}b_1^{(2)}c_1} \end{array}\right]\left[\begin{array}{l} bc^{(2)}b_1\mathbf{c_1}v \\ c_1\pi^{(5)} \end{array}\right] = -2\left[\begin{array}{l} a^{(3)}b^{(2)}c \\ a_1^{(3)}b_1c_1^{(2)} \end{array}\right]\left[\begin{array}{l} bc^{(2)}b_1^{(2)}v \\ c_1\pi^{(5)} \end{array}\right]$$

implies

$$\left[\begin{array}{l} a^{(3)}b^{(2)}c \\ a_1^{(3)}b_1^{(2)}c_1 \end{array}\right] \begin{array}{l} bc^{(2)}b_1c_1 \\ c_1 \end{array} \propto 2\left[\begin{array}{l} a^{(3)}b^{(2)}c \\ a_1^{(3)}b_1^{(2)}c_1 \end{array}\right] \begin{array}{l} bc^{(2)}c_1^{(2)} \\ b_1 \end{array}.$$

Thus, we obtain

$$\begin{array}{l} \alpha^{(3)}\beta^{(2)} \\ \beta \end{array} \propto 3\left[\begin{array}{l} a^{(3)}b^{(2)}c \\ a_1^{(3)}b_1^{(2)}c_1 \end{array}\right] \begin{array}{l} bc^{(2)}c_1^{(2)} \\ b_1 \end{array}, \tag{13}$$

which is equal to 0 by the first result. ■

5.2 Invariants

Let a, b, c, ... ($\subset L$) be equivalent letters, representing a tensor of step 3. In the umbral bracket algebra $\mathrm{UB}_6[L]$, let $\mathrm{M}([\cdot])$ denote the set of bracket monomials over these letters such that the content of each letter in each monomial is either 0 or 3. The bracket polynomial ring generated by all elements of $\mathrm{M}([\cdot])$ is denoted by $\mathrm{P}([\cdot])$. This bracket polynomial ring is really the same thing as what is classically understood to be the ring of invariants by polarizing the positive letters to the corresponding skew-symmetric tensors they represent (see [7], [3]).

To prove that the trivectors in a 6-D space have a unique invariant (Theorem 2), we shall require several technical lemmas.

Lemma 5 *A bracket monomial in* $\mathrm{M}([\cdot])$*, having the form*

$$\begin{bmatrix} a^{(3)}b^{(2)}c \\ bc^{(2)}d^{(2)}. \\ d\cdot\cdot\cdot\cdot\cdot \\ \vdots \end{bmatrix}, \tag{14}$$

is a constant multiple of a bracket monomial with the following form:

$$\begin{bmatrix} a^{(3)}b^{(2)}c \\ bc^{(2)}d^{(3)} \\ \vdots \end{bmatrix}. \tag{15}$$

The proof follows immediately from equality (12).

Lemma 6 *A bracket monomial in* $\mathrm{M}([\cdot])$*, having the form*

$$\begin{bmatrix} a^{(3)}b^{(2)}c \\ d^{(3)}e^{(2)}f \end{bmatrix}\begin{bmatrix} bc^{(2)}e\cdot\cdot \\ \vdots \end{bmatrix} \tag{16}$$

or the form

$$\begin{bmatrix} a^{(3)}b^{(2)}c \\ d^{(3)}e^{(2)}f \end{bmatrix}\begin{bmatrix} bc^{(2)}f\cdot\cdot \\ \vdots \end{bmatrix}, \tag{17}$$

may be written as a linear combination of bracket monomials, each having the form of (15).

Proof. Suppose that the bracket monomial has the form of (16). Let x and y be two arbitrary vectors, then

$$\begin{aligned}
&\begin{bmatrix} a^{(3)}b^{(2)}c \\ \mathbf{d}^{(3)}\mathbf{e}^{(2)}f \end{bmatrix}\begin{bmatrix} bc^{(2)}\mathbf{exy} \\ \vdots \end{bmatrix} \\
= \; & -2\begin{bmatrix} a^{(3)}b^{(2)}c \\ d^{(3)}exf \end{bmatrix}\begin{bmatrix} bc^{(2)}e^{(2)}y \\ \vdots \end{bmatrix} - \begin{bmatrix} a^{(3)}b^{(2)}c \\ d^{(2)}e^{(2)}xf \end{bmatrix}\begin{bmatrix} bc^{(2)}edy \\ \vdots \end{bmatrix} \\
& -2\begin{bmatrix} a^{(3)}b^{(2)}c \\ d^{(3)}eyf \end{bmatrix}\begin{bmatrix} bc^{(2)}e^{(2)}x \\ \vdots \end{bmatrix} - \begin{bmatrix} a^{(3)}b^{(2)}c \\ d^{(2)}e^{(2)}yf \end{bmatrix}\begin{bmatrix} bc^{(2)}edx \\ \vdots \end{bmatrix} \\
& -3\begin{bmatrix} a^{(3)}b^{(2)}c \\ d^{(3)}xyf \end{bmatrix}\begin{bmatrix} bc^{(2)}e^{(3)} \\ \vdots \end{bmatrix} - \begin{bmatrix} a^{(3)}b^{(2)}c \\ de^{(2)}xyf \end{bmatrix}\begin{bmatrix} bc^{(2)}ed^{(2)} \\ \vdots \end{bmatrix} \\
& -2\begin{bmatrix} a^{(3)}b^{(2)}c \\ d^{(2)}exyf \end{bmatrix}\begin{bmatrix} bc^{(2)}de^{(2)} \\ \vdots \end{bmatrix}.
\end{aligned}$$

Applying Lemma 1 and Lemma 5, it is immediately seen that any bracket monomial with form of (16) can be expanded as a linear combination of bracket monomials with form (15).

To show the theorem when the bracket monomial has form (17), let us first show that a bracket monomial in $\mathrm{M}([\cdot])$ having the following form:

$$\begin{bmatrix} a^{(3)}b^{(2)}c \\ d^{(3)}e^{(2)}f \end{bmatrix}\begin{bmatrix} bc^{(2)}f\cdot\cdot \\ ef\cdot\cdot\cdot\cdot \\ \vdots \end{bmatrix}, \tag{18}$$

may be written as a linear combination of bracket monomials with form (15). Let $t^{(3)} = [a^{(3)}b^{(2)}c]bc^{(2)}$ and $t_1^{(3)} = [d^{(3)}e^{(2)}f]ef^{(2)}$, then

$$\begin{bmatrix} t^{(3)}t_1\cdot\cdot \\ t_1^{(2)}\cdot\cdot\cdot\cdot \\ \vdots \end{bmatrix} = \begin{bmatrix} a^{(3)}b^{(2)}c \\ d^{(3)}e^{(2)}f \end{bmatrix}\begin{bmatrix} bc^{(2)}f\cdot\cdot \\ ef\cdot\cdot\cdot\cdot \\ \vdots \end{bmatrix} + \begin{bmatrix} a^{(3)}b^{(2)}c \\ d^{(3)}e^{(2)}f \end{bmatrix}\begin{bmatrix} bc^{(2)}e\cdot\cdot \\ f^{(2)}\cdot\cdot\cdot\cdot \\ \vdots \end{bmatrix}. \tag{19}$$

By formula (7), the left hand side is a constant multiple of a bracket monomial of the form:

$$\begin{bmatrix} t^{(3)}t_1^{(2)}\cdot \\ t_1\cdot\cdot\cdot\cdot\cdot \\ \vdots \end{bmatrix},$$

which, by formula (13), is equal to

$$3\begin{bmatrix} a^{(3)}b^{(2)}c \\ d^{(3)}e^{(2)}f \end{bmatrix}\begin{bmatrix} bc^{(2)}f^{(2)}\cdot \\ e\cdot\cdot\cdot\cdot\cdot \\ \vdots \end{bmatrix}.$$

Thus the left hand side is a constant multiple of a bracket monomial with form (15) by Lemma 5. The second term on the right hand side of (19) has the form of (16). Therefore, a bracket monomial with form (18) is a linear combination of bracket monomials with form (15).

Now the only non-trivial case left for a bracket monomial with form (17) is the following special one:

$$\begin{bmatrix} a^{(3)}b^{(2)}c \\ d^{(3)}e^{(2)}f \end{bmatrix} \begin{bmatrix} bc^{(2)}f\cdot\cdot \\ e\cdot\cdot\cdot\cdot\cdot \\ f\cdot\cdot\cdot\cdot\cdot \\ \vdots \end{bmatrix}. \tag{20}$$

Interchanging e and f, we get

$$\begin{bmatrix} a^{(3)}b^{(2)}c \\ d^{(3)}e^{(2)}f \end{bmatrix} \begin{bmatrix} bc^{(2)}f\cdot\cdot \\ e\cdot\cdot\cdot\cdot\cdot \\ f\cdot\cdot\cdot\cdot\cdot \\ \vdots \end{bmatrix} = -\begin{bmatrix} a^{(3)}b^{(2)}c \\ d^{(3)}f^{(2)}e \end{bmatrix} \begin{bmatrix} bc^{(2)}e\cdot\cdot \\ f\cdot\cdot\cdot\cdot\cdot \\ e\cdot\cdot\cdot\cdot\cdot \\ \vdots \end{bmatrix}. \tag{21}$$

However, by applying the exchange identity (Prop. 11) we also get

$$\begin{aligned} &\begin{bmatrix} a^{(3)}b^{(2)}c \\ \mathbf{d}^{(3)}\mathbf{e}^{(2)}\mathbf{f} \end{bmatrix} \begin{bmatrix} bc^{(2)}\mathbf{f}\cdot\cdot \\ e\cdot\cdot\cdot\cdot\cdot \\ f\cdot\cdot\cdot\cdot\cdot \\ \vdots \end{bmatrix} \\ = &-\begin{bmatrix} a^{(3)}b^{(2)}c \\ d^{(3)}ef^{(2)} \end{bmatrix} \begin{bmatrix} bc^{(2)}e\cdot\cdot \\ \mathbf{e}\cdot\cdot\cdot\cdot\cdot \\ \mathbf{f}\cdot\cdot\cdot\cdot\cdot \\ \vdots \end{bmatrix} - \begin{bmatrix} a^{(3)}b^{(2)}c \\ d^{(2)}e^{(2)}f^{(2)} \end{bmatrix} \begin{bmatrix} bc^{(2)}d\cdot\cdot \\ e\cdot\cdot\cdot\cdot\cdot \\ f\cdot\cdot\cdot\cdot\cdot \\ \vdots \end{bmatrix} \\ = &\begin{bmatrix} a^{(3)}b^{(2)}c \\ d^{(3)}ef^{(2)} \end{bmatrix} \begin{bmatrix} bc^{(2)}e\cdot\cdot \\ f\cdot\cdot\cdot\cdot\cdot \\ e\cdot\cdot\cdot\cdot\cdot \\ \vdots \end{bmatrix} + \pi_1 \\ &-\begin{bmatrix} a^{(3)}b^{(2)}c \\ d^{(2)}e^{(2)}f^{(2)} \end{bmatrix} \begin{bmatrix} bc^{(2)}d\cdot\cdot \\ e\cdot\cdot\cdot\cdot\cdot \\ f\cdot\cdot\cdot\cdot\cdot \\ \vdots \end{bmatrix}. \end{aligned} \tag{22}$$

where π_1 is a sum of bracket monomials, each having the form of (18).

Subtracting (22) from (21), we get

$$\begin{bmatrix} a^{(3)}b^{(2)}c \\ d^{(3)}ef^{(2)} \end{bmatrix} \begin{bmatrix} bc^{(2)}e\cdot\cdot \\ f\cdot\cdot\cdot\cdot\cdot \\ e\cdot\cdot\cdot\cdot\cdot \\ \vdots \end{bmatrix} = \frac{1}{2}\begin{bmatrix} a^{(3)}b^{(2)}c \\ d^{(2)}e^{(2)}f^{(2)} \end{bmatrix} \begin{bmatrix} bc^{(2)}d\cdot\cdot \\ e\cdot\cdot\cdot\cdot\cdot \\ f\cdot\cdot\cdot\cdot\cdot \\ \vdots \end{bmatrix} - \pi_1. \tag{23}$$

Now let us again apply the exchange identity on the left hand side of (23). We obtain

$$\begin{aligned} &\begin{bmatrix} a^{(3)}b^{(2)}c \\ \mathbf{d^{(3)}ef^{(2)}} \end{bmatrix} \begin{bmatrix} bc^{(2)}e\cdot\cdot \\ f\cdot\cdot\cdot\cdot\cdot \\ \mathbf{e}\cdot\cdot\cdot\cdot\cdot \\ \vdots \end{bmatrix} \\ = \ &-\begin{bmatrix} a^{(3)}b^{(2)}c \\ d^{(2)}e^{(2)}f^{(2)} \end{bmatrix} \begin{bmatrix} bc^{(2)}e\cdot\cdot \\ f\cdot\cdot\cdot\cdot\cdot \\ d\cdot\cdot\cdot\cdot\cdot \\ \vdots \end{bmatrix} + \pi_2, \end{aligned}$$

where the bracket monomial π_2 has the form of (16). By interchanging d and e then e and f of the first term on the right hand side of the above equation, we obtain

$$\begin{aligned} &\begin{bmatrix} a^{(3)}b^{(2)}c \\ d^{(3)}ef^{(2)} \end{bmatrix} \begin{bmatrix} bc^{(2)}e\cdot\cdot \\ f\cdot\cdot\cdot\cdot\cdot \\ e\cdot\cdot\cdot\cdot\cdot \\ \vdots \end{bmatrix} \\ = \ &-\begin{bmatrix} a^{(3)}b^{(2)}c \\ d^{(2)}e^{(2)}f^{(2)} \end{bmatrix} \begin{bmatrix} bc^{(2)}d\cdot\cdot \\ e\cdot\cdot\cdot\cdot\cdot \\ f\cdot\cdot\cdot\cdot\cdot \\ \vdots \end{bmatrix} + \pi_2. \end{aligned} \tag{24}$$

Combining equations (23) and (24), we get

$$\frac{3}{2}\begin{bmatrix} a^{(3)}b^{(2)}c \\ d^{(3)}ef^{(2)} \end{bmatrix} \begin{bmatrix} bc^{(2)}e\cdot\cdot \\ f\cdot\cdot\cdot\cdot\cdot \\ e\cdot\cdot\cdot\cdot\cdot \\ \vdots \end{bmatrix} = -\pi_1 + \frac{1}{2}\pi_2.$$

This means that a bracket monomial with form (20) can be expanded as a linear combination of bracket monomials with form (15). ■

Lemma 7 *A bracket monomial having the form*

$$\begin{bmatrix} a^{(3)}b^{(2)}c \\ bc^{(2)}de\cdot \\ d^{(2)}\cdots\cdot \\ e^{(2)}\cdots\cdot \\ \vdots \end{bmatrix}$$

may be written as a linear combination of bracket monomials of the following form:

$$\begin{bmatrix} a^{(3)}b^{(2)}c \\ bc^{(2)}de\cdot \\ d^{(2)}e\cdots \\ e\cdots\cdots \\ \vdots \end{bmatrix}.$$

Proof. The proof follows immediately from the following bracket identity:

$$\begin{bmatrix} a^{(3)}b^{(2)}c \\ bc^{(2)}de\cdot \\ \mathbf{d}^{(2)}\cdots\cdot \\ \mathbf{e}e\cdots\cdot \\ \vdots \end{bmatrix} = -\begin{bmatrix} a^{(3)}b^{(2)}c \\ bc^{(2)}de\cdot \\ de\cdots\cdot \\ de\cdots\cdot \\ \vdots \end{bmatrix} - \sum \begin{bmatrix} a^{(3)}b^{(2)}c \\ bc^{(2)}de\cdot \\ d^{(2)}e\cdots \\ e\cdots\cdots \\ \vdots \end{bmatrix}.$$

The first term on the right hand side disappears by Lemma 1 since d and e are interchangeable. ■

Lemma 8 *A bracket monomial having the form*

$$\begin{bmatrix} a^{(3)}b^{(2)}c \\ bc^{(2)}de\cdot \\ d^{(2)}ef^{(2)}\cdot \\ \vdots \end{bmatrix} \tag{25}$$

may be written as a linear combination of bracket monomials with form (15).

Proof. Let x and y be two vectors. For a bracket monomial with form (15) we only need to consider the following special case:

$$\begin{bmatrix} a^{(3)}b^{(2)}c \\ bc^{(2)}dex \\ d^{(2)}ef^{(2)}y \\ f\cdots\cdots \\ \vdots \end{bmatrix}, \tag{26}$$

where we may assume that both x and y are not equal to e or f (otherwise the case is trivial). By applying the bracket exchange identity, we get

$$\begin{bmatrix} a^{(3)}b^{(2)}c \\ bc^{(2)}dex \\ \mathbf{d^{(2)}ef^{(2)}y} \\ \mathbf{f}\cdots\cdots \end{bmatrix}$$

$$= -\begin{bmatrix} a^{(3)}b^{(2)}c \\ bc^{(2)}dex \\ d^{(2)}ef^{(3)} \\ y\cdots\cdots \end{bmatrix} - \begin{bmatrix} a^{(3)}b^{(2)}c \\ bc^{(2)}dex \\ def^{(3)}y \\ d\cdots\cdots \end{bmatrix} - \begin{bmatrix} a^{(3)}b^{(2)}c \\ bc^{(2)}dex \\ d^{(2)}f^{(3)}y \\ e\cdots\cdots \end{bmatrix}.$$

The first term on the right hand side is a linear combination of bracket monomials with form (15) using Lemma 6. Applying the bracket exchange identity and Lemma 1 to the last two terms, we get

$$-\begin{bmatrix} a^{(3)}b^{(2)}c \\ bc^{(2)}\mathbf{dex} \\ \mathbf{def^{(3)}y} \\ d\cdots\cdots \end{bmatrix}$$

$$= \begin{bmatrix} a^{(3)}b^{(2)}c \\ bc^{(2)}exy \\ d^{(2)}ef^{(3)} \\ d\cdots\cdots \end{bmatrix} + 2\begin{bmatrix} a^{(3)}b^{(2)}c \\ bc^{(2)}e^{(2)}x \\ d^{(2)}f^{(3)}y \\ d\cdots\cdots \end{bmatrix} + \begin{bmatrix} a^{(3)}b^{(2)}c \\ bc^{(2)}efx \\ d^{(2)}ef^{(2)}y \\ d\cdots\cdots \end{bmatrix}$$

$$= \begin{bmatrix} a^{(3)}b^{(2)}c \\ bc^{(2)}exy \\ d^{(2)}ef^{(3)} \\ d\cdots\cdots \end{bmatrix} + 2\begin{bmatrix} a^{(3)}b^{(2)}c \\ bc^{(2)}e^{(2)}x \\ d^{(2)}f^{(3)}y \\ d\cdots\cdots \end{bmatrix} - \begin{bmatrix} a^{(3)}b^{(2)}c \\ bc^{(2)}dex \\ d^{(2)}ef^{(2)}y \\ f\cdots\cdots \end{bmatrix}$$

and

$$-\begin{bmatrix} a^{(3)}b^{(2)}c \\ bc^{(2)}\mathbf{dex} \\ \mathbf{d^{(2)}f^{(3)}y} \\ e\cdots\cdots \end{bmatrix}$$

$$= 2\begin{bmatrix} a^{(3)}b^{(2)}c \\ bc^{(2)}d^{(2)}x \\ def^{(3)}y \\ e\cdots\cdots \end{bmatrix} + \begin{bmatrix} a^{(3)}b^{(2)}c \\ bc^{(2)}dfx \\ d^{(2)}ef^{(2)}y \\ e\cdots\cdots \end{bmatrix} + \begin{bmatrix} a^{(3)}b^{(2)}c \\ bc^{(2)}dxy \\ d^{(2)}ef^{(3)} \\ e\cdots\cdots \end{bmatrix}.$$

The second term of the above identity disappears by Lemma (1) since d and

f are interchangeable. Thereby, we obtain

$$
2\begin{bmatrix} a^{(3)}b^{(2)}c \\ bc^{(2)}dex \\ d^{(2)}ef^{(2)}y \\ f\cdots\cdots \end{bmatrix}
= -\begin{bmatrix} a^{(3)}b^{(2)}c \\ bc^{(2)}dex \\ d^{(2)}ef^{(3)} \\ y\cdots\cdots \end{bmatrix} + \begin{bmatrix} a^{(3)}b^{(2)}c \\ bc^{(2)}exy \\ d^{(2)}ef^{(3)} \\ d\cdots\cdots \end{bmatrix} + 2\begin{bmatrix} a^{(3)}b^{(2)}c \\ bc^{(2)}e^{(2)}x \\ d^{(2)}f^{(3)}y \\ d\cdots\cdots \end{bmatrix}
+2\begin{bmatrix} a^{(3)}b^{(2)}c \\ bc^{(2)}d^{(2)}x \\ def^{(3)}y \\ e\cdots\cdots \end{bmatrix} + \begin{bmatrix} a^{(3)}b^{(2)}c \\ bc^{(2)}dxy \\ d^{(2)}ef^{(3)} \\ e\cdots\cdots \end{bmatrix}.
$$

By the previous lemmas, Lemma (8) follows. ■

Corollary 1 *A bracket monomial, having the form*

$$
\begin{bmatrix} a^{(3)}b^{(2)}c \\ bc^{(2)}def \\ d^{(2)}ef\cdot\cdot \\ \vdots \end{bmatrix}, \tag{27}
$$

can be written as a linear combination of bracket monomials with form (15).

Proof. Applying the bracket exchange identity, we get

$$
\begin{bmatrix} a^{(3)}b^{(2)}c \\ bc^{(2)}d\mathbf{ef} \\ \mathbf{d}^{(2)}\boldsymbol{ef}\cdot\cdot \\ \vdots \end{bmatrix} = \sum\cdots,
$$

where all the terms of the sum on the right hand side have either the form of (26) or the form of (14) by interchanging equivalent letters. So, the theorem follows immediately. ■

Lemma 9 *Let integer $k \geq 2$. Consider the following bracket monomial $\alpha \in \mathrm{M}([\cdot])$:*

$$\begin{bmatrix} w_{-I} \\ \vdots \\ w_1 \\ w_2 \\ \vdots \\ w_k \end{bmatrix}.$$

For every j satisfying $2 \leq j \leq k$, let

$$w_j = w_{j1}w_{j2},$$

where each letter in w_{j1} appears in some w_i for $i \in \{-I, \ldots, 1, \ldots, j-1\}$ and no letters in w_{j2} appear in any w_i for all $i \leq j-1$. If the following conditions (i) and (ii) or (i) and (iii) hold:

(i) *for every j satisfying $2 \leq j \leq k-1$, $\mathrm{length}(w_{j1}) = 3$ and w_{j2} is a monomial consisting of three distinct letters;*

(ii) $\mathrm{length}(w_{k1}) \geq 4$ *and for every j satisfying $2 \leq j \leq k-2$, at least two letters in w_{j2} do not appear in $w_{(j+1)1}$;*

(iii) $\mathrm{length}(w_{k1}) = 3$ *but $w_{k2} = \pi^{(m)} \cdots$ with $m \geq 2, \pi \in L$, and for every j satisfying $2 \leq j \leq k-1$, at least two letters in w_{j2} do not appear in $w_{(j+1)1}$,*

then α may be written as a linear combination of bracket monomials with the following form:

$$\begin{bmatrix} w_{-I} \\ \vdots \\ w_1 \\ w'_2 \\ \vdots \\ w'_k \end{bmatrix} \tag{28}$$

($w'_2 = w'_{21}w'_{22}$: each letter in w'_{21} appears in some w_i for $i \in \{-I, \ldots, 1\}$ and no letters in w'_{22} appear in any w_i for all $i \leq 1$), where either $\mathrm{length}(w'_{21}) \geq 4$ *or $w'_{21} = w_{21}$ but $w'_{22} = \pi'^{(m')} \cdots$ with $m' \geq 2$, $\pi' \in L$.*

Proof. We proceed by induction on k. When $k = 2$, the theorem obviously holds. Suppose that the theorem holds for all k with $k \leq j$. Let $k = j + 1$.

Case 1. (length$(w_{(j+1)1}) \geq 4$) Let $u = w_{j2}w_{(j+1)1}$. Since length(u) is greater than 6 we get the following equality by applying the bracket exchange identity:

$$\sum_u \left[\begin{array}{c} w_{j1}u_{(1)} \\ u_{(2)}w_{(j+1)2} \end{array} \right] = 0. \tag{29}$$

Except for the term $\left[\begin{array}{c} w_{j1}w_{j2} \\ w_{(j+1)1}w_{(j+1)2} \end{array} \right]$, every other term of the sum on the left hand side has the following form:

$$\left[\begin{array}{c} w'_{j1}w'_{j2} \\ w'_{(j+1)1}w'_{(j+1)2} \end{array} \right] \tag{30}$$

(each letter in w'_{j1} appears in some w_i for $i \in \{-I, \ldots, 1, \ldots, j-1\}$ and no letters in w'_{j2} appear in any w_i for all $i \leq j-1$), where either length$(w'_{j1}) \geq 4$ or $w'_{j1} = w_{j1}$ but $w'_{j2} = \pi'^{(m')} \cdots$ with $m' \geq 2$ and $\pi' \in L$. By the induction hypothesis, the theorem follows.

Case 2. (length$(w_{(j+1)1}) = 3$ but $w_{(j+1)2} = \pi^{(3)}$ or $w_{(j+1)2} = \pi^{(2)}u$ with $\pi,\ u \in L$) Let $w_{j2} = xyz$, where y, z are two letters in w_{j2} such that both of them do not appear in $w_{(j+1)1}$ and let $v = yzw_{(j+1)1}\pi^{(3)}$ if $w_{(j+1)2} = \pi^{(3)}$ or $v = yzw_{(j+1)1}\pi^{(2)}$ if $w_{(j+1)2} = \pi^{(2)}u$. Then under the first condition we get

$$\sum_v \left[\begin{array}{c} w_{j1}xv_{(1)} \\ v_{(2)} \end{array} \right] = 0,$$

and under the second condition we get

$$\sum_v \left[\begin{array}{c} w_{j1}xv_{(1)} \\ v_{(2)}u \end{array} \right] = 0.$$

Under either condition, except for the term $\left[\begin{array}{c} w_{j1}w_{j2} \\ w_{(j+1)1}w_{(j+1)2} \end{array} \right]$ every other term of the sum has either the form of (30) or the following form:

$$\left[\begin{array}{c} w''_{j1}w''_{j2} \\ w''_{(j+1)1}w''_{(j+1)2} \end{array} \right]$$

(where each letter in w''_{j1} appears in some w_i for $i \in \{-I, \ldots, 1, \ldots, j-1\}$ and no letters in w''_{j2} appear in any w_i for all $i \leq j-1$), such that

$$w''_{j1} = w_{j1},$$

$$w''_{j2} = \pi \cdots,$$

and

$$w''_{(j+1)1} = w_{(j+1)1}\pi.$$

The bracket monomial is

$$\begin{bmatrix} w_{-I} \\ \vdots \\ w_1 \\ \vdots \\ w_{(j-1)} \\ w''_{j1}w''_{j2} \\ w''_{(j+1)1}w''_{(j+1)2} \end{bmatrix}$$

thereby reduced to the above case 1. By the induction hypothesis, the theorem follows. ■

Lemma 10 *Let integer $k \geq 3$. Consider the bracket monomial $\alpha \in \mathrm{M}([\cdot])$:*

$$\begin{bmatrix} w_{-I_1} \\ \vdots \\ w_{-I_2} \\ \vdots \\ w_1 \\ w_2 \\ \vdots \\ w_k \end{bmatrix}.$$

For every j satisfying $-I_2 \leq j \leq k$, let

$$w_j = w_{j1}w_{j2}$$

where each letter in w_{j1} appears in some w_i for $i \in \{-I_1, \ldots, 1, \ldots, j-1\}$ and no letters in w_{j2} appear in any w_i for all $i \leq j-1$. If the following conditions (i) *and* (ii) *or* (i) *and* (iii) *hold:*

(i) *first, for every j satisfying $-I_2 \leq j \leq 1$, $\mathrm{length}(w_{j1}) = 3$ and w_{j2} is a monomial consisting of three distinct letters, of which at least two do not appear in $w_{(j+1)1}$; secondly, $\mathrm{length}(w_{21}) = 3$ and w_{22} is a monomial consisting of three distinct letters, of which none appears in w_{31}; thirdly for every j satisfying $3 \leq j \leq k-1$, $\mathrm{length}(w_{j1}) = 2$ and w_{j2} consisting of four distinct letters,*

(ii) $\text{length}(w_{k1}) \geq 3$, *and for every j satisfying $3 \leq j \leq k-2$, at least three letters in w_{j2} do not appear in $w_{(j+1)1}$,*

(iii) $\text{length}(w_{k1}) = 2$ *but $w_{k2} = \pi^{(m)} \cdots$ with $m \geq 2$, $\pi \in L$, and for every j satisfying $3 \leq j \leq k-1$, at least three letters in w_{j2} do not appear in $w_{(j+1)1}$,*

then α may be written as a linear combination of bracket monomials, each having the form of (28) or the following form:

$$\begin{bmatrix} w_{-I_1} \\ \vdots \\ w_1 \\ w''_{21}w''_{22} \\ w''_{31}w''_{32} \\ w''_4 \\ \vdots \\ w''_k \end{bmatrix} \tag{31}$$

(each letter in w''_{21} appears in some w_i for $i \in \{-I_1, \ldots, 1\}$ and no letters in w''_{22} appear in any w_i for all $i \leq 1$; each letter in w''_{31} appears in either $w''_{21}w''_{22}$ or some w_i for $i \in \{-I_1, \ldots, 1\}$ and no letters in w''_{32} appear in either $w''_{21}w''_{22}$ or any w_i for all $i \leq 1$), where $(w''_{21}) = w_{21}$, w''_{22} is a monomial consisting of three distinct letters, of which at least two do not appear in w''_{31}, and $\text{length}(w''_{31}) \geq 3$.

The proof of this lemma is similar to the proof of Lemma 9, and is omitted.

Theorem 2 *Bracket monomials in* $\mathrm{M}([\cdot])$ *are universally generated by the following bracket monomial:*

$$I \stackrel{def}{=} \begin{bmatrix} a^{(3)}b^{(2)}c \\ bc^{(2)}d^{(3)} \end{bmatrix}.$$

That is, for a bracket monomial $B \in \mathrm{M}([\cdot])$, it is true that

$$B = \begin{cases} \alpha I_1 I_2 \cdots I_n & \text{if the number of brackets in } B \text{ is even,} \\ 0 & \text{Otherwise,} \end{cases}$$

where I_1, I_2, ..., I_n are equivalent to I and α, which is independent of the tensor t that $a^{(3)}$ represents, is a constant in the field K.

Proof. Let us show that B is a linear combination of bracket monomials of the following form:

$$\begin{bmatrix} a^{(3)}b^{(2)}c \\ bc^{(2)}d^{(3)} \\ \vdots \end{bmatrix}. \tag{32}$$

Let $a < b < c < \cdots$. We may assume that B is standard. Then B is a monomial with one of the following four standard forms:

$$\begin{bmatrix} a^{(3)}\cdots \\ b^{(3)}\cdots \\ \vdots \end{bmatrix}, \quad \begin{bmatrix} a^{(3)}b\cdot\cdot \\ b^{(2)}\cdots\cdot \\ \vdots \end{bmatrix}, \quad \begin{bmatrix} a^{(3)}b^{(2)}\cdot \\ b\cdot\cdot\cdot\cdot\cdot \\ \vdots \end{bmatrix}, \quad \text{and} \quad \begin{bmatrix} a^{(3)}b^{(3)} \\ \vdots \end{bmatrix}.$$

The fourth form vanishes by Lemma 1 and the second form may be reduced to the third one by formula (7) and the straightening algorithm. Since interchanging two equivalent letters in an umbral bracket monomial only changes the sign of the bracket monomial (Lemma 1), the first form is reduced to a linear combination of the other three forms by interchanging the letter next to a in the first row with b and then applying the straightening algorithm. Thus we may assume B has the third standard form. By interchanging equivalent letters and applying the straightening algorithm, B is a linear combination of bracket monomials, each having one of the following three standard forms:

$$\begin{bmatrix} a^{(3)}b^{(3)} \\ \vdots \end{bmatrix}, \quad \begin{bmatrix} a^{(3)}b^{(2)}c \\ bc^{(2)}\cdots \\ \vdots \end{bmatrix}, \quad \text{and} \quad \begin{bmatrix} a^{(3)}b^{(2)}c \\ bc\cdot\cdot\cdot\cdot \\ c\cdot\cdot\cdot\cdot\cdot \\ \vdots \end{bmatrix}.$$

Similarly, the first vanishes by Lemma 1, the third is reduced to the second by formula (9) and the straightening algorithm. So, we may assume that B is a bracket monomial with the standard form

$$\begin{bmatrix} a^{(3)}b^{(2)}c \\ bc^{(2)}\cdots \\ \vdots \\ \cdots\cdots \end{bmatrix}. \tag{33}$$

Again, by interchanging equivalent letters and applying the straightening algorithm, the form of (33) is a linear combination of the following three standard forms:

$$\begin{bmatrix} a^{(3)}b^{(2)}c \\ bc^{(2)}d^{(3)} \\ \vdots \end{bmatrix}, \quad \begin{bmatrix} a^{(3)}b^{(2)}c \\ bc^{(2)}d^{(2)}. \\ d\cdot\cdot\cdot\cdot\cdot \\ \vdots \end{bmatrix}, \quad \text{and} \quad \begin{bmatrix} a^{(3)}b^{(2)}c \\ bc^{(2)}d\cdot\cdot \\ d^{(2)}\cdots\cdot \\ \vdots \end{bmatrix}.$$

Obviously the first form has the form of (32). The second one may be reduced to the first one by Lemma 5 and the straightening algorithm. The third one, by interchanging equivalent letters, and by Lemma 1, Lemma 5, Lemma 6, Lemma 7 as well as by the straightening algorithm, may be restricted to the special standard form

$$\begin{bmatrix} a^{(3)}b^{(2)}c \\ bc^{(2)}def \\ d^{(2)}e\cdots \\ e\cdots\cdot \\ \vdots \end{bmatrix}$$

(those forms other than the special one are reduced to the form of (32)). By Lemma 8, Corollary 1 and by the straightening algorithm, the above special standard form can be further restricted as

$$\begin{bmatrix} a^{(3)}b^{(2)}c \\ bc^{(2)}def \\ d^{(2)}eghi \\ e\cdots\cdot \\ \vdots \end{bmatrix}.$$

Thereby, without loss of generality we may finally assume that B is a bracket monomial with the standard form

$$\begin{bmatrix} a^{(3)}b^{(2)}c \\ bc^{(2)}def \\ d^{(2)}eghi \\ \omega_1 \\ \vdots \\ \omega_m \end{bmatrix}. \tag{34}$$

The last letters in ω_m should be $z^{(3)}$ if z is the largest letter in B, or $i^{(2)}$ if i is the largest letter in B.

It is not difficult to see that there exists a positive integer $m_1 \leq m$ such that the bracket monomial

$$\begin{bmatrix} a^{(3)}b^{(2)}c \\ bc^{(2)}def \\ d^{(2)}eghi \\ \omega_1 \\ \vdots \\ \omega_{m_1} \end{bmatrix} \tag{35}$$

satisfies either the conditions in Lemma 9 or the conditions in Lemma 10. In the first case, by Lemma 9, Lemma 8 and Corollary 1, the bracket monomial B with the form of (34) may be written into a linear combination of bracket monomials with the desired form of (32). Let us now consider the second case.

$\star$ By Lemma 10, the result of the previous case, and by interchanging equivalent letters, we may assume that B is a bracket monomial with the following form:

$$\begin{bmatrix} a^{(3)}b^{(2)}c \\ bc^{(2)}def \\ d^{(2)}eghi \\ \omega'_{31}\omega'_{32} \\ \omega'_4 \\ \vdots \\ \omega'_m \end{bmatrix} \tag{36}$$

where ω'_{31} is a monomial over the set $\{a, b, \ldots, i\}$ with length$(\omega'_{31}) \geq 3$, and ω'_{32} is a monomial over the set $\{j, k, \ldots\}$. By Lemma 9, we may assume that length$(\omega'_{31}) = 3$ and $\omega'_{32} = jkl$. Furthermore, by the standard basis theorem, we may assume the bracket submonomial

$$\begin{bmatrix} \omega'_4 \\ \vdots \\ \omega'_{m2} \end{bmatrix}$$

is standard.

It is easy to see that the B with form (36) again satisfies either the conditions in Lemma 9 or the conditions in Lemma 10 but with the increased value of I_1. Repeating the process starting from $\star$ a finite number of times, we may eventually assume that B satisfies the conditions in Lemma 9 and has the form

$$\begin{bmatrix} a^{(3)}b^{(2)}c \\ bc^{(2)}def \\ d^{(2)}eghi \\ \vdots \end{bmatrix}.$$

By the result of the first case, the proof follows immediately. ■

5.3 The Canonical Forms

We shall use the positive notation to classify trivectors of a 6-dimensional space under the action of the general linear group.

Theorem 3 *In the umbral module* $\mathrm{U}_6(L)$ *consider the following covariants, where all English letters* a, b, c, ... *are equivalent and represent a tensor of step* 3*:*

$$\mathrm{C}_1 : \quad \begin{matrix} a^{(3)}b^{(2)} \\ b; \end{matrix}$$

$$\mathrm{C}_2 : \quad \begin{matrix} a^{(3)}b^{(2)}c \\ bc^{(2)}; \end{matrix}$$

$$\mathrm{C}_3 : \quad \begin{matrix} a^{(3)}b^{(2)}c \\ bc^{(2)}d^{(3)}. \end{matrix}$$

Let x_1,x_2, ... *be vectors. Consider the following canonical forms:*

$$\begin{aligned} \mathrm{I}: &\quad x_1x_2x_3; \\ \mathrm{II}: &\quad x_1x_2x_3 + x_1x_4x_5; \\ \mathrm{III}: &\quad x_1x_2x_3 + x_1x_4x_5 + x_2x_4x_6; \\ \mathrm{IV}: &\quad \text{the generic form.} \end{aligned}$$

Every tensor of step 3 *in a* 6-*D vector space has one of the given canonical forms. Furthermore,* $C_1 \propto 0$ *iff* $a^{(3)}$ *has the canonical form* I (*decomposable*)*; if* $C_1 \not\propto 0$*, but* $C_2 \propto 0$*, then* $a^{(3)} \propto \mathrm{II}$*; if* $C_2 \not\propto 0$*, but* $C_3 \propto 0$*, then* $a^{(3)} \propto \mathrm{III}$*; if* $C_3 \not\propto 0$*, then* $a^{(3)}$ *is generic.*

For each case the sufficiency is easy to check, therefore the proof is omitted. We only prove the necessity here.

5.4 Type I: $x_1x_2x_3$.

By Proposition 17, we know that $\mathrm{rank}(a^{(3)}) \geq 3$. Thus there exist at least three linearly independent vectors in $\mathrm{span}(a^{(3)})$, say, $[a^{(2)}\beta^{(4)}]a$, $[b^{(2)}\alpha^{(4)}]b$, and $[c^{(2)}\gamma^{(4)}]c$. Choose a tensor $\pi^{(3)}$ of step 3 such that

$$\begin{bmatrix} a^{(2)}\beta^{(4)} \\ b^{(2)}\alpha^{(4)} \\ c^{(2)}\gamma^{(4)} \\ abc\pi^{(3)} \end{bmatrix} \neq 0.$$

Let

$$t^{(6)} = \begin{bmatrix} a^{(2)}\beta^{(4)} \\ b^{(2)}\alpha^{(4)} \\ c^{(2)}\gamma^{(4)} \end{bmatrix} abc\pi^{(3)}.$$

Then applying the bracket exchange identity, we get

$$[t^{(3)}d^{(3)}]t^{(3)} = -[t^{(6)}]d^{(3)}.$$

Since $\begin{matrix} a^{(3)}b^{(2)} \\ b \end{matrix} \propto 0$ implies $\begin{matrix} a^{(3)}b \\ b^{(2)} \end{matrix} \propto 0$ by Lemma 2, the left hand side is reduced to only one term

$$\begin{bmatrix} a^{(2)}\beta^{(4)} \\ b^{(2)}\alpha^{(4)} \\ c^{(2)}\gamma^{(4)} \\ \pi^{(3)}d^{(3)} \end{bmatrix} abc,$$

which is

$$[\pi^{(3)}d^{(3)}][a^{(2)}\beta^{(4)}]a \cdot [b^{(2)}\alpha^{(4)}]b \cdot [c^{(2)}\gamma^{(4)}]c.$$

Thus $d^{(3)}$ is decomposable.

5.5 Type II: $x_1x_2x_3 + x_1x_4x_5$

We shall show that $a^{(3)}$ is divisible by a vector. Since $\begin{matrix} a^{(3)}b^{(2)} \\ b \end{matrix} \not\propto 0$, there exists a vector α and a tensor $\pi^{(5)}$ of step 5 such that

$$\begin{bmatrix} a^{(3)}b^{(2)}\alpha \\ b\pi^{(5)} \end{bmatrix} \not\propto 0.$$

Let $t^{(6)} = [a^{(3)}b^{(2)}\alpha]b\pi^{(5)}$. Consider

$$[t^{(3)}d^{(3)}]t^{(3)} = -[t^{(6)}]d^{(3)}.$$

Since $\begin{matrix} a^{(3)}b^{(2)}c \\ bc^{(2)} \end{matrix} \propto 0$ implies $\begin{matrix} a^{(3)}b^{(2)} \\ bc^{(3)} \end{matrix} \propto 0$ by Lemma 3, the left hand side of the above identity is reduced to only one term

$$\begin{bmatrix} a^{(3)}b^{(2)}\alpha \\ \pi^{(3)}d^{(3)} \end{bmatrix} b\pi^{(2)},$$

which is equal to the vector $[a^{(3)}b^{(2)}\alpha]b$ times the bivector $[\pi^{(3)}d^{(3)}]\pi^{(2)}$. Since every bivector, in a vector space of dimension 5, has the canonical form $x_2x_3 + x_4x_5$, the result follows immediately.

5.6 Type III: $x_1x_2x_3 + x_1x_4x_5 + x_2x_4x_6$.

Let us construct the canonical form of $a^{(3)}$. Since $\begin{matrix} a^{(3)}b^{(2)}c \\ bc^{(2)} \end{matrix} \not\propto 0$, there exists a tensor $\pi^{(3)}$ of step 3 such that

$$\begin{bmatrix} a^{(3)}b^{(2)}c \\ bc^{(2)}\pi^{(3)} \end{bmatrix} \not\propto 0.$$

Let $\alpha^{(3)} = [a^{(3)}b^{(2)}c]bc^{(2)}$, and $t^{(6)} = \alpha^{(3)}\pi^{(3)}$. Then

$$\begin{aligned} & [t^{(3)}d^{(3)}]t^{(3)} \\ = \ & [\alpha^{(3)}d^{(3)}]\pi^{(3)} + [\alpha^{(2)}\pi d^{(3)}]\alpha\pi^{(2)} + \end{aligned}$$

$[\alpha\pi^{(2)}d^{(3)}]\alpha^{(2)}\pi + [\pi^{(3)}d^{(3)}]\alpha^{(3)}$. The first term on the right hand side vanishes because we assume $[a^{(3)}b^{(2)}c][bc^{(2)}d^{(3)}] = 0$. The second term, by the bracket exchange identity, is equal to

$$-3[\alpha^{(3)}d^{(3)}]\pi^{(3)} - [\alpha^{(3)}\pi d^{(2)}]d\pi^{(2)},$$

which is equal to

$$-[\alpha^{(3)}\pi d^{(2)}]d\pi^{(2)}.$$

It vanishes according to Lemma 4. Thus

$$[t^{(3)}d^{(3)}]t^{(3)} = [\alpha\pi^{(2)}d^{(3)}]\alpha^{(2)}\pi + [\pi^{(3)}d^{(3)}]\alpha^{(3)}.$$

Since $\alpha^{(3)}$ is decomposable by Lemma 4, let $\alpha^{(3)} = x_1x_2x_4$ where x_1, x_2, x_4 are three linearly independent vectors. Then

$$\begin{aligned} & [t^{(3)}d^{(3)}]t^{(3)} \\ = \ & [x_1\pi^{(2)}d^{(3)}]\pi x_2x_4 + [x_2\pi^{(2)}d^{(3)}]\pi x_1x_4 + [x_4\pi^{(2)}d^{(3)}]\pi x_1x_2 + [\pi^{(3)}d^{(3)}]x_1x_2x_4. \end{aligned}$$

Let

$$x_6 = [x_1\pi^{(2)}d^{(3)}]\pi + [\pi^{(3)}d^{(3)}]x_1,$$

$$x_5 = [x_2\pi^{(2)}d^3]\pi,$$

and

$$x_3 = [x_4\pi^{(2)}d^{(3)}]\pi,$$

then

$$[t^{(3)}d^{(3)}]t^{(3)} = x_1x_2x_3 + x_1x_4x_5 + x_2x_4x_6.$$

5.7 The Generic Form

If the bracket monomial

$$I = \begin{bmatrix} a^{(3)}b^{(2)}c \\ bc^{(2)}d^{(3)} \end{bmatrix} \tag{37}$$

does not vanish at a tensor t of step 3, then we say that t has the **generic form** of trivectors in a 6-D vector space. The representation of this generic form can be found in [7].

Recall that $\mathrm{P}([\cdot])$ ($\subset \mathrm{UB}_6[L]$) denotes the bracket polynomial ring over the equivalent letters a, b, c, ..., where the arity of each letter is 3 and the content of each letter in $\mathrm{P}([\cdot])$ is either 0 or 3. The bracket monomial (37) is the generator of this ring according to Theorem 2. Any element in $\mathrm{P}([\cdot])$, which is a linear combination of products of k brackets, is called a homogeneous bracket polynomial of degree k. If the bracket monomial (37) does not vanish at a tensor t of step 3, then there is no homogeneous bracket polynomial p vanishing at t except that p is identically zero (i.e., p vanishes at every tensor of step 3). In other words, there are no invariants vanishing at t. This is why we call t the generic form.

Conjecture: *Consider a covariant C_λ over the equivalent letters a, b, c, ... with a fixed shape $\lambda = (\lambda_1, \lambda_2, \ldots, \lambda_k)$:*

$$C_\lambda : \begin{matrix} w_1 \\ w_2 \\ \vdots \\ w_k \end{matrix} ,$$

where the content of each letter in C_λ is either 0 or 3. If C_λ is not identically equal to zero (i.e., there exists a tensor of step 3, at which C_λ does not vanish), then the covariant could be obtained as the subtableau of invariants.

If the above conjecture is true and the bracket monomial (37) does not vanish at tensor t, then for any fixed shape there is no covariant vanishing at t. Otherwise, by theorem 2 we obtain

$$P = \alpha I_1 I_2 \cdots I_n,$$

where α is a non-zero element in $\mathrm{UB}_6[L]$, which implies $I = 0$ at the tensor t (a contradiction with the assumption).

Acknowledgment. This paper is part of author's thesis. I thank Professor Rota for stimulating discussions and generous help.

References

[1] A. Brini, A. Palareti, A. Teolis, Proc. Natl. Acad. Sci. USA 85, 1988, pp. 1330–1333.

[2] A. Brini, Grassmann's progressive and regressive products and GC coalgebras, Proceedings of the Curaçao Conference: Invariant Theory in Discrete and Computational Geometry, June 1994, to appear.

[3] W. Chan, G.C. Rota, J. Stein, The power of positive thinking, Proceedings of the Curaçao Conference: Invariant Theory in Discrete and Computational Geometry, June 1994, to appear.

[4] Wendy Chan, Thesis, 1995.

[5] Désarménien, Kung and Rota, Invariant theory, Young bitableau, and combinatorics. Adv. in Math., 27, 1978, pp. 63–92.

[6] Doubilet, Rota and Stein, On the foundations of combinatorial theory: IX–combinatorial methods in invariant theory. Studies in Applied Mathematics, No 3, Vol. LIII, September 1974.

[7] Grosshans, Rota and Stein, Invariant theory and supersymmetric algebras. Conference board of the mathematical sciences, No. 69, Am. Math. Soc., Providence, RI.

[8] Michael Hawrylycz, Thesis, MIT, 1994.

[9] Rosa Huang, Thesis, MIT, 1990.

[10] R. Huang, G.-C. Rota, J. Stein, Supersymmetric bracket algebra and invariant theory, Centro Matematico V. Volterra, Universita Degli Studi Di Roma II, 1990.

[11] G. C. Rota, J.A. Stein, (1986) Proc.Natl.Acad.Sci. USA 83, 844–847.

[12] G.C. Rota, J.A. Stein, (1989) Proc.Natl.Acad.Sci. USA 86, 2521–2524.

[13] G.C. Rota, Combinatorial theory and invariant theory (Bowdoin College, Maine, 1971).

[14] E. B. Vinberg, A. G. Èlashvili, Classification of trivectors of a 9 dimensional space. Originally published in Trudy Sem. Vektor. Tenzor. Anal. 18 (1978), 197-233, Translated by B. A. Datskovsky.

[15] N. White, The bracket ring of a combinatorial geometry I and II, Trans. Ameer. Math. Soc. 202 (1975a) 79–95, 214 (1975b) 233–48.

GE Corporate R & D
Schenectady, NY 12301

Parameter Augmentation for Basic Hypergeometric Series, I

William Y. C. Chen and Zhi-Guo Liu

Dedicated to Professor Gian-Carlo Rota

Abstract

This paper is motivated by the umbral calculus approach to basic hypergeometric series as initiated by Goldman–Rota, Andrews, and Roman, et al. We develop a method of deriving hypergeometric identities by parameter augmentation, which means that a hypergeometric identity with multiple parameters may be derived from its special case obtained by reducing some parameters to zero. Many classical results on basic hypergeometric series easily fall into this framework.

1. Introduction and notation

This paper is motivated by the umbral calculus approach to basic hypergeometric series as initiated by Goldman–Rota [19, 20], Andrews [2], and Roman [25], et al. The main result is the following scheme. Taking an example of ${}_6\psi_6$ with five parameters α, a, b, c, d, what if we say Bailey's summation formula is more or less equivalent to its very special case obtained by letting b, c, d vanish. The idea is to reach the general case by adding parameters to the special case. We introduce an exponential operator defined in terms of the usual q-differential operator. Many classical results on basic hypergeometric series easily fall into this framework, including the Rogers identity — the limiting case of Jackson's theorem, Bailey's ${}_2\psi_2$ transformation formula, Bailey's ${}_3\psi_3$ summation formula, q-Dixon's formula, our generalization to a ${}_3\psi_3$ transformation formula, the Askey beta integral, and an extension of Ramunujan's beta integral.

We follow the notation and terminology in [18], and we always assume that $|q| < 1$. The q-shifted factorial is defined by

$$(a;q)_0 = 1, \quad (a;q)_n = \prod_{j=0}^{n-1}(1-aq^j), \quad n = 1, 2, \ldots, \infty,$$

$$(a_1, a_2, \ldots, a_m; q)_n = (a_1; q)_n (a_2; q)_n \cdots (a_m; q)_n,$$
$$(a_1, a_2, \ldots, a_m; q)_\infty = (a_1; q)_\infty (a_2; q)_\infty \cdots (a_m; q)_\infty.$$

For any integer n, we define $(a; q)_n = (a; q)_\infty / (aq^n; q)_\infty$. It is easy to verify that

$$(q/a; q)_n = (-a)^{-n} q^{\binom{n+1}{2}} (q^{-n}a; q)_\infty / (a; q)_\infty. \tag{1.1}$$

The basic hypergeometric series ${}_{r+1}\phi_r$ and the bilateral basic hypergeometric series ${}_r\psi_r$ are defined by

$${}_{r+1}\phi_r\left(\begin{matrix} a_1, & a_2, & \ldots, & a_{r+1} \\ b_1, & b_2, & \ldots, & b_r \end{matrix}; q, x\right) = \sum_{n=0}^{\infty} \frac{(a_1, a_2, \ldots, a_{r+1}; q)_n \, x^n}{(q, b_1, b_2, \ldots, b_r; q)_n}, \tag{1.2}$$

$${}_r\psi_r\left(\begin{matrix} a_1, & a_2, & \ldots, & a_r \\ b_1, & b_2, & \ldots, & b_r \end{matrix}; q, x\right) = \sum_{n=-\infty}^{\infty} \frac{(a_1, a_2, \ldots, a_r; q)_n \, x^n}{(b_1, b_2, \ldots, b_r; q)_n}. \tag{1.3}$$

2. An exponential operator

The usual q-differential operator, or Euler derivative, is defined as

$$D_q\{f(a)\} = \frac{f(a) - f(aq)}{a}.$$

By convention, D_q^0 is understood as the identity. The q-shift operator, denoted by η in the literature [2, 25], is defined by

$$\eta\{f(a)\} = f(aq) \quad \text{and} \quad \eta^{-1}\{f(a)\} = f(aq^{-1}).$$

Next we recall an operator built on the q-differential operator and the q-shift operator which appeared in the work of Roman [25] and will be denoted by θ in this paper:

$$\theta = \eta^{-1} D_q.$$

The key operator introduced in this paper is an exponential operator constructed from θ:

$$E(b\theta) = \sum_{n=0}^{\infty} \frac{(b\theta)^n q^{\binom{n}{2}}}{(q; q)_n}, \tag{2.1}$$

where b is a parameter. Using the Euler identity, $E(b\theta)$ can also be written as $(-b\theta; q)_\infty$. We first present the Leibniz formula for the operator θ, which is equivalent to a formula of Hahn [22]. Our proof is based on the following standard q-binomial theorem [17, 21], and seems to be new.

Lemma 2.2. *Let A and B be two noncommutative indeterminates satisfying $BA = qAB$, then we have*

$$(A+B)^n = \sum_{k=0}^{n} \begin{bmatrix} n \\ k \end{bmatrix} A^k B^{n-k}, \tag{2.3}$$

where

$$\begin{bmatrix} n \\ k \end{bmatrix} = \frac{(q;q)_n}{(q;q)_k (q;q)_{n-k}}$$

is the q-binomial or Gaussian coefficient.

Theorem 2.4 (Leibniz formula for θ). *For $n \geq 0$, we have*

$$\theta^n\{f(a)g(a)\} = \sum_{k=0}^{n} \begin{bmatrix} n \\ k \end{bmatrix} \theta^k\{f(a)\}\, \theta^{n-k}\{g(aq^{-k})\}. \tag{2.5}$$

Proof. By definition, it follows that

$$\begin{aligned} \theta\{f(a)g(a)\} &= f(a)\frac{g(aq^{-1}) - g(a)}{aq^{-1}} + \frac{f(aq^{-1}) - f(a)}{aq^{-1}} g(aq^{-1}) \\ &= (\theta_g + \theta_f \eta_g^{-1})\{f(a)\, g(a)\}, \end{aligned}$$

where θ_f signifies the action only on f, θ_g, η_g are understood similarly. It is easy to verify $(\theta_f \eta_g^{-1})\theta_g = q\theta_g(\theta_f \eta_g^{-1})$, and it follows that

$$\begin{aligned} \theta^n\{f(a)g(a)\} &= (\theta_g + \theta_f \eta_g^{-1})^n \{f(a)\, g(a)\} \\ &= \sum_{k=0}^{n} \begin{bmatrix} n \\ k \end{bmatrix} \theta_g^k \theta_f^{n-k} \eta_g^{-(n-k)} \{f(a)\, g(a)\} \\ &= \sum_{k=0}^{n} \begin{bmatrix} n \\ k \end{bmatrix} \theta^k\{g(aq^{k-n})\}\, \theta^{n-k}\{f(a)\} \\ &= \sum_{k=0}^{n} \begin{bmatrix} n \\ k \end{bmatrix} \theta^k\{f(a)\}\, \theta^{n-k}\{g(aq^{-k})\}, \end{aligned}$$

which completes the proof. ∎

A fundamental property of the operator θ is the following:

Theorem 2.6. *We have*

$$\theta\{(at;q)_\infty\} = (-t)(at;q)_\infty, \tag{2.7}$$

$$\theta^k\{(at;q)_\infty\} = (-t)^k(at;q)_\infty. \tag{2.8}$$

An important property of the exponential operator $E(b\theta)$ follows from the above theorem:

Theorem 2.9. *We have*

$$E(b\theta)\{(at;q)_\infty\} = (at, bt;q)_\infty. \tag{2.10}$$

The above theorem reflects the differential property or the eigenfunction property of the operator θ, whereas the following theorem is in essence an application of the Leibniz formula for the operator θ. Because of (2.10), we call $E(b\theta)$ an augmentation operator, which is a counterpart of a shift operator E^a: $f(x) \to f(x+a)$ in the classical umbral calculus [26].

Theorem 2.11. *We have*

$$E(b\theta)\{(as, at;q)_\infty\} = \frac{(as, at, bs, bt;q)_\infty}{(abst/q;q)_\infty}. \tag{2.12}$$

Proof. By definition, $E(b\theta)\{(as, at;q)_\infty\}$ equals

$$\begin{aligned}
&\sum_{n=0}^{\infty} \frac{b^n q^{\binom{n}{2}}}{(q;q)_n} \theta^n\{(as, at;q)_\infty\} \\
= &\sum_{n=0}^{\infty} \frac{b^n q^{\binom{n}{2}}}{(q;q)_n} \sum_{k=0}^{n} \begin{bmatrix} n \\ k \end{bmatrix} \theta^k\{(as;q)_\infty\}\theta^{n-k}\{(atq^{-k};q)_\infty\} \\
= &\sum_{n=0}^{\infty} \frac{b^n q^{\binom{n}{2}}}{(q;q)_n} \sum_{k=0}^{n} \begin{bmatrix} n \\ k \end{bmatrix} (-s)^k (as;q)_\infty (-tq^{-k})^{n-k} (atq^{-k};q)_\infty \\
= &(as, at;q)_\infty \sum_{n=0}^{\infty} \frac{b^n q^{\binom{n}{2}}}{(q;q)_n} \sum_{k=0}^{n} \begin{bmatrix} n \\ k \end{bmatrix} (-as)^k (-t)^n q^{\binom{k}{2}-nk} (q/at;q)_k \\
= &(as, at;q)_\infty \sum_{k=0}^{\infty} \frac{(q/at;q)_k}{(q;q)_k} (abst/q)^k \sum_{m=0}^{\infty} \frac{q^{\binom{m}{2}}(-bt)^m}{(q;q)_m} \\
= &(as, at;q)_\infty \frac{(bs;q)_\infty}{(abst/q;q)_\infty} (bt;q)_\infty \\
= &\frac{(as, at, bs, bt;q)_\infty}{(abst/q;q)_\infty},
\end{aligned}$$

as desired. ∎

Note that Theorem 2.11 specializes to Theorem 2.9 by setting $s = 0$. The rest of this paper will be concerned with applications of the operator $E(b\theta)$ along with Theorem 2.9 and Theorem 2.11.

3. The Rogers identity – the limiting case of Jackson's theorem

The Rogers identity [24], or the limiting case of Jackson's Theorem is stated as follows.

Theorem 3.1. *We have*

$$\begin{aligned} &{}_6\phi_5\left(\begin{matrix} \alpha, & q\sqrt{\alpha}, & -q\sqrt{\alpha}, & q/a, & q/b, & q/c \\ \sqrt{\alpha}, & -\sqrt{\alpha}, & \alpha a, & \alpha b, & \alpha c & \end{matrix} ; q, \alpha abc/q^2\right) \\ =& \frac{(\alpha ab/q, \alpha bc/q, \alpha ca/q, \alpha q; q)_\infty}{(\alpha a, \alpha b, \alpha c, \alpha abc/q^2; q)_\infty}. \end{aligned} \tag{3.2}$$

There are four parameters α, a, b, c in the above formula. It should be noted that from the point of view of this paper the parameter α plays a role as a constant, so there are really three parameters in the above formula. The aim of this section is to show that (3.2) can be recovered from its special case for $b = c = 0$. This special case has appeared in the literature, and we will give a simple proof for completeness.

Lemma 3.3. *We have*

$$\sum_{n=0}^{\infty} \frac{(1-\alpha q^{2n})(\alpha, q/a; q)_n}{(q, \alpha a; q)_n} (\alpha a)^n q^{n(n-1)} = \frac{(\alpha; q)_\infty}{(\alpha a; q)_\infty}. \tag{3.4}$$

Proof. Let $f(\alpha)$ denote the left-hand side of (3.4). Since

$$(1-\alpha q^{2j}) = (1-q^j) + q^j(1-\alpha q^j),$$

$f(\alpha)$ equals

$$\begin{aligned} & \sum_{n=1}^{\infty} \frac{(\alpha, q/a; q)_n}{(\alpha a; q)_n (q; q)_{n-1}} (\alpha a)^n q^{n(n-1)} + \sum_{n=0}^{\infty} \frac{(\alpha; q)_{n+1} (q/a; q)_n}{(q, \alpha a; q)_n} (\alpha a)^n q^{n^2} \\ =& \sum_{n=0}^{\infty} \frac{(\alpha, q/a; q)_{n+1}}{(\alpha a; q)_{n+1} (q; q)_n} (\alpha a)^{n+1} q^{n(n+1)} + \sum_{n=0}^{\infty} \frac{(\alpha; q)_{n+1} (q/a; q)_n}{(q, \alpha a; q)_n} (\alpha a)^n q^{n^2} \\ =& \sum_{n=0}^{\infty} \frac{(\alpha; q)_{n+1} (q/a; q)_n}{(q, \alpha a; q)_n} (\alpha a)^n q^{n^2} \left\{ \frac{\alpha a q^n (1-q^{n+1}/a)}{1-\alpha a q^n} + 1 \right\} \\ =& \sum_{n=0}^{\infty} \frac{(1-\alpha q^{2n+1})(\alpha; q)_{n+1} (q/a; q)_n}{(q; q)_n (\alpha a; q)_{n+1}} (\alpha a)^n q^{n^2} \\ =& \frac{(1-\alpha)}{(1-\alpha a)} f(\alpha q) = \frac{(1-\alpha)(1-\alpha q)}{(1-\alpha a)(1-\alpha a q)} f(\alpha q^2) = \cdots = \frac{(\alpha; q)_\infty}{(\alpha a; q)_\infty}, \end{aligned}$$

as required. ∎

Using (1.1), the above lemma can be rewritten as

$$\sum_{n=0}^{\infty} A_n(\alpha a q^n, a q^{-n}; q)_\infty = (\alpha, a; q)_\infty, \tag{3.5}$$

where

$$A_n = (-1)^n \frac{(1-\alpha q^{2n})(\alpha;q)_n}{(q;q)_n} \alpha^n q^{n(3n-1)/2}.$$

Applying the operator $E(b\theta)$ to both sides of (3.5), we get

$$\sum_{n=0}^{\infty} A_n E(b\theta)\{(\alpha a q^n, a q^{-n}; q)_\infty\} = (\alpha;q)_\infty E(b\theta)\{(a;q)_\infty\}. \tag{3.6}$$

By Theorem 2.9 and Theorem 2.11, we have

$$E(b\theta)\{(a;q)_\infty\} = (a,b;q)_\infty, \tag{3.7}$$

$$E(b\theta)\{(\alpha a q^n, a q^{-n}; q)_\infty\} = \frac{(\alpha a q^n, q^{-n}a, \alpha b q^n, q^{-n} b; q)_\infty}{(\alpha a b/q; q)_\infty}. \tag{3.8}$$

Substituting the above identities into (3.4), we get

$$\sum_{n=0}^{\infty} A_n(\alpha b q^{-n}, q^{-n}b, \alpha a q^n, q^{-n}a; q)_\infty = (\alpha, b, a, \alpha a b/q; q)_\infty. \tag{3.9}$$

One more step to reach (3.2) is to apply $E(c\theta)$ to both sides of (3.9):

$$\begin{aligned} &\sum_{n=0}^{\infty} A_n(\alpha b q^{-n}, q^{-n}b; q)_\infty E(c\theta)\{(\alpha a q^n, a q^{-n}; q)_\infty\} \\ = \; &(\alpha, b; q)_\infty E(c\theta)\{(a, \alpha a b/q; q)_\infty\}. \end{aligned} \tag{3.10}$$

Using

$$E(c\theta)\{(\alpha a q^n, a q^{-n}; q)_\infty\} = \frac{(\alpha a q^n, q^{-n}a, \alpha c q^n, q^{-n}c; q)_\infty}{(\alpha a c/q; q)_\infty},$$

$$E(c\theta)\{(a, \alpha a b/q; q)_\infty\} = \frac{(a, \alpha a b/q, c, , \alpha b c/q; q)_\infty}{(\alpha a b c/q^2; q)_\infty},$$

we obtain

$$\begin{aligned} &\sum_{n=0}^{\infty} A_n(\alpha a q^n, \alpha b q^n, \alpha c q^n, q^{-n}a, q^{-n}b, q^{-n}c; q)_\infty \\ = \; &\frac{(\alpha, a, b, c, \alpha a b/q, \alpha b c/q, \alpha c a/q; q)_\infty}{(\alpha a b c/q^2; q)_\infty}, \end{aligned} \tag{3.11}$$

which can be restated as Theorem 3.1. ■

4. Bailey's summation for ${}_6\psi_6$

Bailey's summation formula for ${}_6\psi_6$ is stated as follows.

Theorem 4.1. *We have*

$$
\begin{aligned}
&{}_6\psi_6\left(\begin{matrix} q\sqrt{\alpha}, & -q\sqrt{\alpha}, & q/a, & q/b, & q/c, & q/d \\ \sqrt{\alpha}, & -\sqrt{\alpha}, & \alpha a, & \alpha b, & \alpha c, & \alpha d \end{matrix}; q, \alpha^2 abcd/q^3\right) \\
&= \frac{(q, \alpha q, q/\alpha, \alpha ab/q, \alpha ac/q, \alpha ad/q, \alpha bc/q, \alpha bd/q, \alpha cd/q; q)_\infty}{(\alpha a, a, \alpha b, b, \alpha c, c, \alpha d, d, \alpha^2 abcd/q^3; q)_\infty}. \qquad (4.2)
\end{aligned}
$$

As noted by Andrews, Jacobi's triple product identity (see also [21]) is a very special case of the above theorem [3], so are many other identities. Many proofs of this theorem can be found in the literature, see for example Andrews [3], Askey [12], Askey and Ismail [9], Bailey [14], Slater and Lakin [27]. From the point of view of this paper, one sees the other way around: Bailey's summation formula is not so distant from Jacobi's triple product identity. To be more specific, by setting $b = c = d = 0$ in Theorem 4.1 we get the following special case which is slightly more general than the Jacobi's identity, but it can be also easily derived from Jacobi's identity:

Lemma 4.3. *We have*

$$
\sum_{n=-\infty}^{\infty} \frac{(1-\alpha q^{2n})(q/a;q)_n}{(1-\alpha)(\alpha a;q)_n}(-\alpha^2 a)^n q^{3n(n-1)/2} = \frac{(q, \alpha q, q/\alpha; q)_\infty}{(a, \alpha a; q)_\infty}. \qquad (4.4)
$$

Proof. Let $f(a,\alpha)$ denote the left-hand side of (4.4) and let $g(a,\alpha) = (1-\alpha)f(a,\alpha)$. Since

$$
1 - \alpha q^{2n} = (1 - \alpha a q^{n-1}) + \alpha a q^{n-1}(1 - q^{n+1}/a),
$$

we get

$$
\begin{aligned}
f(a,\alpha) &= \sum_{n=-\infty}^{\infty} \frac{(q/a;q)_n}{(1-\alpha)(\alpha a;q)_{n-1}}(-\alpha^2 a)^n q^{3n(n-1)/2} \\
&\quad + \sum_{n=-\infty}^{\infty} (-1)^n \frac{(q/a;q)_{n+1}}{(1-\alpha)(\alpha a;q)_n} \alpha^{2n+1} a^{n+1} q^{n(3n-1)/2-1}. \qquad (4.5)
\end{aligned}
$$

Substituting $n-1$ with n in the first summation in (4.5), we obtain

$$
\begin{aligned}
g(a,\alpha) &= \sum_{n=-\infty}^{\infty} (-1)^n \frac{(q/a;q)_{n+1}}{(\alpha a;q)_n} \alpha^{2n+1} a^{n+1} (1-\alpha q^{2n+1}) q^{n(3n-1)/2-1} \\
&= (1-q/a)\alpha a q^{-1} g(a/q, \alpha q).
\end{aligned}
$$

Substituting a with aq, it follows that

$$g(aq,\alpha) = (1-a^{-1})\alpha a\, g(a,\alpha q). \tag{4.6}$$

Substituting $n+1$ with n in the second summation in (4.5), it follows that $g(a,\alpha) = q\alpha^{-1}(1-\alpha a/q)\, g(a,\alpha/q)$, which can be rewritten as

$$g(a,\alpha q) = -\alpha^{-1}(1-\alpha a)\, g(a,\alpha). \tag{4.7}$$

Combining (4.6) and (4.7), we obtain

$$\begin{aligned} f(a,\alpha) &= \frac{1}{(1-a)(1-\alpha a)} f(aq,\alpha) \\ &= \frac{1}{(1-a)(1-aq)(1-\alpha a)(1-\alpha a q)} f(aq^2,\alpha) \\ &= \cdots = \frac{1}{(a,\alpha a;q)_\infty} f(0,\alpha). \end{aligned}$$

Using Jacobi's triple product identity, we have

$$\begin{aligned} f(0,\alpha) &= \frac{1}{(1-\alpha)} \sum_{n=-\infty}^{\infty} (1-\alpha q^{2n}) q^{2n^2-n} \alpha^{2n} \\ &= \frac{1}{(1-\alpha)} \sum_{n=-\infty}^{\infty} (-\alpha)^n q^{\binom{n}{2}} = (q,\alpha q, q/\alpha;q)_\infty, \end{aligned}$$

which gives

$$f(a,\alpha) = \frac{(q,\alpha q,q/\alpha;q)_\infty}{(a,\alpha a;q)_\infty}.$$

This completes the proof. ∎

For notational clarity, let us use B_n to denote the coefficient in the above Lemma:

$$B_n = \frac{(1-\alpha q^{2n})}{(1-\alpha)} \alpha^{2n} q^{2n^2-n}.$$

Equation (4.4) can now be expressed as follows:

$$\sum_{n=-\infty}^{\infty} B_n(\alpha a q^n, q^{-n}a;q)_\infty = (q,\alpha q,q/\alpha;q)_\infty. \tag{4.8}$$

Taking the action of $E(b\theta)$ on both sides of (4.8) and using

$$E(b\theta)\{(\alpha a q^n, q^{-n}a;q)_\infty\} = \frac{(\alpha a q^n, q^{-n}a, \alpha b q^n, q^{-n}b;q)_\infty}{(\alpha a b/q;q)_\infty},$$

we obtain

$$\sum_{n=-\infty}^{\infty} B_n(\alpha a q^n, q^{-n}a, \alpha b q^n, q^{-n}b; q)_\infty = (q, \alpha q, q/\alpha, \alpha ab/q; q)_\infty.$$

Taking further action of $E(c\theta)$ on the above equation, we obtain

$$\sum_{n=-\infty}^{\infty} B_n(\alpha a q^n, q^{-n}a, \alpha b q^n, q^{-n}b, \alpha c q^n, q^{-n}c; q)_\infty$$

$$= (q, \alpha q, q/\alpha, \alpha ab/q, \alpha ac/q, \alpha bc/q; q)_\infty.$$

The final step is to take the action of $E(d\theta)$ on the above equation, leading to the following identity:

$$\sum_{n=-\infty}^{\infty} B_n(\alpha a q^n, q^{-n}a, \alpha b q^n, q^{-n}b, \alpha c q^n, q^{-n}c, \alpha d q^n, q^{-n}d; q)_\infty$$

$$= \frac{(q, \alpha q, q/\alpha, \alpha ab/q, \alpha ac/q, \alpha ad/q, \alpha bc/q, \alpha bd/q, \alpha cd/q; q)_\infty}{(\alpha^2 abcd/q^3; q)_\infty},$$

which can be rewritten in the form of Theorem 4.1.

5. Bailey's ${}_2\psi_2$formula and its extension

The following is an identity on ${}_2\psi_2$ due to Bailey [1, 3, 15], which has important applications to mock theta function identities and can be used to derive the Rogers–Ramanujan identities.

Theorem 5.1. *We have*

$$\sum_{n=-\infty}^{\infty} \frac{(1-\alpha q^{2n})(q/a, q/b, q/c, q/d; q)_n}{(1-\alpha)(\alpha a, \alpha b, \alpha c, \alpha d; q)_n} (\alpha^3 abcd)^n q^{n^2-3n}$$

$$= \frac{(q\alpha, q/\alpha, \alpha ab/q, \alpha cd/q; q)_\infty}{(\alpha c, \alpha d, a, b; q)_\infty} {}_2\psi_2\left(\begin{matrix} q/c, & q/d \\ \alpha a, & \alpha b \end{matrix}; q, \alpha cd/q\right). \tag{5.2}$$

In this section, we shall give a treatment of the above theorem in terms of parameter augmentation, and give the following generalization of Bailey's formula:

Theorem 5.3. *We have*

$$\sum_{n=-\infty}^{\infty} \frac{(1-\alpha q^{2n})(q/a,q/b,q/c,q/d,q/e;q)_n}{(1-\alpha)(\alpha a,\alpha b,\alpha c,\alpha d,\alpha e;q)_n}(-\alpha^3 abcde)^n q^{n^2/2-7n/2}$$
$$= \frac{(q\alpha,q/\alpha,\alpha ab/q,\alpha ae/q,\alpha be/q,\alpha cd/q;q)_\infty}{(\alpha c,\alpha d,a,b,e,\alpha^2 abe/q^2;q)_\infty}$$
$$\times {}_3\psi_3\left(\begin{matrix} q/c, & q/d, & \alpha^2 abe/q^2 \\ \alpha a, & \alpha b, & \alpha e \end{matrix};q,\alpha cd/q\right). \tag{5.4}$$

Our point of departure to reach Theorem 5.1 is the following specialization:

Lemma 5.5. *We have*

$$\sum_{n=-\infty}^{\infty} \frac{(q/c,q/d;q)_n}{(\alpha a;q)_n}(\alpha cd/q)^n = \frac{(\alpha c,\alpha d,a;q)_\infty}{(\alpha cd/q,\alpha q,q/\alpha;q)_\infty}$$
$$\times \sum_{n=-\infty}^{\infty} \frac{(1-\alpha q^{2n})(q/a,q/c,q/d;q)_n}{(1-\alpha)(\alpha a,\alpha c,\alpha d;q)_n}(-\alpha^3 acd)^n q^{3n^2/2-5n/2}. \tag{5.6}$$

In order to deal with the above bilateral summation, we need the following lemma which is a special case of [18, p. 54, Ex. 2.22].

Lemma 5.7. *We have*

$$\sum_{n=0}^{\infty} \frac{(q/c,q/d;q)_n}{(q,\alpha a;q)_n}(\alpha cd/q)^n = \frac{(\alpha c,\alpha d;q)_\infty}{(\alpha,\alpha cd/q;q)_\infty}$$
$$\times \sum_{n=0}^{\infty} \frac{(1-\alpha q^{2n})(\alpha,q/a,q/c,q/d;q)_n}{(q,\alpha a,\alpha c,\alpha d;q)_n}(\alpha^2 acd)^n q^{n^2-2n}. \tag{5.8}$$

By the above lemma, we have

$$\sum_{n=-m}^{\infty} \frac{(q/c,q/d;q)_n}{(\alpha a,q^{m+1};q)_n}(\alpha cd/q)^n$$
$$= \sum_{k=0}^{\infty} \frac{(q/c,q/d;q)_{k-m}}{(\alpha a,q^{m+1};q)_{k-m}}(\alpha cd/q)^{k-m}$$
$$= \frac{(q/c,q/d;q)_{-m}}{(\alpha a,q^{m+1};q)_{-m}}\left(\frac{q}{\alpha cd}\right)^m \times \sum_{k=0}^{\infty} \frac{(q^{1-m}/c,q^{1-m}/d;q)_k}{(q,\alpha a q^{-m};q)_k}\left(\frac{\alpha cd}{q}\right)^k$$
$$= \frac{(q/c,q/d;q)_{-m}}{(\alpha a,q^{m+1};q)_{-m}}\left(\frac{q}{\alpha cd}\right)^m \frac{(\alpha cq^{-m},\alpha dq^{-m};q)_\infty}{(\alpha q^{-2m},\alpha cd/q;q)_\infty}$$

$$\times \sum_{k=0}^{\infty} \frac{(1-\alpha q^{2k-2m})(\alpha q^{-2m}, q^{1-m}/a, q^{1-m}/c, q^{1-m}/d; q)_k}{(q, \alpha a q^{-m}, \alpha c q^{-m}, \alpha d q^{-m}; q)_k}$$
$$\times (\alpha^2 acdq^{-m})^k q^{k^2-2k}$$
$$= \frac{(\alpha c, \alpha d; q)_\infty (a;q)_m}{(\alpha, \alpha cd/q; q)_\infty (q/\alpha; q)_m}$$
$$\times \sum_{n=-m}^{\infty} \frac{(1-\alpha q^{2n})(\alpha q^{-m}, q/a, q/c, q/d; q)_n}{(q^{m+1}, \alpha a, \alpha c, \alpha d; q)_n} \left(\frac{\alpha^2 acd}{q^2}\right)^n q^{n^2+nm}.$$

Therefore, Lemma 5.5 follows by taking the limit $m \to \infty$. ∎

Once we have Lemma 5.5, Theorem 5.1 easily follows from the action of $E(b\theta)$. Further application of $E(e\theta)$ to Theorem 5.1 leads to Theorem 5.3.

6. The Askey beta integral

The q-integral is defined by

$$\int_0^a f(t) d_q t = a(1-q) \sum_{n=0}^{\infty} f(aq^n) q^n$$

and

$$\int_a^b f(t) d_q t = \int_0^b f(t) d_q t - \int_0^a f(t) d_q t.$$

The Askey beta integral is stated as follows [11]:

Theorem 6.1. *We have*

$$\int_{-\infty}^{\infty} \frac{(at, bt; q)_\infty}{(-dt, et; q)_\infty} d_q t$$
$$= \frac{2(1-q)(q^2;q^2)_\infty^2 (de, q/de, a/e, -a/d, b/e, -b/d; q)_\infty}{(q;q)_\infty (d^2, e^2, q^2/d^2, q^2/e^2; q^2)_\infty (-ab/deq; q)_\infty}. \tag{6.2}$$

Before we give a parameter augmentation derivation of the Askey beta integral, we present an extension, which is a consequence of applying the operator $E(c\theta)$ to (6.2).

Theorem 6.3. *We have*

$$\int_{-\infty}^{\infty} \frac{(at, bt, ct; q)_\infty}{(-dt, et, -abct/deq^2; q)_\infty} d_q t =$$
$$\frac{2(1-q)(q^2;q^2)_\infty^2 (de, q/de, a/e, -a/d, b/e, -b/d, c/e, -c/d; q)_\infty}{(q;q)_\infty (d^2, e^2, q^2/d^2, q^2/e^2; q^2)_\infty (-ab/deq, -ac/deq, -bc/deq; q)_\infty}. \tag{6.4}$$

Setting $b = 0$ in (6.2), we are led to the following lemma:

Lemma 6.5. *Let*

$$\Delta = \frac{2(1-q)(q^2;q^2)_\infty^2(de,q/de;q)_\infty}{(q;q)_\infty(d^2,e^2,q^2/d^2,q^2/e^2;q^2)_\infty}.$$

Then we have

$$\int_{-\infty}^{\infty} \frac{(at;q)_\infty d_q t}{(-dt,et;q)_\infty} = \Delta(a/e,-a/d;q)_\infty. \tag{6.6}$$

Proof. The following integral is the $f = 0$ case of [18, p. 44, Eq. (2.10.18)] and the starting point of Askey's original proof:

$$\int_a^b \frac{(qt/a,qt/b;q)_\infty}{(dt,et;q)_\infty} d_q t = b(1-q)\frac{(q,bq/a,a/b,abde;q)_\infty}{(ad,ae,bd,be;q)_\infty}. \tag{6.7}$$

Substituting a with $-q^{-n}$, b with q^{-n} and d with $-d$ in the above equation, we obtain

$$\int_{-q^{-n}}^{q^{-n}} \frac{(-q^{n+1}t,q^{n+1}t;q)_\infty}{(-dt,et;q)_\infty} d_q t = \frac{2(1-q)(-q;q)_\infty^2(q,de;q)_\infty(q/de;q)_{2n}}{(q^2/d^2,q^2/e^2;q^2)_n(d^2,e^2;q^2)_\infty}.$$

Taking the limit $n \to +\infty$, we obtain

$$\int_{-\infty}^{\infty} \frac{d_q t}{(-dt,et;q)_\infty} = \frac{2(1-q)(q^2;q^2)_\infty^2(de,q/de;q)_\infty}{(q;q)_\infty(d^2,e^2,q^2/d^2,q^2/e^2;q^2)_\infty}.$$

Using the identity

$$(at;q)_\infty = (a/e;q)_\infty \sum_{n=0}^{\infty} \frac{(et;q)_n}{(q;q)_n}(a/e)^n,$$

we have

$$\begin{aligned}
&\int_{-\infty}^{\infty} \frac{(at;q)_\infty d_q t}{(-dt,et;q)_\infty} \\
&= (a/e;q)_\infty \sum_{n=0}^{\infty} \frac{(a/e)^n}{(q;q)_n} \int_{-\infty}^{\infty} \frac{d_q t}{(-dt,eq^n t;q)_\infty} \\
&= (a/e;q)_\infty \sum_{n=0}^{\infty} \frac{(a/e)^n 2(1-q)(q^2;q^2)_\infty^2(deq^n,q^{1-n}/de;q)_\infty}{(q;q)_n(q;q)_\infty(d^2,e^2q^{2n},q^2/d^2,q^{2-2n}/e^2;q^2)_\infty} \\
&= \frac{2(1-q)(q^2;q^2)_\infty^2(a/e,de,q/de;q)_\infty}{(q;q)_\infty(d^2,e^2,q^2/d^2,q^2/e^2;q^2)_\infty} \sum_{n=0}^{\infty} \frac{(a/d)^n}{(q;q)_n} q^{\binom{n}{2}} \\
&= \frac{2(1-q)(q^2;q^2)_\infty^2(a/e,-a/d,de,q/de;q)_\infty}{(q;q)_\infty(d^2,e^2,q^2/d^2,q^2/e^2;q^2)_\infty} \\
&= \Delta(a/e,-a/d;q)_\infty,
\end{aligned}$$

as desired. ∎

Taking the action of $E(b\theta)$ on (6.6), it follows that

$$\int_{-\infty}^{\infty} \frac{E(b\theta)\{(at;q)_\infty\}d_q t}{(-dt, et;q)_\infty} = \Delta E(b\theta)\{(a/e, -a/d;q)_\infty\},$$

which leads to the Askey beta integral.

7. Some of Ramanujan's identities

The Ramanujan beta integral is stated as follows [10]:

$$\int_0^\infty t^{x-1}\frac{(-at;q)_\infty}{(-t;q)_\infty}dt = \frac{\pi}{\sin \pi x}\frac{(q^{1-x}, a;q)_\infty}{(q, aq^{-x};q)_\infty}, \tag{7.1}$$

where $0 < q < 1$, $x > 0$, and $0 < a < q^x$.

Here we give the following extension:

Theorem 7.2. *We have*

$$\int_0^\infty t^{x-1}\frac{(-at, -bt;q)_\infty}{(-t, -abq^{-x-1}t;q)_\infty}dt = \frac{\pi}{\sin \pi x}\frac{(q^{1-x}, a, b;q)_\infty}{(q, aq^{-x}, bq^{-x};q)_\infty}. \tag{7.3}$$

Eq. (7.3) can be derived from (7.1) by multiplying $(aq^{-x};q)_\infty$ on both sides and then taking the action of $E(b\theta)$.

The following is a formula in Ramanujan's Lost Notebook, and its proofs are given by Andrews [5, 6]. Here we give a treatment in terms of parameter augmentation.

Theorem 7.4. *We have*

$$\sum_{n=0}^\infty \frac{q^n}{(-aq, -bq;q)_n} =$$
$$\frac{-a^{-1}\sum_{m=0}^\infty (-1)^m q^{\binom{m+1}{2}}(b/a)^m}{(-aq, -bq;q)_\infty} + (1+a^{-1})\sum_{m=0}^\infty \frac{q^{\binom{m+1}{2}}(-b/a)^m}{(-bq;q)_m}. \tag{7.5}$$

Setting $b = 0$ in the above theorem, we are led to the following lemma of considerable simplicity, which can be easily verified by the Euler expansion of $(a, q)_\infty$:

Lemma 7.6. *We have*

$$\sum_{n=0}^\infty (-aq^{n+1};q)_\infty q^n = -a^{-1} + a^{-1}(-a;q)_\infty. \tag{7.7}$$

We are now ready to give a derivation of (7.5) by parameter augmentation. First, we note that

$$\theta^n\{a^{-1}\} = (-q)^n(q;q)_n a^{-(n+1)}.$$

Therefore, we have

$$E(b\theta)\{-a^{-1}\} = -a^{-1}\sum_{n=0}^{\infty}(-1)^n q^{\binom{n+1}{2}} b^n a^{-n}. \tag{7.8}$$

Using the Leibniz formula for θ^n, we get

$$\theta^n\{a^{-1}(-a;q)_\infty\} \;=\; (-a;q)_\infty a^{-(n+1)} q^n \sum_{j=0}^{n}\frac{(q;q)_n}{(q;q)_j}(-1)^{n-j}a^j.$$

Then it can be verified that

$$E(b\theta)\{a^{-1}(-a;q)_\infty\} =$$

$$a^{-1}(-a;q)_\infty \sum_{m=0}^{\infty}(-b/a)^m q^{\binom{m+1}{2}}(-bq^{m+1};q)_\infty. \tag{7.9}$$

With (7.8) and (7.9), the action of $E(b\theta)$ on (7.7) leads to (7.5).

8. Two formulas for bilateral series

In this section we give a parameter augmentation treatment of Bailey's ${}_3\psi_3$ summation formula [18, p. 239], and remark that it is equivalent to q-Dixon's formula [4, 16, 18, 28].

Theorem 8.1. *We have*

$${}_3\psi_3\left(\begin{matrix} q/a, & q/b, & q/c \\ a, & b, & c\end{matrix}\;;q, abc/q^2\right) = \frac{(q, ab/q, bc/q, ac/q;q)_\infty}{(a,b,c,abc/q^2;q)_\infty}. \tag{8.2}$$

Setting $c=0$ in the above formula, we have the following lemma:

Lemma 8.3. *We have*

$$\sum_{n=-\infty}^{\infty}(-1)^n q^{\binom{n}{2}}\frac{(q/a,q/b;q)_n}{(a,b;q)_n}(ab/q)^n = \frac{(q,ab/q;q)_\infty}{(a,b;q)_\infty}. \tag{8.4}$$

Proof. Using the following identity [23]:

$$\sum_{k=-\infty}^{\infty} (-1)^k q^{\binom{k}{2}} \begin{bmatrix} 2n \\ n+k \end{bmatrix} = \delta_{0,n},$$

and by Heine's q-analogue of Gauss' summation formula, we have

$$\begin{aligned}
1 &= \sum_{n=0}^{\infty} \frac{(q/a, q/b; q)_n}{(q;q)_{2n}} (ab/q)^n \delta_{0,n} \\
&= \sum_{n=0}^{\infty} \frac{(q/a, q/b; q)_n}{(q;q)_{2n}} (ab/q)^n \sum_{k=-\infty}^{\infty} (-1)^k q^{\binom{k}{2}} \begin{bmatrix} 2n \\ n+k \end{bmatrix} \\
&= \sum_{k=-\infty}^{\infty} (-1)^k q^{\binom{k}{2}} \frac{(q/a, q/b; q)_k}{(q;q)_{2k}} (ab/q)^k \, {}_2\phi_1 \left(\begin{matrix} q^{k+1}/a, \; q^{k+1}/b \\ q^{2k+1} \end{matrix} ; q, ab/q \right) \\
&= \frac{(a,b;q)_\infty}{(q, ab/q; q)_\infty} \sum_{k=-\infty}^{\infty} (-1)^k q^{\binom{k}{2}} \frac{(q/a, q/b; q)_k}{(a,b;q)_k} (ab/q)^k,
\end{aligned}$$

which gives (8.4). ■

Rewriting (8.4) as follows:

$$\sum_{n=-\infty}^{\infty} q^{n^2-n} b^n \frac{(q/b;q)_n}{(b;q)_n} (q^{-n}a, q^n a; q)_\infty = \frac{(q;q)_\infty}{(b;q)_\infty} (a, ab/q; q)_\infty,$$

then (8.2) can be easily obtained by taking the action of $E(c\theta)$ on both sides of (8.4).

Using (1.1), we may rewrite (8.2) into the following form:

$$\begin{aligned}
&\sum_{n=-\infty}^{\infty} (-1)^n q^{(3n^2+n)/2} (q^{-n}a, q^n a, q^{-n}b, q^n b, q^{-n}c, q^n c; q)_\infty \\
= \; & \frac{(q, ab/q, bc/q, ac/q, a, b, c; q)_\infty}{(abc/q^2; q)_\infty}.
\end{aligned}$$

Substituting a, b, c with $q^{a+1}, q^{b+1}, q^{c+1}$ in the above identity, and then dividing both sides by $(q;q)_\infty^6$, we get

$$\begin{aligned}
&\sum_{n=-\infty}^{\infty} (-1)^n q^{(3n^2+n)/2} \frac{1}{(q;q)_{a+n}(q;q)_{a-n}(q;q)_{b-n}(q;q)_{b+n}(q;q)_{c-n}(q;q)_{c+n}} \\
&\qquad = \frac{(q;q)_{a+b+c}}{(q;q)_{a+b}(q;q)_{b+c}(q;q)_{a+c}(q;q)_a(q;q)_b(q;q)_c},
\end{aligned}$$

which can be rewritten as the q-Dixon's formula:

Theorem 8.5. *We have*

$$\sum_{n=-\infty}^{\infty} (-1)^n q^{(3n^2+n)/2} \begin{bmatrix} b+c \\ c+n \end{bmatrix} \begin{bmatrix} a+c \\ a+n \end{bmatrix} \begin{bmatrix} a+b \\ b+n \end{bmatrix} = \frac{(q;q)_{a+b+c}}{(q;q)_a\,(q;q)_b\,(q;q)_c}.$$

Starting with the finite form of Euler's pentagonal number theorem [23]:

$$\sum_{k=-\infty}^{\infty} (-1)^k q^{k(3k-1)/2} \begin{bmatrix} 2n \\ n+k \end{bmatrix} = \frac{(q;q)_{2n}}{(q;q)_n},$$

we may obtain the following lemma:

Lemma 8.6. *We have*

$$\sum_{n=-\infty}^{\infty} \frac{(q/a,q/b;q)_n}{(a,b;q)_n}(-ab/q)^n q^{n(3n-1)/2} = \frac{(q,ab/q;q)_\infty}{(a,b;q)_\infty}\sum_{n=0}^{\infty}\frac{(q/a,q/b;q)_n}{(q;q)_n}(ab/q)^n. \tag{8.7}$$

Using Heine's q-analogue of Gauss' summation formula at the last step, the proof of (8.7) is given below:

$$\begin{aligned}
&\sum_{n=0}^{\infty}\frac{(q/a,q/b;q)_n}{(q;q)_n}(ab/q)^n \\
&= \sum_{n=0}^{\infty}(q/a,q/b;q)_n\,(ab/q)^n \sum_{k=-\infty}^{\infty}(-1)^k\frac{q^{k(3k-1)/2}}{(q;q)_{n+k}(q;q)_{n-k}} \\
&= \sum_{k=-\infty}^{\infty}(-1)^k\frac{(q/a,q/b;q)_k}{(q;q)_{2k}}\,q^{k(3k-1)/2}\,(ab/q)^k \\
&\qquad \times {}_2\phi_1\left(\begin{matrix} q^{k+1}/a,\ q^{k+1}/b \\ q^{2k+1}\end{matrix};q,ab/q\right).
\end{aligned}$$

Using (1.1), the above identity can be rewritten as

$$\sum_{n=-\infty}^{\infty}\frac{(q/b;q)_n}{(b;q)_n}(b/q)^n q^{2n^2}(q^{-n}a,aq^n;q)_\infty,$$

$$= \frac{(q;q)_\infty}{(b;q)_\infty}\sum_{n=0}^{\infty}\frac{(q/b;q)_n}{(q;q)_n}(b/q)^n(-1)^n q^{\binom{n+1}{2}}(ab/q,q^{-n}a;q)_\infty.$$

Taking the action of $E(c\theta)$, we obtain the following transformation formula.

Theorem 8.8. *We have*

$$\sum_{n=-\infty}^{+\infty} \frac{(q/a, q/b, q/c; q)_n}{(a, b, c; q)_n} (abc)^n q^{n^2-2n}$$
$$= \frac{(q, ab/q, ac/q, bc/q; q)_\infty}{(a, b, c, abc/q^2; q)_\infty} \sum_{n=0}^{\infty} \frac{(q/a, q/b, q/c; q)_n}{(q, q^3/abc; q)_n} q^n. \tag{8.9}$$

Acknowledgements. This work was done under the auspices of the U.S. Department of Energy, the National Science Foundation of China, and the Stroock Foundation. The authors would like to thank G.E. Andrews, W.C. Chu, Q.H. Hou, M.E.H. Ismail, J.D. Louck and G.-C. Rota for valuable discussions. We are grateful for the referees for many important suggestions leading to improvement of an earlier version of this paper.

References

[1] G. E. Andrews, On a transformation of bilateral series with applications, *Proc. Amer. Math. Soc.* **25** (1970), 554–558.

[2] G. E. Andrews, On the foundations of combinatorial theory V. Eulerian differential operators, *Studies in Appl. Math.* **50** (1971), 345–375.

[3] G. E. Andrews, Applications of basic hypergeometric function, *SIAM Rev.* **16** (1974), 441–484.

[4] G. E. Andrews, "Problems and prospects for basic hypergeometric functions," in: *The Theory and Application of Special Functions*, R. Askey, ed., Academic Press, New York, 1975, pp. 191–224.

[5] G. E. Andrews, An introduction to Ramanujan's "Lost" Notebook, *Amer. Math. Monthly* **86** (1979), 89–108.

[6] G. E. Andrews, Ramanujan's "Lost" Notebook I, Partial θ-function, *Adv. in Math.* **41** (1981), 137–172.

[7] G. E. Andrews, Ramanujan's "Lost" Notebook III, The Rogers–Ramanujan continued fraction, *Advances in Math.* **41** (1981), 186–208.

[8] G. E. Andrews and R. Askey, A simple proof of Ramanujan's summation of ${}_1\psi_1$, *Aequationes Math.* **18** (1978), 333–337.

[9] R. Askey and M. E. H. Ismail, The very well poised ${}_6\psi_6$, *Proc. Amer. Math. Soc.* **77** (1979), 218–222.

[10] R. Askey, Ramanujan's extensions of the gamma and beta functions, *Amer. Math. Monthly* **87** (1980), 346–359.

[11] R. Askey, q-extension of Cauchy's form of the beta integral, *Quart. J. Math. Oxford* **32** (2), (1981), 255–266.

[12] R. Askey, The very well poised ${}_6\psi_6$ II, *Proc. Amer. Math. Soc.* **90** (1984), 575–579.

[13] R. Askey and J. Wilson, Some basic hypergeometric orthogonal polynomials that generalize Jacobi polynomials, *Memoirs. Amer. Math. Soc.* **319** (1985).

[14] W. N. Bailey, Series of hypergeometric type which are infinite in both directions, *Quart. J. Math.* **7** (1936), 105–115.

[15] W. N. Bailey, On the basic bilateral hypergeometric series ${}_2\psi_2$, *Quart. J. Math. Oxford* **1** (2), (1950), 194–198.

[16] W. N. Bailey, On the analogue of Dixon's theorem for bilateral basic hypergeometric series, *Quart. J. Math. Oxford* **1** (2), (1950), 318–320.

[17] J. Cigler, Operatormethoden für q-Identitäten, *Monatsh. für Math.* **88** (1979), 87–105.

[18] G. Gasper and M. Rahman, *Basic Hypergeometric Series*, Cambridge University Press, Cambridge, 1990.

[19] J. Goldman and G.-C. Rota, "The number of subspaces of a vector space," in: *Recent Progress in Combinatorics*, W. Tutte, ed., Academic Press, New York, 1969, pp. 75–83.

[20] J. Goldman and G.-C. Rota, On the foundations of combinatorial theory IV: Finite vector spaces and Eulerian generating functions, *Studies in Appl. Math.* **49** (1970), 239–258.

[21] I. P. Goulden and D. M. Jackson, *Combinatorial Enumeration*, John Wiley & Sons, 1983.

[22] W. Hahn, Über orthogonalpolynome, die q-differenzengleichungen genügen, *Math. Nachr.* **2** (1949), 4–34.

[23] P. Paule, On identities of the Rogers–Ramanujan type, *J. Math. Anal. Appl.* **107** (1985), 255–284.

[24] L. J. Rogers, Third memoir on the expansion of certain infinite product, *Proc. London Math. Soc.* **26** (1895), 15–32.

[25] S. Roman, More on the umbral calculus, with emphasis on the q-umbral calculs, *J. Math. Anal. Appl.* **107** (1985), 222–254.

[26] G.-C. Rota, *Finite Operator Calculus*, Academic Press, New York, 1975.

[27] L. J. Slater and A. Lakin, Two proofs of the ${}_6\psi_6$ summation theorem, *Proc. Edin. Math. Soc.* **9** (1956), 116–121.

[28] H. S. Wilf and D. Zeilberger, An algorithmic proof theory for hypergeometric (ordinary and "q") multisum/integral identities, *Invent. Math.* **108** (1992), 575–633.

William Y.C. Chen
T-7, Mail Stop B284
Los Alamos National Laboratory
Los Alamos, NM 87545
USA
email: chen@t7.lanl.gov
and
Nankai Institute of Mathematics
Nankai University
Tianjin 300071
P. R. China
email: chen@public.tpt.tj.cn

Zhi-Guo Liu
Department of Mathematics
Xinxiang Education College
Xinxiang, Henan 453000
P. R. China

Unities and Negation

Henry Crapo and Claude Le Conte de Poly-Barbut

For Gian-Carlo Rota,
on the occasion of his 2^nth birthday,
for some n

1. Introduction

We define a 'unity' as a preordered set with a certain separation property, imitating the preorder induced on the irreducible elements of a finite lattice. As proved in [C], any finite lattice L has a representation $L \simeq \underline{\mathbf{2}}^I$ as the set of all maps from an appropriate preordered structure I, which we call the *negation of the unity of irreducibles* of L, into a small unity that we call $\underline{\mathbf{2}}$. In this paper, we characterize unities of irreducibles of lattices, show how arbitrary unities can be constructed from such unities of irreducibles, and discuss possible definitions of maps between unities. In this way we lay some groundwork necessary for a categorical study of unities, of the negation operator, and of functors between the category $\mathcal{U}$ of unities, the category $\mathcal{P}$ of partially ordered sets, and the category $\mathcal{L}$ of lattices.

2. Representation theorems

We recall that a finite Boolean algebra is representable as the set $\underline{\mathbf{2}}^P$ of functions from a finite set P to the two-element set $\underline{\mathbf{2}} = \{0,1\}$, in the pointwise order; that is, functions f, g are in the relation $f \leq g$ if and only if

$$(\forall a \in P) \;\; f(a) = 1 \;\Rightarrow\; g(a) = 1.$$

The set P has natural representations in the Boolean algebra $\underline{\mathbf{2}}^P$ both as its set of atoms and as its set of coatoms (see Figure 1). Recall the well-known analogue for distributive lattices, due to Garrett Birkoff, with subsequent work by M. H. Stone [S] and H. Priestley [P]. A finite distributive lattice is representable as the set of *order-preserving* functions from a finite partially-ordered set P to the two-element set $\underline{\mathbf{2}} = \{0, 1\}$, again in the pointwise order. In this representation $D \simeq \underline{\mathbf{2}}^P$ of a finite distributive lattice D, the *opposite* of the partially-ordered set P occurs naturally as a sub-poset of D both as the set of join-irreducible elements of D and as the set of meet-irreducible elements of D (see Figure 2, drawn as a sublattice of the Boolean algebra). A map $f \in \underline{\mathbf{2}}^P$ is meet-irreducible if and only if $f^{-1}(0)$ is a principal order ideal of P, and is join-irreducible if and only if $f^{-1}(1)$ is a principal order filter of P.

In order to construct such a representation theorem $L \simeq \underline{\mathbf{2}}^P$ for arbitrary finite lattices, the appropriate domain P is obtained by inverting the sets of join-irreducible and meet-irreducible elements of L and combining them into a single preordered set, often with considerable overlap. To this end, we suggest a generalization of the notion of partially-ordered set, called a *unity* [C], or L-space [U], which will take into account the influence of both join- and meet-irreducible elements of L. Figure 3 will provide a rough visual idea of what is involved; the sections which follow will gradually unfold its meaning.

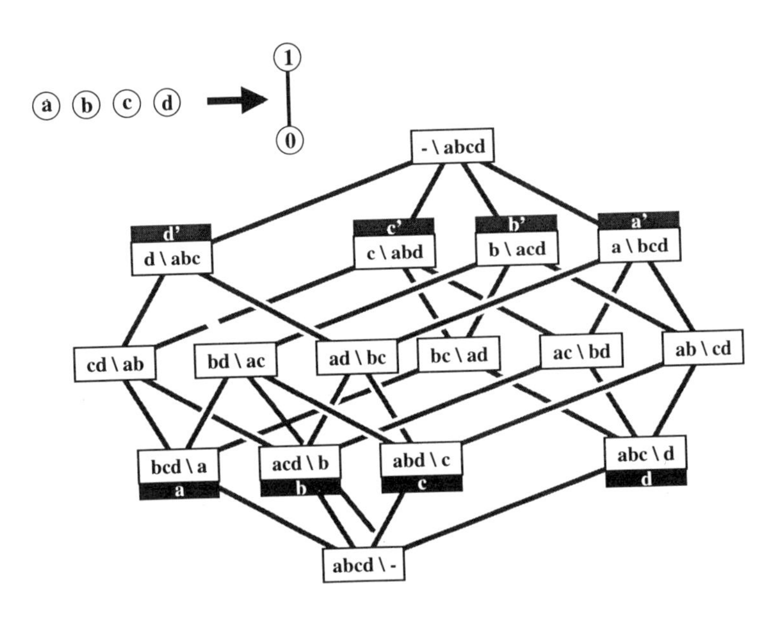

Figure 1. Representation of a finite Boolean algebra.

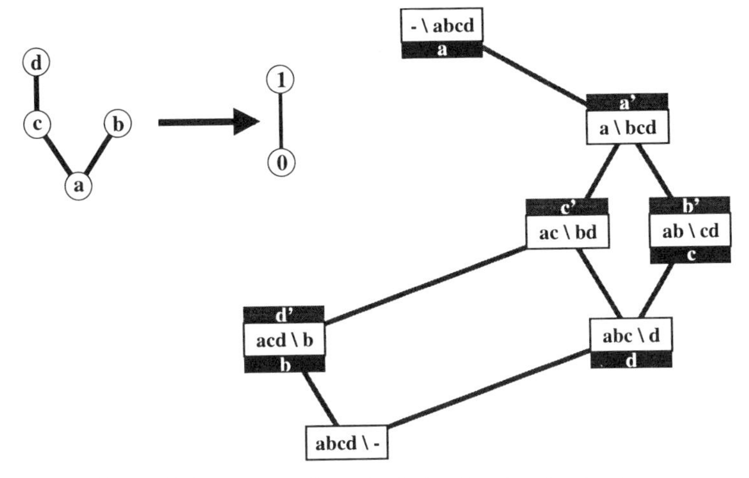

Figure 2. Representation of a finite distributive lattice.

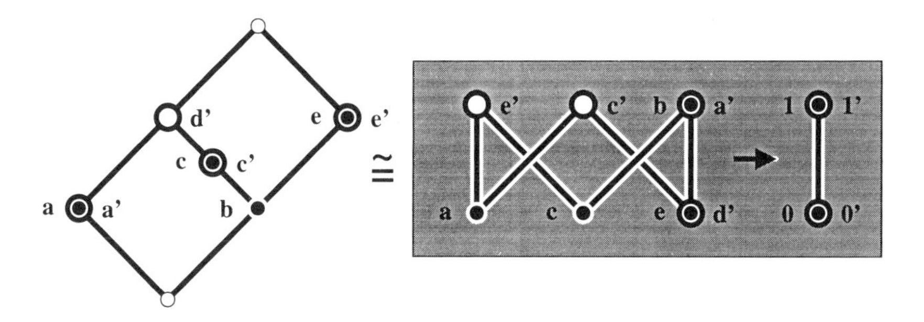

Figure 3. An example of the representation theorem for finite lattices.

3. Unities

A *unity* I is any preorder $\leq$ on the formally-disjoint union $I = J \oplus M$ of two sets J and M, such that the preorders induced on J and on M are antisymmetric (partial orders), and with the following separation property (see Figure 4):

$$\text{for any elements } x, y \in I, \text{ if } x \not\leq y, \text{ then there exist elements} \\ a \in J, \;\; a' \in M \text{ such that } a \leq x, \;\; y \leq a', \;\; a \not\leq a'. \tag{1}$$

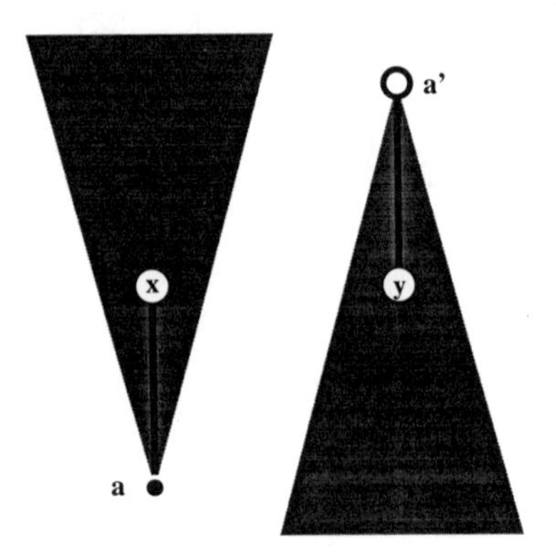

Figure 4. The separation property for a unity.

The separation property assures that every element of a unity is a supremum of elements in J and an infimum of elements in M, as we show below in Proposition 2.

Any preorder induces a partial order on its set of equivalence classes of elements. In the case of a unity I, we call this associated poset Supp (I), the *support* of the unity.

Examples of unities are shown in Figures 5 and 6. We use unprimed letters $a, b, \ldots$ to denote elements of J, primed letters $a', b', \ldots$ to denote elements of M. The 'prime' is *not* intended to represent an operator; there is no necessary connection between a and a'. In our figures, elements of J are shown as solid black dots, elements of M as larger circles.

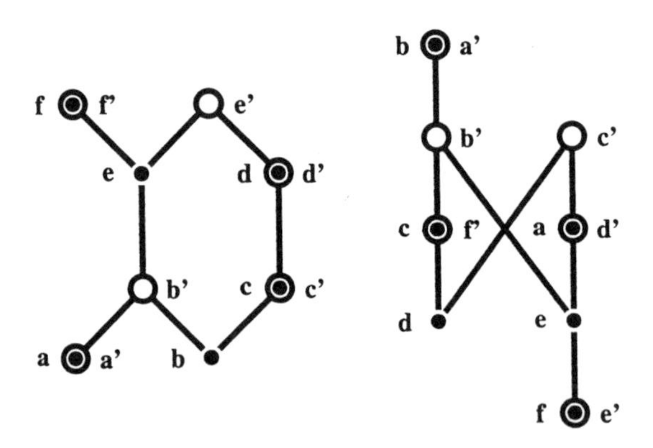

Figure 5. A pair of unities.

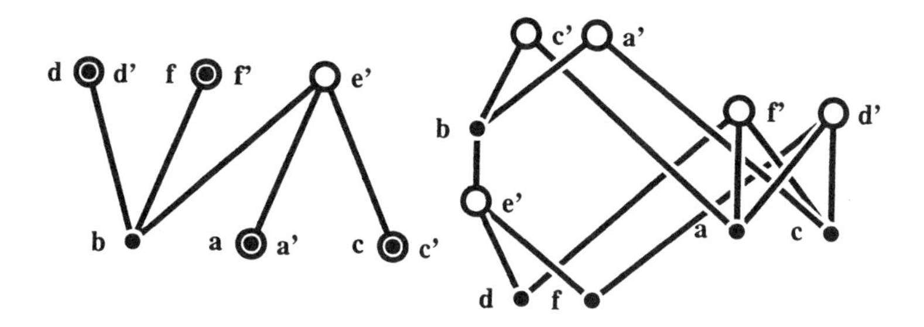

Figure 6. Another pair of unities.

In Sections 4 and 5 we shall characterize those unities which are unities of irreducibles of finite lattices, and shall show how all unities may be constructed.

4. Irreducible elements

An element x in a partially ordered set P is *join-irreducible* if and only if x is not the supremum (least upper bound) of the set $\{y \in P;\ y < x\}$ of elements of P strictly less than x. Dually, x is *meet-irreducible* if and only if x is not the infimum (greatest lower bound) of the set of elements of P strictly greater than x. (Of course if the set $\{y \in P;\ y < x\}$ has *no* supremum, then x is join-irreducible.)

Proposition 1. *An element x in a partially ordered set P with at least two elements is join-irreducible if and only if there is an element $x' \in P$ such that $x' \not\geq x$ but $x' \geq y$ for all elements y such that $y < x$. A dual condition holds for meet-irreducibility.*

Proof. Any element x is an upper bound for the set $P_x = \{y \in P;\ y < x\}$. If x is join-irreducible, x is not the join of the set P_x. There must be another upper bound x' for P_x, with $x \not\leq x'$. Conversely, if there is an element $x' \not\geq x$, an upper bound for the set P_x, x is not the join of any subset $A \subseteq P_x$ in P, and is thus not the join of any subset not containing x. ■

Proposition 2. *Given a poset P, and a pair of subsets J and M of P such that $P = J \cup M$, then the preorder induced on $J \oplus M$ is a unity if and only if every element of P is a meet of elements in M, and is a join of elements in J.*

Proof. An element $x \in P$ is not a join of elements of J if and only if x is not the join of the set $K = \{a \in J; a \leq x\}$. That is, if and only if there is an upper bound $y \in P$ for K such that $x \not\leq y$. In that event, the separation of x from y is not assured: there is no element $a \in J$ with $a \leq x, a \not\leq y$. An analogous argument proves that the separation property is violated if x is not a meet of elements of M.

Conversely, if every element of P is a join of a subset of J then for any pair of elements $x \not\leq y$ in P, $x = \bigvee K$ for some subset $K \subseteq J$, and y is not an upper bound for K. So there is an element $a \in K$ with $a \leq x, a \not\leq y$. An analogous argument shows that since y is a meet of elements of M, there is an element $a' \in M$ such that $a \not\leq a', a \leq x, y \leq a'$. ■

For an element x in a finite lattice, the set $\{y \in P;\ y < x\}$ has a join, say x', with $x' \leq x$. If $x' = x$, then x is join-reducible. Otherwise, x is join-irreducible, and x' is the unique element covered by x in L. (Note that the least element 0 in a finite lattice is join-reducible, being the join of the empty set, and that the greatest element 1 is meet-reducible.) There is an analogous statement which holds for any finite poset.

Theorem 3. *Every element x in a finite poset P is the join of the set J_x, where*

$$J_x = \{y \in P;\ y \text{ join-irreducible, and } y \leq x\}.$$

Proof. Assume that the statement is false for some poset P and that $x \in P$ is a minimal element such that x is not the join of the corresponding subset J_x. Since x is an upper bound for J_x, there exists an element $x' \in P$ such that $x \not\leq x'$, x' an upper bound for J_x. If x' were also an upper bound for the set $P_x = \{y \in P;\ y < x\}$, then x would be join-irreducible, and would be the join of the set J_x. Since this is not the case, there is an element $y \in P$ with $y < x, y \not\leq x'$. By the minimality of x, $y = \bigvee J_y$. Since $J_y \subseteq J_x$, x' is also an upper bound for J_y, and $y \leq x'$, a contradiction. The proposition is true. ■

Proposition 4. *For any finite poset P, the formally disjoint union of the set J of join-irreducible elements of P and the set M of meet-irreducible elements of P, in the preorder induced by the order on P, is a unity.*

Proof. Let x and y be elements of P such that $x \not\leq y$. Since $x = \bigvee J_x$, where J_x is the set of join-irreducible elements $a \in P$ such that $a \leq x$, it follows that y is not an upper bound for J_x, and there is a join-irreducible element $a \in J$ such that $a \not\leq y$. Similarly, a is not a lower bound for M_y, so there is an element $a' \in M$, with $y \leq a'$ but $a \not\leq a'$. Thus $J \oplus M$, in the preorder induced by the order on P, satisfies the separation property (1), and is a unity. ■

Definition 5. The unity defined by Proposition 4, we denote by Irred (P), the *unity of irreducibles* of the poset P.

Proposition 6. *In a unity $I = (J \oplus M, \leq)$, every join-irreducible element of Supp (I) is in J, every meet-irreducible element is in M.*

Proof. Assume an element x is join-irreducible in Supp (I). Then there is an element $y \in I, x \not\leq y$, with $z \leq y$ for all elements $z < x$. By the separation property of I, there is an element $a \in J$, with $a \leq x, a \not\leq y$. This element a cannot be strictly less than x, so x is itself in J. Dually, every meet-irreducible element is in M. ■

Definition 7. For any finite poset P, let $P^{[2]}$ be the *double* of P, the unity obtained by setting $J = M = P$.

Proposition 8. *A unity $I = (J \oplus M, \leq)$ has support poset P if and only if Irred $P \subseteq I$ and for all $x \in P, x \in J \cup M$.*

Proof. We know by Proposition 6 that all join- and meet-irreducible elements of P are in J and M, respectively. For any larger choice of subsets J', M' with $J \subseteq J' \subseteq P$ and $M \subseteq M' \subseteq P$, the pair J', M' will satisfy the separation property (1), and will form a unity with support P. ■

Corollary 9. *For any unity $I = (J \oplus M, \leq)$ with support poset P,*

$$\textit{Irred}\,(P) \subseteq I \subseteq P^{[2]}.$$

There exist unities which are not representable as unities of irreducibles of a poset. Figure 7 lists the unities supported on posets of 0, 1, and of 2 elements. Only three of these unities are unities of irreducibles of posets, and one of these is empty!

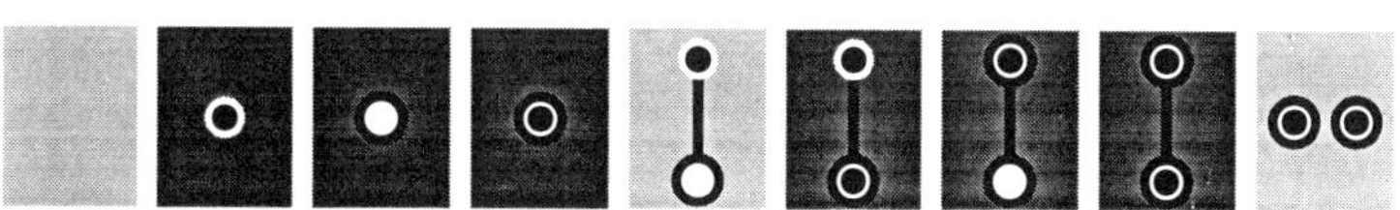

Figure 7. Unities supported on small posets: the 1st, 5th and 9th are unities of irreducibles.

Proposition 10. *A unity I is the unity of irreducibles of a poset if and only if $I = \mathit{Irred}\,(\mathit{Supp}\,(I))$.*

Proof. Say $I = \text{Irred}\,(P)$ for some poset P, and let $C = \text{Supp}\,(I)$. We show that P and C have the same irreducible elements. An element x is join-irreducible in C if and only if there is an element $x' \in C$, with $x \not\leq x'$, but $y \leq x'$ for all $y \in C$, $y < x$. Since every element $z < x$ in P is a join of join-irreducible elements $< x$, it is also true that $z \leq x'$, so x is join-irreducible in P. Conversely, if x is join-irreducible in P, there is an element $x' \in P$ such that $y \leq x'$ for all $y < x$, and thus for all irreducible elements $y < x$. Since x' is a meet of meet-irreducible elements a', by the dual of Theorem 3, and since $x \not\leq x'$, there is a meet-irreducible element $a' \in C$ such that $x' \leq a', x \not\leq a'$. So x is join-irreducible in C. ∎

5. Completion

To every unity $I = I(J, M, \leq)$ we associate its lattice *completion* Latt (I) via the usual Galois connection.

A pair (A, B) of subsets $A \subseteq J$, $B \subseteq M$ are *closed* , and the pair (A, B) is said to be a *node*, if and only if

$$B = \sigma(A), \quad A = \tau(B);$$

in the Galois connection

$$\begin{aligned} \sigma: \; A \subseteq J &\mapsto \{y \in M \mid (\forall x \in A) \;\; x \leq y\}, \\ \tau: \; B \subseteq M &\mapsto \{x \in J \mid (\forall y \in B) \;\; x \leq y\}. \end{aligned}$$

A node (A, B) is thus a maximal totally-related pair of subsets:

$$\begin{gathered} \forall x \in A, \forall y \in B, x \leq y \\ \forall x \notin A, \exists y \in B, x \not\leq y \\ \forall y \notin B, \exists x \in A, x \not\leq y\,. \end{gathered}$$

For any subset $C \subseteq J$, the passage via σ, then τ, with $B = \sigma(C), A = \tau(B)$ yields a node (A, B). The map $\tau \circ \sigma$ is a closure operator on 2^J, making the set of nodes on I a complete lattice in the order

$$(A, B) \leq (C, D) \quad \text{iff} \quad A \subseteq C \quad \text{iff} \quad D \subseteq B.$$

Definition 11. This lattice is called the *completion* Latt (I) of the unity I.

Note that if a unity I is a doubled poset $= P^{[2]}$, then the completion Latt (I) is simply the Dedekind–MacNeille completion of the poset P. Furthermore, any finite lattice is the Dedekind–MacNeille completion of the support of its unity of irreducibles (see Theorem 17). These principles are illustrated in Figures 8 and 9.

Proposition 12. *Any unity I has a natural embedding in its completion Latt (I), given by the map $x \longrightarrow N_x = (A_x, B_x)$, where*

$$A_x = \{a \in J;\ a \leq x\},\ \textit{and}\ B_x = \{a' \in M;\ x \leq a'\}.$$

Proof. If $x \leq y$ in I, then $A_x \subseteq A_y$, so $N_x \leq N_y$. If $x \not\leq y$, then there is an element $a \in J$ with $a \leq x, a \not\leq y$, so $A_x \not\subseteq A_y$, and $N_x \not\leq N_y$. ■

Definition 13. The image N_x of an element x in this embedding is called its *principal node.*

Proposition 14. *A node (A, B) in the completion $L =$ Latt (I) of a unity I is join-irreducible in L if and only if it is the principal node of a join-irreducible element x in Supp (I). The analogous property holds for meet-irreducibles.*

Proof. Say (A, B) is a node of I, join-irreducible in L. Then (A, B) covers a unique node (A', B') in L, with $A' \subset A$ and $B \subset B'$ strict inclusions of subsets. Let x be any element of the difference set $A \backslash A'$. For the principal node $N_x = (A_x, B_x)$, $N_x \leq (A, B)$, but $N_x \not\leq (A', B')$, so $N_x = (A, B)$. For any element $a \in J$ with $a < x$, $N_a < N_x$, so $N_a \leq (A', B')$. On the other hand, $A' \subseteq \{a \in J;\ a < x\}$, so $\bigvee_{a \in A'} N_a = (A', B')$. Since B is a proper subset of B', there in an element y of M, with $x \not\leq y$, that is an upper bound for A', and thus for all join-irreducible elements $a < x$. It follows that the element x itself is join-irreducible. Conversely, if an element $x \in I$ is join-irreducible, x is an upper bound, but not a least upper bound, for the set $A' = \{a \in J;\ a < x\}$. It follows that A' is closed, forming a node (A', B'), say, that is the unique node covered by the principal node N_x in Irred (L). ■

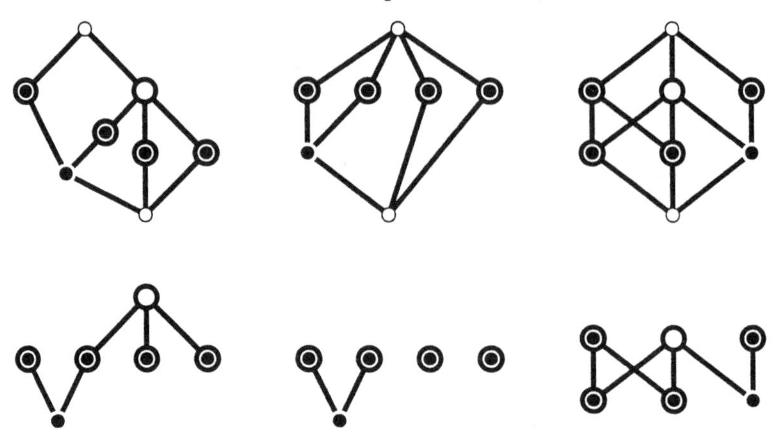

Figure 8. Two lattices and a non-lattice poset, with their unities of irreducibles.

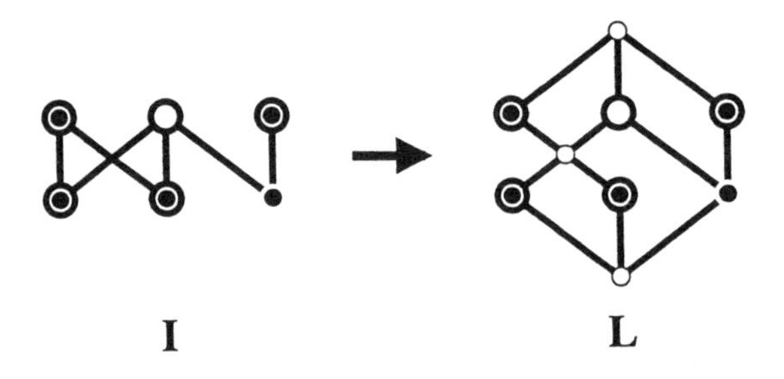

Figure 9. A unity and its lattice completion.

Corollary 15. *For any finite poset P, Latt $(P^{[2]})$ = Latt (Irred (P)).* ■

Theorem 16. *A unity I is isomorphic to the unity of irreducibles of a finite lattice if and only if it is the unity of irreducibles of a finite poset, if and only if I = Irred (Supp (I)).*

Proof. By Proposition 10, it suffices to show that if I = Irred (P) for some poset P, then I = Irred (Latt (P)). But by Proposition 14, Irred (P) = Irred (Latt (P)). ■

All unities with a given completion L are included between the smallest such unity, Irred (L), and the largest such unity, the preorder $L^{[2]}$ obtained by doubling the lattice L.

Theorem 17. *Let P be a finite partially ordered set, and I = Irred (P) its unity of irreducibles. Then the completion L = Latt (I) is isomorphic to the Dedekind–MacNeille completion of the poset P, and I is isomorphic to the unity Irred (L) of irreducibles of the lattice Latt (I).*

Proof. The Dedekind–MacNeille completion of P is the completion Latt $(P^{[2]})$, which is in turn equal to Latt (Irred (P)), by Corollary 15. ■

Corollary 18. *Two posets P and Q have isomorphic MacNeille completions if and only if their unities of irreducibles are isomorphic, if and only if Q is isomorphic to a subposet $P' \subseteq$ Latt (Irred (P)) satisfying*

$$\mathit{Supp}\,(\mathit{Irred}\,(P)) \subseteq P' \subseteq \mathit{Latt}\,(\mathit{Irred}\,(P)) \simeq \mathit{MacNeille}\,(P).$$

6. Unities and relations

We define the unity of a relation. We shall assume in what follows that every relation $\rho \subseteq J \times M$ in question is *reduced*, that is:

(a) no pair a, b in J are related via ρ to exactly the same elements of M,

(b) no pair a', b' in M are related via ρ^{-1} to exactly the same elements of J.

This assumption is not essential, but it assures that the unity induces an order, rather than simply a preorder, on the sets J and M.

For any pair of sets J, M, the product $(J \oplus M) \times (J \oplus M)$ is expressible as a disjoint union:

$$(J \oplus M) \times (J \oplus M) \;=\; (J \times J) \oplus (J \times M) \oplus (M \times M) \oplus (M \times J)\,.$$

Thus any preorder on $J \oplus M$ is the sum of (1) a preorder on J, (2) a preorder on M, (3) a relation from J to M, and (4) a relation from M back to J.

Definition 19. For any binary relation $\rho \subseteq J \times M$, the *unity of the relation* ρ is the largest preorder $I(\rho)$ on $J \oplus M$ compatible with the restriction

$$I(\rho) \mid_{J \times M} \;=\; \rho.$$

It is not clear apriori that such a largest preorder exists. We prove this below in Theorem 21. An example of this construction is in Figure 10. Starting with the relation on the left, imagine that the elements $x \in J$ will rise as far as possible, being blocked only by those elements $y' \in M$ to which they are related in the relation ρ. Thus the element c' descends, merging with the element b, its only relatum in J, while the element c rises, being unrelated to elements of M, to become the top element of the unity.

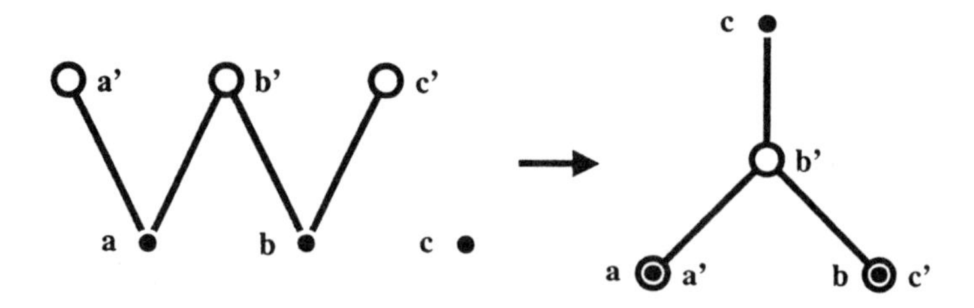

Figure 10. Passing from a relation ρ to its unity $I(\rho)$.

In attempting to build a unity $I \subseteq (J \oplus M)^2$ such that $I(\rho) \mid_{J \times M} = \rho$, we encounter the following constraints:

(1) In $J \times J$, an ordered pair $a \leq b$ can be in the preorder on J only if the required transitivity of the expression $a \leq b \leq b'$, for some $b \leq b'$ in ρ, does not force a pair $a \leq b'$ not in ρ.
(2) In $M \times M$, an ordered pair $a' \leq b'$ can be in the preorder on M only if the required transitivity of the expression $a \leq a' \leq b'$, for some $a \leq a'$ in ρ, does not force a pair $a \leq b'$ not in ρ.
(3) In $M \times J$, an ordered pair $a' \leq a$ can be in the relation from M back to J only if no pairs $b \leq a'$ and $a \leq b'$ in ρ force a pair $b \leq b'$ not in ρ, by transitivity of $b \leq a' \leq a \leq b$.

Surprisingly, if a pair is not explicitly *excluded* by one of the above conditions, it can be *included*, and the resulting preorder is clearly maximum among those with the given restriction ρ to $J \times M$.

Definition 20. (Equivalent to Definition 19) For any binary relation $\rho \subseteq J \times M$, let $I(\rho)$ be defined as follows, in the products:

$$\begin{aligned} J \times M: &\quad a \leq b' \text{ iff } a\rho b'; \\ J \times J: &\quad a \leq b \text{ iff } (\forall b' \in M)\; b\rho b' \Rightarrow a\rho b'; \\ M \times M: &\quad a' \leq b' \text{ iff } (\forall a \in J)\; a\rho a' \Rightarrow a\rho b'; \\ M \times J: &\quad a' \leq b \text{ iff } (\forall a \in J, b' \in M)\; (a\rho a' \;\&\; b\rho b') \Rightarrow a\rho b'. \end{aligned}$$

Theorem 21. *$I(\rho)$ is a unity.*

Proof. This is proven in [C]. First, we verify eight cases of transitivity. For example, to prove that for all $a, c \in J$ and all $b' \in M$,

$$a \leq b' \leq c \;\Rightarrow\; a \leq c,$$

we observe that:

$$\begin{array}{ll} & a \leq b' \leq c \\ \Rightarrow & (a\rho b' \text{ and } ((\forall c' \in M)\ (a\rho b' \text{ and } c\rho c') \Rightarrow a\rho c')) \\ \Rightarrow & ((\forall c' \in M)\ c\rho c' \Rightarrow a\rho c') \\ \Rightarrow & a \leq c \end{array}$$

If $x \not\leq y$ for a pair of elements x, y in $I(\rho)$, then by negating the corresponding line of the definition, we obtain the required separating elements in J and M. ■

Theorem 22. $\forall I\ \exists \rho \quad I = I(\rho)$ *That is, every unity is the unity of a relation.*

Proof. Given I, let ρ be the restriction of $I \subseteq (J \oplus M)^2$ to the product $J \times M$. The details of the proof are given in [C]. ■

7. Negation

In the representation theorems $L \simeq \underline{\mathbf{2}}^P$ for different classes of finite lattices L, the appropriate domain P is obtained by inverting the sets of join- and meet-irreducibles of L and combining them into a single ordered set, often with considerable overlap. We now describe this general procedure. Since every unity I on $J \oplus M$ is the unity of the relation obtained by restricting I to the product $J \times M$, any operation on relations will induce an operation on unities. Thus, the operations *union* and *intersection* of unities are inherited from the set-theoretic operations of union and intersection of relations. Of particular significance is the operation inherited from complementation of relations, which we call 'negation' of unities. We now describe this operation in some detail. See Figure 11.

Definition 23. The *negation* I^o of a unity I is the unity of the *complementary* relation

$$J \times M \setminus (I \mid_{J \times M}).$$

Since $I^{oo} = I$ for any unity I, we may speak of I and I^o as *opposite* unities, being negatives of one another.

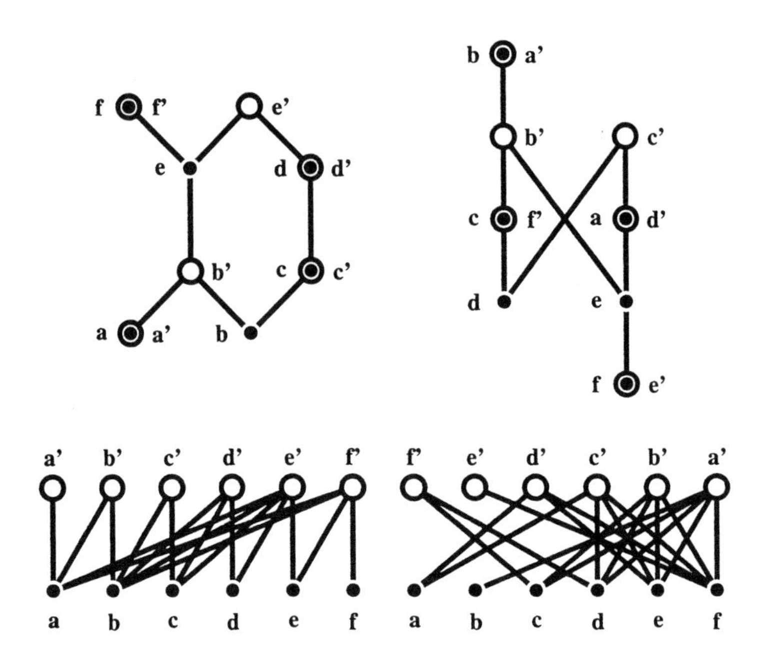

Figure 11. Opposite unities, with their complementary relations.

Two examples of pairs of opposite unities are shown in Figures 5 and 6. In these figures, the four unities are unities of irreducibles of lattices. But the opposite of a unity Irred (L) of a poset L need not be a unity of irreducibles. See Figure 13.

Proposition 24. *For any unity I, the partial orders $I|_{J\times J}$ and $I^o|_{J\times J}$ are anti-isomorphic via the identity map on J, as are $I|_{M\times M}$ and $I^o|_{M\times M}$ via the identity map on M.*

Proof. By negation of the relation in Definition 20. ■

Proposition 25. *For any unity I, and any pair (a', a) in $M \times J$,*

$$\begin{aligned} a' \leq^o a \quad & \textit{iff} \quad \forall c \in J,\ a \leq c \textit{ or } c \leq a' \\ & \textit{iff} \quad \forall c' \in M,\ a \leq c' \textit{ or } c' \leq a' \end{aligned}$$

Proof. Both conditions expand to

$$\begin{aligned} & \forall c \in J, c' \in M,\quad c \not\leq c' \text{ or } a \leq c' \text{ or } c \leq a' \\ \Leftrightarrow\ & \forall c \in J, c' \in M,\quad c \leq^o c' \text{ or } a \not\leq^o c' \text{ or } c \not\leq^o a'\,. \end{aligned}$$

This is equivalent to $a' \leq^o a$ because

$$a' \leq^o a \Leftrightarrow \forall c \in J, c' \in M \ \ ((c \leq^o a' \text{ and } a \leq^o c') \Rightarrow c \leq^o c')$$

■

Definition 26. A pair of elements $a \in J, a' \in M$ are *pseudocomplementary* in a unity I if and only if $a \not\leq a'$ and a, a' satisfy the condition stated in Proposition 25.

Corollary 27. *A pair of elements $a \in J, a' \in M$ are equivalent in the negation I^o of a unity I if and only if they are pseudocomplementary in I.*

8. Dual lattices

The following notion of duality is in the sense of Brian Davey *et al* [D1, D2], and is not to be confused with a simple inversion of order in a lattice.

Definition 28. The *dual* L^* of a lattice L is the completion Latt $(I^o(L))$ of the opposite of the unity of irreducibles of L.

Figure 12 illustrates such a dual pair of lattices. A finite lattice L need not be isomorphic to its double dual L^{**}. In Figure 13, the elements c, c' are reducible in L^*. So only the upper diamond-shaped region of L will remain in L^{**}. The unity I^o is of course *not* a unity of irreducibles.

Problem 29. For any finite lattice L, are there natural embeddings of L^{**} in L?

Theorem 30. *Duality of lattices exchanges products of chains with sums of chains.*

Proof. Let C_i be a chain of $k_i + 1$ elements $p_{i,0} < \cdots < p_{i,k_i}$, length k_i, for $i = 1, \cdots, n$. Let D be the distributive lattice $C_1 \times \cdots \times C_n$, with elements of the form

$$(p_{1,t_1}, \ \cdots, \ p_{n,t_n}).$$

The join-irreducible elements of D are of the form

$$a_{i,t} \ := \ (p_{1,0}, \ \cdots, \ p_{i-1,0}, \ p_{i,t}, \ p_{i+1,0}, \ \cdots, \ p_{n,0})$$

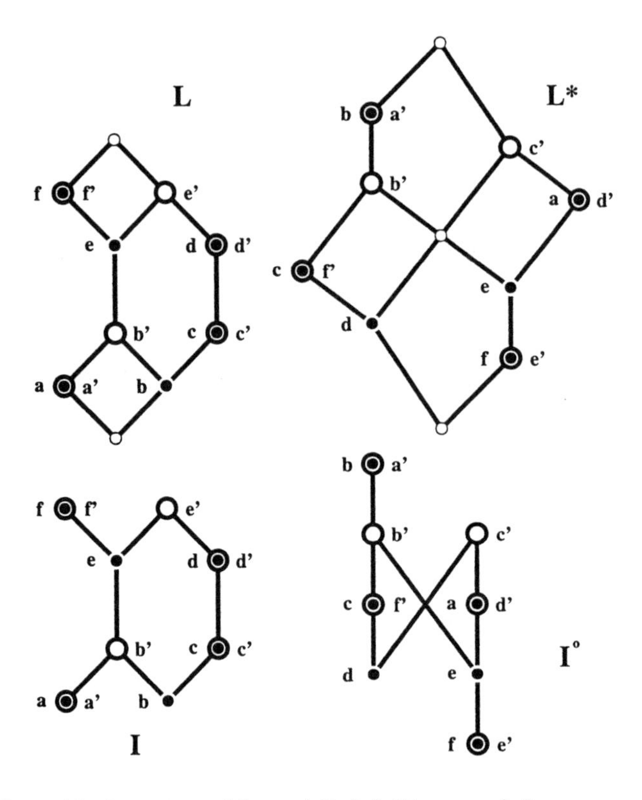

Figure 12. Opposite unities and their lattice completions

for $1 \leq i \leq n$ and for $0 < t \leq k_i$, while the meet-irreducible elements are of the form

$$a'_{i,t} := (p_{1,k_1}, \cdots, p_{i-1,k_{i-1}}, p_{i,t}, p_{i+1,k_{i+1}}, \cdots, p_{n,k_n})$$

for $1 \leq i \leq n$ and for $0 \leq t < k_i$. In the summand $J \times M$ of the unity I of irreducibles of the lattice D,

$$a_{i,t} \leq a'_{j,r} \text{ if and only if } i \neq j \text{ or } t \leq r$$

so

$$a_{i,t} \leq^o a'_{j,r} \text{ if and only if } i = j \text{ and } r < t$$

in the negation I^o of I. This means $a_{i,t}$ is equivalent to $a'_{i,t-1}$ in I^o,

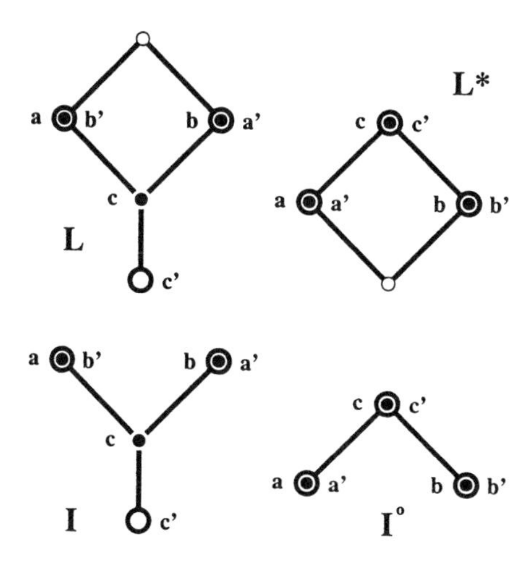

Figure 13. A lattice L for which $L \not\simeq L^{**}$

for $0 < t \leq k_i$, and $I^o = P^{[2]}$, where P is the disjoint union of n chains $a_{i,k_i} < \cdots < a_{i,2} < a_{i,1}$ of k_i elements, for $i = 1, \cdots, n$. ■

For examples, see Figures 14 and 15. Recall also that for the *unordered* set P, the double $P^{[2]}$ has as completion the lattice L_n consisting of $0, 1$, and n atoms a_i, with $0 < a_i < 1$. The dual of this lattice is the Boolean algebra B_n, the product of n chains of length 2. Since the negation of the unity of irreducibles of a distributive lattice is such a simple object, one might well expect the same behavior for modular lattices. Figures 16 and 17 show the free modular lattice on three generators, together with its dual lattice, a simpler structure, indeed.

Problem 31. Characterize those lattices L that are isomorphic to their own double-duals: $L \simeq L^{**}$. As we have seen, the lattices in question are those for which Irred $(L)^o$ is a unity of irreducibles.

9. Coverings and chains

The negation of the unity of irreducibles of a lattice arises naturally from a canonical labelling of the *steps* (covering pairs) of the lattice by sets of irreducibles. See Figure 14.

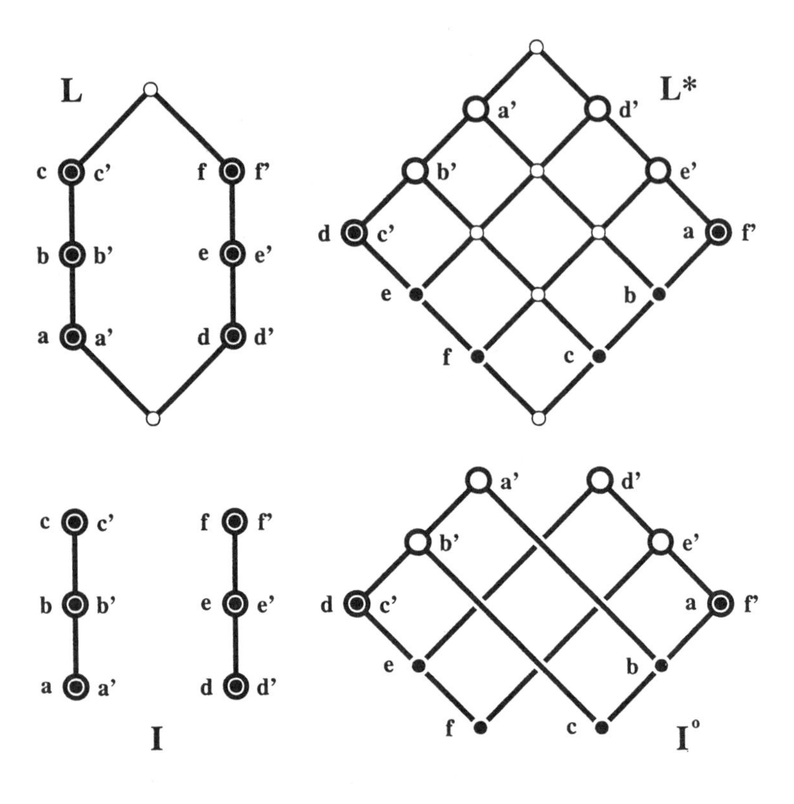

Figure 14. A sum of chains — dual to a product of chains.

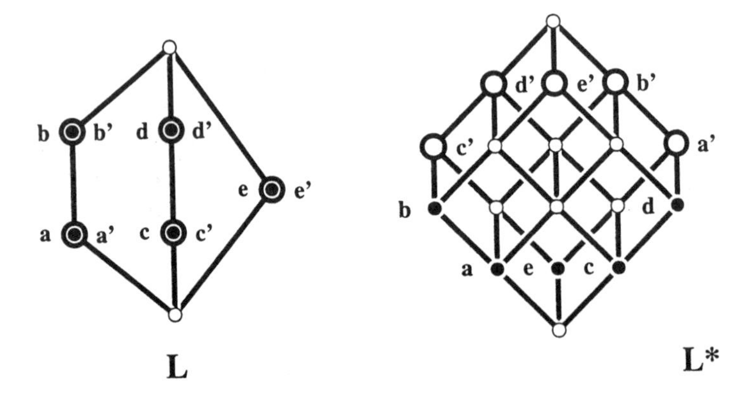

Figure 15. Dual lattices, the sum and product of three chains.

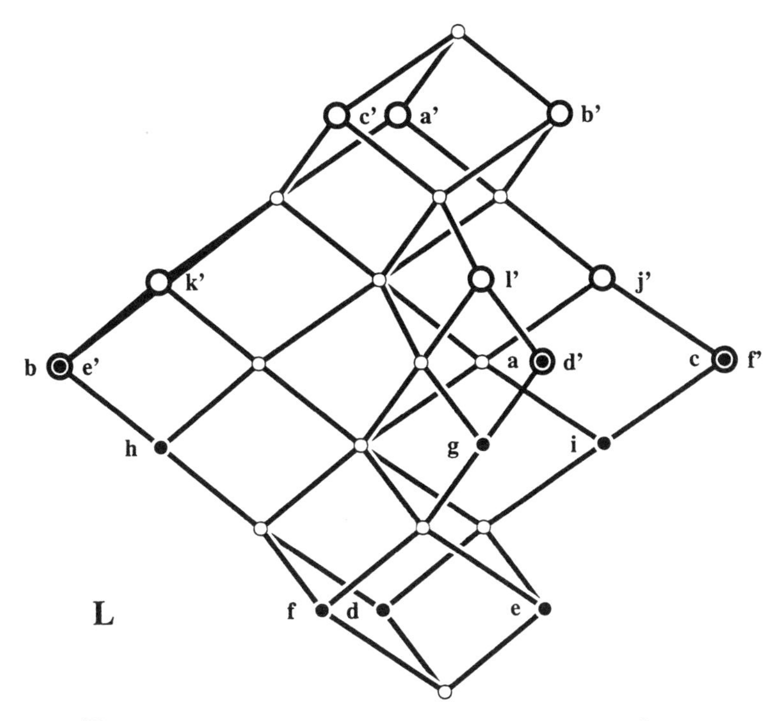

Figure 16. The free modular lattice on 3 generators a, b, c.

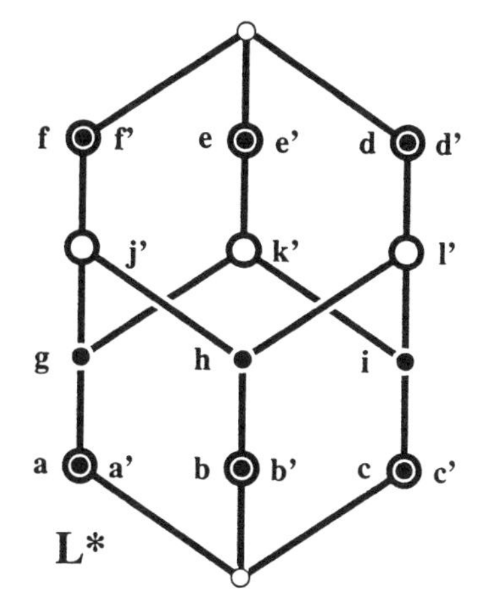

Figure 17. The dual of the free modular lattice on 3 generators a, b, c.

Definition 32. Let $I = (J \oplus M, \leq)$ be the unity of irreducibles of a finite lattice L. To any covering pair $x < y$ in L we assign the sets (S, T) of *labels*, where

$$\begin{aligned} S &= \{a \in J;\ a \not\leq x, a \leq y\} \\ T &= \{a' \in M;\ x \leq a', y \not\leq a'\}. \end{aligned}$$

We define a unity I_C, using the opposite preorder of the labels along a maximal chain C, as follows:

Definition 33. To any maximal chain C from 0 to 1 in a finite lattice L we assign the unity I_C for which $x \leq_C y$ if and only if the label x occurs higher in the chain C than the label y.

There is equally well a unity associated to each *element* of a lattice.

Definition 34. Let L be a finite lattice, and z an element of L. The unity I_z is the opposite $I^o(\rho_z)$ of the unity of the relation ρ_z, where

$$a\rho_z a' \text{ iff } a \leq z \leq a'.$$

for a join-irreducible in L, a' meet-irreducible in L.

Each unity I_z has just two equivalence classes of elements, with

$$\begin{gathered} \{a \in J \mid a \leq z\} \cup \{a' \in M \mid z \not\leq a'\} \\ \text{above} \\ \{a \in J \mid a \not\leq z\} \cup \{a' \in M \mid z \leq a'\} \end{gathered}$$

Theorem 35. *The opposite Irred* $^o(L) = (\mathit{Irred}\,(L))^o$ *of the unity of irreducibles of a lattice is represented as the intersection of unities:*

$$\bigcap_{\text{chains } C} I_C = I^o = \bigcap_{z \in L} I_z$$

Proof. Proven in [C]. See Figure 19. ∎

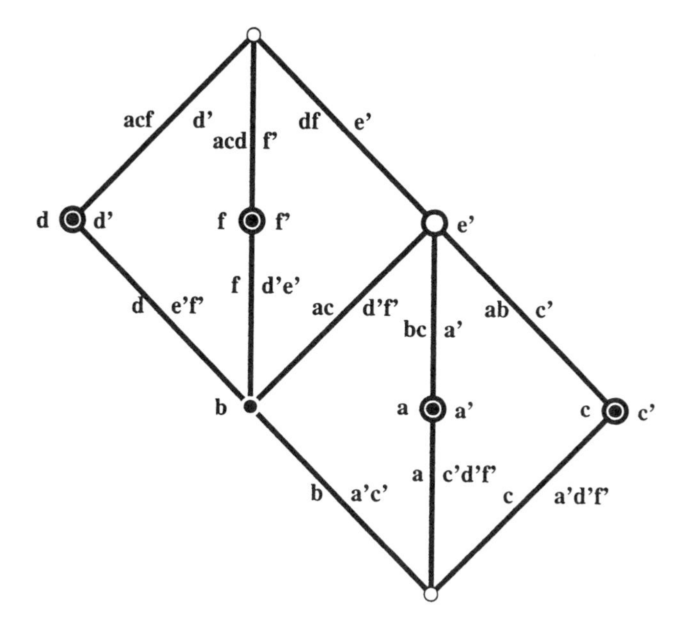

Figure 18. The labelling of the steps of a lattice.

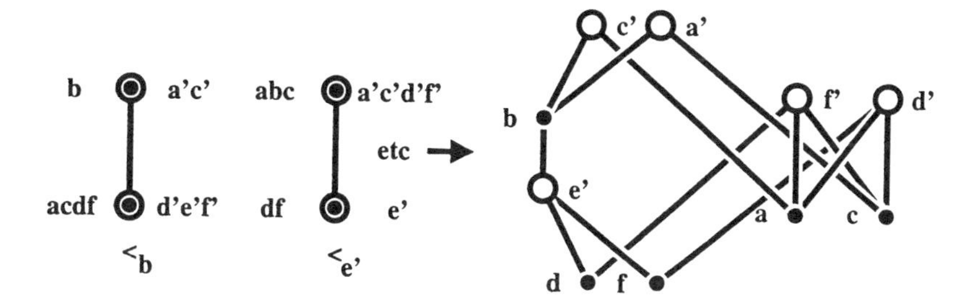

Figure 19. Two of the unities I_z, and the unity $I^o = \bigcap_{z \in L} I_z$ for the lattice in Figure 18.

10. Splits

Definition 36. Given two subsets S, T of a preordered set P, the set T is *final* in S if and only if

$$\forall s \in S \ \exists t \in S \cap T \quad \text{such that } s \leq t,$$

and the set T is *initial* in S if and only if

$$\forall s \in S \;\; \exists t \in S \cap T \quad \text{such that } t \leq s.$$

Definition 37. A *split* of a unity $I \subseteq (J \oplus M)^2$ is an ordered pair (C, D) such that

$$M \text{ is final in } C, \text{ an order ideal of } I,$$
$$J \text{ is initial in the complementary order filter } D = I \backslash C.$$

A most attentive referee here points out that this condition is equivalent to: an element $x \in I$ is an element of C if and only if every element a such that $a \leq x, a \in J$ is in C, and similarly for D.

Split (I) is the set of splits of a unity I, in the pointwise order, as order-preserving functions from I to $\underline{\mathbf{2}}$. The connection between splits of a unity and nodes of its opposite are illustrated in Figure 20, and are the subject of our main theorem, below.

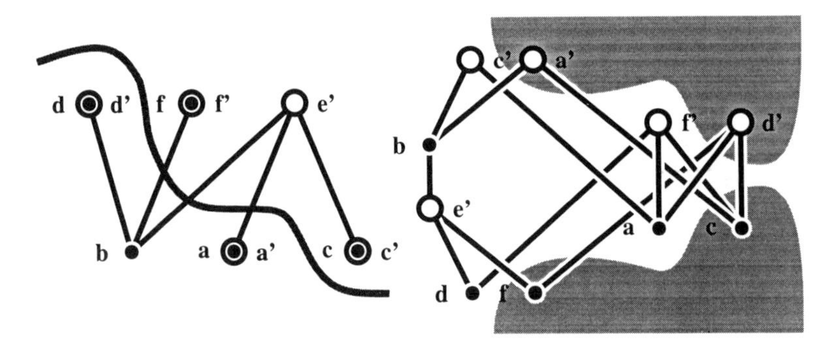

Figure 20. A split of I, and the corresponding node of I^o.

Main Theorem 38. *For any unity I with negation* I^o,

$$\textit{Split } (I) \simeq \textit{Completion } (I^o).$$

Proof. If (C, D) is a split of I, let

$$A = J \cap D, \;\; B = M \cap C\,.$$

This is a bijection, with

$$C = B \cup (J \backslash A), \;\; D = A \cup (M \backslash B).$$

Assume (C, D) is a split of I. For every $x \in A$, $x' \in B$, we have $x \nleq x'$, so $x \leq^o x'$ in I^o. If $y \in J \backslash A$, then $y \in C$, so there is an element $x' \in M \cap C$ with $y \leq x'$. But then $x' \in B$, and $y \nleq^o x'$, as required. Similarly, for any $y' \in M \backslash B$ there is an element $x \in A$ with $x \nleq^o y'$. So (A, B) is a node of I^o.

Conversely, to see that $C = B \cup (J \backslash A)$ is an order ideal of I, we consider a pair of elements $x \leq y$ in I, with $y \in C$. First assume $y \in B$. If $x \in M$, then $x \leq y$ in $M \times M$, so $y \leq^o x$ in I^o by Proposition 24, and $x \in B \subseteq C$. If $x \in J$, then $x \leq y$ in $J \times M$, so $x \nleq^o y$ in I^o, and x is not a lower bound for B. That is, $x \notin A$, and $x \in C$. Second, assume $y \in J \backslash A$. If $x \in J$, then $x \leq y$ in $J \times J$, so $y \leq^o x$ in I^o. Since $y \notin A$, and since A is an order ideal of J, $x \notin A$, and $x \in J \backslash A \subseteq C$. If $x \in M$, then $x \leq y$ in $M \times J$, so for all $c \in J$, $y \leq^o c$ or $c \leq^o x$, by Proposition 25. Consequently, for all elements $c \in A$, where $y \nleq^o c$, it must be the case that $c \leq^o x$. That is, x is an upper bound for A in I^o, and $x \in B \subseteq C$.

Assume (A, B) is a node of I^o, and let y' be an element of $M \cap D$. Then $y' \notin B$, so there is an element $x \in A$ with $x \nleq^o y'$. So $x \in D$ and $x \leq y'$, and J is initial in D. Similarly, M is final in C, and (C, D) is a split of I. ■

Theorem 39. *A finite lattice is distributive if and only if the negation of its unity of irreducibles is a doubled poset.*

Proof. For a finite lattice L, if $(\text{Irred}\,(L))^o = P^{[2]}$, then every decomposition (A, B) of P into complementary sets A, B, with A an order filter, B an order ideal of P, is a split of $P^{[2]}$. The lattice of splits is thus the lattice of order ideals of P, and is distributive. Conversely, if L is distributive, then for every join-irreducible element a in L there is a pseudo-complementary meet-irreducible element a', the maximal element x of L such that $x \ngeq a$. In the negation of the unity of irreducibles of L, a is equivalent with a'. ■

11. Maps

A reasonable definition for maps between unities should possess most, if not all, of the following properties:

(a) Composition of maps is well-defined and associative.
(b) For each unity I there is a unique identity map ι_I such that for any

map $\alpha : I \longrightarrow K$ of unities,

$$\iota_I \, \alpha = \alpha = \alpha \, \iota_K$$

as maps form I to K.

(c) The operations Opp, Supp, Irred, Latt, Doub, and the like, extend naturally to functors

$$\begin{gathered}\text{Opp } : \mathcal{I} \longrightarrow \mathcal{I}, \ \text{Supp } : \mathcal{I} \longrightarrow \mathcal{P}, \ \text{Irred } : \mathcal{L} \longrightarrow \mathcal{I}, \\ \text{Latt } : \mathcal{I} \longrightarrow \mathcal{L}, \ \text{Doub } : \mathcal{P} \longrightarrow \mathcal{I}, \ldots\end{gathered}$$

with properties established in the present article.

(d) The functor Opp is an anti-isomorphism of the category $\mathcal{I}$.

(e) For any unity I, there is a natural correspondence between splits of I and maps $I \longrightarrow \underline{\mathbf{2}}$, as well as between nodes of I and maps $\underline{\mathbf{2}}^o \longrightarrow I$, where $\underline{\mathbf{2}}$ is the double $C_2^{[2]}$ of the two-element chain C_2.

(The first two properties assure that maps of unities form a category.)

It is first necessary to decide whether a map $\alpha : I \longrightarrow K$ of unities sends elements of I to *elements* of K, or to *sets of elements* of K. The latter approach, albeit the more complicated of the two, is more likely to have a reasonable behavior relative to the completion functor Latt. (Maps which preserve irreducibility of elements play no natural role in lattice theory.)

In [C] we showed that it is not entirely trivial to give a definition of maps between unities that behaves well under completion and converts splits into maps to $\underline{\mathbf{2}}$. For a given split σ of a unity I and for most nodes $\alpha(A, B)$ of I, it is clear how to define $\sigma(\alpha)$, but there may exist nodes α for which no natural choice is possible. We called this the question of *determinacy*.

For the moment it seems possible only to suggest a reasonable analogue for unities of join- or of meet-preserving maps of lattices.

Definition 40. A J-map from a unity I_1 to a unity I_2 is a function α mapping subsets of J_1 to subsets of M_2 and satisfying

$$\begin{aligned}\alpha(\overline{A}) &= \alpha(A) = \overline{\alpha(A)} \\ \alpha(A) &= \bigcap_{a \in A} \alpha(a)\end{aligned}$$

The *composite* of a map $\alpha : I_1 \longrightarrow I_2$ with a map $\beta : I_2 \longrightarrow I_3$ is obtained as the composite of functions

$$2^{J_1} \longrightarrow 2^{M_2} \longrightarrow 2^{J_2} \longrightarrow 2^{M_3},$$

where $2^{M_2} \longrightarrow 2^{J_2}$ is the Galois connection of the unity I_2.

A definition of *M-maps* is given by the dual. J-maps and M-maps form categories, and are mapped contravariantly to one another by the negation functor Opp. But these categories are too small to admit all splits as maps to $\underline{\mathbf{2}}$. Indeed, it is still a temptation to believe that splits are exactly the maps from a unity to $\underline{\mathbf{2}}$ in some category, with a particularly simple definition of maps which uses only the notion of separation.

In closing, we wish to thank the referees for their most careful reading of the paper, and for their exceptionally helpful suggestions. It is a pleasure to have crossed their paths.

References

[B] B. Banaschewski and G. Bruns, *Categorical characterization of the MacNeille completion*, Arch. Math **18** (1967), 369–377.

[C] Henry Crapo, *Unities and Negation: representation of finite lattices*, J. of Pure and Applied Algebra **23** (1982), 109–135.

[D1] Brian A. Davey and H. Werner, *Dualities and equivalences: for varieties of algebras* La Trobe University, Pure Maths Res Paper 81-1, 1981.

[D2] Brian A. Davey and Dwight Duffus, *Exponentiation and Duality,* La Trobe University, Pure Maths Res Paper 81-12, 1981.

[L] F. William Lawvere, *Continuously variable sets: Algebraic Geometry = Geometric Logic*, in Logic Colloquium '73, North-Holland, Amsterdam, 1975, pages 135–156.

[M] George Markowsky, *The factorization and representation of distributive lattices*, Proc. Lond. Math. Soc. (1972), 507–530.

[P] H. Priestley, *Ordered topological spaces and representation of lattices,* Trans. Amer. Math. Soc. (1975), 185–200.

[S] M. H. Stone, *Topological characterization of distributive lattices and Brouwerian logics*, Casopis Pest. Math. Fys. **67** (1937), 1–25.

[U] Alisdair Urquhart, *A topological representation for lattices*, Algebra Universalis **8** (1978), 45–58.

Henry Crapo and Claude Le Conte de Poly-Barbut
C.A.M.S., 54 bd Raspail, 75270 Paris Cedex 06, France
electronic mail: *crapo@ehess.fr, barbut@ehess.fr*

The Would-Be Method of Targeted Rings

Ottavio M. D'Antona

Abstract

We describe the notion of an algebraic structure underlying a combinatorial identity or an inequality. In so doing, we supply an algebraic motivation for symbolic substitutions like $x_n \to x^n$, $x_n \to (x)_n$ and others.

1. Introduction

In this paper we describe the notion of an algebraic structure (the targeted ring) underlying a combinatorial identity, an inequality or the symbolic structure of a formula.

In order to prove a numerical identity $\alpha = \beta$, for example, the would-be method of targeted rings proceeds as follows.

(1) Find a structure R, for example, where the identity $A = B$ holds.

(2) Find a (linear) functional ν that maps the structure into the real numbers such that $\nu(A) = \alpha$ and $\nu(B) = \beta$.

This way, applying ν to both sides of $A = B$ proves the desired identity. We then say that R is the structure *underlying* the identity $\alpha = \beta$, or equivalently that R is a *targeted ring* associated to the identity.

In the following we will describe a few instances of this procedure that will allow us to give new proofs of the Inclusion-Exclusion Principle, Bonferroni's and Fréchet's inequalities, a relationship between ordinary and factorial moments of a discrete random variable and other identities involving Stirling numbers. Unfortunately, at present, there is no systematic way of deriving the (multiplicative) structure of the ring from the features of the relationship at point.

Besides being a method of proof, our work can be seen from another point of view. Given a sequence of real numbers a_n $(n = 0, 1, \ldots)$ and a sequence of polynomials of degree n, $p_n(x)$ $(n = 0, 1, \ldots)$, there is a unique linear functional L on the vector space of polynomials with real coefficients such that $a_n = L(p_n(x))$ for $n = 0, 1, \ldots$. In the classical umbral calculus, expressions like $x_n = x^n$ or $x_n = (x)_n$ were often seen as "symbolic identities" allowing us to describe (and to work with) numerical coefficients in terms of polynomials. This paper will show that some of these representations are associated to ring structures. This way we provide explicit algebraic motivation for the symbolic substitutions $x_n \to x^n$, $x_n \to (x)_n$ and others. (For a further description of this concept the reader is referred to the beautiful introduction by H. Crapo to Rota's umbral calculus [4].)

2. The Inclusion-Exclusion Principle

In this section we describe three rings underlying the Inclusion-Exclusion Principle. The relationship among two of them will be discussed in Section 4.1. We start by setting some notation that will be used throughout the paper. Let $\Sigma = \{\sigma_1, \sigma_2, \ldots, \sigma_n\}$ be a non-empty collection of (not necessarily distinct) subsets of an N-element set W. Let, for positive k, S_k be the sum of the number of elements of W lying in the sets $\sigma_{i_1} \cap \sigma_{i_2} \cap \cdots \cap \sigma_{i_k}$, for all increasing k-tuples $i_1, i_2, \ldots, i_k$. In symbols, we set

$$S_k = \sum_{1 \leq i_1 < i_2 < \ldots i_k \leq n} |\sigma_{i_1} \cap \sigma_{i_2} \cap \ldots \cap \sigma_{i_k}| \tag{1}$$

for positive k, and $S_0 = N$. Let, for $k = 0, 1, \ldots, n$, $p(k)$ denote the number of elements of W belonging to exactly k subsets in Σ. The Inclusion-Exclusion Principle

$$p(k) = \sum_{i=k}^{n} (-1)^{i-k} \binom{i}{k} S_i \tag{2}$$

spells out the relationship between $p(k)$ and the coefficients $S_k, S_{k+1}, \ldots$.

2.1. A polynomial ring underlying the Inclusion-Exclusion Principle.

In his treatment of the Inclusion-Exclusion Principle, J. Riordan [11] suggests introducing a "formal" variable S and writing the right-hand side of (2) as a polynomial in S. In this way one obtains the symbolic form of the Inclusion-Exclusion Principle:

$$p(k) = \frac{S^k}{(1+S)^{k+1}}.$$

Here the substitution $x_n \to x^n$ replaces the (numerical) coefficients $S_k, S_{k+1}, \ldots$ with the powers of the "variable" S. We describe below a *targeted ring* [9] giving an algebraic motivation for this symbolic substitution.

Let $\mathbb{Q}\langle x_1, x_2, \ldots, x_n\rangle$ be the associative ring of polynomials in the variables $x_1, x_2, \ldots, x_n$ over the rationals. Let I be the ideal generated by all monomials $x_{i_1}x_{i_2}\ldots x_{i_k}$ in which at least two indexes are not in strictly increasing order. Let $R = \mathbb{Q}\langle x_1, x_2, \ldots, x_n\rangle/I$, and let ϕ be the canonical homomorphism of $\mathbb{Q}\langle x_1, x_2, \ldots, x_n\rangle$ onto $\mathbb{Q}\langle x_1, x_2, \ldots, x_n\rangle/I$. We call the ring R the *segment ring* in n variables, and write f_i for the image of x_i under ϕ. The multiplication on R is determined by the rule

$$f_i f_j = \begin{cases} f_i f_j & \text{if } i < j, \\ 0 & \text{otherwise.} \end{cases}$$

A non-zero monomial $f_{i_1}f_{i_2}\ldots f_{i_k}$ of R will be called a *segment*, and the integer k its *length*. We define next two distinguished elements of R. Namely, we set $A = f_1 + f_2 + \ldots + f_n$, and $S = -1 + (1+f_1)(1+f_2)\ldots(1+f_n)$. It is easy to see that, for positive k, A^k is the sum of all segments of length k, and that

$$S = A + A^2 + \ldots + A^n = (1+S)A \tag{3}$$

holds. Moreover, since $A^m = 0$ for any $m > n$, we have $(1-A)^{-1} = 1 + S$. Thus, recalling (3) one can easily derive that, for $k \geq 0$,

$$A^k - A^{k+1} = \frac{S^k}{(1+S)^{k+1}}. \tag{4}$$

Now we take another step peculiar to a targeted ring, that is defining a counting functional on the ring. Specifically, we introduce the linear functional $\nu : R \to \mathbb{Q}$ such that, for any segment of length $k > 0$, $\nu(f_{i_1}f_{i_2}\cdots f_{i_k})$ equals the number of elements of W belonging to the set $\sigma_{i_1} \cap \sigma_{i_2} \cdots \cap \sigma_{i_k}$ but not belonging to any set $\sigma_{j_1} \cap \sigma_{j_2} \cdots \cap \sigma_{j_k}$, for any k-tuple $(j_1, j_2, \ldots, j_k)$ lexicographically less than $(i_1, i_2, \ldots, i_k)$. We set $\nu(1) = N$. Since any element of R may be written as a unique linear combination of segments, it follows that ν is well defined. For instance we have $\nu(f_1f_2) = |\sigma_1 \cap \sigma_2|$, $\nu(f_1f_3) = |\sigma_1 \cap \sigma_3 \cap \overline{\sigma_2}|$, and $\nu(f_2f_3) = |\sigma_2 \cap \sigma_3 \cap \overline{\sigma_1}|$. Thus, if $n = 3$, we have that $\nu(A^2) = \nu(f_1f_2 + f_1f_3 + f_2f_3)$ equals the number of elements lying in two or in three subsets of the family Σ. But for non-negative k, $\nu(A^k)$ equals the number of elements lying in at least k subsets of Σ. This proves the following:

$$\nu(A^k - A^{k+1}) = p(k).$$

On the other hand, with a mild additional effort (the reader is referred to [9] for computational details) one can show that

$$\nu(S^k) = S_k. \tag{5}$$

Therefore, by applying ν to both sides of (4), one obtains a proof of the Inclusion-Exclusion Principle. Identity (5) provides the algebraic motivation for the symbolic substitution $x_n \to x^n$.

2.2. A ring of matrices underlying the Inclusion-Exclusion Principle.

Let $X = \{x_1, x_2, \ldots\}$ be an infinite alphabet, and let T' be the following matrix

$$T' = \begin{pmatrix} 0 & x_1 & x_2 & x_4 & x_7 & \cdots \\ 0 & 0 & x_3 & x_5 & x_8 & \cdots \\ 0 & 0 & 0 & x_6 & x_9 & \cdots \\ 0 & 0 & 0 & 0 & x_{10} & \cdots \\ 0 & 0 & 0 & 0 & 0 & \cdots \\ \vdots & \vdots & \vdots & \vdots & \vdots & \vdots \end{pmatrix} = t(i,j)$$

$$t(i,j) = \begin{cases} x_{i+\binom{j-1}{2}} & \text{if } i < j \\ 0 & \text{otherwise.} \end{cases} \tag{6}$$

For a positive integer n, we define n infinite matrices $F_1, F_2, \ldots, F_n$ over $\mathbb{Q}[x_1, x_2, \ldots]$ as follows. For $h = 1, 2, \ldots, n$ the matrix F_h is an infinite matrix all of whose entries equal zero except for the $(h+1)$-st column, which equals the $(h+1)$-st column of T'. For example we have

$$F_1 = \begin{pmatrix} 0 & x_1 & \cdots \\ 0 & 0 & \cdots \\ \vdots & \vdots & \cdots \end{pmatrix}, \quad F_2 = \begin{pmatrix} 0 & 0 & x_2 & \cdots \\ 0 & 0 & x_3 & \cdots \\ 0 & 0 & 0 & \cdots \\ \vdots & \vdots & \vdots & \cdots \end{pmatrix},$$

$$F_3 = \begin{pmatrix} 0 & 0 & 0 & x_4 & \cdots \\ 0 & 0 & 0 & x_5 & \cdots \\ 0 & 0 & 0 & x_6 & \cdots \\ 0 & 0 & 0 & 0 & \cdots \\ \vdots & \vdots & \vdots & \vdots & \cdots \end{pmatrix}, \quad F_4 = \begin{pmatrix} 0 & 0 & 0 & 0 & x_7 & \cdots \\ 0 & 0 & 0 & 0 & x_8 & \cdots \\ 0 & 0 & 0 & 0 & x_9 & \cdots \\ 0 & 0 & 0 & 0 & x_{10} & \cdots \\ 0 & 0 & 0 & 0 & 0 & \cdots \\ \vdots & \vdots & \vdots & \vdots & \vdots & \cdots \end{pmatrix}.$$

More precisely, for $1 \le h \le n$,

$$F_h(i,j) = \begin{cases} x_{i+\binom{h}{2}} & \text{if } i \leq h, \text{ and } j = h+1 \\ 0 & \text{otherwise.} \end{cases} \tag{7}$$

Let F be the ring generated by the matrices $F_1, F_2, \ldots, F_n$. The product of two generators is characterized as follows. If $h \geq k$, then $F_h F_k = 0$. Otherwise the only non-zero entries of $F_h F_k$ are found in the first h positions of column $k+1$. Specifically, for $i = 1, 2, \ldots, h$, entry $(i, k+1)$ of the product $F_h F_k$ is

$$x_{i+\binom{h}{2}} x_{h+1+\binom{k}{2}}.$$

Conversely, suppose we are given a matrix M whose only non-zero entries are the first h entries of column $k+1$. Suppose such entries are $x_a x_b, x_{a+1} x_b, \ldots, x_{a+h} x_b$. Then M is the product of two generators F_h, and F_k, say. But indeed h, and k can be derived by taking the positive solution of the system of equations

$$\begin{cases} a &= 1 + \binom{h}{2} \\ b &= h + 1 + \binom{k}{2}. \end{cases}$$

This possibility of uniquely factoring a non-zero product of two generators can be generalized. If the sequence $1 \leq h_1, h_2, \ldots, h_t \leq n$ is not in strictly increasing order, the product $F_{h_1} F_{h_2} \cdots F_{h_t}$ vanishes. Otherwise, the only non-zero entries of the product are the first h_1 elements of column $h_t + 1$. Specifically, for $i = 1, 2, \ldots, h_1$, entry $(i, h_t + 1)$ of the product $F_{h_1} F_{h_2} \cdots F_{h_t}$ is

$$x_{i+\binom{h_1}{2}} x_{h_1+1+\binom{h_2}{2}} \cdots x_{h_{t-1}+1+\binom{h_t}{2}}.$$

In conclusion, one can devise an algorithm that, when input with any non-zero element M, for example, of F outputs the (unique) polynomial in the generators which equals M. Such an algorithm performs a rightward scan of the first row of M, and to each monomial in the $x's$ writes down the corresponding monomial in the generators by solving certain second degree equations.

The above discussion proves the following.

Proposition 2.1.

The segment ring R and F *are isomorphic under the linear map $\psi : \mathrm{F} \to R$ that sends monomial $F_{h_1} F_{h_2} \cdots F_{h_t}$ to the monomial $f_{h_1} f_{h_2} \cdots f_{h_t}$.*

Two distinguished elements of F are $T = F_1 + F_2 + \ldots + F_n$ and $Z = -1 + (1 + F_1)(1 + F_2) \ldots (1 + F_n)$. In particular, $\psi(T^k) = A^k$ and $\psi(Z^k) = S^k$.

Thus, in F we have

$$T^k - T^{k+1} = \frac{Z^k}{(1+Z)^{k+1}}. \tag{8}$$

Finally we define a linear functional ξ on F by setting $\xi = \nu \circ \psi$. Applying ξ to both sides of (8) we again obtain the Inclusion-Exclusion Principle.

We close the section with a question. The matrices here introduced are a representation of the generators of the segment ring R. Indeed, one could define an analogous set of matrices by choosing other shapes for the non-zero region non-zero. For instance, suitable triangles will do. What is the minimum amount of information these new generators should display in order to be able to factor a given matrix of the ring?

2.3. Yet another ring underlying the Inclusion-Exclusion Principle.

Let $\mathbb{Q}[y_1, y_2, \ldots, y_n]$ be the associative ring of polynomials in the variables $y_1, y_2, \ldots, y_n$ over the rationals. Let I be the ideal generated by the elements $y_1^2, y_2^2, \ldots, y_n^2$. Let $G = \mathbb{Q}[y_1, y_2, \ldots, y_n]/I$, and let ϕ be the canonical homomorphism of $\mathbb{Q}[y_1, y_2, \ldots, y_n]$ onto $\mathbb{Q}[y_1, y_2, \ldots, y_n]/I$. We call the ring G the *square-free ring* in n variables, and write g_i for the image of y_i under ϕ. The multiplication on G is determined by

$$g_i g_j = \begin{cases} g_j g_i & \text{if } i \neq j, \\ 0 & \text{otherwise.} \end{cases}$$

Again, following [9] we consider a distinguished element in our targeted ring by setting

$$T = g_1 + g_2 + \cdots + g_n.$$

For $k = 0, 1, \ldots, n$ the kth power of T has the following expression:

$$T^k = k! \sum_{i_1 < i_2 < \cdots < i_k} g_{i_1} g_{i_2} \cdots g_{i_k}.$$

In other words, $T^k/k!$ is the sum of all monomials of degree k. We define a linear functional $\mu : G \to \mathbb{Q}$ such that, any k-tuple $i_1, i_2, \ldots, i_k$ with distinct indexes $\mu(g_{i_1} g_{i_2} \cdots g_{i_k})$ equals the number of elements of W belonging to the set $\sigma_{i_1} \cap \sigma_{i_2} \cap \cdots \cap \sigma_{i_k}$ and to no other subset of Σ. For instance, we have $\mu(g_1 g_2) = |\sigma_1 \cap \sigma_2 \cap \overline{\sigma_3}|$, $\mu(g_1 g_3) = |\sigma_1 \cap \sigma_3 \cap \overline{\sigma_2}|$, and $\mu(g_2 g_3) = |\sigma_2 \cap \sigma_3 \cap \overline{\sigma_1}|$. Moreover, since any element of G can be written as a unique linear combination of monomials, μ is well defined. Thus one easily sees that, for $k = 0, 1, \ldots, n$

$$\mu(\frac{T^k}{k!}) = p(k). \tag{9}$$

Now, by writing the (finite) sum

$$e^T = 1 + \frac{T}{1!} + \frac{T^2}{2!} + \cdots + \frac{T^n}{n!} + 0 + 0 + \cdots$$

we deduce that $\mu(e^T)$ equals N. Note that e^T can also be written as

$$e^T = (1 + g_1)(1 + g_2) \cdots (1 + g_n).$$

Again from [9] we recall that, for any k-tuple $i_1 < i_2 < \cdots < i_k$, one can write

$$g_{i_1} g_{i_2} \cdots g_{i_k} e^T = \sum g_{j_1} g_{j_2} \cdots$$

where the sum ranges over all monomials which are divisible by $g_{i_1} g_{i_2} \cdots g_{i_k}$. (For instance, for $n = 4$ we have $g_1 g_2 e^T = g_1 g_2 + g_1 g_2 g_3 + g_1 g_2 g_4 + g_1 g_2 g_3 g_4$.) Specifically, we have

$$\frac{T^k}{k!} e^T = \sum_{i=k}^{n} \binom{i}{k} \frac{T^i}{i!}. \tag{10}$$

Since, by definition of μ,

$$\mu(g_{i_1} g_{i_2} \cdots g_{i_k} e^T) = |\sigma_{i_1} \cap \sigma_{i_2} \cap \cdots \cap \sigma_{i_k}|$$

then it follows that

$$\mu(\frac{T^k}{k!} e^T) = S_k. \tag{11}$$

By multiplying both sides of

$$e^{-T} = 1 - \frac{T}{1!} + \frac{T^2}{2!} - \cdots + (-1)^n \frac{T^n}{n!} + 0 + 0 + \cdots$$

by $\frac{T^k}{k!}$ one obtains the identity

$$\frac{T^k}{k!} e^{-T} = \sum_{i \geq k} (-1)^{i-k} \binom{i}{k} \frac{T^i}{i!}.$$

Applying the linear functional μ to both sides of the above identity and recalling (9) and (11) we again obtain the Inclusion-Exclusion Principle:

$$
\begin{aligned}
p(k) &= \mu(\frac{T^k}{k!}) = \\
&= \sum_{i\geq k}(-1)^{i-k}\binom{i}{k}\mu(\frac{T^i}{i!}e^T) = \\
&= \sum_{i\geq k}(-1)^{i-k}\binom{i}{k}S_i.
\end{aligned}
$$

In this way, we have obtained another algebraic motivation to the symbolic substitution $x_n \longrightarrow \frac{x^n}{n!}e^T$.

3. The lower-factorial ring

Let M_n be the n-th (ordinary) moment of a finitely valued random variable X, and $M_{(n)}$ be its n-th factorial moment (a concept introduced by J. F. Steffensen in 1923), that is to say the expected value of the random variable $(X)_n$. In symbols we have

$$
\begin{aligned}
M_n &= E[X^n] = \sum_{k\geq 0} k^n p_k, \\
M_{(n)} &= E[(X)_n] = \sum_{k\geq 0}(k)_n p_k,
\end{aligned}
$$

where $p_k = P[X = k]$. Recall that the *Stirling numbers of the first kind* $s(n,k)$ are defined as the connection constants between powers and lower factorial polynomials, that is

$$(x)_n = \sum_{k=0}^{n} s(n,k)x^k.$$

Note that $s(n,0) = 0$ whenever n is positive and $s(0,0) = 1 = (x)_0$. Thus, the relationship between moments and factorial moments is easily seen to be

$$M_{(n)} = \sum_{k=0}^{n} s(n,k)M_k.$$

In his 1947 book [10] M. Fréchet remarks that the latter formula could also be obtained by introducing a "formal" variable M and writing

$$M_{(n)} = (M)_n.$$

After expanding the expression on the right, the "symbolic method" would simply replace each power M^k with the moment M_k. The targeted ring we develop below is meant to give an algebraic justification for such a symbolic substitution $x_n \to (x)_n$.

Let, as in [5], $\mathbb{Q}[y_1, y_2, \ldots, y_n]$ be the associative ring of polynomials in the variables $y_1, y_2, \ldots, y_n$ over the rationals. Denote by I the ideal generated by elements $y_1^2 - y_1$, and by elements $y_i y_j$ where $i \neq j$. Let $L = \mathbb{Q}[y_1, y_2, \ldots, y_n]/I$ and ϕ be the canonical homomorphism of P onto P/I. We call L the *lower-factorial ring* in n variables, and write h_i for the image of y_i under ϕ. The multiplication on L is determined by the following

$$h_i h_j = \begin{cases} h_i & \text{if } i = j, \\ 0 & \text{otherwise.} \end{cases}$$

As usual we start by identifying an element of our targeted ring. Set $M = h_1 + 2h_2 + \cdots + nh_n$. A basic property enjoyed by the powers of M is the following.

Proposition 3.1. *For any positive $m \leq n$,*

$$M^m = \sum_{k=1}^{n} k^m h_k. \tag{12}$$

Proof. In the product $(h_1 + 2h_2 + \cdots + nh_n)^m$ every monomial $i_1 h_{i_1} i_2 h_{i_2} \cdots i_m h_{i_m}$ vanishes unless $i_1 = i_2 = \cdots = i_n$. In such an event, writing k for the common value of the indexes, we have $i_1 h_{i_1} i_2 h_{i_2} \cdots i_m h_{i_m} = k^m h_k$. ∎

We extended this result by linearity.

Proposition 3.2. *If $p(u)$ is any m-degree polynomial ($1 \leq m \leq n$) of $\mathbb{Q}[u]$ such that $p(0) = 0$, then*

$$p(M) = \sum_{k=1}^{n} p(k) h_k. \tag{13}$$

Proof. Let $p(u) = \sum_{k=1}^{m} a_i u^i$ and $p(M) = \sum_{k=1}^{m} c_i h_i$ for some $a_i, c_i \in \mathbb{R}$. The proof sums up to show that $c_k = p(k)$ for $k = 1, 2, \ldots, n$. By (12) we can write

$$\begin{aligned} p(M) &= \sum_{k=1}^{m} a_i M^i = \sum_{k=1}^{m} a_i \sum_{k=1}^{n} k^i h_k = \\ &= \sum_{k=1}^{n} (\sum_{k=1}^{m} a_i k^i) h_k = \sum_{k=1}^{n} p(k) h_k. \end{aligned}$$

∎

In particular, setting $p(u) = (u)_m$, we find

$$(M)_m = \sum_{k=1}^{n} (k)_m h_k. \tag{14}$$

Define a linear functional $\nu : L \to [0,1]$ such that, for $k = 1, 2, \ldots, n$, $\nu(h_k) = p_k = P[X = k]$. Then it is immediate that, for $m = 1, 2, \ldots, n$,

$$\begin{aligned} \nu(M^n) &= E[X^n] = M_n \\ \nu((M)_n) &= E[(X)_n] = (M)_n. \end{aligned}$$

Thus, we have provided an algebraic interpretation of the symbolic expression $M_{(n)} = (M)_n$.

To close the section it is worth noting that in [6] we showed that any element of the lower factorial ring can be uniquely expressed as a linear combination of the elements $(M)_0 = 1, (M)_1, (M)_2, \ldots, (M)_n$. Define a linear functional $\gamma : L \to \mathbb{R}$ by setting $\gamma((M)_n) = 1$. Then one can prove [6] that $\lambda(M^n) = B_n$ (as usual B_n denotes the number of partitions of an n-set) by reproducing the steps made in [13]. In turns, defining, for $k = 0, 1, \ldots, n$ the linear functionals $\gamma_k : L \to \mathbb{R}$ by setting $\gamma_k((M)_n) = \delta_{n,k}$ allows us to prove the classical recurrence of the Stirling numbers of the second kind by reproducing the steps made in [1].

4. Final remarks

Here we report a few additional results and comments on the rings we have discussed so far.

4.1. Relating the segment ring R and the square-free ring G

Since G is commutative and R is not, there is no hope of finding any isomorphism between the two rings. Instead, we will give a relationship between R and G in terms of the two counting functionals μ and ν. To this end, following [8], we first introduce the linear map $\phi : R \to G$ such that, for any segment $f_{i_1} f_{i_2} \cdots f_{i_k}$,

$$\phi(f_{i_1} f_{i_2} \cdots f_{i_k}) =$$

$$= g_{i_1} g_{i_2} \cdots g_{i_k} \left(1 + \sum_{i_k < j_1} g_{j_1} + \sum_{i_k < j_1 < j_2} g_{j_1} g_{j_2} + \ldots + g_{i_k+1} g_{i_k+2} \cdots g_n \right).$$

For example, for $n = 4$, we have $\phi(f_2) = g_2 + g_2 g_3 + g_2 g_4 + g_2 g_3 g_4$. In this way, $\mu(\phi(f_2))$ is easily seen to coincide with $\nu(f_2)$. This remark can be generalized. For any $z \in R$,

$$\nu(z) = \mu(\phi(z)).$$

Next, we define a linear map $\psi : G \to R$ by setting

$$\psi(g_{i_1} g_{i_2} \dots g_{i_k}) = f_{i_1} f_{i_2} \dots f_{i_k} - \sum_{i_k < j} f_{i_1} f_{i_2} \dots f_{i_k} f_j$$

for any monomial $g_{i_1} g_{i_2} \dots g_{i_k}$ of G. For instance, $\psi(g_2) = f_2 - f_2 f_3 - f_2 f_4$. Again, one can prove this, for any $y \in G$,

$$\mu(y) = \nu(\psi(y)).$$

4.2. An order relation on the square-free ring

Here we use the square-free ring G to prove the inequalities of Fréchet and Bonferroni. As in [2], we define a partial order relation on G by setting, for any $x, y \in G$, $x \leq_G y$ whenever $x - y$ can be written as a linear combination of monomials with non-negative coefficients. The remarkable property of this order relation is that the counting functional μ we introduced in Section 2.3 to prove the Inclusion-Exclusion Principle is a *non-decreasing monotone function.*

Fréchet inequality [3], that is

$$\frac{S_{k-1}}{\binom{n}{k-1}} \geq \frac{S_k}{\binom{n}{k}} \tag{15}$$

can now be derived in its simpler expression

$$(n - k + 1) S_{k-1} \geq k S_k.$$

Proposition 4.1. *For $k = 1, 2, \dots, n$*

$$(n - k + 1) \frac{T^{k-1}}{(k-1)!} e^T \geq_G k \frac{T^k}{k!} e^T \tag{16}$$

Proof. Recall from Section 2.3 the identity (10), that is

$$\frac{T^k}{k!} e^T = \sum_{i \geq k} \binom{i}{k} \frac{T^i}{i!}.$$

Observe that in the expression

$$(n - k + 1) \frac{T^{k-1}}{(k-1)!} + \sum_{i=k+1}^{n} [(n - k + 1) \binom{i}{k-1} - k \binom{i}{k}] \frac{T^i}{i!}$$

the coefficients of $T^i/i!$ are non-negative and that $T^i/i!$ is a *sum* of monomials. ∎

Since $\mu((T^k/k!)e^T) = S_k$, applying the monotone counting functional μ to both sides of (16) proves Fréchet's inequality. Bonferroni inequality [3] states that, for $h = 0, 1, \ldots, n$ the sum $-p(0) + S_0 - S_1 + S_2 - \cdots + (-1)^h S_h$ alternates in sign according to the parity of h.

Proposition 4.2. *Let h be a non-negative integer not exceeding n. If h is even, then*

$$\sum_{i=0}^{h}(-1)^h \frac{T^i}{i!} e^T \geq_G 1, \tag{17}$$

otherwise

$$1 \geq_G \sum_{i=0}^{h}(-1)^h \frac{T^i}{i!} e^T. \tag{18}$$

Proof. Let c_j be the coefficient of $T^i/i!$ in the sum

$$\sum_{i=0}^{h}(-1)^h \frac{T^i}{i!} e^T = \sum_{i=0}^{h}(-1)^h \sum_{j\geq i} \binom{j}{i} \frac{T^i}{i!}.$$

If $j = 0$, then $c_j = 1$. If $0 < j \leq h$, from [14] we know that $c_j = 0$. If $j > h$, a short manipulation shows that $c_j = (-1)^h \binom{j-1}{h}$. Now we can write

$$\sum_{i=0}^{h}(-1)^h \frac{T^i}{i!} e^T = \sum_{j=0}^{n} c_j \frac{T^j}{j!} = 1 + (-1)^h \sum_{j\geq h} \binom{j-1}{h} \frac{T^j}{j!}.$$

We have proved that

$$\sum_{i=0}^{h}(-1)^h \frac{T^i}{i!} e^T - 1 \geq_G 0$$

when h is even, and that

$$-1 + \sum_{i=0}^{h}(-1)^h \frac{T^i}{i!} e^T \geq_G 0$$

when h is odd. ∎

Applying the monotone counting functional μ to both sides of (17) and (18), and recalling that $\mu(1) = p(0)$ gives an algebraic proof of Bonferroni's inequality.

Finally, we notice that one could define in the segment ring R an analogous order relation and then rework the proofs of the above inequalities. However, it is worth noting that these proofs can be made even simpler by exploiting the maps ψ and ϕ introduced in 4.1.

4.3. A ring of formal series

Here we describe a structure (introduced in [7]) that displays several analogies with $\mathbb{R}[[u]]$, the ring of formal power series on the indeterminate u. Our ring is the inductive limit of the lower factorial ring introduced in Section 3.

Let $h_1, h_2, \dots$ be an infinite set of variables satisfying

$$h_i h_j = \begin{cases} 0 & \text{if } i \neq j, \\ h_i & \text{if } i = j. \end{cases}$$

Infinite series of the form $a_0 + \sum_{i>0} a_i h_i$, where the a_i are real numbers, make an associative and commutative ring, here called H, with respect to the componentwise sum and the following multiplication

$$(a_0 + \sum_{i>0} a_i h_i)(b_0 + \sum_{i>0} b_i h_i) = a_0 b_0 + \sum_{i>0} (a_0 b_i + a_i b_0 + a_i b_i) h_i.$$

The additive identity is the series in which $a_i = 0$ for every i, and will be denoted by 0. The multiplicative identity is the series in which $a_0 = 1$, $a_i = 0$ for every positive i, and will be denoted by 1.

An element of H is invertible if and only if $a_0 \neq 0$, and $a_i \neq a_0$ for every positive i. The inverse of $a_0 + \sum_{i>0} a_i h_i$ is the series

$$\frac{1}{a_0} + \sum_{i>0} \frac{-a_i}{a_0(a_0 + a_i)} h_i.$$

It is worth noting that H is not a local ring. To see this, simply consider that neither f_i nor $1 - f_i$ are invertible, but their sum has an inverse.

We introduce a linear operator D on H to be called *(pseudo) derivative* by setting

$$D(h_i) = \frac{h_{i-1}}{i}.$$

Moreover $D(h_1) = 1$ and $D(1) = D(0) = 0$. Thus, for any series $\alpha = a_0 + \sum_{i>0} a_i h_i$ we have

$$D(a_0 + \sum_{i>0} a_i h_i) = a_1 + \sum_{i>0} a_{i+1} D(h_i).$$

We set $D^0(\alpha) = \alpha$ and for any positive m, $D^m(\alpha) = D(D^{m-1}(\alpha))$.

To show some properties of our derivative we define the series

$$M! = 1 + \sum_{i>0} i!h_i = 0! + 1!h_1 + 2!h_2 + 3!h_3 + \cdots$$

and note that $D(M!) = M!$. In other words, in H the series $M!$ plays the role of the exponential series e^t in $\mathbb{R}[[u]]$. To make this statement precise we introduce a linear functional $\tau : \mathbb{R}[[u]] \to H$ such that $\tau(t^i/i!) = i!h_i$, for $i = 1, 2, \ldots$. Then we have

$$\tau(e^u) = M!.$$

Moreover, denoting by $\frac{d}{du}$ the formal derivative of power series, we have the following.

Proposition 4.3. *For any formal power series* $F(u)$

$$D(\tau(u^n)) = \tau(\frac{d}{du}F(u)).$$

Proof. Write for any positive n

$$D(\tau(F(u))) = D(\tau(n!)^2 h_n) = n!(n-1)h_{n-1} = \tau(n\ u^{n-1}) = \tau(\frac{d}{du}u^n).$$

Moreover $D(\tau(1)) = 0 = \tau(\frac{d}{du}1)$. By linearity the result follows. ∎

Finally we describe a result very similar to a basic fact of the Roman umbral calculus [12]. Given a formal power series

$$F(u) = \sum_{k\geq 0} \frac{a_k}{k!} u^k$$

one can define a linear functional on P, the algebra of polynomials in the single variable x and complex coefficients by writing $\langle F(u)|x^n\rangle = a_n$ for $n = 0, 1, \ldots$. In particular one has $\langle u^k|x^n\rangle = n!\delta_{n,k}$. In the same vein, to any series

$$\alpha = \sum_{k\geq 0} a_k k! h_k,$$

we can associate a linear functional on H by setting $\langle \alpha|x^n\rangle = a_n$ for $n = 0, 1, \ldots$. In particular one has

$$\langle h_k|x^n\rangle = \frac{1}{n!}\delta_{n,k}.$$

Finally we derive the analog of Theorem 2.1.10 of [12].

Proposition 4.4. *For any series $\alpha \in H$ and polynomial $p(x)$,*

$$\langle \alpha | xp(x) \rangle = \langle D(\alpha) | p(x) \rangle.$$

Proof. Once again it suffices to prove the result for $p(x) = x^n$. Let $\alpha = a_0 + \sum_{i>0} a_i i! h_i$. Then we write

$$\begin{aligned} \langle D(\alpha) | x^n \rangle &= \langle \sum_{i>0} a_i (i-1)! h_{i-1} | x^n \rangle = \\ &= \sum_{i>0} a_i (i-1)! \langle h_{i-1} | x^n \rangle = \\ &= \sum_{i>0} a_i (i-1)! \frac{1}{n!} \delta_{n,i-1} = a_{n+1} = \langle \alpha | x^{n+1} \rangle. \end{aligned}$$

■

Acknowledgments. The author wishes to thank C. Mereghetti and F. Regonati for several useful discussions on the preliminary drafts of the paper.

References

[1] M. Aigner, *Combinatorial Theories*, Springer, New York/Berlin, 1979.

[2] S. Bertoluzza and O. D'Antona, Considerazioni di tipo algebrico sulla diseguaglianza di Bonferroni, Technical Report RIDIS 28-87, Dipartimento di Informatica e Sistemistica, Università di Pavia, 1987.

[3] L. Comtet, *Advanced Combinatorics*, Birkäuser, 1974.

[4] H. H. Crapo, Rota's "combinatorial theory," in *Gian-Carlo Rota on Combinatorics*, J.P.S. Kung, ed., xix–xliii, Birkäuser, Boston 1995.

[5] O. D'Antona, A ring underlying probabilistic identities, *J. Math. Anal. Appl.*, **108**:211–215, 1985.

[6] O. D'Antona, Combinatorial properties of the factorial ring, *J. Math. Anal. Appl.*, **117**:303–309, 1986.

[7] O. D'Antona, Pseudo power series, Technical Report RIDIS 27-87, Dipartimento di Informatica e Sistemistica, Università di Pavia, 1987.

[8] O. D'Antona and A. Pesci, Connecting two rings underlying the inclusion-exclusion principle, *J. Combin. Theory Ser. A*, **40**:439–443, 1985.

[9] O. D'Antona and G.-C. Rota, Two rings connected with the inclusion-exclusion principle, *J. Combin. Theory Ser. A*, **24**:65–72, 1978.

[10] M. Fréchet, *Les probabilités associées à un système d'événements compatible et dépendants, Vol. I*, Herman, Paris, 1940.

[11] J. Riordan, *An Introduction to Combinatorial Analysis*, J. Wiley & Sons, 1958.

[12] S. Roman, *Umbral Calculus*, Reidel, Dordrecht, 1984.

[13] G.-C. Rota, The number of partitions of a set, *Amer. Math. Monthly*, **71**:498–505, 1964.

[14] R. P. Stanley, *Enumerative Combinatorics*, Wadworth & Brooks, Monterey, 1986.

Dipartimento di Scienze dell'Informazione
Università degli Studi di Milano
via Comelico, 39
20135 Milano
Italy
E-mail: `dantona@dsi.unimi.it`

Lattice Walks and Primary Decomposition

Persi Diaconis, David Eisenbud, and Bernd Sturmfels *

1. Introduction

This paper shows how primary decompositions of an ideal can give useful descriptions of components of a graph arising in problems from combinatorics, statistics, and operations research. We begin this introduction with the general formulation. Then we give the simplest interesting example of our theory, followed by a statistical example similar to that which provided our original motivation. Later on we study the primary decompositions corresponding to some natural combinatorial problems.

Let $\mathcal{B}$ be a set of vectors in $\mathbf{Z}^n$. Define a graph $G_\mathcal{B}$ on whose vertices are the non-negative n-tuples $\mathbf{N}^n$ as follows: $\boldsymbol{u}, \boldsymbol{v} \in \mathbf{N}^n$ are connected by an edge of $G_\mathcal{B}$ if and only if $\boldsymbol{u} - \boldsymbol{v}$ is in $\pm\mathcal{B}$. We say $\boldsymbol{u}$ and $\boldsymbol{v}$ in $\mathbf{N}^n$ can be *connected via* $\mathcal{B}$ if they are in the same connected component of $G_\mathcal{B}$. We shall consider the problem of characterizing the components of $G_\mathcal{B}$.

The simplest sort of characterization we know is by linear functionals. For example, if $n = 2$ and $\mathcal{B} = \{(1,-1)\}$ then two vectors $\boldsymbol{u} = (u_1, u_2)$ and $\boldsymbol{u}' = (u_1', u_2')$ are in the same component of $G_\mathcal{B}$ if and only if $u_1 + u_2 = u_1' + u_2'$. However for $\mathcal{B} = \{(2,-2), (3,-3)\}$ there is no such characterization.

To treat such cases, we connect our problem with commutative algebra. For any vector $\boldsymbol{u} = (u_1, \ldots, u_n)$ of non-negative integers we define a monomial $\boldsymbol{x}^{\boldsymbol{u}} = x_1^{u_1} x_2^{u_2} \ldots x_n^{u_n} \in k[x_1, \ldots, x_n]$. Every vector $\boldsymbol{u} \in \mathbf{Z}^n$ can be written uniquely as $\boldsymbol{u} = \boldsymbol{u}_+ - \boldsymbol{u}_-$ where $\boldsymbol{u}_+, \boldsymbol{u}_-$ are non-negative vectors with disjoint support. For example $(1,-2) = (1,0) - (0,2)$. To a vector $\boldsymbol{u} \in \mathbf{Z}^n$ we associate the binomial difference $\boldsymbol{x}^{\boldsymbol{u}_+} - \boldsymbol{x}^{\boldsymbol{u}_-}$. Let k be any field. To the subset $\mathcal{B} \subset \mathbf{Z}^n$ we associate the ideal generated by the corresponding binomials:

$$I_\mathcal{B} = \langle \boldsymbol{x}^{\boldsymbol{u}_+} - \boldsymbol{x}^{\boldsymbol{u}_-} : \boldsymbol{u} \in \mathcal{B} \rangle \;\subset\; k[x_1, \ldots, x_n].$$

* The authors are grateful to the NSF for partial support during the preparation of this paper. The third author is also supported by a David and Lucile Packard Fellowship.

The following theorem shows that this is a good encoding scheme:

Theorem 1.1. *Two vectors* $\boldsymbol{u}, \boldsymbol{v} \in \mathbf{N}^n$ *are in the same component of* $G_{\mathcal{B}}$ *if and only if* $\boldsymbol{x}^{\boldsymbol{u}} - \boldsymbol{x}^{\boldsymbol{v}} \in I_{\mathcal{B}}$.

Theorem 1.1 has been rediscovered many times. One early reference is [MM]. See also [Stu, §5]. Applications of the graphs $G_{\mathcal{B}}$ to integer programming can be found in [Tho].

As explained further in Section 2, there are various decompositions

$$I_{\mathcal{B}} \quad = \quad J_1 \cap J_2 \cap \ldots \cap J_r,$$

where the J_i correspond to other, and in some ways simpler, combinatorial problems. For example, we might take a primary decomposition. Since $\boldsymbol{u}$ and $\boldsymbol{v}$ are in the same component of $G_{\mathcal{B}}$ if and only if $\boldsymbol{x}^{\boldsymbol{u}} - \boldsymbol{x}^{\boldsymbol{v}} \in J_i$, $1 \leq i \leq r$, such a decomposition of ideals allows a decomposition of the original problem. The following examples suggest how the theory will go:

Example 1.2.

The simplest typical example is provided by the set

$$\mathcal{B} = \{(2,-2),(3,-3)\} \ \subset \ \mathbf{Z}^2$$

when the characteristic of the field k is different from 2. If $\boldsymbol{u}, \boldsymbol{v} \in \mathbf{N}^2$ are connected via $\mathcal{B}$ then clearly $u_1 + u_2 = v_1 + v_2$. We shall see that the converse is true provided that $u_1 + u_2 \geq 3$. When this inequality is not satisfied, the situation is more delicate: Of the vectors $(2,0), (1,1), (0,2)$ with sum 2, only $(0,2)$ and $(2,0)$ are connected. Each of the vectors $(1,0), (0,1)$, and $(0,0)$ is isolated.

These statements are all easy, but we will now derive them using the general method of this paper. We first compute the primary decomposition

$$I_{\mathcal{B}} \quad = \quad \langle x^2 - y^2, x^3 - y^3 \rangle \quad = \quad \langle x - y \rangle \cap \langle x + y, x^3, x^2 y, x y^2, y^3 \rangle.$$

From this we see that (u_1, u_2) is connected to (v_1, v_2) via $\mathcal{B}$ if and only if it satisfies two conditions, corresponding to the two ideals on the right-hand side of the equation. The first condition is that $x^{u_1} y^{u_2} - x^{v_1} y^{v_2} \in \langle x - y \rangle$, that is, $x - y$ divides $x^{u_1} y^{u_2} - x^{v_1} y^{v_2}$, or equivalently $u_1 + u_2 = v_1 + v_2$. The second condition is harder to interpret combinatorially. Note that $\langle x + y, x^3, x^2 y, x y^2, y^3 \rangle$ contains all monomials of degree ≥ 3. Thus if $u_1 + u_2 = v_1 + v_2 \geq 3$, then $\boldsymbol{u}$ and $\boldsymbol{v}$ are connected via $\mathcal{B}$. Since $x^2 - y^2$

is divisible by $x + y$ it is also in the second ideal, and $(2,0)$ is connected with $(0,2)$. Inspection shows no other difference of monomials is in $I_{\mathcal{B}}$, completing the proof.

Example 1.3. (Poisson regression).

Here is a small example from statistics. Suppose that a chemical to control insects is sprayed on successive equally infested plots in increasing concentrations 0,1,2,3,4 (in some units). After the spraying the numbers of insects left alive on the plots are 44,25,21,19,11. Roughly: Greater concentration leads to fewer insects.

To extrapolate we need a model. One standard model postulates that the number of insects at concentration i has a Poisson distribution with mean parameter e^{a+bi} where a and b are parameters to be fitted from the data. If $\hat{a}$ and $\hat{b}$ are estimates of a and b, and $\hat{\lambda}_i = e^{\hat{a}+i\hat{b}}$, then a test for goodness of fit of the Poisson model can be based on the chi-square statistic

$$\sum_{i=1}^{5} \frac{(\hat{\lambda}_i - N_i)^2}{\hat{\lambda}_i}.$$

Asymptotic theory predicts an approximate chi-square distribution on 3 degrees of freedom. In this example, $\hat{a} = 3.707$ $\hat{b} = -.3125$, the 5 fitted values are $\hat{\lambda}_i = (40.8, 29.8, 21.8, 16.0, 11.7)$ and the chi-square statistic is 1.7.

Does this value of chi-square show that the data were well fitted? Poorly fitted? Calibrating the chi-square test leads to a combinatorial problem of the type considered above. Let $\mathcal{X}$ be the set of all non-negative 5-tuples $x = (x_0, \ldots, x_4)$ with $S(x) = x_0 + \ldots + x_4 = 120$ and $T(x) = 0x_0 + x_1 + 2x_2 + 3x_3 + 4x_4 = 168$ matching the data above. $\mathcal{X}$ is a finite set, a component of the graph G defined by the linear functionals S and T. We want to know what proportion of the 5-tuples in $\mathcal{X}$ have chi-square greater than 1.7. We do not want to enumerate all 5-tuples to find out (indeed, in realistic problems of this kind there are simply too many to enumerate) and we know no general theory that will solve this problem accurately. Thus we will approximate a solution by choosing examples uniformly from the set, and seeing what proportion of the examples chosen have chi-square greater than 1.7.

To make these choices, we might run a random walk starting at the original data vector $x^* = (44, 25, 21, 19, 11)$. As a first approximation, it might seem reasonable to take, for basic moves in the walk, any set of elements that span the sublattice of $\mathbf{Z}^n$ defined by the vanishing of the functionals S and T. For example, we might take

$$\mathcal{B} = \{(1,-1,-1,1,0),(1,-1,0,-1,1),(0,1,-1,-1,1)\}.$$

At each step of our walk we randomly choose $\pm$ one of the vectors in $\mathcal{B}$ and then add it to the current vector in $\mathcal{X}$. If the entry of the result is positive, we step to the sum, which is again a vector in $\mathcal{X}$. Otherwise we discard it. Thus, the walk might go

$$(44,25,21,19,11) \to (45,24,20,20,11) \to (45,23,21,21,10) \to \dots .$$

This walk generates a symmetric process which leads to the uniform distribution on the component of $G_{\mathcal{B}}$ containing x^*.

It turns out in this example that that the observed chi-square distribution for this walk is rather close to the true distribution. Unfortunately, not every pair of vectors in $\mathcal{X}$ can be connected by steps in $\mathcal{B}$, so the set of 5-tuples over which we are averaging is not quite the same as the set we want! For example, the vector $(36,0,84,0,0)$ cannot be connected to x^* by steps in $\mathcal{B}$ keeping all entries positive. The primary decomposition exhibited below shows that two non-negative integer vectors (i_1,j_1,h_1,l_1,m_1) and (i_2,j_2,h_2,l_2,m_2) in $\mathcal{X}$ are connected via $\mathcal{B}$ if and only if

$$\begin{aligned} i_r + j_r + h_r &\geq 1 \text{ and} \\ i_r + j_r + l_r &\geq 1 \text{ and} \\ j_r + l_r + m_r &\geq 1 \text{ and} \\ h_r + l_r + m_r &\geq 1. \end{aligned} \tag{1.1}$$

The set $\mathcal{B}$ can be enlarged to

$$\mathcal{B}' = \mathcal{B} \cup \{(1,-2,1,0,0),(0,1,-1,1,0),(0,0,1,-2,1)\},$$

and we shall see that this enlargement is sufficient. However, the distribution observed for chi-square coming from a random walk based on $\mathcal{B}$ is close to one based on $\mathcal{B}'$. In particular, using $\mathcal{B}$, the proportion of samples with chi-squared < 1.7 is .0031, whereas using $\mathcal{B}'$ the proportion is .0046. See the histograms below, each of which is based on a walk of 90,000 steps, an initial 10,000 having been discarded.

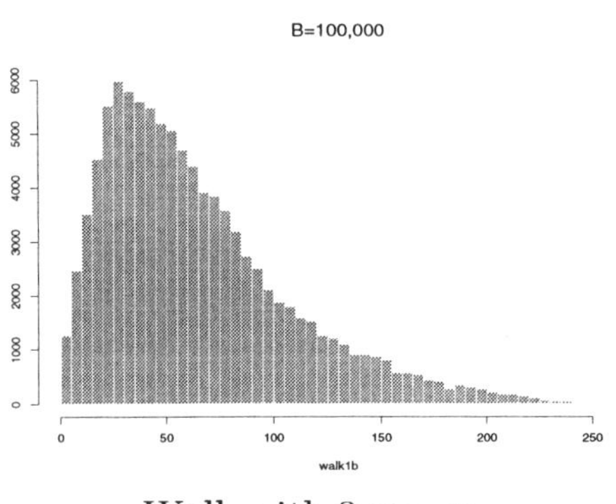

Walk with 3 moves

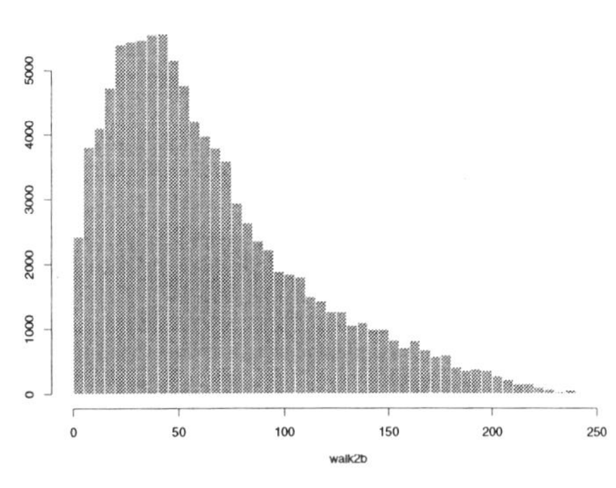

Walk with 6 moves

To understand the situation we turn to primary decomposition. We work in the polynomial ring $k[x_1, x_2, x_3, x_4, x_5]$. The set $\mathcal{B}$ is encoded by the binomial ideal $I_\mathcal{B} = \langle x_2x_3 - x_1x_4, x_2x_4 - x_1x_5, x_3x_4 - x_2x_5 \rangle$. Two 5-tuples are connected via $\mathcal{B}$ if and only if $I_\mathcal{B}$ contains

$$x_1^{i_1} x_2^{j_1} x_3^{k_1} x_4^{l_1} x_5^{m_1} - x_1^{i_2} x_2^{j_2} x_3^{k_2} x_4^{l_2} x_5^{m_2}. \tag{1.2}$$

The primary decomposition of $I_\mathcal{B}$ is found to be

$$I_\mathcal{B} = I \cap \langle x_1, x_2, x_3 \rangle \cap \langle x_1, x_2, x_4 \rangle \cap \langle x_2, x_4, x_5 \rangle \cap \langle x_3, x_4, x_5 \rangle, \tag{1.3}$$

where I is the prime ideal generated by the 2×2-minors of the matrix as is

$$\begin{pmatrix} x_1 & x_2 & x_3 & x_4 \\ x_2 & x_3 & x_4 & x_5 \end{pmatrix}.$$

Note that (1.2) is in $I_\mathcal{B}$ if and only if it is in each ideal on the right of (1.3). It is in I if and only if $i_1 + j_1 + h_1 + l_1 + m_1 = i_2 + j_2 + k_2 + l_2 + m_2$ and $j_1 + 2k_1 + 3l_1 + 4m_1 = j_2 + 2k_2 + 3l_2 + 4m_2$. The remaining ideals on the right-hand side of (1.3) are generated by monomials. A polynomial is in a monomial ideal J if and only if each term in J. This gives the remaining characterizing relations claimed in (1.1).

The ideal I encodes a lattice walk that connects all of $\mathcal{X}$. The reader may wonder why we didn't simply begin with the set $\mathcal{B}'$, corresponding to the six generators of I, in constructing the random walk above. In larger problems, it is computationally quite taxing to find connecting sets of moves. It is natural to take a smaller set of moves such as a lattice basis and "hope for the best;" indeed, this approach is taken in several published studies. The goal of the theory initiated in this paper is to understand the nature of such approximations.

The rest of this paper is laid out as follows. In Section 2 we set up the algebraic technique which comes from the work on binomial ideals [ES]. Section 3 treats contingency tables; $a \times b$ arrays of non-negative integers with given row and column sums. Section 4 describes a different basis $\mathcal{B}$ for the problem of contingency tables, the "adjacent minors." In the final section we discuss a systematic way of making a relatively small choice of basis for any lattice walk problem. We call these circuit walks.

2. Lattice ideals

With notation as above, let $\mathcal{B}$ be a set of vectors in $\mathbf{Z}^n$. Let $\mathcal{L}$ be the subgroup of $\mathbf{Z}^n$ generated by $\mathcal{B}$. Call $\boldsymbol{u}, \boldsymbol{v} \in \mathbf{N}^n$ *equivalent* if $\boldsymbol{u} - \boldsymbol{v} \in \mathcal{L}$. In Example 1.2, $\mathcal{L}$ is the set $(u_1, u_2) \in \mathbf{Z}^2$ with $u_1 + u_2 = 0$ and $(u_1, u_2), (v_1, v_2) \in \mathbf{N}^2$ are equivalent if and only if $u_1 + u_2 = v_1 + v_2$. In Example 1.3, the equivalence classes generated by $\mathcal{L}$ are the set of all $\boldsymbol{u} \in \mathbf{N}$ with $u_1 + u_2 + u_3 + u_4 + u_5$, $0 \cdot u_1 + 1u_2 + 2u_3 + 3u_4 + 4u_5$ having fixed values. In applied problems the equivalence classes are often the basic objects of interest. One wants to construct a set of edges (that is a choice of $\mathcal{B}$) that connects elements within an equivalence class by a path along which all components stay non-negative. In [DS], it is shown how to find a finite set $\mathcal{B}$ by finding a Gröbner basis for $I_{\mathcal{L}}$. The division algorithm gives an effective algorithm for finding a connecting path.

If $\boldsymbol{u}$ and $\boldsymbol{v}$ lie in the same connected component of $G_{\mathcal{B}}$ they are equivalent but not conversely. It can be shown that two vectors $\boldsymbol{u}, \boldsymbol{v} \in \mathbf{N}^n$ are equivalent if and only if $\boldsymbol{u} + \boldsymbol{w}$ can be connected to $\boldsymbol{v} + \boldsymbol{w}$ by a move in $\mathcal{B}$ for some (sufficiently large) $\boldsymbol{w} \in \mathbf{N}^n$. It is an interesting problem to give useful bounds on $\boldsymbol{w}$. Theorem 1.1 gives the following condition for equivalence.

Corollary 2.1. *Let $\mathcal{B} \subseteq \mathbf{Z}^n$ generate $\mathcal{L} = \mathbf{Z}\mathcal{B}$. Every pair of $\mathcal{L}$ equivalent vectors is connected via $\mathcal{B}$ if and only if $I_{\mathcal{L}} = I_{\mathcal{B}}$.*

The ideals considered here are all generated by binomials $\boldsymbol{x}^{\boldsymbol{u}_+} - \boldsymbol{x}^{\boldsymbol{u}_-}$. We shall need a few results from the recently developed theory of binomial ideals [ES], and we briefly review now. General references for commutative algebra with emphasis on computational aspects are [CLO], [Stu]. Thorough treatments of primary decomposition can be found in [AM], [ES].

A *binomial* in $k[x_1, \ldots, x_n]$ is a polynomial with at most two terms. A *binomial ideal* is an ideal generated by binomials. Thus, monomial ideals are also binomial ideals. The following theorem is proved in [ES].

Theorem 2.2. *Every binomial ideal has a binomial primary decomposition.*

In [ES, §9] there is an explicit algorithm which expresses a given binomial ideal as an intersection of primary binomial ideals. A primary decomposition algorithm for general polynomial ideals is given in [BW §8]. Specializing to the situation of the paper we get

$$I_{\mathcal{B}} = J_1 \cap J_2 \cap \ldots \cap J_r,$$

where each J_i is primary and generated by binomials $\alpha \boldsymbol{x}^{\boldsymbol{u}} - \beta \boldsymbol{x}^{\boldsymbol{v}}$.

Let $\mathcal{L}$ be a lattice in $\mathbf{Z}^n$. Call $\mathcal{L}$ *saturated* if for each $r \in \mathbf{Z}, \boldsymbol{u} \in \mathbf{Z}^n, r \cdot \boldsymbol{u} \in \mathcal{L}$ implies $\boldsymbol{u} \in \mathcal{L}$. Equivalently, $\mathcal{L}$ is saturated if and only if the quotient group $\mathbf{Z}^n/\mathcal{L}$ is free abelian. All of the lattices that appear in the examples of this paper are saturated. In [ES] it is proved that the binomial ideal $I_{\mathcal{L}}$ is prime if and only if $\mathcal{L}$ is saturated.

If $\mathcal{L} = \mathbf{Z}\mathcal{B}$ is saturated then [ES] show that the prime $I_{\mathcal{L}}$ appears among the J_i. Otherwise, $I_{\mathcal{L}}$ equals the intersection of some of the J_i. All other J_i's must contain monomials by [ES §2].

Theorem 1.1 shows that $\boldsymbol{u}$ and $\boldsymbol{v}$ are connected via $\mathcal{B}$ if and only if $\boldsymbol{x}^{\boldsymbol{u}} - \boldsymbol{x}^{\boldsymbol{v}}$ lies in J_i for all i. If J_i is a monomial ideal, the corresponding combinatorial condition is easy: suppose

$$J_i \quad = \quad \langle \boldsymbol{x}^{\boldsymbol{a}}, \boldsymbol{x}^{\boldsymbol{b}}, \ldots, \boldsymbol{x}^{\boldsymbol{c}} \rangle,$$

then $\boldsymbol{x}^{\boldsymbol{u}} - \boldsymbol{x}^{\boldsymbol{v}} \in J_i$ if and only if $\boldsymbol{x}^{\boldsymbol{u}}, \boldsymbol{x}^{\boldsymbol{v}} \in J_i$. Further, $\boldsymbol{x}^{\boldsymbol{u}} \in J_i$ if and only if $\boldsymbol{u} \geq \boldsymbol{a}$ or $\boldsymbol{u} \geq \boldsymbol{b}$ or ... or $\boldsymbol{u} \geq \boldsymbol{c}$. Even if the J_i's are not monomial ideals, artful choices may allow neat necessary and sufficient conditions or neat necessary conditions. Examples appear in Section 3 and 4 below. It is often convenient to combine the J_i in groups, so that the intersection of each group is generated by monomials $\boldsymbol{x}^{\boldsymbol{w}}$ and pure binomials $\boldsymbol{x}^{\boldsymbol{u}} - \boldsymbol{x}^{\boldsymbol{v}}$. Such a regrouping facilitates the combinatorial translation of containment in J_i. See [ES, Cor. 8.2] for further details.

3. Corner minors

The prototype of the problems considered here is that of generating random "contingency tables" — tables of nonnegative integers of given size with fixed row and column sums. The statisticians J. Darroch and G. Glonek (see [Gl]) introduced a random walk technique: Start at a given table and take steps that do not change the nonnegativity or the row and column sums. For example, consider the following procedure: at each step, a position l in the first row and m in the first column are chosen randomly. The current table is changed to a new table by altering the four entries in positions $(1,1)$, $(l,1)$, $(1,m)$, (l,m) by adding or subtracting 1 following either the pattern of signs $\begin{pmatrix} + & - \\ - & + \end{pmatrix}$ or $\begin{pmatrix} - & + \\ + & - \end{pmatrix}$, the choice being random as well. For instance, for $a = b = 3$ this basis is

$$\mathcal{B}_{\text{cor}} = \left\{ \begin{pmatrix} +1 & -1 & 0 \\ -1 & +1 & 0 \\ 0 & 0 & 0 \end{pmatrix}, \begin{pmatrix} +1 & 0 & -1 \\ -1 & 0 & +1 \\ 0 & 0 & 0 \end{pmatrix}, \begin{pmatrix} +1 & -1 & 0 \\ 0 & 0 & 0 \\ -1 & +1 & 0 \end{pmatrix}, \begin{pmatrix} +1 & 0 & -1 \\ 0 & 0 & 0 \\ -1 & 0 & +1 \end{pmatrix} \right\}.$$

A change is suppressed if it would lead to a table with negative entries. This process defines a symmetric random walk with a uniform stationary distribution on the set of tables connected to the starting table by the given moves. It turns out, however, that these moves may not connect all the non-negative tables with the given row and column sums. Glonek showed that they do suffice to connect these tables when the row and column sums are all ≥ 2. In this section we derive a strengthening of Glonek's result by describing the primary decomposition of the ideal corresponding to the set of chosen moves.

More formally, we are concerned with the lattice $\mathbf{Z}^{a \times b}$ of $a \times b$-integer matrices, and $\mathcal{L}$ is the sublattice of matrices with zero row sums and zero column sums. We begin by considering a random walk with a larger set $\mathcal{B}_{all}$ of possible moves: Again, the walk is over all non-negative integer $a \times b$-matrices with fixed row and column sums. To describe a move in $\mathcal{B}_{all}$ we select the positions in a 2×2-submatrix and alter them by adding

$$\begin{pmatrix} +1 & -1 \\ -1 & +1 \end{pmatrix} \quad \text{or} \quad \begin{pmatrix} -1 & +1 \\ +1 & -1 \end{pmatrix}. \tag{3.1}$$

The following result appears in [Gl] but was probably known before.

Lemma 3.1. *The moves in (3.1) are necessary and sufficient to connect any pair of non-negative integer $a \times b$-matrices with the same row and column sums.*

Proof. It is known (see e.g., [Stu, Prop. 5.4]) that $I_{\mathcal{L}}$ is a prime ideal which is minimally generated by the 2×2-minors of an $a \times b$-matrix of indeterminates:

$$\begin{pmatrix} x_{11} & x_{12} & \cdots & x_{1b} \\ x_{21} & x_{22} & \cdots & x_{2b} \\ \vdots & \vdots & \ddots & \vdots \\ x_{a1} & x_{a2} & \cdots & x_{ab} \end{pmatrix}. \tag{3.2}$$

Lemma 3.1 now follows from Corollary 2.1. ∎

The number of moves (3.1) is $\binom{a}{2}\binom{b}{2}$, which is much larger than $\operatorname{rank}(\mathcal{L}) = (a-1)(b-1)$, the size of the collection of moves $\mathcal{B}_{\text{cor}}$ described at the beginning of this section — quartic rather than quadratic in the size of the tables. It turns out that though the moves $\mathcal{B}_{\text{cor}}$ do not connect all possible tables with the given row and column sums, they come rather close. The following primary decomposition result allows an analysis of the situation.

Theorem 3.2. *Let $I_{\mathcal{L}}$ be the prime ideal generated by all 2×2-minors of (3.2), where $a, b \geq 2$. Let $R := \langle x_{11}, \ldots, x_{1b} \rangle$ and $C := \langle x_{11}, \ldots, x_{a1} \rangle$. The ideal of "corner minors" $I_{\mathcal{B}_{\text{cor}}} := \langle\, x_{11}x_{ij} - x_{1j}x_{i1} \;:\; 2 \leq i \leq a,\, 2 \leq j \leq b \,\rangle$ has the primary decomposition*

$$I_{\mathcal{B}_{\text{cor}}} \quad = \quad I_{\mathcal{L}} \cap R \cap C \cap (I_{\mathcal{B}_{\text{cor}}} + R^2 + C^2). \tag{3.3}$$

If $a, b > 2$ then this primary decomposition is minimal, whereas if $b = 2$, the last two terms can be dropped, and similarly if $a = 2$.

From (3.3) we can read off the connectivity properties of the basis $\mathcal{B}_{\text{cor}}$.

Corollary 3.3. *Two non-negative integer $a \times b$-matrices U, V are connected via $\mathcal{B}_{\text{cor}}$ if*

(a) U and V have the same row and column sums, and
(b) U, V have positive first row sum,
(c) U, V have positive first column sum,
(d) U, V have either row sum ≥ 2 or column sum ≥ 2.

Proof. The primary decomposition (3.3) implies

$$I_{\mathcal{B}_{\text{cor}}} \quad \supseteq \quad I_{\mathcal{L}} \cap R \cap C \cap (R^2 + C^2).$$

This inclusion of ideals is equivalent to the assertion of Corollary 3.3. ∎

Proof of Theorem 3.2. We will deal only with the case $a, b > 2$, leaving the (easy) remaining cases to the reader.

The left-hand side of (3.3) is clearly contained in the right-hand side of (3.3). Order the variables row-wise $x_{11} < x_{12} < \ldots < x_{1b} < x_{21} < \ldots < x_{ab}$ and let $<$ denote the resulting reverse lexicographic term order. We shall prove the equality

$$in_<(I_\mathcal{L}) \cap in_<(R) \cap in_<(C) \cap in_<(I_{\mathcal{B}_{\text{cor}}} + R^2 + C^2) \quad = \quad in_<(I_{\mathcal{B}_{\text{cor}}}) \quad (3.4)$$

This implies (3.3) because the left-hand side of (3.4) contains the initial ideal of the right-hand side of (3.3); hence both sides of (3.3) have the same initial ideal and are thus equal.

In order to evaluate the constituents in (3.4) we introduce the ideals

$$R' := \langle x_{12}, x_{13} \ldots, x_{1b} \rangle \qquad \text{and} \qquad C' := \langle x_{21}, x_{31} \ldots, x_{a1} \rangle \qquad \text{and}$$
$$I_{s,t} := \langle\, x_{ij} x_{kl} \mid s \leq i < k,\ j > l \geq t \,\rangle \qquad \text{for } 1 \leq s \leq a \text{ and } 1 \leq t \leq b.$$

Since the 2×2-minors are a Gröbner basis (see e.g., [Stu, Prop. 5.4]), we have

$$in_<(I_\mathcal{L}) \quad = \quad I_{1,1}. \qquad (3.5)$$

Note that $in_<(R) = R$ and $in_<(C) = C$. We next derive the identity

$$in_<(I_{\mathcal{B}_{\text{cor}}} + R^2 + C^2) \quad = \quad (R + C)^2. \qquad (3.6)$$

It is evident that $R'C' \subseteq in_<(I_{\mathcal{B}_{\text{cor}}})$ and it follows that $(R + C)^2 \subseteq in_<(I_{\mathcal{B}_{\text{cor}}} + R^2 + C^2)$. For the reverse inclusion it suffices to show that the minimal generators of $I_{\mathcal{B}_{\text{cor}}} + R^2 + C^2$ are a Gröbner basis. Using Buchberger's first criterion [BW, Theorem 5.68], this reduces to a few easily checked cases, such as

$$\text{s-pol}\big(\, x_{k1}x_{1j} - x_{11}x_{kj}\,,\, x_{k1}x_{1l} - x_{11}x_{kl} \,\big) \quad = \quad x_{11}x_{1j}x_{kl} - x_{11}x_{1l}x_{kj} \in R^2. \qquad (3.7)$$

For the definition of *s-pol(ynomial)* see [BW, pp. 211] or [CLO, pp. 82]. Having thus verified (3.6), we now claim the following more complicated identity:

$$\begin{aligned} in_<(I_{\mathcal{B}_{\text{cor}}}) \quad = \quad & R'C' \;+\; x_{11}^2 I_{2,2} \;+\; \sum_{t \geq 2} x_{11}x_{1t} I_{2,t} \;+\; \sum_{s \geq 2} x_{11}x_{s1} I_{s,2} \\ & + x_{11} \cdot \langle\, x_{ij}x_{kl} \mid i < k \text{ and } j > l \text{ and } (i = 1 \text{ or } l = 1) \rangle. \end{aligned} \qquad (3.8)$$

Here each summand is a product of ideals. We abbreviate the last summand by M. We first show that $in_<(I_{\mathcal{B}_{\text{cor}}})$ contains the right-hand side of (3.8). To see that $in_<(I_{\mathcal{B}_{\text{cor}}})$ contains M, it suffices (by symmetry) to check that it contains $x_{11}x_{1j}x_{kl}$ for $1 < k, j > l$. This is clear from (3.7). Next let $s \leq i < k$, $j \geq l \geq 2$ and consider the following $\mathcal{B}_{\text{cor}}$-walk:

$$\begin{aligned} &x_{11}x_{s1}x_{ij}x_{kl} \to x_{1j}x_{s1}x_{i1}x_{kl} \to x_{sj}x_{11}x_{i1}x_{kl} \to x_{sj}x_{1l}x_{i1}x_{k1} \\ &\to x_{sj}x_{11}x_{il}x_{k1} \to x_{s1}x_{1j}x_{il}x_{k1} \to x_{s1}x_{11}x_{il}x_{kj}. \end{aligned}$$

This shows that $x_{11}x_{s1} \cdot (x_{ij}x_{kl} - x_{il}x_{kj}) \in I_{\mathcal{B}_{\text{cor}}}$. Applying (3.5) to $(a-s) \times (b-1)$-matrices, we conclude that $x_{11}^2 I_{2,2} \subset in_<(I_{\mathcal{B}_{\text{cor}}})$ and $x_{11}x_{s1}I_{s,2} \subset in_<(I_{\mathcal{B}_{\text{cor}}})$ for $s \geq 2$. By symmetry, we also obtain $x_{11}x_{1t}I_{2,t} \subset in_<(I_{\mathcal{B}_{\text{cor}}})$ for $t \geq 2$. It is evident that $R'C' \subset in_<(I_{\mathcal{B}_{\text{cor}}})$. We have shown that the right-hand side of (3.8) lies in the left-hand side.

The inclusion $\supseteq$ in (3.4) is clear. To prove equality in both (3.4) and (3.8), it suffices to show that the right-hand side of (3.8) contains the left-hand side in (3.4), i.e.,

$$\begin{aligned} &I_{1,1} \cap R \cap C \cap (R+C)^2 \quad \subseteq \\ &\qquad R'C' + x_{11}^2 I_{2,2} + \sum_{t \geq 2} x_{11}x_{1t}I_{2,t} + \sum_{s \geq 2} x_{11}x_{s1}I_{s,2} + M. \end{aligned}$$

Let m be a monomial in $I_{1,1} \cap R \cap C \cap (R+C)^2$. If m is not divisible by x_{11}, then since $m \in R \cap C$ we must have $m \in R' \cap C' = R'C'$. Thus we may suppose that x_{11} divides m. If m is not divisible by x_{11}^2, then since $m \in (R+C)^2$ we must have $m \in R'$ or $m \in C'$, say m is divisible by $x_{11}x_{1t}$ for some t. Since $m \in I_{1,1}$ as well, we see that either:

- m is also divisible by x_{su} for some $s > 1$ and $u < t$, in which case $m \in M$, or else
- m is also divisible by $x_{su}x_{vw}$ with $s > v > 1$, $t \leq u < w$, in which case $m \in x_{11}x_{1t}I_{2,t}$.

In either case m lies in the desired sum. Now consider the case where x_{11}^2 divides m. Since $m \in I_{1,1}$, m is also divisible by a product of the form $x_{ij}x_{kl}$ with $i < j$, $k > l$. If $i = l = 1$, then $m \in R'C'$. If exactly one of i or l equals 1, then $m \in M$. If both $i > 1$ and $l > 1$ then $m \in x_{11}^2 I_{2,2}$, and we are done.

Finally, we show that the intersection in (3.3) is irredundant. It suffices to show this for $a = b = 3$. In this special case the monomial ideal (3.8) equals

$$in_{\prec}(I_{\mathcal{B}_{\text{cor}}}) = \langle x_{12}x_{21}, x_{13}x_{21}, x_{12}x_{31}, x_{13}x_{31}, x_{11}^2 x_{23}x_{32}, x_{11}x_{12}x_{23}x_{32}, \\ x_{11}x_{21}x_{23}x_{32}, x_{11}x_{22}x_{31}, x_{11}x_{23}x_{31}, x_{11}x_{13}x_{22}, x_{11}x_{13}x_{32} \rangle. \tag{3.9}$$

The following "witnesses" show that each of the four primary components in (3.3) is needed:

$$x_{11}^2, \quad x_{31}\cdot(x_{21}x_{32}-x_{22}x_{31}), \quad x_{13}\cdot(x_{13}x_{22}-x_{12}x_{23}), \quad x_{11}\cdot(x_{22}x_{33}-x_{23}x_{32}).$$

The intersection of any three of the ideals on the right-hand side of (3.3) contains one the four polynomials listed. But none of these four is in $I_{\mathcal{B}_{\text{cor}}}$ because none of their terms lies in (3.9). This completes the proof of Theorem 3.2. ∎

Remark 3.4. In general when we have a primary decomposition $I = \cap_j I_j$ in a polynomial ring with a term order $<$, then $in_>(I) \subseteq \cap_j in_>(I_j)$, but the two sides will usually not be equal. For a simple example, suppose that $x < y$ are indeterminates, and consider

$$\langle x^2 \rangle = in_<\langle x^2 - y^2 \rangle \quad \neq \quad \langle x \rangle \cap \langle x \rangle = in_<\langle x - y \rangle \cap in_<\langle x + y \rangle.$$

But in the setting of Theorem 3.1 a small miracle occurs and the corresponding intersections of initial ideals are equal for the correct choice of term order. It would be interesting to understand when such things happen in general.

Remark 3.5. The referee suggested an alternate approach to the proof of Theorem 3.2 which simplifies the computations but yields a little less: Write J for the ideal on the right-hand side of (3.3). As J is obviously contained in $I_{\mathcal{B}_{\text{cor}}}$, it suffices to prove $I_{\mathcal{B}_{\text{cor}}} \subseteq J$. It is easy to see that the two ideals become equal after adding (x_{11}) to both sides, so it suffices to show that

$$x_{11}(J : x_{11}) \subseteq I_{\mathcal{B}_{\text{cor}}}.$$

Using the computation of the initial ideal of $I_{\mathcal{B}_{\text{cor}}} + R^2 + C^2$ in the proof above one can prove that $(I_{\mathcal{B}_{\text{cor}}} + R^2 + C^2) : x_{11} = R + C$, and it follows that $(J : x_{11}) = I_{\mathcal{L}} \cap (R + C)$. Thus it suffices to show that $x_{11}(I_{\mathcal{L}} \cap (R + C)) \subseteq I_{\mathcal{B}_{\text{cor}}}$....

4. Adjacent minors

Another natural basis for the lattice $\mathcal{L}$ in Section 3 is the set $\mathcal{B}_{adj}$ of adjacent 2×2-minors. Here the situation is more complicated than before. Let us examine the case $a = b = 4$ in detail. The adjacent 2×2-moves for 4×4-matrices are encoded by the ideal

$$\begin{aligned} I \;:=\; I_{\mathcal{B}_{adj}} \;=\; & \langle x_{12}x_{21} - x_{11}x_{22},\, x_{13}x_{22} - x_{12}x_{23},\, x_{14}x_{23} - x_{13}x_{24}, \\ & x_{22}x_{31} - x_{21}x_{32},\, x_{23}x_{32} - x_{22}x_{33},\, x_{24}x_{33} - x_{23}x_{34}, \\ & x_{32}x_{41} - x_{31}x_{42},\, x_{33}x_{42} - x_{32}x_{43},\, x_{34}x_{43} - x_{33}x_{44}\rangle. \end{aligned}$$

Two nonnegative integer 4×4-matrices (a_{ij}) and (b_{ij}) with the same row and column sums can be connected by a sequence of adjacent 2×2-moves if and only if the binomial

$$\prod_{1 \le i,j \le 4} x_{ij}^{a_{ij}} \;-\; \prod_{1 \le i,j \le 4} x_{ij}^{b_{ij}}$$

lies in the ideal I, by Corollary 2.1.

Proposition 4.1. *Two non-negative integer 4×4-matrices with the same row and column sums can be connected by a sequence of adjacent 2×2-moves if both of them satisfy the following six inequalities:*

(i) $a_{21} + a_{22} + a_{23} + a_{24} \ge 2$;
(ii) $a_{31} + a_{32} + a_{33} + a_{34} \ge 2$;
(iii) $a_{12} + a_{22} + a_{32} + a_{42} \ge 2$;
(iv) $a_{13} + a_{23} + a_{33} + a_{43} \ge 2$;
(v) $a_{12} + a_{22} + a_{23} + a_{24} + a_{31} + a_{32} + a_{33} + a_{43} \ge 1$;
(vi) $a_{13} + a_{21} + a_{22} + a_{23} + a_{32} + a_{33} + a_{34} + a_{42} \ge 1$.

We remark that these sufficient conditions remain valid if (at most) one of the four inequalities "≥ 2" is replaced by "≥ 1." No further relaxation of the conditions (i)–(vi) is possible, as is shown by the following two pairs of matrices, which are disconnected:

$$\begin{pmatrix} 0 & 0 & 0 & 0 \\ 0 & 1 & 1 & 0 \\ 0 & 1 & 0 & 0 \\ 0 & 0 & 0 & 1 \end{pmatrix} \leftrightarrow \begin{pmatrix} 0 & 0 & 0 & 0 \\ 0 & 0 & 1 & 1 \\ 0 & 1 & 0 & 0 \\ 0 & 1 & 0 & 0 \end{pmatrix}$$

and

$$\begin{pmatrix} 0 & 0 & 1 & 0 \\ 1 & 1 & 0 & 0 \\ 0 & 0 & 0 & 2 \\ 0 & 0 & 0 & 0 \end{pmatrix} \leftrightarrow \begin{pmatrix} 0 & 0 & 0 & 1 \\ 0 & 0 & 1 & 1 \\ 1 & 1 & 0 & 0 \\ 0 & 0 & 0 & 0 \end{pmatrix}.$$

The necessity of conditions (v) and (vi) is seen from the disconnected matrices

$$\begin{pmatrix} n & n & 0 & n \\ 0 & 0 & 0 & n \\ n & 0 & 0 & 0 \\ n & 0 & n & n \end{pmatrix} \leftrightarrow \begin{pmatrix} n & 0 & n & n \\ n & 0 & 0 & 0 \\ 0 & 0 & 0 & n \\ n & n & 0 & n \end{pmatrix} \qquad \text{for any integer} \quad n \geq 0.$$

Proof of Proposition 4.1. Let $I_{\mathcal{L}}$ be the prime ideal generated by all 36 2×2-minors of a 4×4-matrix (x_{ij}) of indeterminates. Define also the primes

$$\begin{aligned} C_1 &:= \langle x_{12}, x_{22}, x_{23}, x_{24}, x_{31}, x_{32}, x_{33}, x_{43} \rangle \quad \text{and} \\ C_2 &:= \langle x_{13}, x_{21}, x_{22}, x_{23}, x_{32}, x_{33}, x_{34}, x_{42} \rangle. \end{aligned}$$

Using a computer algebra system – such as MACAULAY – it can be verified easily that

$$\begin{aligned} I_{\mathcal{L}} \cap C_1 \cap C_2 \cap \langle x_{21}, x_{22}, x_{23}, x_{24} \rangle \cap \langle x_{31}, x_{32}, x_{33}, x_{34} \rangle^2 \\ \cap \langle x_{12}, x_{22}, x_{32}, x_{42} \rangle^2 \cap \langle x_{13}, x_{23}, x_{33}, x_{43} \rangle^2 \subseteq I_{\mathcal{B}_{adj}}. \end{aligned} \tag{4.1}$$

This containment of ideals implies Proposition 4.1. ■

For completeness we describe the primary decomposition of $I = I_{\mathcal{B}_{adj}}$. This is a good test case for implementations of (binomial) primary decomposition. Consider the prime ideals

$$A \quad := \quad \langle x_{12}x_{21} - x_{11}x_{22}, x_{13}, x_{23}, x_{31}, x_{32}, x_{33}, x_{43} \rangle \qquad \text{and}$$

$$\begin{aligned} B := \langle\, & x_{11}x_{22} - x_{12}x_{21}, x_{11}x_{23} - x_{13}x_{21}, x_{11}x_{24} - x_{14}x_{21}, \\ & x_{12}x_{23} - x_{13}x_{22}, x_{12}x_{24} - x_{14}x_{22}, x_{13}x_{24} - x_{14}x_{23}, \\ & x_{31}, x_{32}, x_{33}, x_{34} \rangle. \end{aligned}$$

Rotating and reflecting the matrix (x_{ij}), we find eight ideals $A_1, A_2, \ldots, A_8$ equivalent to A and four ideals B_1, B_2, B_3, B_4 equivalent to B. Note that A_i has codimension 7 and degree 2, B_j has codimension 7 and degree 4, and C_k has codimension 8 and degree 1, while P has codimension 9 and degree 20.

Proposition 4.2. *The minimal associated primes of the binomial ideal I are the 15 primes A_i, B_j, C_j and P. Each of these occurs with multiplicity one in I, so that*

$$\mathrm{Rad}(I) \quad = \quad A_1 \cap A_2 \cap \cdots \cap A_8 \cap B_1 \cap B_2 \cap B_3 \cap B_4 \cap C_1 \cap C_2 \cap P.$$

In particular, both I and its radical Rad(I) *have codimension* 7 *and degree* 32.

We next present the list of all the embedded components of I. We are grateful to Serkan Hosten and Jay Shapiro for pointing out some errors in a previous version of this list. Each of the following five ideals D, E, F, F' and G was shown to be primary by using Algorithm 9.4 in [ES].

Our first primary ideal is

$$\begin{aligned} D \quad := \quad & \langle x_{13}, x_{23}, x_{33}, x_{43} \rangle^2 \; + \; \langle x_{31}, x_{32}, x_{33}, x_{34} \rangle^2 \; + \\ & \langle\, x_{ik}x_{jl} - x_{il}x_{jk} \; : \; \min\{j,l\} \leq 2 \;\text{ or }\; (3,3) \in \{(i,k),(j,l),(i,l),(j,k)\} \,\rangle \end{aligned}$$

Its radical Rad(D) is a prime of codimension 10 and degree 5. (Commutative algebra experts will notice that Rad(D) is a *ladder determinantal ideal.*) Up to symmetry, there are four such ideals D_1, D_2, D_3, D_4.

Our second type of primary ideal is

$$\begin{aligned} E \quad := \quad & \big(\, [I + \langle x_{12}^2, x_{21}^2, x_{22}^2, x_{23}^2, x_{24}^2, x_{32}^2, x_{33}^2, x_{34}^2, x_{42}^2, x_{43}^2 \rangle] \\ & \qquad\qquad\qquad : (x_{11}x_{13}x_{14}x_{31}x_{41}x_{44})^2 \, \big). \end{aligned}$$

Its radical Rad(E) is a monomial prime of codimension 10. Up to symmetry, there are four such primary ideals E_1, E_2, E_3, E_4.

Our third type of primary ideal has codimension 10 as well. It equals

$$\begin{aligned} F \quad := \quad & \big(\, [I + \langle x_{12}^3, x_{13}^3, x_{22}^3, x_{23}^3, x_{31}^3, x_{32}^3, x_{33}^3, x_{34}^3, x_{42}^3, x_{43}^3 \rangle] \\ & \qquad\qquad : (x_{11}x_{14}x_{21}x_{24}x_{41}x_{44})^2 (x_{11}x_{24} - x_{21}x_{14}) \, \big) \end{aligned}$$

Its radical Rad(F) is a monomial prime. Up to symmetry, there are four such primary ideals F_1, F_2, F_3, F_4. Note how *Rad*(F) differs from *Rad*(E).

Our fourth type of primary is the following ideal of codimension 11:

$$F' \quad := \quad \big(\, [I + \langle x_{12}^3, x_{13}^3, x_{22}^3, x_{23}^3, x_{31}^3, x_{32}^3, x_{33}^3, x_{34}^3, x_{42}^3, x_{43}^3 \rangle] \\ : (x_{11}x_{14}x_{21}x_{24}x_{41}x_{44})(x_{21}x_{44} - x_{41}x_{24}) \, \big)$$

Up to symmetry, there are four such primary ideals F_1', F_2', F_3', F_4'. Note that $Rad(F') = Rad(F) + \langle x_{14}x_{21} - x_{11}x_{24} \rangle$. In particular, the ideals F and F' lie in the same *cellular component* of I; see [ES, Section 6].

Our last primary ideal has codimension 12. It is unique up to symmetry.

$$G \quad := \quad \big(\, [I + \langle x_{12}^5, x_{13}^5, x_{21}^5, x_{22}^5, x_{23}^5, x_{24}^5, x_{31}^5, x_{32}^5, x_{33}^5, x_{34}^5, x_{42}^5, x_{43}^5 \rangle] \\ : (x_{11}x_{14}x_{41}x_{44})^5 (x_{11}x_{44} - x_{14}x_{41}) \, \big).$$

In summary, we have the following theorem.

Theorem 4.3. *The ideal I of adjacent 2×2-minors of a generic 4×4-matrix has 32 associated primes, 15 minimal and 17 embedded. Using the decomposition in Proposition 4.2, we get the minimal primary decomposition*

$$I \quad = \quad \mathrm{Rad}(I) \cap D_1 \cap \cdots \cap D_4 \cap E_1 \cap \cdots \cap E_4 \cap F_1 \cap \cdots \cap F_4 \cap F_1' \cap \cdots \cap F_4' \cap G.$$

The correctness of the intersection can be checked by MACAULAY.

It remains an open problem to find a primary decomposition for the ideal of adjacent 2×2-minors for larger sizes. We do not even have a reasonable conjecture for generalizing the result in Proposition 4.2.

In the special case $a = 2$ the ideal $I_{\mathcal{B}_{adj}}$ is radical and has a nice explicit prime decomposition. We shall present this decomposition using a slightly simplified notation. We write I as the ideal generated by the following n binomials in $2n + 2$ variables:

$$x_{i-1} \cdot y_i \quad - \quad x_i \cdot y_{i-1} \qquad\qquad (i = 1, 2, \ldots, n). \tag{4.2}$$

Let $f(n)$ denote the n-th *Fibonacci number*, which is defined recursively by $f(0) = f(1) = 1$ and $f(n) = f(n-1) + f(n-2)$.

Theorem 4.4. *The ideal I of adjacent 2×2-minors of a generic $2 \times (n+1)$-matrix is the intersection of $f(n)$ prime ideals; in particular, I is radical.*

Proof. The ideal I is a complete intersection, which means I has codimension n and degree 2^n. (To see this, note that the left-hand terms in (4.2) are pairwise relatively prime. They are the leading terms in the lexicographic order.) By Macaulay's Unmixedness Theorem [Eis, Corollary 18.14], every associated prime of I is minimal and has codimension n.

Let $\mathcal{D}(n)$ denote the set of all subsets of $\{1,2,\ldots,n-1\}$ which do not contain two consecutive integers. The cardinality of $\mathcal{D}(n)$ equals the Fibonacci number $f(n)$. For instance, $\mathcal{D}(4) = \{\emptyset,\{1\},\{2\},\{3\},\{4\},\{1,3\},\{1,4\},\{2,4\}\}$. For each element S we define a binomial ideal I_S in $k[x_0,\ldots,x_n,y_0,\ldots,y_n]$. The generators of I_S are the variables x_i and y_i for all $i \in S$, and the binomials $x_j y_k - x_k y_j$ for all $j,k \notin S$ such that no element of S lies between j and k. It is easy to see that I_S is a prime ideal of codimension n. Moreover, I_S contains I, and therefore I_S is a minimal prime of I. We claim that

$$I \quad = \quad \bigcap_{S\in\mathcal{D}(n)} I_S. \tag{4.3}$$

In view of Macaulay's Unmixedness Theorem, it suffices to prove the identity

$$\sum_{S\in\mathcal{D}(n)} \text{degree}(I_S) \quad = \quad 2^n. \tag{4.4}$$

First note that $I_\emptyset$ is the determinantal ideal $\langle x_i y_j - x_i x_j \,:\, 0 \leq i < j \leq n\rangle$. It is known (see e.g., [Har, Example 19.10]) that the degree of $I_\emptyset$ equals $n+1$. Using the same fact for matrices of smaller size, we find that, for S non-empty, the degree of I_S equals the product

$$i_1\cdot(i_2-i_1+1)\cdot(i_3-i_2+1)\cdots(i_r-i_{r-1}+1)\cdot i_r \text{ where } S = \{i_1 < i_2 < \cdots < i_r\}. \tag{4.5}$$

Consider the surjection $\phi : 2^{\{1,\ldots,n\}} \to \mathcal{D}(n)$ defined by

$$\phi(\{j_1<j_2<\cdots<j_r\}) = \{j_{r-1}, j_{r-3}, j_{r-5}, \ldots\}.$$

The product in (4.5) is the cardinality of the inverse image $\phi^{-1}(S)$. This proves $\sum_{S\in\mathcal{D}(n)} \#(\phi^{-1}(S)) = 2^n$, which implies (4.4) and hence Theorem 4.4 is proved. ∎

5. Circuit walks

Let $\mathcal{A}$ be a $d \times n$-integer matrix of rank d. The integer kernel of $\mathcal{A}$ is a sublattice $\mathcal{L}$ in $\mathbf{Z}^n$ of rank $n-d$. In this case $\mathcal{L}$ is saturated and hence $I_\mathcal{L}$ is a prime ideal. A non-zero vector $\mathbf{u} = (u_1, \ldots, u_n)$ in $\mathcal{L}$ is called a *circuit* if its coordinates u_i are relatively prime and its support $\mathrm{supp}(\mathbf{u}) = \{ i : u_i \neq 0\}$ is minimal with respect to inclusion. In this section we discuss the walk defined by the set $\mathcal{C}$ of all circuits in $\mathcal{L}$. This makes sense for two reasons:

- The lattice $\mathcal{L}$ is generated by the circuits, i.e., $\mathbf{Z}\mathcal{C} = \mathcal{L}$ (see e.g., [ES, Lemma 8.8]).
- The circuits can be computed easily from the matrix $\mathcal{A}$.

Here is a simple algorithm for computing $\mathcal{C}$. Initialize $\mathcal{C} := \emptyset$. For any $(d+1)$-subset $\tau = \{\tau_1, \ldots, \tau_{d+1}\}$ of $\{1, \ldots, n\}$ form the vector

$$C_\tau \quad = \quad \sum_{i=1}^{d+1} (-1)^i \cdot det(\mathcal{A}_{\tau \backslash \{\tau_i\}}) \cdot \mathbf{e}_{\tau_i},$$

where $\mathbf{e}_j$ denotes the j-th unit vector and $\mathcal{A}_\sigma$ denotes the submatrix of $\mathcal{A}$ with column indices σ. If C_τ is non-zero then remove common factors from its coordinates. The resulting vector is a circuit and all circuits are obtained in this manner (see e.g., [Stu, §4]).

Example 5.1. Let $d - 2, n = 4$ and $\mathcal{A} = \begin{pmatrix} 0 & 2 & 5 & 7 \\ 7 & 5 & 2 & 0 \end{pmatrix}$. Then the set of circuits equals

$$\mathcal{C} \quad = \quad \pm\{\, (3,-5,2,0),\, (5,-7,0,2),\, (2,0,-7,5),\, (0,2,-5,3) \,\}. \qquad (5.1)$$

It is instructive to check that the $\mathbf{Z}$-span of $\mathcal{C}$ equals $\mathcal{L} = \ker_{\mathbf{Z}}(\mathcal{A})$. (For instance, try to write $(1,-1,-1,1) \in \mathcal{L}$ as a $\mathbf{Z}$-linear combination of $\mathcal{C}$). We shall derive the following result: *Two $\mathcal{L}$-equivalent non-negative integer vectors (A,B,C,D) and (A',B',C',D') can be connected by the circuits in (5.1) if both of them satisfy the following inequality*

$$\min\Big\{ \max\{A,B,C,D\},\ \max\{B, \frac{9}{4}C, \frac{9}{4}D\},\ \max\{\frac{9}{4}A, \frac{9}{4}B, C\} \Big\} \ \geq\ 9 \qquad (5.2)$$

We remark that the following two $\mathcal{L}$-equivalent pairs cannot be connected by circuits:

$$(4,9,0,2) \leftrightarrow (5,8,1,1) \quad \text{and} \quad (1,6,6,1) \leftrightarrow (3,4,4,3). \qquad (5.3)$$

To analyze circuit walks in general, we consider the *circuit ideal* $I_{\mathcal{C}}$ generated by the binomials $\mathbf{x}^{\mathbf{u}_+} - \mathbf{x}^{\mathbf{u}_-}$ where $\mathbf{u} = \mathbf{u}_+ - \mathbf{u}_-$ runs over all circuits in $\mathcal{L}$. The primary decomposition of circuit ideals was studied in [ES, §8]. We summarize the relevant results. Let $\mathrm{pos}(\mathcal{A})$ denote the d-dimensional convex polyhedral cone in $\mathbf{R}^d$ spanned by the column vectors of $\mathcal{A}$. Each face of $\mathrm{pos}(\mathcal{A})$ is identified with the subset $\sigma \subset \{1,\ldots,n\}$ consisting of all indices i such that the i-th column of $\mathcal{A}$ lies on that face. If σ is a face of $\mathrm{pos}(\mathcal{A})$ then the ideal $I_\sigma := \langle x_i : i \notin \sigma \rangle + I_\mathcal{L}$ is prime. Note that $I_{\{1,\ldots,n\}} = I_\mathcal{L}$ and $I_{\{\}} = \langle x_1, x_2, \ldots, x_n \rangle$.

Theorem 5.2. [ES, Theorem 8.3, Example 8.6 and Proposition 8.7]

$$\mathrm{Rad}(I_\mathcal{C}) \;=\; I_\mathcal{L} \quad \textit{and} \quad Ass(I_\mathcal{C}) \subseteq \{ I_\sigma : \sigma \textit{ is a face of } \mathrm{pos}(\mathcal{A}) \}.$$

Theorem 8.3 in [ES] gives a procedure for computing a binomial primary decomposition of the circuit ideal $I_\mathcal{C}$. This enables us to analyze the connectivity of the circuit walk in terms of the faces of the polyhedral cone $\mathrm{pos}(\mathcal{A})$.

Example 5.1. *(continued)* We choose variables a, b, c, d for the four columns of $\mathcal{A}$. The cone $\mathrm{pos}(\mathcal{A}) = \mathrm{pos}\{(7,0),(5,2),(2,5),(0,7)\}$ equals the positive orthant in $\mathbf{R}^2$. It has one 2-dimensional face, labeled $\{a,b,c,d\}$, two 1-dimensional faces, labeled $\{a\}$ and $\{d\}$ and one 0-dimensional face, labeled $\{\}$. The lattice ideal of $\mathcal{L} = \ker_\mathbf{Z}(\mathcal{A})$ is the prime ideal

$$I_\mathcal{L} \;=\; \langle\, ad - bc,\; ac^4 - b^3d^2,\; a^3c^2 - b^5,\; b^2d^3 - c^5,\; a^2c^3 - b^4d \,\rangle.$$

The circuit ideal equals

$$I_\mathcal{C} \;=\; \langle\, a^3c^2 - b^5,\; a^5d^2 - b^7,\; a^2d^5 - c^7,\; b^2d^3 - c^5 \,\rangle.$$

It has the minimal primary decomposition

$$\begin{aligned} I_\mathcal{C} \;=\; I_\mathcal{L} & \cap \langle b^9, c^4, d^4, b^2d^2, c^2d^2, b^2c^2 - a^2d^2, b^5 - a^3c^2 \rangle \\ & \cap \langle a^4, b^4, c^9, a^2b^2, a^2c^2, b^2c^2 - a^2d^2, c^5 - b^2d^3 \rangle \\ & \cap \big(\langle a^9, b^9, c^9, d^9 \rangle + I_\mathcal{C} \big). \end{aligned}$$

Here the second ideal is primary to $I_{\{a\}} = \langle b, c, d \rangle$ and the third ideal is primary to $I_{\{d\}} = \langle a, b, c \rangle$. The given primary decomposition implies (5.2) because

$$\langle a^9, b^9, c^9, d^9 \rangle \cap \langle b^9, c^4, d^4 \rangle \cap \langle a^4, b^4, c^9 \rangle \cap I_\mathcal{L} \;\subset\; I_\mathcal{C}. \qquad (5.2')$$

Returning to our general discussion, Theorem 5.2 implies that for each face σ of the polyhedral cone $\text{pos}(\mathcal{A})$ there exists a non-negative integer M_σ such that

$$I_{\mathcal{L}} \;\cap \bigcap_{\substack{\sigma \text{ face} \\ \text{of } \text{pos}(\mathcal{A})}} \langle x_i : i \notin \sigma \rangle^{M_\sigma} \quad \subset \quad I_{\mathcal{C}}. \tag{5.4}$$

Corollary 5.3. *For each proper face σ of the cone* $\text{pos}(\mathcal{A})$ *there is an integer M_σ such that any two $\mathcal{L}$-equivalent vectors $(a_1, \ldots, a_n)$ and $(b_1, \ldots, b_n)$ in $\mathbf{N}^n$ with the property*

$$\sum_{i \notin \sigma} a_i \geq M_\sigma \quad \textit{and} \quad \sum_{i \notin \sigma} b_i \geq M_\sigma \quad \textit{for all proper faces } \sigma \textit{ of } \text{pos}(\mathcal{A}) \tag{5.4'}$$

can be connected by circuits.

This suggests the following research problem.

Problem 5.4. *Find good bounds for the integers M_σ in terms of the matrix $\mathcal{A}$.*

The optimal value of M_σ seems to be related to the singularity of the toric variety defined by $I_{\mathcal{L}}$ along the torus orbit labeled σ: The worse the singularity is, the higher the value of M_σ. It would be very interesting to understand these geometric aspects.

In Example 5.1 we can choose the integers M_σ as follows:

$$M_{\{\}} \;=\; 15 \quad \text{and} \quad M_{\{a\}} \;=\; 11 \quad \text{and} \quad M_{\{d\}} \;=\; 11.$$

These choices are optimal. This is seen from the disconnected pairs in (5.3).

References

[AM] M. F. Atiyah and I. G. Macdonald, *Introduction to Commutative Algebra,* Addison Wesley, Reading, MA, 1969.

[BW] T. Becker, V. Weispfenning, *Gröbner Bases: A Computational Approach to Commutative Algebra,* Graduate Texts in Mathematics, Springer, New York, 1993.

[CLO] D. Cox, J. Little, and D. O'Shea, *Ideals, Varieties, and Algorithms,* Springer-Verlag, NY, 1992.

[DS] P. Diaconis and B. Sturmfels, Algebraic algorithms for sampling from conditional distributions, to appear in *Annals of Statistics.*

[Eis] D. Eisenbud, *Commutative Algebra with a View Toward Algebraic Geometry*, Springer-Verlag, NY, 1995.

[ES] D. Eisenbud and B. Sturmfels, Binomial ideals, *Duke Mathematical Journal* **84** (1996), 1–45.

[Gl] G. Glonek, Some aspects of log linear models. Thesis, School of Math. Sci., Flinders Univ. of S. Australia, 1987.

[Har] J. Harris, *Algebraic Geometry*, Graduate Texts in Math., Springer, New York, 1992.

[MM] E. Mayr and A. Meyer, The complexity of the word problem for commutative semigroups and polynomial ideals, *Advances in Mathematics* **46** (1982) 305–329.

[Stu] B. Sturmfels, *Gröbner Bases and Convex Polytopes*, American Mathematics Society, Providence, RI, 1995.

[Tho] R.R. Thomas, A geometric Buchberger algorithm for integer programming, *Mathematics of Operations Research* **20** (1995), 864–884.

Persi Diaconis
Department of Mathematics
Cornell University
Ithaca, NY 14850

David Eisenbud
Mathematical Sciences Research Institute
1000 Centennial Drive
Berkeley, CA 94720
de@msri.org

Bernd Sturmfels
Department of Mathematics
University of California
Berkeley, CA 94720
bernd@math.berkeley.edu

Natural Exponential Families and Umbral Calculus

*A. Di Bucchianico** and D. E. Loeb*†

Dedicated to our friend Gian-Carlo on the occasion of his 64th birthday

Abstract

We use the Umbral Calculus to investigate the relation between natural exponential families and Sheffer polynomials. As a corollary, we obtain a new transparent proof of Feinsilver's theorem which says that natural exponential families have a quadratic variance function if and only if their associated Sheffer polynomials are orthogonal.

AMS Classification: 05A40, 33C45, 60E05, 62E10

Keywords: natural exponential family, variance function, umbral calculus, Sheffer polynomials, orthogonal polynomials, approximation operators.

1. Introduction

Exponential families of probability measures play a traditional role in statistics (dating back to the thirties) because of their nice estimation properties (see e.g. [25]). However, recently exponential families appear as the cornerstone of the important class of generalized linear models (see [6] for an excellent introduction). In [18], Morris studied natural exponential families on the real line. He showed that there are six classes of natural exponential families with quadratic variance function (i.e., where the variance is a polynomial of degree at most two). We study natural exponential families in light of polynomial expansions of their density function. These polynomials belong to the class of Sheffer polynomials. We study these polynomials and their relation to natural exponential families from an Umbral Calculus

*Author supported NATO CRG 930554.

†Author partially supported by URA CNRS 1304, EC grant CHRX-CT93-0400, the PRC Maths-Info, and NATO CRG 930554.

viewpoint. In particular, the polynomials expansions associated with a natural exponential family are Sheffer and their delta operator is related to their variance function. Using slightly different terminology, Feinsilver proved [7, Chapter 4] that a natural exponential family has a quadratic variance function if and only if the corresponding Sheffer polynomials are orthogonal. This result immediately follows from our approach to natural exponential families. It is interesting to note (as pointed out to us by one of the referees) that the Morris classification was discovered a few years earlier in approximation theory by May (see [16] and for generalizations [10]). We will discuss the relation between exponential families and exponential approximation operators in Section 5. We will indicate how our approach differs from the approach in [16, 10].

Before presenting our Umbral Calculus approach to exponential families, we begin with brief introductions to these two subjects.

2. Natural exponential families

We begin by recalling the definition of a natural exponential family. Our notation closely follows [15].

Let ν be a measure on the real line. We assume that ν is not concentrated in one point.

Let the Laplace transform of ν be given by

$$L(\theta) = \int_{-\infty}^{\infty} e^{x\theta}\, d\nu(x). \tag{1}$$

We define Θ to be the interior of the set $\{\theta \in \mathbb{R} \mid L(\theta) < \infty\}$. If Θ is non-empty, then the **natural exponential family** generated by ν is the set of probability distributions of the form

$$P_\theta(A) = \int_A e^{x\theta - k(\theta)}\, d\nu(x), \tag{2}$$

where k is the **cumulant** of ν, i.e., $k(\theta) := \log L(\theta)$, and $\theta \in \Theta$. We will see later that different ν may generate the same natural exponential family. Since $k(\theta) = \log L(\theta)$, we have

$$e^{k(\theta)} = \int_{-\infty}^{\infty} e^{x\theta}\, d\nu(x). \tag{3}$$

It follows by differentiating (3) with respect to θ that

$$k'(\theta) = \int_{-\infty}^{\infty} x\, dP_\theta(x). \tag{4}$$

Differentiating (3) twice with respect to θ and using (4), we obtain

$$k''(\theta) = \int_{-\infty}^{\infty} (x - k'(\theta))^2 \, dP_\theta(x). \tag{5}$$

Since k is strictly convex on Θ by the Hölder inequality,[1] it follows that the map $\theta \mapsto k'(\theta)$ is a bijection of Θ on its range, which we will denote by M_ν. The inverse of this map will be denoted by

$$\psi : M_\nu \mapsto \Theta.$$

This means that we may reparametrize the densities with respect to ν in (2) as

$$\varphi(m, x) = e^{x\psi(m) - k(\psi(m))}. \tag{6}$$

Using the reparametrization of (6), we now come to the following important definition.

Definition 2.1. Let $\{P_\theta \,|\, \theta \in \Theta\}$ be the natural exponential family generated by a measure ν. The function $V_\nu : M_\nu \to \mathbb{R}$ defined by $V_\nu(m) = \int_{-\infty}^{\infty} (x - m)^2 \, \varphi(m, x) \, d\nu(x)$ is called the *variance function* of $\{P_\theta \,|\, \theta \in \Theta\}$.

A natural exponential family is uniquely determined by its variance function together with the domain of the variance function [18]. In the theory of generalized linear models, the variance function is called the **link function** [6]. The link function is essential for estimating purposes.

Before we continue, we give an example in order to illustrate the notions introduced above.

Example. (Poisson family) Consider a Poisson distribution with parameter θ, i.e., $\Pr(n) = \dfrac{e^{-\theta}\,\theta^n}{n!}$ for $n = 0, 1, 2, \ldots$. Writing $\dfrac{e^{-\theta}\,\theta^n}{n!} = \dfrac{e^{n \log \theta}}{n! \, e^{e^{\log \theta}}}$, we see that $\{P_\theta \,|\, \theta \in (0, \infty)\}$ is a natural exponential family generated by the discrete measure $\nu\{n\} = 1/n!$, $n = 0, 1, 2, \ldots$ where P_θ is Poisson($\log \theta$) distributed. An easy calculation shows that $k(\theta) = e^{\theta}$, $\Theta = \mathbb{R}$, $\psi(m) = \log m$, $M_\nu = (0, \infty)$, and $V_\nu(m) = m$. We see that the standard change from θ to the so-called natural parameter $\log \theta$ (cf. [6]) is nothing but our reparametrization (6).

The following lemma is crucial to our approach.

---

[1]If $0 < \lambda < 1$, then $k(\lambda\theta + (1 - \lambda)\xi) = \log\left(\int_{-\infty}^{\infty} e^{\lambda\theta x} e^{(1-\lambda)\xi x} \, d\nu(x)\right) < \log\left(\left(\int_{-\infty}^{\infty} e^{\theta x} \, d\nu(x)\right)^{\lambda} \left(\int_{-\infty}^{\infty} e^{\xi x} \, d\nu(x)\right)^{1-\lambda}\right) = \lambda k(\theta) + (1 - \lambda) k(\xi)$. Note that the inequality is strict, since ν is not concentrated in one point.

Lemma 2.2. *If $(P_\theta \mid \theta \in \Theta)$ is a natural exponential family, then there exist a real number t and a natural exponential family $\{\widetilde{P}_\theta \mid \theta \in \widetilde{\Theta}\}$ generated by a measure μ such that*

1. $P_\theta(A) = \widetilde{P}_\theta(A+t)$

2. $\int_{-\infty}^{\infty} x \, d\mu(x) = 0$

3. $0 \in \widetilde{\Theta}$

4. $\widetilde{k}'(0) = 0$

5. $V_\mu(m) = V_\nu(m+t)$.

Proof. First note that $\{P_\theta \mid \theta \in \Theta\}$ is also generated by the measure $e^{\theta_0 x} \, d\nu(x)$ for any $\theta_0 \in \Theta$ and that the corresponding parameter set $\widetilde{\Theta}$ equals $\Theta - \theta_0$. In particular, $0 \in \widetilde{\Theta}$. Now define the measure μ by $d\mu(x) = d\nu(x + k'(0))$. It follows from (4) that $\widetilde{k}'(0) = 0$. Moreover, easy calculations show that *1.* and *5.* hold with $t = k'(0)$. ■

Example. (Poisson family continued) Let μ be the measure obtained by shifting the generating measure ν one $(= k'(0))$ to the left, i.e., $\mu\{n\} = 1/(n+1)!$, $n = -1, 0, 1, 2, \ldots$. An easy calculation yields that μ is of mean zero, $\widetilde{\Theta} = \mathbb{R}$, $\widetilde{k}'(\theta) = e^{\theta} - 1$, $M_\mu = (-1, \infty)$, $\widetilde{k}'(0) = 0$, and $V_\mu(m) = m + 1$.

Thus we may and will assume without loss of generality that $(P_\theta \mid \theta \in \Theta)$ satisfies the extra conditions of the above Lemma. By well-known properties of Laplace transforms, k and ψ are analytic functions in a neighbourhood of zero. Hence, we may expand (6) into a power series in m for $m \in M_\nu$. It follows from (2) and (4) that $k'(0) = 0$. Moreover, since ν is not concentrated in one point, it follows from (5) that $k''(0) \neq 0$. Thus $\psi(0) = 0$ and $\psi'(0) \neq 0$, which implies that s_n is a polynomial of degree exactly n. The associated Sheffer polynomials of a natural exponential family are the polynomials $(s_n)_{n \geq 0}$ defined by

$$\varphi(m, x) = \sum_{n=0}^{\infty} s_n(x) \, m^n, \tag{7}$$

where φ is defined by (6). In the following section we study Sheffer polynomials. The relation of exponential families with approximation theory will be discussed in Section 5.

3. Sheffer polynomials

In this section we give an introduction to Sheffer polynomials. In the spirit of the Umbral Calculus, we give an algebraic definition. Our definitions come from [27] and differ slightly from the orginal approach as presented in [22].

Definition 3.1. A sequence $(p_n)_{n\geq 0}$ is a *polynomial sequence* if $p_0 \neq 0$ and p_n is a polynomial of degree exactly n for all n.

Definition 3.2. A polynomial sequence $(s_n)_{n\geq 0}$ is a *Sheffer sequence* if there exists a polynomial sequence $(q_n)_{n\geq 0}$ such that

$$s_n(x+y) = \sum_{k=0}^{n} s_k(x)\, q_{n-k}(y) \tag{8}$$

holds for all n, x, and y.

Note that this definition differs a factor $n!$ from [27]. If $g(z)$ is a power series in z such that $g(0) = 0$ and $g'(0) \neq 0$, then the polynomial sequence $(s_n)_{n\geq 0}$ defined by the generating function

$$\sum_{n=0}^{\infty} s_n(x)\, z^n = A(z)\, e^{xg(z)}$$

is a Sheffer sequence. The advantage of Definition 3.2 is that it is purely algebraic and avoids the notion of formal power series. As we will see (Theorem 3.8) these definitions are essentially equivalent.

The Umbral Calculus is based on linear operators on the vector space of polynomials.

Definition 3.3. A linear operator T on the vector space of polynomials is said to be *shift-invariant* if $TE^a = E^aT$ for all a, where the shift operators E^a are defined by $(E^a p)(x) = p(x+a)$. If moreover Tx is a non-zero constant, then T is said to be a *delta operator*.

The following Theorem relates Sheffer sequences to delta operators.

Theorem 3.4. *The sequence $(s_n)_{n\geq 0}$ is a Sheffer sequence if and only if there exists a delta operator Q such that $Qs_n = s_{n-1}$ for $n \geq 1$. In both cases, we also have $Qq_n = q_{n-1}$ for all $n \geq 1$, where $(q_n)_{n\geq 0}$ is as in (8). Moreover, $q_0 = 1$ and $q_n(0) = 1$ for $n \geq 1$.*

Proof. See [27, Theorem 1.1]. ∎

Remark 3.5.

1. In the terminology of the Umbral Calculus, the polynomial sequence $(q_n)_{n\geq 0}$ in Theorem 3.4 is the **basic sequence** of the delta operator Q (apart from a factor $n!$). Conversely, for each delta operator Q there exists a basic set $(q_n)_{n\geq 0}$, i.e., a polynomial sequence such that $q_0 = 1$, $q_n(0) = 0$ and $Qq_n = q_{n-1}$ for $n \geq 1$. It follows from [22, Proposition 3, p. 688] and [22, Theorem 1, p. 689] that

$$q_n(x+y) = \sum_{k=0}^{n} q_k(x)\, q_{n-k}(y) .$$

2. Let $(s_n)_{n\geq 0}$ be a Sheffer sequence with delta operator Q. Define the linear operator A by $As_n = q_n$, where $(q_n)_{n\geq 0}$ is the basic set of Q. Then both A and A^{-1} are invertible shift-invariant operators (use [22, Proposition 1, p. 698] and the fact that the inverse of a shift-invariant operator is shift-invariant).

The importance of the last remark is shown by the powerful operator methods which lead to the following expansion theorems. Since these results are not explicitly mentioned in [22], we include the proofs.

Theorem 3.6. *Let $(s_n)_{n\geq 0}$ be a Sheffer sequence with delta operator Q and invertible operator A. If p is a polynomial, then*

$$p = \sum_{k=0}^{\infty} \left[AQ^k p\right]_{x=0} s_k .$$

Proof. Since $(s_n)_{n\geq 0}$ is a basis for all polynomials in one variable, it suffices to show that the formula holds for $p = s_n$ $(n \geq 0)$. Using $Q^k s_n = s_{n-k}$ for $0 \leq k \leq n$, we see that the right-hand side with $p = s_n$ equals $\sum_{k=0}^{n} q_{n-k}(0)\, s_k$, which equals s_n by the definition of a Sheffer sequence. ∎

Taylor's Theorem is a special case of Theorem 3.6 where $s_k(x) = x^k/k!$ and hence $Q = D$ and $A = I$.

Theorem 3.7. *Let $(s_n)_{n\geq 0}$ be a Sheffer sequence with delta operator Q and invertible operator A. If T is a shift-invariant operator, then*

$$T = \sum_{k=0}^{\infty} \left[T\, s_k\right]_{x=0} A\, Q^k .$$

Proof. Let p be an arbitrary polynomial of degree n. Applying Theorem 3.6 to $E^y p$, where $E^y p(x) := p(x+y)$, we obtain $TE^y p = \sum_{k=0}^{n} (A\,Q^k E^y p)(0)\,T s_k = \sum_{k=0}^{n} (A\,Q^k p)(y)\,T s_k$. Hence, $(Tp)(y) = (E^y Tp)(0) = (TE^y p)(0) = \sum_{k=0}^{n} (Ts_k)(0)\left(A\,Q^k p\right)(y)$ for all y. Since Q reduces the degree of a polynomial by one, we may write the last expression as $\sum_{k=0}^{\infty} (Ts_k)(0)\left(A\,Q^k p\right)(y)$. ■

In particular, if we take $Q = D$ and $s_k = x^k/k!$, then it follows that each delta operator can be expanded into a power series in D. Note that there are no convergence problems, since all infinite sums reduce to finite sums when applied to a polynomial.

The following theorem relates these expansions to the generating function of a Sheffer sequence.

Theorem 3.8. *A polynomial sequence $(s_n)_{n\geq 0}$ is Sheffer if and only if the following formal generating function identity holds:*

$$\sum_{n=0}^{\infty} s_n(x)\,z^n = \frac{1}{f(g(z))}\,e^{xg(z)}. \tag{9}$$

In this case, the delta operator Q of $(s_n)_{n\geq 0}$ equals $g^{(-1)}(D)$, where $g^{(-1)}$ denotes the compositional inverse of the formal power series g and the invertible operator A, defined by $As_n = q_n$, satisfies $A = f(D)$.

Proof. See [22, Proposition 5, p. 702]. ■

Theorem 3.8 is fundamental to our approach, since we will use it in the following section to relate the variance function of a natural exponential family to the delta operator and the invertible operator of the associated Sheffer sequence.

4. Main results

In this section we combine the results from the previous section which yield our Umbral Calculus approach to natural exponential families. The following theorem relates the variance function of a natural exponential family to the delta operator of its associated Sheffer sequence.

Theorem 4.1. *Let $\{P_\theta \mid \theta \in \Theta\}$ be a natural exponential family generated by a measure ν with associated Sheffer sequence $(s_n)_{n\geq 0}$ (thus we assume without loss of generality that the extra conditions of Lemma 2.2 hold). Let $Q = q(D)$ be the delta operator and $A = f(D)$ be the invertible operator of $(s_n)_{n\geq 0}$.*

Then $q'(D) = V_\nu(q(D))$ and $f'(D) = q(D)\, f(D)$. Moreover, f is the Laplace transform of ν.

Proof. It follows from (6) and (7) and Theorem 3.8 that $q(D) = \psi^{-1}(D) = k'(D)$. Thus, by (5) and the definition of variance function, we arrive at $q'(D) = k''(D) = V_\nu(k'(D))$. For the second statement, note that by Theorem 3.8 we have $f(D) = e^{k(D)}$. Hence, $f'(D) = k'(D)\, e^{k(D)} = q(D)\, f(D)$. The last statement follows from $k(\theta) = \log L(\theta)$ and (1). ■

Remark 4.2. The differential equations in Theorem 4.1 are formal differential equations, i.e., the solutions must be a formal power series. By a theorem of Borel, this is equivalent to saying that the solutions must be functions that are infinitely differentiable at zero.

A crucial role is played by the **Pincherle derivative** of a delta operator. The Pincherle derivative of a shift-invariant operator T is defined by $T' := T\,\mathbf{x} - \mathbf{x}\,T$, where $\mathbf{x}$ denotes the multiplication by x operator. By Theorem 3.7, there exists a formal power series t in D such that $T = t(D)$. A convenient way to calculate T' is given by the formula $T' = t'(D)$ (see e.g., [22, Proposition 2, p. 695]). This also shows that the Pincherle derivative of a shift-invariant operator is again a shift-invariant operator, and that the Pincherle derivative obeys usual rules of differentiation such as the product rule for derivatives (see [22, Section 4]).

We now set out to prove that a natural exponential family has a quadratic variance function if and only if the associated Sheffer polynomials are orthogonal. By Favard's Theorem (see e.g., [4, Theorem 4.4]), a polynomial sequence $(s_n)_{n\geq 0}$ is orthogonal if and only if it satisfies the following three-term recurrence relation for $n \geq 0$:

$$s_{n+1}(x) = (a_n\, x - b_n)\, s_n(x) - c_n\, s_{n-1}(x), \tag{10}$$

where $s_1 = 0$, s_0 is a non-zero constant, and $c_n\, a_n\, a_{n-1} > 0$ for $n \geq 1$. We must be careful with normalizations when dealing with Sheffer sequences, because if $(s_n)_{n\geq 0}$ is Sheffer, then $(\lambda_n s_n)_{n\geq 0}$ need not be Sheffer. Hence, we must use the non-monic form (10).

The following Theorem shows the relation between the three-term recurrence relation (10) and differential equations for the delta operator and shift-invariant operator of a Sheffer sequence. These differential equations are to be considered in the Pincherle sense (cf. Remark 4.2).

Theorem 4.3. *Let $(s_n)_{n\geq 0}$ be a Sheffer sequence with delta operator Q and invertible operator A. If $(s_n)_{n\geq 0}$ satisfies the three-term recurrence relation*

(10), then

$$Q' = \frac{1}{a_0} I + \left(\frac{b_1}{a_1} - \frac{b_0}{a_0}\right) Q + \left(\frac{b_2}{a_2} - \frac{c_1}{a_1}\right) Q^2 \tag{11}$$

$$A' = \frac{b_0}{a_0} A + \frac{c_1}{a_0} A Q. \tag{12}$$

Conversely, if $Q' = d_1 + d_2 Q + d_3 Q^2$ *and* $A' = d_4 A + d_5 AQ$, *then* $(s_n)_{n \geq 0}$ *satisfies the following three-term recurrence relation:*

$$(n+1)\, d_1\, s_{n+1}(x) = (x - n\, d_2 - d_4)\, s_n(x) - ((n-1)\, d_3 + d_5)\, s_{n-1}(x).$$

Proof. We make extensive use of the Operator Expansion Theorem (Theorem 3.7) and of the fact that $q_n(0) = 0$ for $n \geq 1$. In particular,

$$T' = \sum_{k=0}^{\infty} \left[T' s_k\right]_{x=0} A Q^k,$$

where $T' = t'(D) = T\mathbf{x} - \mathbf{x}T$ is the Pincherle derivative of the operator $T = t(D)$. Notice that

$$\left[T'p\right]_{x=0} = \left[Txp - xTp\right]_{x=0} = \left[T\mathbf{x}p\right]_{x=0}. \tag{13}$$

It will be convenient to adopt the convention that $q_k = 0$ and $s_k = 0$ for $k < 0$.

Assume that $(s_n)_{n \geq 0}$ satisfies the three-term recurrence relation (10). We first apply (13) to $T = A'$.

$$\begin{aligned} A' &= \sum_{k=0}^{\infty} \left[Axs_k\right]_{x=0} A\, Q^k \\ &= \sum_{k=0}^{\infty} \frac{1}{a_k} \left[A\left(s_{k+1} + b_k\, s_k + c_k s_{k-1}\right)\right]_{x=0} AQ^k \\ &= \sum_{k=0}^{\infty} \frac{1}{a_k} \left(q_{k+1}(0) + b_k\, q_k(0) + c_k\, q_{k-1}(0)\right) A\, Q^k \\ &= \frac{b_0}{a_0} A + \frac{c_1}{a_0} A\, Q. \end{aligned}$$

We now apply (13) to $T = (AQ)'$.

$$AQ' + A'Q = (AQ)'$$

$$
\begin{aligned}
&= \sum_{k=0}^{\infty} \left[(AQ)'\, s_k\right]_{x=0} A\, Q^k \\
&= \sum_{k=0}^{\infty} \left[AQ\, x s_k\right]_{x=0} A\, Q^k \\
&= \sum_{k=0}^{\infty} \frac{1}{a_k} \left[AQ\,(s_{k+1} + b_k s_k + c_k s_{k-1})\,\right]_{x=0} AQ^k \\
&= \frac{1}{a_0} A + \sum_{k=1}^{\infty} \frac{1}{a_k} \left(q_k(0) + b_k q_{k-1}(0) + c_k q_{k-2}(0)\right) AQ^k \\
&= \frac{1}{a_0} A + \frac{b_1}{a_1} AQ + \frac{c_2}{a_2} AQ^2.
\end{aligned}
$$

Now we eliminate A' in the last equation by using (12):

$$
\begin{aligned}
AQ' + A'Q &= \frac{1}{a_0} A + \frac{b_1}{a_1} AQ + \frac{c_2}{a_2} AQ^2 \\
AQ' &= \frac{1}{a_0} A + \left(\frac{b_1}{a_1} - \frac{b_0}{a_0}\right) AQ + \left(\frac{b_2}{a_2} - \frac{c_1}{a_1}\right) AQ^2 \\
Q' &= \frac{1}{a_0} I + \left(\frac{b_1}{a_1} - \frac{b_0}{a_0}\right) Q + \left(\frac{b_2}{a_2} - \frac{c_1}{a_1}\right) Q^2.
\end{aligned}
$$

In the last step of the proof, we used that A is invertible .

Conversely, assume that $Q' = d_1 + d_2 Q + d_3 Q^2$ and $A' = d_4 A + d_5 AQ$. We apply the Polynomial Expansion Theorem 3.6 to $x s_n$, which yields

$$
\begin{aligned}
x s_n &= \sum_{k=0}^{n+1} \left[AQ^k x s_n\right]_{x=0} s_k \\
&= \sum_{k=0}^{n+1} \left[(AQ^k)' s_n\right]_{x=0} s_k \\
&= \sum_{k=0}^{n+1} \left\{\left(A'Q^k + k\, AQ^{k-1}Q'\right) s_n(0)\right\} s_k \\
&= \sum_{k=0}^{n+1} \left\{A' s_{n-k}(0) + k\, AQ' s_{n-k+1}(0)\right\} s_k \\
&= \sum_{k=0}^{n+1} \{(d_4 A + d_5 AQ)\, s_{n-k}(0) + \\
&\qquad k\left(d_1 A + d_2 AQ + d_3 AQ^2\right) s_{n-k+1}(0)\Big\}\, s_k \\
&= \sum_{k=0}^{n+1} \{d_4\, q_{n-k}(0) + d_5\, q_{n-k-1}(0) +
\end{aligned}
$$

$$\begin{aligned} & k\,d_1\,q_{n-k+1}(0) + k\,d_2\,q_{n-k}(0) + k\,d_3\,q_{n-k-1}(0)\}\,s_k \\ = \;& d_4 s_n + d_5\,s_{n-1} + (n+1)\,d_1\,s_{n+1} + n\,d_2\,s_n + (n-1)\,d_3\,s_{n-1} \\ = \;& ((n-1)\,d_3 + d_5)\,s_{n-1} + (n\,d_2 + d_4)\,s_n + (n+1)\,d_1 s_{n+1}\,. \end{aligned}$$

■

Since $d_1 \neq 0$, this means that $(s_n)_{n\geq 0}$ satisfies the three-term recurrence relation of the form (10).

Solving the differential equations for Q and A, we find all orthogonal Sheffer polynomials (see Section 6). Thus we have a new proof of the Meixner classification of orthogonal Sheffer polynomials (see [17] for the original proof, other proofs can be found in [8, 12, 13, 23, 24]). The advantage of our proof is that it is a constructive proof based on first principles of the Umbral Calculus. A generalization of the Meixner classification was obtained by Al-Salam in [2], where it is shown that the Meixner result remains true even if we consider the more general class of polynomials with generating function $e^{Q(x,t)}$, where $Q(x,t)$ is a polynomial in x and a power series in t.

We now are ready to prove the classification result mentioned in the introduction. The original proof is in [7]. Another proof of this result can be found in [14, Theorem 4.1]. The merit of our proof is that it explains why the result is true.

Theorem 4.4 (Feinsilver). *A natural exponential family has quadratic variance function if and only if the associated Sheffer polynomials are orthogonal.*

Proof. Combine Theorems 4.1 and 4.3. ■

5. Natural exponential families and approximation theory

In this section we will show that exponential families appear in disguise in approximation theory. An important consequence of this is that the results of [16, 9, 10] are of importance for the statistics literature (in particular, it turns out that many of the results in [18] were predated by the above-mentioned papers). We will also see how our approach differs from the approach in [16, 9, 10] (apart from different terminology).

We begin with recalling the basics of exponential-type approximation operators, following the exposition in [16] (see also [10, 9]). We slightly change the notation in order to be able to compare directly.

Let $W(\lambda, m, x)$ be the kernel of an exponential-type operator, i.e., $W(\lambda, m, x)$ is a generalized function[2] such that

$$\begin{aligned} W(\lambda, m, x) &\geq 0 && (14) \\ \int_{-\infty}^{\infty} W(\lambda, m, x)\, dx &= 1 && (15) \\ \frac{\partial}{\partial m} W(\lambda, m, x) &= \frac{\lambda}{p(m)} W(\lambda, m, x)\,(x - m), && (16) \end{aligned}$$

where p is analytic and positive on an interval on the real line. The corresponding positive approximation operator is defined by

$$(S_\lambda f)(t) = \int_{-\infty}^{\infty} W(\lambda, t, x)\, f(x)\, dx. \tag{17}$$

It is shown in [10, Corollary 3.2] that any solution of the partial differential equation (16) (together with the normalization condition (15)) is of the form

$$W(\lambda, m, x) = \exp\left(\lambda \int_c^m \frac{x - y}{p(y)}\, dy\right) C(\lambda, x). \tag{18}$$

The normalization condition (15) yields that $\exp\left(\lambda \int_c^{g(m)} y/p(y)\, dy\right)$ is the Laplace transform of $C(\lambda, x)$, where $g(m) = \int_c^m 1/p(y)\, dy$. In other words, for fixed λ, the $W(\lambda, m, x)$ form a natural exponential family generated by $d\nu(x) = C(\lambda, x)$ (cf. formulas (2) and (6)) such that $\psi(m) = \int_c^m \lambda/p(y)\, dy$.

Conversely, let $\{P_\theta \,|\, \theta \in \Theta\}$ be the natural exponential family generated by a measure ν. Consider the reparametrization (6). Define the functions $W(\lambda, m, x)$ by

$$W(\lambda, m, x) := e^{x\psi(\lambda m) - k(\psi(\lambda m))}. \tag{19}$$

Since $\psi = k^{-1}$, it follows immediately that $\psi'(m) = 1/k''(\psi(m))$ and $k(\psi(m))' = m/k''(\psi(m))$. Hence, the functions $W(\lambda, m, x)$ defined by (19) satisfy

$$\frac{\partial}{\partial m} W(\lambda, m, x) = \frac{\lambda}{k''(\psi(m))} (x - m).$$

Note that $k''(\psi)$ is the variance function of $\{P_\theta \,|\, \theta \in \Theta\}$ by (6) (i.e., it is the p appearing in (16)).

[2]In fact, one would like to say that $W(\lambda, m, x)$ is the density function of a random variable. However, since we don't want to exclude random variables with discrete parts, we have to resort to generalized functions.

We have thus obtained a complete correspondence between kernels of exponential-type approximation and natural exponential families. Hence, we have shown that the classification problems for exponential-type approximation operators and natural exponential families are equivalent. However, we now want to point out some differences between the approach in [16, 10, 9] and our approach. Cast in our terminology, Ismail and May expand the moment generating function of ν and invert Laplace transforms, while we expand the densities with respect to ν in (2) and solve differential equations. As a consequence, the polynomial sequences that correspond to approximation operators as in [9] differ from our polynomial sequences. Thus although the classifications yield the same probability distributions, they yield different associated polynomial sequences.

6. Quadratic variance functions

In this section we will use the results of the previous sections to present the classification for natural exponential families with quadratic variance function in full detail.

Theorem 4.1 tells us that we must solve the differential equations (11) and (12) in order to obtain all exponential families with quadratic variance function. Note that since $Q = q(D)$ is a delta operator, we must have $q(0) = 0$ and $q'(0) \neq 0$. Since $A' = AQ$ and $A = a(D)$ is invertible, it follows that $\log(A^{-1}A') = Q$. Hence,

$$a(D) = \exp(\int q(D)\, dD). \tag{20}$$

Note that the integration constant must be equal to zero, since $a(t)$ is the Laplace transform of a mean zero distribution.

Under these conditions, we find the following natural exponential families.

6.1. Normal distribution.

If the variance function is constant, then $Q' = \alpha$ with $\alpha > 0$. Hence, $Q = \alpha D$. Now (20) yields that $A = e^{\alpha D^2/2}$. Thus the corresponding natural exponential family is generated by a normal distribution with mean zero and variance α for $\alpha > 0$. The associated Sheffer polynomials are the Hermite polynomials of variance α [23, p. 87 ff.].

6.2. Poisson distribution.

If the variance function is a polynomial of degree one, then $Q' = \alpha + \beta Q$. Thus, $Q = \alpha\left(\frac{e^{\beta D} - 1}{\beta}\right)$ and $A = \exp\left(\frac{\alpha}{\beta}\left(\frac{1}{\beta}e^{\beta D} - D\right)\right)$. Thus the corresponding family is the Poisson family. The associated Sheffer polynomials are the Poisson–Charlier polynomials [23, p. 119 ff.].

6.3. Gamma distribution.

If the variance function is a polynomial of degree two with two identical roots, then $Q' = \alpha(Q-\beta)^2$. Hence, $Q = \beta \dfrac{D}{D + 1/(\alpha\beta)}$ and $A = e^{\beta D}\,(1+\alpha\beta D)^{1/\alpha}$. Thus the corresponding natural exponential family is the gamma distribution family. The associated Sheffer polynomials are the Laguerre polynomials of variance α [23, p. 108 ff.].

6.4. Binomial distribution.

If the variance function has two different positive roots, then the corresponding natural exponential family is the binomial distribution family. The associated Sheffer polynomials are the Krawtchouk polynomials [23, p. 125–126].

6.5. Negative binomial distribution.

If the variance function has two different negative roots, then the corresponding natural exponential family is the negative binomial distribution family. The associated Sheffer polynomials are a subclass of the Meixner polynomials of the first kind [23, p. 125-126].

6.6. Hyperbolic distribution.

If the variance function has two complex conjugate roots, then the associated Sheffer polynomials are a subclass of the Meixner polynomials of the second kind [23, p. 126]. The corresponding natural exponential family is generated by the hyperbolic distribution (see [15]).

7. Conclusion

A few final remarks on generalizations are in order. Our approach is not restricted to natural exponential families with quadratic variance function. For example, it could be used to obtain a classification of natural exponential families with cubic variance function as in [15] (in [10] no attempt is being made to obtain a complete classification). Letac and Mora[15] state that it seems hard to obtain classifications of natural exponential families with higher order polynomial variance functions. In light of our approach, this is probably related to the fact that the differential equations $Q' = V(Q)$ are hard (resp. impossible) to solve explicitly when V is a polynomial of degree more than three (resp. four).

A more interesting direction is to generalize our approach to natural exponential families generated by multivariate distributions [3, 5, 11, 14, 20, 21]. Together with Gian-Carlo Rota, the authors are developing a basis-free multivariate Umbral Calculus [1], one of the hoped for applications is the above mentioned multivariate case.

Acknowledgement. We would like to thank one of the referees for drawing our attention to the references on approximation operators. This put our paper in a proper context and led us to write the section on the relation between exponential approximation operators and natural exponential families.

References

[1] A. Di Bucchianico, D.E. Loeb, and G.-C. Rota, Umbral calculus in Hilbert space, to appear in Rota Festschrift.

[2] W.A. Al-Salam, On a characterization of Meixner's polynomials, *Quart. J. Math. Oxford* **2** 17 (1966), 7–10.

[3] S.K. Bar-Lev, D. Bshouty, P. Enis, G. Letac, I.-L. Lu, D. Richards, The diagonal multivariate natural exponential families and their classification. *J. Theoret. Probab.* **7** (1994), 883–929.

[4] T.S. Chihara, An Introduction to Orthogonal Polynomials, Gordon and Breach, 1978.

[5] Y. Chikuse, Multivariate Meixner classes of invariant distributions, *Lin. Alg. Appl.* **82** (1986), 177–200.

[6] A.J. Dobson, An introduction to generalized linear models, Chapman and Hall, 1994.

[7] P.J. Feinsilver, Special Functions, Probability Semigroups, and Hamiltonian Flows, *Lect. Notes in Math.* **696**, Springer, Berlin, 1978.

[8] J.M. Freeman, Orthogonality via transforms, *Stud. Appl. Math.* **77** (1987), 119–127.

[9] M.E.H. Ismail, Polynomials of binomial type and approximation theory, *J. Approx. Theor.* **23** (1978), 177–186.

[10] M.E.H. Ismail and C.P. May, On a family of approximation operators, *J. Math. Anal. Appl.* **63** (1978), 446–462.

[11] S.A. Joni, Multivariate exponential operators, *Stud. Appl. Math.* **62** (1980), 175–182.

[12] A.N. Kholodov, The umbral calculus and orthogonal polynomials, Acta Appl. Math. 19 (1990), 55–76.

[13] H.O. Lancaster, Joint probability distributions in the Meixner classes, *J. Roy. Stat. Soc. B* **37** (1975), 434–443.

[14] G. Letac, Le problème de la classification des familles exponentielles naturelles de $\mathbb{R}^d$ ayant une fonction variance quadratique, in: Probability measures on groups IX, Lect. Notes in Math. 1379, H. Heyer (ed.), Springer, Berlin, 1989, 192–216.

[15] G. Letac and M. Mora, Natural real exponential families with cubic variance functions, *Ann. Stat.* **18** (1990), 1–37.

[16] C.P. May, Saturation and inverse theorems for combinations of a class of exponential-type operators, *Can. J. Math.* **28** (1976), 1224–1250.

[17] J. Meixner, Orthogonale Polynomsysteme mit einer besonderen Gestalt der erzeugenden Funktion, *J. London Math. Soc.* **9** (1934), 6–13.

[18] C.N. Morris, Natural exponential families with quadratic variance functions, *Ann. Stat.* **10** (1982), 65–80.

[19] C.N. Morris, Natural exponential families with quadratic variance functions: statistical theory, *Ann. Stat.* **11** (1983), 515–529.

[20] D. Pommeret, Familles exponentielles naturelles et algèbres de Lie, *Exposition. Math.* **14** (1996), 353–381.

[21] D. Pommeret, Orthogonal polynomials and natural exponential families, *Test* **5** (1996), 77–111.

[22] G.-C. Rota, D. Kahaner, and A. Odlyzko, On the foundations of combinatorial theory VII. Finite operator calculus, *J. Math. Anal. Appl.* **42** (1973), 684–760.

[23] S.M. Roman, The Umbral Calculus, Academic Press, 1984.

[24] S.M. Roman, P. De Land, R. Shiflett, and H. Schultz, The umbral calculus and the solution to certain recurrence relations, *J. Comb. Inf. Sys. Sci.* **8** (1983), 235–240.

[25] S.D. Silvey, Statistical inference, Monographs on Statistics and Applied Probability 7, Chapman and Hall, 1991.

[26] I.M. Sheffer, Some properties of polynomials of type zero, *Duke Math. J.* **5** (1939), 590–622.

[27] X.-H. Sun, New characteristics of some polynomial sequences in combinatorial theory, *J. Math. Anal. Appl.* **175** (1993), 199–205.

A. Di Bucchianico
Eindhoven University of Technology
Dept. of Mathematics and Computing Science
P.O. Box 513
5600 MB Eindhoven, The Netherlands
sandro@win.tue.nl
http://www.win.tue.nl/ sandro

D.E. Loeb
Laboratoire Bordelais de Recherche en Informatique
Université de Bordeaux I
33405 Talence Cedex, France
loeb@labri.u-bordeaux.fr
http://dept-info.labri.u-bordeaux.fr/ loeb/index.html

Umbral Calculus in Hilbert Space

Alessandro Di Bucchianico[*], *Daniel E. Loeb*[†]
and Gian-Carlo Rota[‡]

Extended Abstract

AMS Classification: 05A40 / 47N50, 33C55

Keywords: Umbral Calculus, Bose Algebra, Binomial Type, Appell Polynomials, Bernoulli Polynomials, Hermite Polynomials, Abel Polynomials, Laplace Operator, Annihilation Operator, Hilbert Space

1. Introduction

Umbral calculus in its modern form [28, 46] is a powerful tool for calculations with polynomials. Applications of the umbral calculus include combinatorics (e.g. [14, 16, 22, 29, 31, 36, 39, 40, 42, 45, 50, 55]), special function theory [10, 21], approximation theory [17, 19, 26, 47], statistics (e.g. [7, 12, 30, 32]), probability theory (e.g. [6, 49, 51]), topology (e.g. [35, 37, 38]), and physics (e.g. [3, 4, 57]).

In light of these applications, several extensions of the umbral calculus to multiple variables have been made (see e.g. [5, 8, 9, 15, 18, 25, 34, 41, 43, 48, 52, 53, 54]) or even infinitely many variables [1]. However, these extensions all suffer from the same drawback: they are not basis-free. In other words, these extensions are made over a polynomial ring $\mathbf{R}[x_1, x_2, \ldots, x_n]$ where $\mathbf{x} = (x_1, x_2, \ldots, x_n)$ is a fixed set of variables taking values; that is, $\mathbf{x}$ is an indeterminate taking values (expressed in terms of the "standard basis") in the vector space $\mathbf{R}^n$. In the present work, we summarize the results we have obtained so far in developing a basis-free umbral calculus over a real Hilbert space $\mathcal{H}$ (of finite or infinite dimension) instead of a fixed set of variables.

[*]Author supported by NATO CRG 930554.

[†]Author partially supported by URA CNRS 1304, EC grant CHRX-CT93-0400, the PRC Maths-Info, and NATO CRG 930554.

[‡]Author supported by NATO CRG 930554.

We believe such a (separable) Hilbert space to be the natural setting for umbral calculus. Our polynomials are sums of products of inner products $\langle x|h\rangle$. By specializing to any fixed orthonormal basis $\{b_1, b_2, \ldots\}$, the inner products $\langle x|h\rangle$ can be expanded in terms of $x_i = \langle x|b_i\rangle$ and we recover the classical theory of polynomials. Most of the classical formulas may be extended to this setting, and most importantly, they acquire an ease and naturalness which even in the one variable case is not visible classically. For example, the special role of *diagonal sets* [43] (called *normal* by Watanabe [53, 54] and *coherent* by Barnabei *et al.* [1]) is seen to be an artifact of having specialized to the wrong basis in the case of the basic sequence of a nondiagonal delta set.

The present Hilbert space setting offers another advantage, that of studying orthogonally invariant special functions (for example, spherical harmonics) by umbral methods. Last but not least, the present notation streamlines and in some cases makes fully rigorous the Boson calculus of second quantization [33].

Let us make one final remark concerning topology. Typically in algebraic combinatorics the topologies employed are quite simple. For example, the sum of operators may be deemed to converge only if finitely many terms are nonzero when applied to any given object. In most papers, questions of convergence are left to the reader since they truly are trivial once one applies all the relevant definitions.

In this paper however there are several infinite processes occuring simultaneously, and a variety of special topologies are applied depending on the circumstances. Passing from polynomial to formal series requires the "combinatorial" topology resulting from the grading of $\mathcal{P}$ into homogeneous components. On the other hand, passing from polynomials to quasi-polynomials requires the use of filters via ascending chains of subspaces. Finally, the Hilbert space has its own topology which is inherited by the inner product on $\mathcal{P}$.

We thus are constrained in each instance to pay particular attention to issues of convergence and topology. Nevertheless, in the interest of space, we defer most complete proofs and serious discussion of topological issues to our sequel paper.

Similarly, we only give three generalizations of classical special functions to a Hilbert space setting: Abel polynomials, Bernoulli polynomials (limited to finite-dimensional Hilbert space), and Hermite polynomials (generalizing the pioneering work of Appell and Kampé de Feriet).

2. Generalized Polynomials

2.1. Polynomials and quasi-polynomials

The first step in any generalization of the umbral calculus is to clearly define the objects in question which play the role of ordinary polynomials. It turns out that in an infinite dimensional Hilbert space two such generalizations are available: polynomials and quasi-polynomials. In a certain sense to be made clear in Section 3, these notions are complementary and neither can be fully understood without the other.

In this paper, $\mathcal{H}$ will denote a real Hilbert space with a countable basis.

Given a Hilbert space $\mathcal{H}$, a *generalized polynomial* $p(x)$ (or simply *polynomial* when there is no risk of confusing) is a finite sum of products of inner products $\langle x|h\rangle$ where $h \in \mathcal{H}$. The **R**-algebra of these polynomials (with the obvious addition and multiplication) is denoted by $\mathcal{P}$. This is to be distinguished from the more general notion of a *quasi-polynomial* which is a continuous function from the Hilbert space $\mathcal{H}$ to the reals **R** whose restrictions to finite dimensional subspaces of $\mathcal{H}$ are polynomial.

By definition, polynomials are quasi-polynomials. However, quasi-polynomials are not in general polynomials. For example, consider the *norm function* $\mathcal{N} : x \mapsto \langle x|x\rangle = \|x\|^2$.

Proposition 2.1. *The norm function $\mathcal{N}$ is a quasi-polynomial, but unless* $\dim(\mathcal{H})$ *is finite, $\mathcal{N}$ is not a polynomial.*

Proof. Given an orthonormal basis $\{b_1, \dots, b_n\}$ of a finite dimensional subspace $\mathcal{H}'$ of $\mathcal{H}$, we have $\mathcal{N}|_{\mathcal{H}'}(x) = \sum_{k=1}^{n} \langle x|b_k\rangle^2$. Thus, $\mathcal{N}$ is a quasi-polynomial.

Suppose $\mathcal{H}$ is infinite dimensional and $\mathcal{N}$ was expressed by some finite sum $p = \sum_{j=1}^{k} c_j \langle x|r_{j,1}\rangle \langle x|r_{j,2}\rangle \dots \left\langle x \middle| r_{j,i_j}\right\rangle$. Now, let h be a nonzero vector in the space $\{r_{j,1} : 1 \leq j \leq k\}^{\perp}$. Clearly, $p(h) = 0 \neq \mathcal{N}(h)$. ∎

The **R**-algebra of polynomials $\mathcal{P}$ can be recognized as isomorphic to the free algebra generated by $\mathcal{H}$. When equipped with the inner product of Subsection 3.2, the latter algebra becomes the Bose algebra $\Gamma_0(\mathcal{H})$ — dear to physicists where $\mathcal{H}$ is called the *base* or *one-particle space* [33, p. 6]. On the other hand, the **R**-algebra $\mathcal{Q}$ of quasi-polynomials does not seem to have appeared in the literature.

At each step in generalizing the umbral calculus, it is helpful to express the results of the classical umbral calculus as a special case of our generalized results in order to better see the relation between the two. Here, if $\mathcal{H} = \mathbf{R}^n$ equipped with the standard basis $\{b_1, \dots, b_n\}$, then we may identify x with $(x_1, x_2, \ldots, x_n)$. Hence, the inner product plays the role of a "variable" $\langle x|b_k\rangle = x_k$, and $\mathcal{P} = \mathcal{Q} \equiv \mathbf{R}[x_1, x_2, \ldots, x_n]$. Thus, our generalized polynomials reduce to ordinary polynomials, and quasi-polynomials are trivially polynomials.

2.2. Multisets

We will use the shorthand x^R to denote *monomials* such as $\langle x|r_1\rangle \langle x|r_2\rangle \ldots \langle x|r_n\rangle$, where $R = \{r_1, r_2, \ldots, r_n\}$ denotes a *multiset* of elements of $\mathcal{H}$. The multiplicity of an element r of a multiset R is denoted by R_r.

In particular, $x^\emptyset$ equals the constant 1 identically, and the map $\mathbf{x} : r \mapsto x^{\{r\}} = \langle x|r\rangle$ is linear.

The *degree* of x^R is defined to be the *cardinality* of $|R|$, in other words $\deg(x^R) = |R| = \sum_{r\in\mathcal{H}} R_r$ is the sum of the *multiplicities* R_r. The degree of a (quasi-)polynomial is the largest degree of its terms.

Given a fixed basis B for $\mathcal{H}$, the monomials x^R for $R \subset B$ will be seen to be a basis for the space $\mathcal{Q}$ of quasi-polynomials.

We reserve the notation $R \subset B$ to indicate that R is a multiset on B, where B is a subset of $\mathcal{H}$. This should not be confused with the notation $\subseteq$ used here to denote the usual inclusion of sets or multisets.

We define the *factorial* as

$$R! = \prod_{h\in\mathcal{H}} R_h!,$$

and the *binomial coefficients* as

$$\binom{R}{S} = \frac{R!}{S!\,(R \setminus S)!}.$$

Sums over submultisets of a given multiset are taken without regard to multiplicity (unless specially indicated as $\sum'$). For example,

$$\sum_{S\subseteq R} \binom{R}{S} c_S = {\sum_{S\subseteq R}}' c_S = \sum_{I\subseteq\{1,2,\ldots,n\}} c_{\{r_k : k\in I\}} \tag{1}$$

where $R = \{r_1, r_2, \ldots, r_n\}$ is a multiset and the last sum is over subsets I of the index set $\{1, 2, \ldots, n\}$.

The *sum* or *product* of any quantity over elements of a multiset is given by their multiplicity

$$\sum_{h\in R} c_h = \sum_{h\in\mathcal{H}} R_h\, c_h \quad \text{and} \quad \prod_{h\in R} c_h = \sum_{h\in\mathcal{H}} c_h^{R_h}.$$

Thus, $\sum_{r\in R}$ is equivalent to $\sum'_{\substack{S\subseteq R\\|S|=1}}$ but not to $\sum_{\substack{S\subseteq R\\|S|=1}}$.

For example, the cardinality of a multiset is given by the sum

$$|R| = \sum_{r\in R} 1,$$

and we have the following generalization of the binomial formula.

Theorem 2.2. (Binomial Theorem) *If R is a multiset on $\mathcal{H}$ and x and y are elements of $\mathcal{H}$, then we have*

$$(x+y)^R = \sum_{S\cup T=R} \binom{R}{S} x^S\, y^T = \sum_{S\cup T=R}{}' x^S\, y^T.$$

Proof. Using the multiset analog of Pascal's formula

$$\binom{R\cup\{u\}}{S} = \binom{R}{S\setminus\{u\}} + \binom{R}{S},$$

we obtain the result by induction on $|R|$. ∎

2.3. Formal and quasi-formal Hilbert series

A (quasi-)polynomial p is said to be *homogeneous* of degree k if it can be expressed so that all its terms are of degree k, that is, $p(\lambda x) = \lambda^k p(x)$ for all $x \in \mathcal{H}$ and $\lambda \in \mathbf{R}$. It is easy to see that the spaces $\mathcal{P}_k$ (resp. $\mathcal{Q}_k$) of homogeneous polynomials (resp. quasi-polynomials) of degree k grade the $\mathbf{R}$-algebras

$$\mathcal{P} = \bigoplus_{k=0}^{\infty} \mathcal{P}_k \qquad \left(\text{resp. } \mathcal{Q} = \bigoplus_{k=0}^{\infty} \mathcal{Q}_k\right).$$

Proposition 2.3.

0. $\mathcal{P}_0 = \mathcal{Q}_0 \cong \mathbf{R}$.

1. $\mathcal{P}_1 = \mathcal{Q}_1 \simeq \mathcal{H}$.

2. $\mathcal{P}_2 \neq \mathcal{Q}_2$ *unless* $\dim(\mathcal{H})$ *is finite.*

Proof.

0. The only monomial of degree zero is $x^{\emptyset} = 1$.

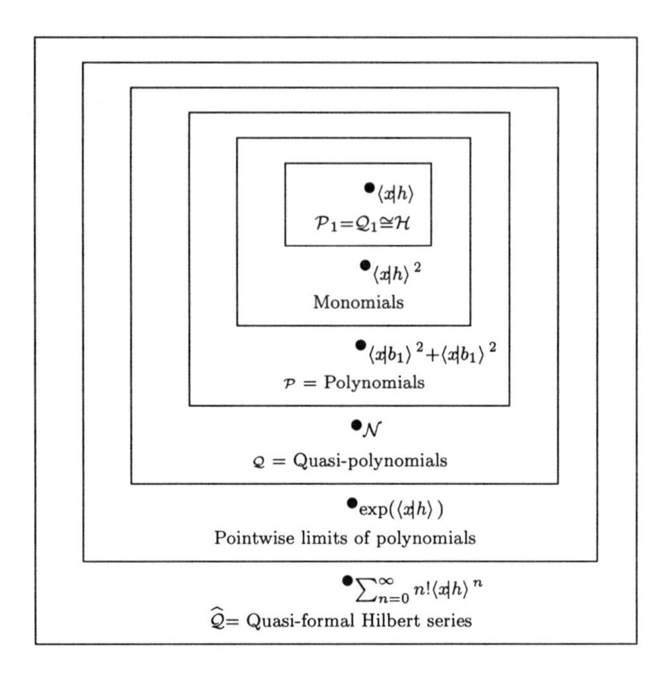

Figure 1: Generalized Polynomials

1. **Polynomials.** By continuity, the additivity of a quasi-polynomial on finite dimensional subspaces extends to all of $\mathcal{H}$. Hence, by the Riesz-Fischer theorem, it is a linear functional, i.e. an element of $\mathcal{P}_1$.

2. The norm function $\mathcal{N}$ is in $\mathcal{Q}_2$, but not in $\mathcal{P}_2$ by Proposition 2.1. ∎

We now define $\widehat{\mathcal{P}}$ to be the completion of $\mathcal{P}$ over this grading. The members of $\widehat{\mathcal{P}}$ are called *formal Hilbert series*. They are formal sums

$$f(t) = \sum_{k=0}^{\infty} f_k(t)$$

of homogeneous polynomials f_k one for each degree k, where for all $k \geq 0$, the coefficient c_R is nonzero for only finitely many multisets $R \subseteq \mathcal{H}$ of cardinality k. That is, f is a formal sum of polynomials. E.g., $\sum_k \langle x|h\rangle^k$ is a member of $\widehat{\mathcal{P}}$ (but not of $\mathcal{Q}$ since it diverges at $x = h$), while the quasi-polynomial $\mathcal{N}$ does not belong to $\widehat{\mathcal{P}}$. Similarly, we define $\widehat{\mathcal{Q}}$ to be the completion of $\mathcal{Q}$. Its members (called *quasi-formal Hilbert series*) are formal sums of quasi-polynomials. As opposed to (quasi-)polynomials, (quasi-)formal Hilbert series are not necessarily representable as a function $\mathcal{H} \to \mathbf{R}$, and we use the letter t to denote their purely formal "variable."

If the Hilbert space $\mathcal{H}$ is the usual finite dimensional Euclidean space $\mathbf{R}^n$, then we have $\widehat{\mathcal{P}} = \widehat{\mathcal{Q}} \cong \mathbf{R}[[x_1, x_2, \ldots, x_n]]$. Here, formal Hilbert series reduce to formal power series.

The *order* of a (quasi-)formal Hilbert series q is the largest n such that q can be written using terms of degree at least n, that is, $q \in \oplus_{k=n}^{\infty} \mathcal{Q}_k$, but $q \notin \oplus_{k=n+1}^{\infty} \mathcal{Q}_k$. Hence, the *order* of 0 is taken to be $+\infty$. Addition and multiplication of (quasi-)polynomials are continuous, and thus can be extended by continuity to (quasi-)formal Hilbert series. A (quasi-)formal Hilbert series is invertible if and only if it is of order zero. In other words, it is invertible if and only if it has a constant term.

3. Operators

The duality between polynomials and their operators has served since around 1970 [28] as the logical foundation of umbral calculus. The key to applying these powerful operator methods in the new context of generalized polynomials or quasi-polynomials over a Hilbert space $\mathcal{H}$ is to seek out the operators which play the role of the familiar derivative d/dx and shift operators $E^y : p(x) \mapsto p(x+y)$. It is essential that the operators which commute with the shift operator (*shift-invariant operators*) are exactly those that can be expressed in terms of derivatives.

In coalgebraic terms [20, Section V.3], the shift operator must play the role of coproduct in a Hopf algebra over $\mathcal{P}$ (resp. $\mathcal{Q}$). (We will not use coalgebra theory or terminology in our paper. However, it is important to recognize the crucial role that coalgebra theory plays in the design of a generalization of umbral calculus.)

The algebra of shift-invariant operators on $\mathcal{P}$ (resp. $\mathcal{Q}$) must then be seen as dual to $\mathcal{Q}$ (resp. $\mathcal{P}$). The resulting inner product is of considerable interest [44].

3.1. Derivative

The *derivative* D_h for $h \in \mathcal{H}$ is the unique derivation on $\mathcal{P}$ (or $\mathcal{Q}$) defined by $D_h \langle x|r\rangle = \langle h|r\rangle$. For example, the derivative of a monomial is

$$D_h\, x^R = \sum_{r \in R} \langle h|r\rangle\, x^{R\setminus\{r\}}, \tag{2}$$

and the derivative of the norm $\mathcal{N}$ is given by

$$D_h \mathcal{N} = 2\,\langle x|h\rangle\,. \tag{3}$$

If u is a unit vector, then D_u can be thought of as the derivative in the direction u. Otherwise, $h = cu$ and $D_h = cD_u$. In particular, D_0 is the zero operator.

Since the derivatives commute $D_{h_1} D_{h_2} = D_{h_2} D_{h_1}$, we may thus define the *multiple derivative* $D^R = \prod_{r \in R} D_r$.

3.2. Inner product

We now endow $\mathcal{P}$ with the following inner product:

$$\left\langle x^R \middle| x^S \right\rangle = \begin{cases} 0 & \text{if } j \neq k, \text{ and} \\ \operatorname{perm} \begin{pmatrix} \langle r_1|s_1\rangle & \cdots & \langle r_j|s_1\rangle \\ \vdots & \ddots & \vdots \\ \langle r_1|s_j\rangle & \cdots & \langle r_j|s_j\rangle \end{pmatrix} & \text{if } j = k \end{cases} \tag{4}$$

where

- $R = \{r_1, r_2, \ldots, r_k\}$ and $S = \{s_1, s_2, \ldots, s_j\}$ are multisets on $\mathcal{H}$,
- the *permanent* of an $n \times n$ matrix is defined as a "signless" determinant

$$\operatorname{perm}(M) = \sum_{\sigma \in \mathcal{S}_n} \left(\prod_{i=1}^{n} M_{i,\sigma(i)} \right) \tag{5}$$

  and
- $\mathcal{S}_n$ is the symmetric group of permutations of n elements.

Proposition 3.1. **([33, Lemma 1.7B])** *Let B be an orthonormal basis of $\mathcal{H}$. Given finite multisets R and S on B, we have*

$$\left\langle x^R \middle| x^S \right\rangle = \begin{cases} S! & \text{if } R = S, \text{ and} \\ 0 & \text{otherwise.} \end{cases}$$

It is a direct corollary of this proposition that the derivative D_h is the adjoint of the operator X_h of multiplication by $\langle x|h\rangle$ with respect to this inner product.

Note that this inner product can not be extended to quasi-polynomials in general. For example, by the above definition, $\langle \mathcal{N}|\mathcal{N}\rangle$ would be given by an infinite divergent sum.

This inner product defines a topology on $\mathcal{P}$ unrelated to that used in Section 2.3 for the definition of the completion $\widehat{\mathcal{P}}$. For example, in this topology $\langle x|h\rangle^n$ does not tend to zero, but $\frac{1}{n}\sum_{j=1}^{n} \langle x|b_j\rangle$ does tend to zero.

Endowed with this topology, $\mathcal{P}$ is called the Bose algebra in physics [33, p. 6].

Let B be an orthonormal basis of $\mathcal{H}$. Since $\mathbf{x} = \{x^R : R \subset B\}$ is an orthogonal basis of $\mathcal{P}$, the following proposition is easily seen to be true.

Proposition 3.2. *Let S be a finite multiset on $\mathcal{H}$, and let B be an orthonormal basis of $\mathcal{H}$. The monomials x^S may then be expanded as*

$$x^S = \sum_{R \subset B} \left\langle x^S \middle| x^R \right\rangle x^R / R!.$$

We have the following extension of Leibniz's rule.

Proposition 3.3. *Let $R, S \subseteq \mathcal{H}$ be finite multisets.*

$$D^S x^R = \sum_{V \cup W = R} \binom{R}{V} \left\langle x^S \middle| x^W \right\rangle x^V = \sum_{V \cup W = R}' \left\langle x^S \middle| x^W \right\rangle x^V.$$

We may now also prove the analog of Taylor's theorem.

Proposition 3.4. *Let q be a quasi-polynomial and $B \subset \mathcal{H}$ be an orthonormal basis. Then q can be expressed as an infinite sum*

$$q(x) = \sum_{R \subset B} c_R x^R / R!, \tag{6}$$

where the coefficients c_R are uniquely given by $c_R = \left[D^R q \right]_{x=0}$.

3.3. Shift-Invariant Operators

For $h \in \mathcal{H}$, the *shift-operator* E^h is defined on $\mathcal{P}$ and $\mathcal{Q}$ as usual by

$$E^h := \exp(D_h) = \sum_{k=0}^{\infty} D_h^k / k!$$

or equivalently $E^h p(x) = p(x+h)$ for a (quasi-)polynomial p. Note that all but finitely many terms vanish when the sum is applied to any given quasi-polynomial.

A linear operator θ on $\mathcal{P}$ or $\mathcal{Q}$ is said to be *shift-invariant* if $\theta E^a = E^a \theta$ for all $a \in \mathcal{H}$. The set of shift-invariant operators over $\mathcal{P}$ or $\mathcal{Q}$ forms an $\mathbf{R}$-algebra denoted SIO or $QSIO$ respectively, where $(\theta + \phi) f = \theta(f) + \phi(f)$ and $(\theta \times \phi) f = \theta(\phi(f))$.

One might ask why we choose to study *both* polynomials and quasi-polynomials rather than focusing our attention on one, say polynomials as do physicists. In fact, shift-invariant operators on one can not be adequately understood without recourse to the other. Shift-invariant operators over polynomials are expressed in terms of derivatives by the use of quasi-formal Hilbert series. Conversely, expressing shift-invariant operators over quasi-polynomials requires formal Hilbert series.

By the above, the multiple derivatives D^R are shift-invariant for all multisets $R \subseteq \mathcal{H}$ as are any linear combinations thereof. Thus, every quasi-formal Hilbert series $f = \sum_{R \subset B} c_R \, t^R$ can be identified with a shift-invariant operator on $\mathcal{P}$ namely $\iota(f) = f(D) = \sum_{R \subset B} c_R \, D^R$, and every formal Hilbert series $f = \sum_{R \subset \mathcal{H}} c_R \, t^R$ can be identified with a shift-invariant operator on $\mathcal{Q}$ namely $\iota(f) = f(D) = \sum_{R \subset \mathcal{H}} c_R \, D^R$.

This identification is one-to-one and onto; that is, the map ι is an isomorphism identifying the spaces $\hat{\mathcal{Q}}$ and $\hat{\mathcal{P}}$ with SIO and $QSIO$, respectively. Hence $QSIO$ can be naturally identified with a subspace of SIO. The inverse image under ι of a shift-invariant operator is known as its indicator.

Theorem 3.5. *Let θ be a shift-invariant operator on either $\mathcal{P}$ or $\mathcal{Q}$, and $B \subset \mathcal{H}$ be an orthonormal basis. Then θ is given by the convergent sum*

$$\theta = \sum_{R \subset B} c_R \, D^R \tag{7}$$

over submultisets of B where $c_R = \left[\theta x^R / R!\right]_{x=0}$.

Thus, considering the more general quasi-polynomials restricts our choice of shift-invariant operators while considering only polynomials allows the choice of shift-invariant operators corresponding to quasi-formal Hilbert series.

The *Laplace operator* $\mathcal{L} = \iota \mathcal{N}$ is an example of a shift-invariant operator on $\mathcal{P}$ which does not extend to $\mathcal{Q}$.

Proposition 3.6. *The action of the Laplace operator $\mathcal{L}$ is for $S \setminus \mathcal{H}$ given by*

$$\mathcal{L} \, x^S = \sum_{r \in S} \sum_{s \in S \setminus \{r\}} \langle r | s \rangle \, x^{S \setminus \{r,s\}}.$$

Proof. By linearity and continuity, we may assume without loss of generality that S is a multiset on some orthonormal basis B. Thus, we can write

$$\begin{aligned} \mathcal{L} &= \iota \mathcal{N} \\ &= \sum_{b \in B} \iota \, \langle x | b \rangle^2 . \\ \mathcal{L} x^S &= \sum_{b \in B} D_b^2 \, x^S \end{aligned}$$

$$= \sum_{b \in B} S_b(S_b - 1) x^{S \setminus \{b,b\}}$$
$$= \sum_{b \in S} \sum_{b' \in S \setminus \{b\}} \langle b | b' \rangle \, x^{S \setminus \{b,b'\}}$$

since $\langle b|b' \rangle = \delta_{bb'}$. ■

3.4. Linear operators

This setting allows us to obtain a notable generalization (Theorem 3.8.) of Kurbanov and Maksimov's result [23] giving an explicit construction of an arbitrary linear operator in terms of the derivative and multiplication by x. Such expansions are useful in that the derivative is easy to calculate both numerically and symbolically. We can thus manipulate arbitrary linear operators with similar ease. For example, the expansion of the integration operator gives an asymptotic formula for integration [11, p. 54].

Let θ be a continuous, linear polynomial (resp. quasi-polynomial) operator. We extend θ by acting on the coefficients of bivariate series. We use two variables x and t to denote quasi-formal Hilbert series whose coefficients are not simply constants but polynomials or quasi-polynomials. The set of such series is thus given by the real tensor product $\mathcal{P} \otimes \widehat{\mathcal{Q}}$ (resp. $\mathcal{Q} \otimes \widehat{\mathcal{Q}}$).

For example, let $B = \{b_1, b_2, \ldots\}$ be an orthonormal basis. Then the inner product

$$\langle x|t \rangle = \sum_{i=1}^{\infty} \langle x|b_i \rangle \, \langle t|b_i \rangle$$

is a member of $\mathcal{P} \otimes_{\mathbf{R}} \widehat{\mathcal{Q}}$. Its exponential is a *generating function* for the monomials x^R

$$\mathrm{Gen}\,(\mathbf{x}) := \sum_{R \subset B} x^R t^R / R! = \sum_{n=0}^{\infty} \frac{\langle x|t \rangle^n}{n!} = \exp \langle x|t \rangle \,.$$

Thus, to compute the derivative $D_h \langle x|t \rangle$, we expand $\langle x|t \rangle$.

$$\begin{aligned} D_h \langle x|t \rangle &= \sum_{i=1}^{\infty} D_h \langle x|b_i \rangle \, \langle t|b_i \rangle \\ &= \sum_{i=1}^{\infty} \langle h|b_i \rangle \, \langle t|b_i \rangle \\ &= \sum_{i=1}^{\infty} h_i \, \langle b_i|t \rangle \\ &= \left\langle \sum_{i=1}^{\infty} h_i \, b_i \middle| t \right\rangle \end{aligned}$$

$$= \langle h|t\rangle .$$

Conversely, we extend the definition of ι (Subsection 3.3) by requiring that $\iota x^R t^S = X^R D^S$ where X^R is the operator of multiplication by x^R. (Note that in general $X^R D^S \neq D^S X^R$.)

For example, we define the *mixed operator* or *number operator* $\mathcal{M} = \iota \langle x|t\rangle$.

Proposition 3.7. *The action of the mixed operator $\mathcal{M}$ is given by $\mathcal{M}\, x^R = |R|\, x^R$ for $R \subset \mathcal{H}$.*

We thus have the following generalization of Kurbanov and Maksimov's expansion formula [23].

Theorem 3.8. *Let θ be a linear operator on $\mathcal{P}$ or $\mathcal{Q}$. Then θ may be uniquely expressed in terms of X^R and D^S as follows:*

$$\theta = \iota \left(\frac{\theta \exp\left(\langle x|t\rangle\right)}{\exp\left(\langle x|t\rangle\right)} \right). \tag{8}$$

That is to say, any linear operator can be expressed in terms of a differentiation and multiplication by monomials as follows.

1. Apply to $\exp \langle x|t\rangle$.
2. Multiply by $\exp(- \langle x|t\rangle)$.
3. Replace t^R by D^R.

3.5. Composition

The fundamental theorem of umbral calculus highlights the interplay between polynomials and operators: the umbral composition of two sequences of polynomials corresponds to the ordinary composition of the indicators of their delta operators.

To this end, we explain what the ordinary composition of polynomials (or quasi-polynomials) is in this context. Classically, the functional composition of multivariate polynomials is a sort of plethysm. Each variable x_n in the polynomial q is replaced by a polynomial p_n. In our context, we must choose a polynomial p_h to replace $\langle x|h\rangle$ for each vector $h \in \mathcal{H}$.

We define an *umbral set* $\mathbf{q}$ of quasi-polynomials to be an injective continuous linear map from $\mathcal{H}$ to $\mathcal{Q}$. For example, $\mathbf{x} : r \mapsto \langle x|r\rangle$ is an umbral set of polynomials. There is a unique continuous algebra endomorphism on $\mathcal{P}$ (or $\mathcal{Q}$) that maps $\langle x|r\rangle$ to q_r. It is called the *umbral operator* of the set $\mathbf{q}$.

The image of $\mathbf{q}$ under this map is denoted $p(\mathbf{q})$. We further define an *umbral sequence* $\mathbf{q}$ of (quasi-)polynomials to be an injective map from finite multisets R on $\mathcal{H}$ to (quasi-)polynomials denoted q_R which is k-linear when R is considered as a k-tuple $R = (r_1, r_2, \ldots, r_k)$. For example, $\mathbf{x} : R \mapsto x^R$ is an umbral sequence[1]. More generally, given any umbral set $\mathbf{q}$ of (quasi-)polynomials, the map

$$\mathbf{q}: R \mapsto q^R = \prod_{r \in R} q_r$$

is an umbral sequence. The map defining an umbral sequence can be uniquely extended to an $\mathbf{R}$-linear map on $\mathcal{P}$ (or $(\mathcal{Q})$. The image of a polynomial p under this map is denoted $p(\mathbf{q})$.

Lemma 3.9. is crucial if we wish to relate our umbral sequences to other results concerning multivariate polynomial sequences which appear in the umbral calculus literature. Indeed, the umbral sequence p_R is determined by its values when $R \subset B$ for any orthonormal basis B. In that case, p_R is a classical sequence of multivariate polynomials. The coefficient of p_R in the polynomial p_S is given by the inner product $\left\langle x^S \middle| x^R \right\rangle$ defined in Section 3.2.

Lemma 3.9. *Let B be an orthonormal basis, and let $\mathbf{p}$ be an umbral sequence. The expansion of the elements of $\mathbf{p}$ is given by*

$$p_S = \sum_{R \subset B} \frac{\left\langle x^S \middle| x^R \right\rangle}{R!} p_R.$$

4. Orthogonal Invariance

4.1. Orthogonally invariant quasi-polynomials

Using ordinary polynomials, the only notion of symmetry available is that of symmetric functions [24] invariant under permutation of variables.

However, here, the natural notion of symmetry is much stricter. Recall that the usual notion of variables *per se* only appears in $\mathcal{P}$ by fixing a particular orthonormal basis for our Hilbert space $\mathcal{H}$. However, a priori, there is no natural choice for such a privileged coordinate system, and moreover by leaving the choice of coordinate system free we avoid diagonality problems for sequences of binomial type. (See Section 6.)

A truly symmetric quasi-polynomial in our context should be the same when written in any coordinate system. Thus, a quasi-polynomial p is said to be *orthogonally invariant* if for all isometries $\phi: \mathcal{H} \to \mathcal{H}$, we have $p = p \circ \phi$.

[1]It should be clear from the context when $\mathbf{x}$ is being used as an *umbral set* and when it is being used as an *umbral sequence*.

Theorem 4.1. *A quasi-polynomial q is orthogonally invariant if and only if it can be written in the form $q = f(\mathcal{N})$ for some ordinary polynomial $f \in \mathbf{R}[n]$.*

Proof. If $\mathcal{H}$ is finite dimensional, then this follows directly from the First Main Theorem of Invariant Theory ([13, 56]). For infinite dimensional $\mathcal{H}$ we need the extra fact that any isometry on a finite dimensional subspace of $\mathcal{H}$ extends to an isometry on all of $\mathcal{H}$. ∎

Thus, over an infinite dimensional Hilbert space, there are no orthogonally invariant polynomials other than constants, but only orthogonally invariant quasi-polynomials.

4.2. Orthogonally invariant shift-invariant operators

Given our practice of treating polynomials and operators acting on them as of equal importance, we now define an analogous orthogonally invariant condition for polynomial operators. An operator on (quasi-)polynomials is *orthogonally invariant* if for all isometries ϕ it commutes with the associated *lifting operator* on (quasi-)polynomials p defined by $\widetilde{\phi} : p \mapsto p \circ \phi$. An umbral set or sequence is orthogonally invariant if and only if its *umbral operator* (Section 3) is orthogonally invariant.

Proposition 4.2. *Let θ be a shift-invariant operator. The following are equivalent:*

0. *θ is orthogonally invariant.*

1. *The indicator of θ is orthogonally invariant.*

2. *θ commutes with the continuous extension $\widetilde{\phi}$ of the lifting operator $p \mapsto p \circ \phi$ for all isometries ϕ.*

3. *$\theta(p \circ \phi) = (\theta p) \circ \phi$ for all isometries ϕ.*

4. *$\theta = \phi\theta$ for all isometries ϕ where we define $\phi D^S = D_{\phi^{-1}s_1} \cdots D_{\phi^{-1}s_k}$, where $S = \{s_1, \ldots, s_k\}$.*

5. *$\theta = f(\mathcal{L})$ for some formal power series $f \in \mathbf{R}[[\ell]]$.*

Examples of orthogonally invariant shift-invariant operators:

- The Laplace Operator $\mathcal{L}$ (Proposition 3.6.),

- the Bernoulli operator J (Section 5.2), and
- the Weierstrass operator $W = \exp(-\mathcal{L}/2)$ (Section 5.3).

Thus, any orthogonally invariant shift-invariant operator can be expressed as a formal power series in the *Laplace operator* $\mathcal{L}$. In general, however to express an orthogonally invariant linear operator one requires not only multiplication by the norm $\mathcal{N}$ and the Laplace operator $\mathcal{L}$ but also the non-shift-invariant mixed or number operator $\mathcal{M}$ where $\mathcal{M}\, x^R = |R|\, x^R$.

Furthermore, note that if $\mathcal{H}$ is infinite dimensional, $\mathcal{L}$ is a quasi-formal Hilbert series, and thus acts only on polynomials, not general quasi-polynomials such as $\mathcal{N}$. On the other hand, if $\mathcal{H}$ is finite dimensional, then orthogonally invariant shift-invariant operators act on orthogonally invariant polynomials.

Proposition 4.3. *If $\mathcal{H}$ is of dimension $n < \infty$, then the action of $\mathcal{L}$ on $\mathcal{N}$ is well-defined. In particular,*

$$\mathcal{L}\, \mathcal{N}^k = 2k\,(2k - 2 + n)\, \mathcal{N}^{k-1}.$$

If a shift-invariant operator is known to be orthogonally invariant, a more powerful version of the expansion theorem (Theorem 3.5.) is available.

Theorem 4.4. (Orthogonally Invariant Expansion) *Let $\mathcal{H}$ be an n-dimensional Hilbert space, and let θ be an orthogonally invariant shift-invariant operator. Then θ is given by the sum*

$$\theta = \sum_{k=0}^{\infty} c_k \mathcal{L}^k,$$

where $c_k = \left[\theta \mathcal{N}^k/(2k)!!\, n(n+2)\ldots(n+2k-2)\right]_{x=0}$ *and the* double factorial *is defined by* $(2n)!! = 2 \times 4 \times 6 \times \cdots \times 2n = 2^n\, n!$.

Similarly, we have a powerful orthogonally invariant version of Proposition 3.4.

Theorem 4.5. (Orthogonally Invariant Taylor) *Let $\mathcal{H}$ be an n-dimensional Hilbert space, and let p be an orthogonally invariant polynomial. Then p is given by the finite sum*

$$p = \sum_{k=0}^{\infty} c_k \mathcal{N}^k,$$

where $c_k = \left[\mathcal{L}^k p/(2k)!!n(n+2)\ldots(n+2k-2)\right]_{x=0}$.

Note that if $n = 1$, then the denominators above simplify to $(2k)!$ and Theorems 4.4. and 4.5. reduce to the usual univariate Expansion and Taylor formulas (restricted to the special case of even operators and even polynomials).

5. Appell sequences

5.1. Results

At this point, the true machinery of umbral calculus can be introduced. We briefly present Appell sequences, the theory of which is quite simple.

Given an invertible (quasi-)shift-invariant operator A, the (quasi-)polynomials $a_R = A^{-1}x^R$ indexed by finite multisets $R \subset \mathcal{H}$ form an umbral sequence which is called an *Appell sequence.*

We give equivalent characterizations of Appell sequences in terms of convolution formulas, operator identities and generating functions.

Proposition 5.1. *Let a_R be an umbral sequence of (quasi-)polynomials. Then the following are equivalent:*

1. *a_R is an Appell sequence,*
2. *$a_\emptyset \neq 0$, and a_R obeys the following "binomial formula"*
$$a_R(x+y) = \sum_{S \subseteq R} \binom{R}{S} a_{R\setminus S}(x)\, y^S = \sum_{S \subseteq R}{}' a_{R\setminus S}(x)\, y^S.$$
3. *$a_\emptyset \neq 0$, and for all $h \in \mathcal{H}$,*
$$D_h\, a_R = \sum_{r \in R} \langle h|r\rangle\, a_{R\setminus\{h\}}.$$
4. *The generating function for* $\mathbf{a}$ *is given by*
$$\operatorname{Gen}(\mathbf{a}) := \sum_{R \subset B} a_R\, t^R/R! = g(t)\, \exp\left(\langle x|t\rangle\right),$$
where g is a fixed formal Hilbert series of order one and B is any orthonormal basis.

In this case, $g = \operatorname{Ind}(A^{-1})$ is the indicator of the invertible operator for a_R.

We develop two examples of Appell sequences in a Hilbert setting: Bernoulli polynomials and Hermite polynomials.

5.2. Example: Bernoulli polynomials

Let $\mathcal{H}$ be an n-dimensional real Hilbert space. Consider the n-dimensional unit ball $B_n = \{y\colon \langle y|y\rangle \leq 1\}$. The volume v_n and surface area ω_n of the unit n-ball B_n are given by [27, Eq. (2)]

$$\omega_n = 2\pi^{n/2}/\Gamma(n/2) \qquad (9)$$
$$v_n = (\pi)^{n/2}/\Gamma((n+2)/2), \qquad (10)$$

where the Gamma function of positive multiples of one-half is given by

$$\Gamma\left(\frac{n}{2}\right) = \begin{cases} \left(\frac{n-2}{2}\right)! & \text{if } n \text{ is even and} \\ (n-2)!!\,\sqrt{\pi}/2^{(n-1)/2} & \text{if } n \text{ is odd.} \end{cases} \tag{11}$$

Classically, the Bernoulli operator is defined to be an integral over the unit cube $[0,1]^n \subset \mathbf{R}^n$. The cube however imposes an artificial coordinate system on the Hilbert space. A more natural, symmetrical and thus orthogonally invariant definition is given by the corresponding integral over the unit n-ball. We thus define the *Bernoulli operator* $J : \mathcal{P} \to \mathcal{P}$ by the integral

$$Jp(x) = \frac{1}{v_n} \int_{y \in B_n} p(x+y)\, dy. \tag{12}$$

By the Orthogonally Invariant Expansion Theorem, this is equivalent to

$$\begin{aligned} J &= 1 + \tfrac{\mathcal{L}}{2\,(n+2)} + \tfrac{\mathcal{L}^2}{4!!\,(n+2)(n+4)} + \tfrac{\mathcal{L}^3}{6!!\,(n+2)(n+4)(n+6)} + \cdots \\ &= \sum_{k=0}^{\infty} \tfrac{\mathcal{L}^k}{(2k)!!\,(n+2)\cdots(n+2k)} \\ &= \Gamma\left(\tfrac{n}{2}+1\right) \left(-\tfrac{4}{\sqrt{\mathcal{L}}}\right)^{n/4} J_{n/2}\left(\sqrt{-\mathcal{L}}\right) \\ &= {}_0F_1\left({\atop n+1} ; \tfrac{\mathcal{L}}{4}\right). \end{aligned} \tag{13}$$

In particular, in one-dimensional space, J is (up to rescaling and shifting) the usual Bernoulli operator. Usually, one integrates over the interval $[0, 1]$. This is equivalent to the antiderivative of the forward difference operator $(\exp(D) - 1)/D$. Since the interval $B_1 = [-1, 1]$ is twice as long, we rescale yielding $(\exp(2D) - 1)/2D$, and then shift by applying the shift operator E^{-1} giving

$$J = \frac{\exp(D) - \exp(-D)}{2D} = \frac{\sinh(D)}{D} = \frac{\sinh\left(\sqrt{\mathcal{L}}\right)}{\sqrt{\mathcal{L}}}$$

as above, where $D = D_u$ is the derivative in the direction of either unit vector u.

On the other hand, we may wish to consider infinite dimensional spaces. The integral defining the Bernoulli operator (Equation 12) is only well-defined over a finite dimensional space. We could however adopt Equation (13) as a definition of J and take the limit as n tends to infinity. Unfortunately, all of the coefficients except the first converge to zero and $J \to I$ as $n \to \infty$. This problem can be avoided by rescaling by a factor of $\sqrt{n}$. Define the *renormalized Bernoulli operator* by

$$J\tilde{}\,p(x) = \frac{1}{v_n} \int_{y \in B_n} p(\sqrt{n}x + y)\, dy.$$

$$\begin{aligned}
B_{\emptyset} &= 1 \\
B_{\{h_1\}} &= \langle x|h_1\rangle \\
B_{\{h_1,h_2\}} &= \tfrac{1}{n+2}\left(\langle x|h_1\rangle\langle x|h_2\rangle + \langle h_1|h_2\rangle\right) \\
B_{\{h_1,h_2,h_3\}} &= x^{\{h_1,h_2,h_3\}} + \\
&\quad \tfrac{1}{n+2}(\langle h_2|h_3\rangle\langle x|h_1\rangle + \langle h_1|h_3\rangle\langle x|h_2\rangle + \langle h_1|h_2\rangle\langle x|h_3\rangle)
\end{aligned}$$

Table 1: Bernoulli polynomials

We then have

$$J^{\sim} = \sum_{k=0}^{\infty} \frac{\mathcal{L}^k}{2^k k! \left(1+\frac{2}{n}\right)\left(1+\frac{4}{n}\right)\cdots\left(1+\frac{2k}{n}\right)}.$$

Taking the limit as the dimension of $\mathcal{H}$ increases, we obtain

$$\begin{aligned}
\lim_{n\to\infty} J^{\sim} &= \sum_{k=0}^{\infty} \frac{\mathcal{L}^k}{2^k!} \\
&= W,
\end{aligned}$$

where $W = \exp(\mathcal{L}/2)$ is the *Weierstrass operator.* (See Section 5.3.)

Applying the ordinary expansion theorem to Equation (13), we can immediately calculate $\left[Jx^R\right]_{x=0}$. In other words, we have a powerful method by which to calculate integrals of monomials with respect to a spherically symmetrical measure.

Proposition 5.2. *Let $n, e_1, \ldots, e_n$ be nonnegative integers, and let $k = \sum_{i=1}^n e_i$. Then*

$$\frac{1}{v_n}\int_{x\in B_n} x_1^{e_1}\cdots x_n^{e_n}\,dx = \frac{\prod_{i=1}^{n}(e_i-1)!!}{(n+2)(n+4)\cdots(n+k)}$$

if all of the e_i are even, and zero otherwise.

The Appell sequence given by J is called the *Bernoulli polynomials* and is denoted B_R. Note that $B_{\{r_1,r_2,\ldots,r_n\}}(x)$ is a symmetric function of $\langle x|r_k\rangle$ in the classical sense [24]. The list of Bernoulli polynomials in Table 1 follows easily from Equation (13) and Proposition 3.6.

$$\begin{aligned}
H_{\emptyset} &= 1 \\
H_{\{h_1\}} &= \langle x|h_1\rangle \\
H_{\{h_1,h_2\}} &= \langle x|h_1\rangle \langle x|h_2\rangle - \langle h_1|h_2\rangle \\
H_{\{h_1,h_2,h_3\}} &= x^{\{h_1,h_2,h_3\}} - \\
&\quad (\langle h_2|h_3\rangle \langle x|h_1\rangle + \langle h_1|h_3\rangle \langle x|h_2\rangle + \langle h_1|h_2\rangle \langle x|h_3\rangle)
\end{aligned}$$

Table 2: Hermite polynomials

5.3. Example: Hermite polynomials

The *Hermite polynomials* H_R are the Appell sequence associated with the orthogonally invariant shift-invariant *Weierstrass operator* $W = \exp(\mathcal{L}/2)$. Over an n-dimensional Hilbert space ($n < \infty$), the integral

$$Wp(x) = (2\pi)^{-n/2} \int_{y\in\mathcal{H}} p(x+y)e^{-\|y\|^2/2}\,dy \tag{14}$$

gives an equivalent definition for the Weierstrass operator.

Let X_r be the operator of multiplication by $\langle x|r\rangle$. The Hermite polynomials are characterized by their particularly simple *Roman operator* $X_r - D_r$, i.e. the operator that takes H_R to $H_{R\cup\{r\}}$. Its expression as a sum of products of inner products (Table 2) is particularly satisfying with all of the possible terms included and coefficients of ± 1 depending on their parity.

Theorem 5.3. *The Hermite polynomial is for $R \subset \mathcal{H}$ given by*

$$H_R = \sum_{k=0}^{\lfloor n/2\rfloor} (-1)^k \left(\sum_{S\subseteq R, |S|=2k} \mathrm{Hf}\,(S) \right) x^{R\setminus S},$$

where the Hafnian $\mathrm{Hf}\,(S)$ *of a finite multiset $S = \{s_1, s_2, \ldots, s_{2k}\} \subset \mathcal{H}$ is given by the sum*

$$\sum_{\sigma} \prod_{\substack{\text{cycles} \\ (i,j)}} \langle s_i|s_j\rangle$$

over involutions σ of $\{1, 2, \ldots, 2k\}$ without fixed points (σ is thus the product of k disjoint transpositions), and the product is over that set of k transpositions (i,j).

The Hermite polynomials are also given by the Rodrigues formula

$$H_S(x) = (-1)^{|S|} \exp(\mathcal{N}/2)\, D^S \exp(-\mathcal{N}/2)$$

and generating function

$$\mathrm{Gen}\,(\mathbf{H}) = \sum_{R \subset B} H_R\, t^R / R! = \exp\left(\frac{\langle t|t\rangle}{2} + \langle x|t\rangle\right),$$

where B is any orthonormal basis.

Let L be the functional associated with the Weierstrass operator

$$Lp = [Wp]_{x=0}\,.$$

Its action on monomials is given by a Hafnian.

$$Lx^R = \mathrm{Hf}\,(R)\,.$$

We also have the following orthogonality relation.

Proposition 5.4. *If R and S are multisets on $\mathcal{H}$, then we have*

$$\begin{aligned} L\,(H_R\, H_S) &= \left\langle x^R \middle| x^S \right\rangle \\ &= \left[D^R\, x^S\right]_{x=0} \\ &= \left[D^S\, x^R\right]_{x=0}. \end{aligned}$$

6. Sequences of binomial type

The main problem in studying classical multivariate sequences of binomial type is that one is faced with two choices.

Either one can admit all sequences $p_{mn\ldots}(x, y, \ldots)$ of classical multivariate polynomials obeying the binomial formula

$$\begin{aligned} &p_{mn\ldots}(x + x', y + y', \ldots) \\ &\quad = \sum_{\substack{0\le j\le m \\ 0\le k\le n \\ \vdots}} \binom{m}{j}\binom{n}{k}\cdots\, p_{jk\ldots}(x, y, \ldots)\, p_{m-j,n-k,\ldots}(x', y', \ldots) \end{aligned} \tag{15}$$

in which case one must require that a certain determinant $\det(A)$ not vanish and this determinant would then appear in the Transfer formula and elsewhere making the theory needlessly opaque. Or else one can make the theory needlessly restrictive by requiring that A is the identity matrix. In this case, $p_{mn\ldots}$ is called a *diagonal sequence* [43] (or *normal* in [53] and *coherent* in [2]). In this case, the leading term of $p_{mn\ldots}$ is $x^m\, y^n \ldots$.

Compare the diagonal Transfer formula of [43, Theorem 6], [53, Theorem 1.3.6] and [2, Theorem 7.2] with the nondiagonal formula [43, Theorem 13]. For example, the bivariate transfer formula is given in the diagonal case by

$$p_{i,j}(x,y) = J(f)(D_x/f_1)^{1+i}(D_y/f_2)^{1+j}x^i y^j,$$

where D_x and D_y are the partial derivatives with respect to x and y, respectively. The transfer formula for the general case is given by

$$p_{i,j}(x,y) = \det(A^{-1}) \sum_{k=0}^{i} \sum_{\ell=0}^{j} \frac{J(f)}{k!\,\ell!}\, g_1^k g_2^\ell\, (b_{11}x + b_{21}y)^{i+k}\, (b_{12}x + b_{22}y)^{j+\ell},$$

where

- $f_1 = a_{11}D_x + a_{12}D_y + g_1$ and $f_2 = a_{21}D_x + a_{22}D_y + g_2$ are delta operators, where the matrix $A = [a_{ij}]$ is required to be the identity in the diagonal case, and simply nonsingular in the general case.
- $J(f)$ is the "Jacobian" operator $f_1^{(x)} f_2^{(y)} - f_2^{(x)} f_1^{(y)}$, where $\theta^{(v)} = \theta v - v\theta$ is the Pincherle derivative.

In our context we have no fixed preordained coordinate system, so a classical multivariate polynomial sequence with $p_{11}(x,y) = x(x+y)$ need not be rejected out of hand for not being "diagonal". Its failure to be diagonal is seen as a relic of having specialized the "wrong" orthonormal basis, for if we define a new variable $z = x + y$ then $p_{11} = xz$ is in fact "diagonal".

Such manipulations are absolutely rigorous from our Hilbert point of view, for there is a unique invertible linear transformation τ such that $p_{\{r\}} = \langle x | \tau r \rangle$. The leading term of p_R is then $x^{\tau R}$. We thus see that the umbral sequence $\mathbf{q}$ can be "reindexed" to form a new umbral sequence $q_R = p_{\tau^{-1}R}$ which truly is diagonal.

We now present an example of a sequence of binomial type that clearly shows the advantages of our basis-free approach (cf. e.g. [53]).

Example (Abel polynomials): Let $a \in \mathcal{H}$ be arbitrary and let $\mathcal{M}$ be the number operator defined in Section 3.4 (cf. Proposition 3.7.). Define the Abel polynomials by $p_R(x) := \frac{1}{n}\mathcal{M}(x - na)^R$, where n is the cardinality of the multiset $R \subset \mathcal{H}$. Using the definition of the shift operator E^b and the commutation rule $ME^b = E^a(M - D_b)$, we may rewrite the Abel polynomials as follows: $p_R(x) = (1/n)ME^{-na}x^R = (1/n)E^{-na}(M + nD_a)x^R$. Using Proposition 3.7. and commutativity of derivatives and shifts, we obtain after some simplifications that

$$p_R(x) = (1 + D_a)(x - na)^R. \tag{16}$$

We now show that the Abel polynomials form a diagonal basic sequence for the delta set $Q_h = D_h E^a$, i.e. $Q_h p_R = \sum_{r \in R} \langle h|r \rangle \, p_{R \setminus \{r\}}$. Using the commutation rules $E^a M = (M + D_a)E^a$ and $MD_a = D_a(M-1)$, we obtain $Q_h p_R(x) = (1/n)(M + D_a + 1)D_h E^{-(n-1)a} x^R$. Using the definition of derivative, we may rewrite this as

$$\sum_{r \in R} \langle h|r \rangle \frac{1}{n} M(x-(n-1)a)^{R \setminus \{r\}} + \sum_{r \in R} \langle h|r \rangle \frac{1}{n}(1 + D_a)(x-(n-1)a)^{R \setminus \{r\}}.$$

It follows from (16) that this simplifies to $\sum_{r \in R} \langle h|r \rangle \, p_{R \setminus \{r\}}(x)$.

Analogously to the classical umbral calculus, it can be shown that basic sequences are of binomial type, i.e. they satisfy

$$p_R(x+y) = \sum_{S \cup T = R}{}' p_S(x) \, p_T(y).$$

Note that the binomial theorem (Theorem 2.2.) is the special case $a = 0$.

Acknowledgements. We thank the referees for many helpful remarks, which lead to a considerable improvement of our paper.

References

[1] M. Barnabei, A. Brini, and G. Nicoletti. A general umbral calculus in infinitely many variables. *Adv. Math.*, 50:49–93, 1983. (MR 85g:05025).

[2] M. Barnabei, A. Brini, and G. Nicoletti. A general umbral calculus. *Adv. Math., Suppl. Stud*, 10:221–244, 1986. (Zbl. 612.05009).

[3] L. Biedenharn, R. Gustafson, M. Lohe, J. Louck, and S. Milne. Special functions and group theory in theoretical physics. In *Special functions: group theoretical aspects and applications*, Math. Appl., pages 129–162. Reidel, Dordrecht, 1984. (MR 86h:22034).

[4] L. Biedenharn, R. Gustafson, and S. Milne. An umbral calculus for polynomials characterizing $U(n)$ tensor products. *Adv. Math.*, 51:36–90, 1984. (MR 86m:05016).

[5] J. Brown. On multivariable Sheffer sequences. *J. Math. Anal. Appl.*, 69:398–410, 1979. (MR 80j:05007).

[6] M. Cerasoli. Enumerazione binomiale e processi stocastici di Poisson composti. *Bollettino U.M.I.*, (5) 16-A:310–315, 1979. (MR 80k:05008).

[7] C. Charalambides and J. Singh. A review of the Stirling numbers, their generalizations and statistical applications. *Comm. Stat. Th. Methods*, (8) 17:2533–2595, 1988. (MR 89d:62017).

[8] W. Chen. The theory of compositionals. *Discrete Math.*, 122:59–87, 1993. (MR 95i:60131).

[9] W. Chen. Compositional calculus. *J. Combin. Theory Ser. A*, 64:149–188, 1993. (MR 95g:05014).

[10] F. Cholewinski. *The finite calculus associated with Bessel functions*, volume 75 of *Contemporary Mathematics*. Amer. Math. Soc., 1988. (MR 89m:05013).

[11] A. Di Bucchianico and D.E. Loeb. Operator expansion in the derivative and multiplication by x. *Int. Transf. Spec. Fun.*, 4:49–68, 1996.

[12] A. Di Bucchianico and D.E. Loeb. Natural exponential families and umbral calculus, in: *Festschrift in Honor of Gian-Carlo Rota*, Birkhäuser Boston, pp. 195–211, 1998.

[13] J.A. Dieudonné and J.B. Carrell. *Invariant Theory, Old and New*. Academic Press, 1971.

[14] H. Domingues. The dual algebra of the Dirichlet coalgebra. *Rev. Mat. Estatist.*, 1:7–13, 1983. (MR 86i:05022).

[15] A. Garsia and S. Joni. Higher dimensional polynomials of binomial type and formal power series inversion. *Comm. Algebra*, 6:1187–1211, 1978. (MR 58#10484).

[16] H. Gzyl. Interpretacion combinatorica de polinomios de tipa binomial. *Acta Cient. Venezolana*, 27:244–246, 1976. (MR 55#118).

[17] M. Ismail. Polynomials of binomial type and approximation theory. *J. Approx. Th.*, 23:177–186, 1978. (MR 81a:41033).

[18] S. Joni. Lagrange inversion in higher dimension and umbral operators. *J. Linear and Multilinear Algebra*, 6:111–121, 1978. (MR 58#10485).

[19] S. Joni. Multivariate exponential operators. *Stud. Appl. Math.*, 62:175–182, 1980. (MR 81c:41050).

[20] S. Joni and G.-C. Rota. Coalgebras and algebras in combinatorics. *Stud. Appl. Math.*, 61:93–139, 1979. (MR 81c:05002).

[21] A. Kholodov. The umbral calculus and orthogonal polynomials. *Acta Appl. Math.*, 19:1–54, 1990. (MR 92b:33022).

[22] G. Kreweras. The number of more or less 'regular' permutations. *Fibonacci Quart.*, 18:226–229, 1980. (MR 82c:05011).

[23] S. Kurbanov and V. Maksimov. Mutual expansions of differential operators and divided difference operators. *Dokl. Akad. Nauk UzSSR*, 4:8–9, 1986. (MR 87k:05021).

[24] I. G. Macdonald. *Symmetric Functions and Hall Polynomials.* Oxford Mathematical Monographs. Oxford University Press, Walton Street, Oxford, England, 1979.

[25] P. Michor. Contributions to finite operator calculus in several variables. *J. Combin. Inform. System Sci.*, 4:39–65, 1979. (MR 81b:05013).

[26] G. Moldovan. Algebraic properties of a class of positive convolution operators. *Studia Univ. Babeş-Bolyai Math.*, 26:9–14, 1981. (MR 83i:41029).

[27] C. Müller. *Spherical Harmonics.* Number 17 in Lecture Notes in Mathematics. Springer-Verlag, Berlin, Heidelberg, 1966.

[28] R. Mullin and G.-C. Rota. On the foundations of combinatorial theory III. Theory of binomial enumeration. In Harris, editor, *Graph theory and its applications*, pages 167–213. Academic Press, 1970. (MR 43#65).

[29] H. Niederhausen. Sheffer polynomials in path enumeration. *Congr. Num.*, 26:281–294, 1980. (MR 82d:05015).

[30] H. Niederhausen. Sheffer polynomials for computing exact Kolmogorov-Smirnov and Rényi type distributions. *Ann. Statist.*, 9:923–944, 1981. (MR 84b:62067).

[31] H. Niederhausen. How many paths cross at least l given lattice points. *Congr. Num.*, 36:161–173, 1982. (MR 85b:05014).

[32] H. Niederhausen. Sheffer polynomials for computing Takács's goodness-of-fit distributions. *Ann. Statist.*, 11:600–606, 1983. (MR 84h:62077).

[33] T. T. Nielsen. *Bose Algebras: The Complex and Real Wave Representation.* Springer-Verlag, 1991.

[34] C. Parrish. Multivariate umbral calculus. *J. Linear and Multilinear Algebra*, 6:93–109, 1978. (MR 58#10487).

[35] N. Ray. Symbolic calculus: a 19th century approach to MU and BP. In *Homotopy theory (Durham 1985)*, volume 117 of *London Math. Soc. Lect. Notes Series*, pages 195–238. Cambridge University Press, 1987. (MR 89k:55007).

[36] N. Ray. Umbral calculus, binomial enumeration and chromatic polynomials. *Trans. Amer. Math. Soc.*, 309:191–213, 1988. (MR 89k:05014).

[37] N. Ray. *Loops on the 3-sphere and umbral calculus*, volume 96 of *Cont. Math.*, pages 297–302. Amer.Math. Soc., 1989. (MR 90i:55006).

[38] N. Ray. *Stirling and Bernoulli numbers for complex oriented homology theory*, volume 1370 of *Lect. Notes in Math.*, pages 362–373. Springer, Berlin, 1989. (MR 90f:55010).

[39] M. Razpet. An application of the umbral calculus. *J. Math. Anal. Appl.*, 149:1–16, 1990. (MR 91i:05018).

[40] M. Razpet. A new class of polynomials with applications. *J. Math. Anal. Appl.*, 150:85–99, 1990. (MR 91i:05020).

[41] D. Reiner. Multivariate sequences of binomial type. *Stud. Appl. Math.*, 57 (2):119–133, 1977. (MR 58#21668).

[42] D. Reiner. The combinatorics of polynomial sequences. *Stud. Appl. Math.*, 58:95–117, 1978. (MR 58#260).

[43] S. Roman. The algebra of formal series. III. several variables. *J. Approx. Theory*, 26:340–381, 1979. (MR 81i:05023b).

[44] S. Roman. *The umbral calculus.* Academic Press, 1984. (MR 87c:05015 = Zbl. 536.33001).

[45] G.-C. Rota. The number of partitions of a set. *Amer. Math. Monthly*, 71:498–504, 1964. (MR 28#5009).

[46] G.-C. Rota, D. Kahaner, and A. Odlyzko. On the foundations of combinatorial theory VII. Finite operator calculus. *J. Math. Anal. Appl.*, 42:684–760, 1973. (MR 49#10556).

[47] E. Shiu. Proofs of central-difference interpolation formulas. *J. Approx. Theory*, 35:177–180, 1982. (MR 84i:41004).

[48] S. Singh and S. Asthana. Multivariate shift invariant operators. *J. Math. Anal. Appl.*, 118:422–442, 1986. (MR 87i:33041).

[49] A. J. Stam. Polynomials of binomial type and renewal sequences. *Stud. Appl. Math.*, 77:183–193, 1987. (MR 90m:60097).

[50] A. J. Stam. Two identities in the theory of polynomials of binomial type. *J. Math. Anal. Appl.*, 122:439–443, 1987. (MR 88b:05015).

[51] A. J. Stam. Polynomials of binomial type and compound Poisson processes. *J. Math. Anal. Appl.*, 130:493–508, 1988. (MR 89d:60134).

[52] K. Ueno. General power umbral calculus in several variables. *J. Pure Appl. Algebra*, 59:299–308, 1989. (MR 90m:05014).

[53] T. Watanabe. On a dual relation for addition formulas of additive groups: I. *Nagoya Math. J.*, 94:171–191, 1984. (MR 86f:05020).

[54] T. Watanabe. On a dual relation for addition formulas of additive groups: II. *Nagoya Math. J.*, 97:95–135, 1985. (MR 86i:05023).

[55] T. Watanabe. On a generalization of polynomials in the ballot problem. *J. Statist. Planning & Inference*, 14:143–152, 1986. (MR 87j:05024).

[56] A. Weyl. *Classical Groups. Their Invariants and Representations.* Number 1 in Princeton mathematical series. Princeton University Press, 1946.

[57] B. Wilson and F. Rogers. Umbral calculus and the theory of multispecies nonideal gases. *Phys. A*, 139:359–386, 1986. (MR 88d:82024).

A. DiBuccianico
Eindhoven University of Technology
Dept. of Mathematics and Computing Science
P.O. Box 513
5600 MB Eindhoven, The Netherlands
sandro@win.tue.nl
http://www.win.tue.nl/math/bs/statistics/bucchianico

G.-C. Rota
Department of Applied Mathematics,
The Massachusetts Institute of Technology,
Cambridge, MA 02138, USA,
rota@math.mit.edu

D.E. Loeb
Laboratoire Bordelais de Recherche en Informatique
Université de Bordeaux I
33405 Talence Cedex, France
loeb@labri.u-bordeaux.fr
http://dept-info.labri.u-bordeaux.fr/~loeb/index.html

A Strategy for Determining Polynomial Orthogonality

J. M. Freeman

The novice attempting to learn about orthogonal polynomials is easily daunted by the lack of a discernable general framework. Things seem so *ad hoc* and obscurely connected. I am sure this impression is not shared by the facile experts who rule the game, but then there are the rest of us who would welcome a more transparent framework.

It is in this spirit that we would like to present a general strategy for determining when a generating function is orthogonal, *i.e.*, the polynomial sequence it generates is orthogonal.

The strategy works smoothly in a variety of substantial cases. Here we apply it to Sheffer generating functions,

$$e(x,t) = \exp[f(t) + xg(t)]\,,$$

those of a general Tschebychev form,

$$e(x,t) = f(t)[1 - xg(t)]^{-1}\,,$$

and to those of the Al-Salam/Carlitz form

$$e(x,t) = f(t)\prod_{n=0}^{\infty}(1 - q^n tx)^{-1}.$$

In the first case we easily obtain Meixner's list [1], and in the second, the generating functions of Tschebychev, types I and II, plus some additional, possibly new, orthogonal ones. In the third case, using a less ambitious Ansatz, we obtain precisely the generating functions of the orthogonal q-polynomials of Al-Salam and Carlitz [2].

The strategy is based on some simple yet fundamental facts about operators on $K[x][[t]]$. Loosely stated, anything you do to a generating function,

$$e(x,t) = \sum_{n=0}^{\infty} p_n(x)t^n \qquad (\deg\ p_n(x) = n),$$

with an x-operator X, you can do with a unique t-operator T, and vice-versa. In fact, you can do anything you want to $e(x,t)$ with an x-operator. That is, for arbitrary $f(x,t)$ in $K[x][[t]]$, there exists a unique x-operator X with $Xe(x,t) = f(x,t)$.

And for yet a little more precision, x-operators are those which are $K[[t]]$-linear and map $K[x]$ to itself. The t-operators do the reverse, being $K[x]$-linear and mapping $K[[t]]$ to itself (t-adic continuously). The algebras of x-operators and t-operators commute with each other.

When X is an x-operator and T a t-operator which has the same effect on $e(x,t)$, *i.e.*, $Xe(x,t) = Te(x,t)$, we write $\hat{X} = T$ and $X = \check{T}$. These are the e-transform of X and the inverse e-transform of T respectively. These transforms are anti-isomorphisms between the two algebras.

Actually for present purposes we will be interested primarily in $\hat{x}$, the e-transform of multiplication by x. We will at this point take our field K to be the reals $\mathbf{R}$.

Now we turn to the orthogonality of $e(x,t)$, *i.e.*, the orthogonality of the polynomial sequence it generates. We are asking about the existence of a functional μ for which

$$\mu[p_k(x)p_n(x)] = \lambda(k)\delta(n,k)\,,$$

where the scalar sequence λ has no zeros. There is only one candidate for μ, namely, the x-operator μ for which

$$\mu e(x,t) = 1.$$

This asserts that 1 is orthogonal to all other $p_n(x)$. The complete orthogonality condition can be harmlessly restated as

$$\mu(x^k e(x,t)) = q_k(t)\,,$$

where $q_k(t)$ is a polynomial of degree k.

Now taking the e-transform of x^k gives $\mu x^k e(x,t) = \hat{x}^k \mu e(x,t) = \hat{x}^k 1$. Thus $\hat{x}^k 1 = q_k(t)$ must be of degree k, and hence the new orthogonality condition,

$$\deg(\hat{x}q(t)) = \deg q(t) + 1 \qquad \text{for } q(t) \text{ in } \mathbf{R}[t].$$

From this our strategy for determining orthogonality emerges.

First compute $\hat{x}$ explicitly, and then test whether $\deg(\hat{x}t^n) = n + 1$ for all n.

For the first application, we select out the orthogonal generating functions from among the Sheffer generating functions.

We compute $\hat{x}$ relative to

$$e(x,t) = \exp[f(t) + xg(t)]\,,$$

where $f(0)=0$, $g(0)=0$, $g'(0)=1$. These generating functions satisfy

$$\frac{\partial}{\partial t}e(x,t)=[f'(t)+xg'(t)]e(x,t),$$

a relation which we solve for $xe(x,t)$;

$$xe(x,t)=[z(t)\frac{\partial}{\partial t}+q(t)]e(x,t)\,,$$

where $z(t)=1/g'(t)$ and $q(t)=-f'(t)/g'(t)$. The bracketed expression is a t-operator so

$$\hat{x}=z(t)\frac{\partial}{\partial t}+q(t).$$

Then $\hat{x}t^n=nz(t)t^{n-1}+t^nq(t)$. Thus it is necessary for orthogonality that $q(t)$ be a polynomial of degree 1, say $q(t)=ct+d$, and $z(t)$ a polynomial of degree ≤ 2 with $z(0)=1$. We write $z(t)=(1-at)(1-bt)$ and here a,b are real or conjugate complex. The final requirement is $abn+c\neq 0$ for any n. All this is necessary and sufficient for $\deg(\hat{x}t^n)=n+1$, *i.e.*, for orthogonality.

To obtain the explicit form of $e(x,t)$ in this case, we simply make the replacements $f'(t)=-q(t)/z(t)$ and $g'(t)=1/z(t)$ in the partial differential equation whose solution is then

$$e\left(x,t\;\middle|\;\begin{matrix}a & b\\ c & d\end{matrix}\right)=\exp\int_0^t\frac{x-cu-d}{(1-au)(1-bu)}du.$$

Varying the four parameters consistent with the above constraints gives Meixner's list [1]: Hermite, confluent hypergeometric (including Laguerre), Charlier, and Meixner types I and II. There are also some perhaps spurious newcomers arising because Meixner's initial constraints on the parameters are tighter, requiring that $ab\ge 0$ and $c>0$.

One such newcomer is

$$e\left(x,t\;\middle|\;\begin{matrix}1 & -1\\ 1/2 & 0\end{matrix}\right)=(1-t^2)^{\frac{1}{4}}\left(\frac{1+t}{1-t}\right)^{\frac{x}{2}}.$$

We now apply our strategy to generating functions of the general Tschebychev form

$$e(x,t)=f(t)[1-xg(t)]^{-1}\,,$$

where $f(0)=1$, $g(0)=0$ and $g'(0)=1$. These satisfy the difference equation

$$\begin{aligned}xe(x,t)&=\frac{e(x,t)-f(t)e(x,0)}{g(t)}\\&=\frac{1}{g(t)}[1-f(t)\tau]e(x,t),\end{aligned}$$

τ being the point evaluation at $t = 0$, a t-operator. Thus

$$\hat{x} = \frac{1}{g(t)}[1 - f(t)\tau],$$

and so

$$\hat{x}1 = \frac{1 - f(t)}{g(t)} = q(t),$$

$$\hat{x}t = \frac{t}{g(t)} = z(t),$$

and

$$\hat{x}t^n = z(t)t^{n-1} \qquad \text{for } n \geq 1.$$

Hence for orthogonality, $q(t)$ and $z(t)$ must be polynomials of degree 1 and 2 respectively, $z(0) = 1$, and this suffices.

Now to get $e(x,t)$ explicitly we first note that $g(t) = t/z(t)$ and $f(t) = (z(t) - tq(t))/z(t)$. The polynomial $w(t) = z(t) - tq(t)$ has degree ≤ 2 and $w(0) = 1$. We write

$$\begin{aligned} z(t) &= (1 - at)(1 - bt) \\ w(t) &= (1 - ct)(1 - dt) \end{aligned}$$

where $ab \neq 0$ and a, b and c, d are real or conjugate complex pairs. The condition $ab \neq cd$ is also required to ensure that $q(t) = [z(t) - w(t)]/t$ is of degree 1. With these constraints on the four parameters, the orthogonal generating functions of the general Tschebychev form are

$$e\left(x, t \,\middle|\, \begin{matrix} a & b \\ c & d \end{matrix}\right) = \frac{(1 - ct)(1 - dt)}{(1 - at)(1 - bt)}\left[1 - \frac{xt}{(1 - at)(1 - bt)}\right]^{-1}.$$

Tschebychev types I and II are

$$e\left(x, t \,\middle|\, \begin{matrix} 1 & -1 \\ i & -i \end{matrix}\right) = (1 - t^2)(1 - xt + t^2)^{-1}$$

$$e\left(x, t \,\middle|\, \begin{matrix} 0 & 0 \\ i & -i \end{matrix}\right) = (1 - xt + t^2)^{-1}.$$

One example of a newcomer is

$$e\left(x, t \,\middle|\, \begin{matrix} 1 & 0 \\ 1 & 1 \end{matrix}\right) = \left[1 - \frac{xt}{1 + t} + t\right]^{-1},$$

for which Mourad Ismail has found the weight function to be $\sqrt{\frac{4-x}{x}}$ on $[0, 4]$.

Finally we test generating functions of the form

$$e(x,t) = f(t)\prod_{n=0}^{\infty}(1-q^n tx)^{-1},$$

where $f(0) = 1$ and $|q| < 1$. We abbreviate the product as

$$e_q(t) = \prod_{n=0}^{\infty}(1-q^n t)^{-1}$$

and have the difference equation

$$xe_q(xt) = \frac{e_q(xt) - e_q(xqt)}{t}.$$

The right-hand side is $Te_q(xt)$ where T is the t-operator

$$Th(t) = [h(t) - h(qt)]/t.$$

Now from the above we have

$$[f(t)Tf(t)^{-1}]f(t)e_q(xt) = xf(t)e_q(xt).$$

Thus the e-transform of x relative to $e(x,t) = f(t)e_q(xt)$ is

$$\hat{x} = f(t)Tf(t)^{-1}.$$

Hence $\hat{x}1 = \frac{1}{t}\left[1 - \frac{f(t)}{f(qt)}\right]$ must be a polynomial of degree 1. Take $p(t) = f(t)/f(qt)$ and $z(t) = 1-p(t)$. Then $z(t)$ must have degree 2 and $z(0) = 0$ and thus $p(t)$ is a polynomial of degree 2 with $p(0) = 1$, so $p(t) = (1-at)(1-bt)$ and $ab \neq 0$.

Iteration gives for all n that

$$f(t) = p(t)f(qt) = p(t)p(qt)\cdots p(q^n t)f(q^{n+1}t)$$

i.e.,

$$\begin{aligned} f(t) &= \prod_{n=0}^{\infty}(1-aq^n t)(1-bq^n t) \\ &= e_q(at)^{-1}e_q(bt)^{-1}. \end{aligned}$$

By direct computation, we have further that

$$\hat{x}t^n = (1-q^n)t^{n-1}z(t)$$

which is of degree $n+1$. Thus the generating functions,

$$e(x,t) = e_q(at)^{-1}e_q(bt)^{-1}e_q(xt),$$

of Al-Salam and Carlitz are precisely the orthogonal ones among those of the form $f(t)e_q(xt)$.

In wrapping up the discussion of our strategy, it is in order to divulge a small secret. In the case of orthogonality the explicit formula for the t-operator $\hat{x}$ is just the classical three term recursion in compact form — or rather, the transform of same. One simply takes the inverse e-transform of the formula and applies it to $p_n(x)$ to get the recursion.

A similar use of transform techniques gives the relevant second order differential equations.

Though the computation of $\hat{x}$ does not always proceed as smoothly as in the examples presented, we feel that the strategy holds broad promise as soon as more tricks for the computation of $\hat{x}$ are discovered.

References

[1] J. Meixner, Orthogonale Polynomsysteme mit einer besonderen Gestalt der erzeugenden Function, Part 1 *J. London Math Soc.* (1934).

[2] T. S. Chihara, "An introduction to orthogonal polynomials," in: *Mathematics and its Applications*, Vol. 13, Gordon and Breach, New York, 1978.

[3] J. M. Freeman, Transforms of operators on $K[x][[t]]$, *Congr. Numer.* **48** (1985), 115–132.

[4] J. M. Freeman, "Orthononality via Transforms," *Stud. Appl. Math.*, **77**:2 (1987).

Florida Atlantic University
777 Glades Road
Boca Raton, FL 33431-0991
U.S.A.

Plethystic Formulas and Positivity for q,t-Kostka Coefficients

A. M. Garsia† *and J. Remmel*††

Abstract

Our results concern the Macdonald q,t-Kostka coefficients $K_{\lambda\mu}(q,t)$. More precisely we work here with the expressions $\tilde{K}_{\lambda\mu}(q,t) = t^{n(\mu)}K_{\lambda\mu}(q,1/t)$ where $n(\mu)$ denotes the sum of the legs of the cells of μ. The $K_{\lambda\mu}(q,t)$ have been conjectured by Macdonald to be polynomials in q,t with positive integer coefficients. We prove here the Macdonald conjecture for arbitrary μ when λ is an augmented hook. Our proof is based on explicit formulas yielding $\tilde{K}_{\lambda\mu}(q,t)$ as a symmetric polynomial plethystically evaluated at $\tilde{K}_{(n-1,1),\mu}(q,t)$. More precisely, it can be shown that for each $\gamma \vdash k$ there is a unique symmetric polynomial $\mathbf{k}_\gamma(x;q,t)$, of degree $\leq k$ in x, such that for any λ of the form $\lambda = (n-k,\gamma)$, we have $\tilde{K}_{\lambda\mu}(q,t) = \mathbf{k}_\gamma[\tilde{K}_{(n-1,1),\mu}(q,t);q,t\,]$. The proof of existence of the polynomials $\mathbf{k}_\gamma(x;q,t)$ is algorithmic. There are now two separate algorithms yielding $\mathbf{k}_\gamma(x;q,t)$. We present here the chronologically first one. The second one is presented in a paper by Garsia–Tesler [12] where it yields that $\mathbf{k}_\gamma(x;q,t)$ is a polynomial with integer coefficients in $x,q,t,1/q,1/t$. Although the first algorithm is not as good with denominators, we present it here since, combined with some recent work of Lapointe–Vinet, it yields what may be the simplest proof that the $K_{\lambda\mu}(q,t)$ are polynomials with integer coefficients.

Introduction

Given a partition μ we shall represent it as customary by a Ferrers diagram. We shall use the French convention here and, given that the parts of μ are $\mu_1 \geq \mu_2 \geq \cdots \geq \mu_k > 0$, we let the corresponding Ferrers diagram have μ_i lattice squares in the ith row (counting from the bottom up). We shall also adopt the Macdonald convention of calling the *arm, leg, coarm* and *coleg* of a lattice square s the parameters $a_\mu(s), l_\mu(s), a'_\mu(s)$ and $l'_\mu(s)$ giving the number of cells of μ that are respectively *strictly* EAST, NORTH, WEST and SOUTH of s in μ.

---

† Work carried with support from NSERC.

†† Work carried out under NSF grant support.

Macdonald defines the q,t-Kostka coefficients $K_{\lambda\mu}(q,t)$ through an expansion, which in λ-ring notation may be written as

$$J_\mu(x;q,t) \;=\; \sum_\lambda S_\lambda[X(1-t)]\, K_{\lambda\mu}(q,t)\ . \tag{I.1}$$

Macdonald calls the $J_\mu(x;q,t)$ "integral forms." They are related to the polynomials $P_\mu(x;q,t)$, $Q_\mu(x;q,t)$ via the formulas

$$J_\mu(x;q,t) \;=\; h_\mu(q,t)\, P_\mu(x;q,t) \;=\; h'_\mu(q,t)\, Q_\mu(x;q,t)\ , \tag{I.2}$$

where we set

$$h_\lambda(q,t) \;=\; \prod_{s\in\lambda}(1-q^{a_\lambda(s)}t^{l_\lambda(s)+1})\ ,\quad h'_\lambda(q,t) \;=\; \prod_{s\in\lambda}(1-q^{a_\lambda(s)+1}t^{l_\lambda(s)})\ . \tag{I.3}$$

For our purposes it will be more convenient to work with the polynomials $\tilde H_\mu(x;q,t)$ which are defined by setting

$$\tilde H_\mu(x;q,t) = t^{n(\mu)}\, J_\mu[\frac{X}{1-1/t};q,1/t] \qquad \left(n(\mu)=\sum_{s\in\mu} l_\mu(s)\right)\ . \tag{I.4}$$

This gives us the Schur function expansion

$$\tilde H_\mu(x;q,t) \;=\; \sum_\lambda S_\lambda(x)\, \tilde K_{\lambda\mu}(q,t)\ ,\qquad \left(\ \tilde K_{\lambda\mu}(q,t) = K_{\lambda\mu}(q,1/t)\, t^{n(\mu)}\ \right)\ . \tag{I.5}$$

The starting point of the investigations that led to the present results is formula 2.13 of [8] which gives that

$$\tilde K_{(n-k,1^k),\mu}(q,t) = e_k[B_\mu(q,t)-1] \qquad \text{for all } \mu\vdash n>k\ , \tag{I.6}$$

where e_k denotes the ordinary elementary symmetric function, and

$$B_\mu(q,t) = \sum_{s\in\mu} q^{a'_\mu(s)}t^{l'_\mu(s)}\ .$$

In (I.6) the brackets indicate that $B_\mu(q,t)-1$ is plethystically substituted into e_k.[1] It develops that formula (I.6) is but a very special instance of a general result that may be stated as follows

[1] We recall that the plethystic substitution of an expression $E(t_1,t_2,\ldots)$ into a symmetric polynomial P is obtained by first expanding P in terms of the power basis and then making the replacements $p_k \to E(t_1^k,t_2^k,\ldots)$.

Theorem I.1. *For any given $\gamma \vdash k$ we can construct a symmetric polynomial $\mathbf{k}_\gamma(x;q,t)$ such that*

$$\tilde{K}_{(n-k,\gamma),\mu}(q,t) \;=\; \mathbf{k}_\gamma[\, B_\mu(q,t)\,;q,t] \qquad (\;\forall \quad \mu \vdash n \geq k + max(\gamma)\;)\;. \quad \text{(I.7)}$$

The polynomials $\mathbf{k}_\gamma(x)$ are uniquely determined by (I.7) and by the condition that their Schur function expansion is of the form

$$\mathbf{k}_\gamma(x;q,t) \;=\; \sum_{|\rho|\leq k} S_\rho\, \mathbf{k}_{\rho\gamma}(q,t)\,, \qquad \text{(I.8)}$$

with coefficients $\mathbf{k}_{\rho\gamma}(q,t)$ rational functions in q, t with integer coefficients.

It is easily seen that the identities in (I.7) overdetermine the symmetric polynomials $\mathbf{k}_\gamma$. In fact, the reader who is in possession of q,t-Kostka tables, without reading any further, is in a position to quickly construct a few instances of the polynomials $\mathbf{k}_{\rho\gamma}(q,t)$ by treating the coefficients $\mathbf{k}_{\rho\gamma}(q,t)$ as unknowns and solving for them, using a system of equations obtained from (I.7) for a sufficiently large number of partitions μ. We should mention that the polynomials $\mathbf{k}_{11}$, $\mathbf{k}_2$, $\mathbf{k}_{111}$, $\mathbf{k}_{21}$, $\mathbf{k}_3$ were first computed in this manner by F. Bergeron[2] when Theorem I.1 was still a conjecture.

At any rate, a look at the first few instances quickly reveals that the coefficients $\mathbf{k}_{\rho\gamma}(q,t)$ appear to be much better than stated above. In fact, in all the cases that have been computed so far (and these are sufficient to give us the $\tilde{K}_{\lambda\mu}(q,t)$ for $\lambda,\mu \vdash n \leq 12$), the $\mathbf{k}_{\rho\gamma}(q,t)$ turn out to be polynomials in $1/q$ and $1/t$ with integer coefficients. We have two distinct algorithms for constructing the the symmetric polynomials $\mathbf{k}_\gamma$. The chronologically first, is the simplest of the two, to program on a computer, but it has the theoretical disadvantage that it uses a recursion which introduces denominator factors $1-q^i$, $1-q^j$ in the course of computation. The second algorithm (given in [12]), takes more effort to construct but it shows that the coefficients $\mathbf{k}_{\rho\gamma}(q,t)$, when reduced to their normal forms, have only denominator factors of the form t^iq^i.

Nevertheless, the integrality of the q,t-Kostka $K_{\lambda\mu}(q,t)$ may be established using only theoretical consequences of the first algorithm. To do this all we need is some other algorithm which produces expressions for the $\tilde{K}_{\lambda\mu}(q,t)$ whose denominator factors are coprime with $1-q^i$, $1-q^j$. It develops that a formula given by Vinet–La Pointe in [16] yields precisely such an algorithm. The details of this derivation are given in the last section. We also show there that some of the polynomials $\mathbf{k}_\gamma$ are in fact sufficient to yield the positive integrality of the corresponding family of $K_{\lambda\mu}(q,t)$.

[2] Personal communication

Before we can present our algorithm we need to state some identities satisfied by the polynomials $\tilde{H}_\mu(x;q,t)$. These identities can be derived, by rewriting in the present notation the corresponding identities satisfied by the Macdonald polynomials $P_\mu(x;,q,t)$. In all cases they follow in a straightforward manner from (I.2), (I.3) and (I.4). We should mention that we will make extensive use of λ-ring notation here. The reader who is not familiar with this device may find a brief introduction to its use in [1], [8].

We should mention that there are presently five preprints containing proofs of polynomiality of $K_{\lambda\mu}(q,t)$. Knop in [14], [15] and Sahi in [21] derive it from a parallel theory of non symmetric [19] non-homogeneous Macdonald polynomials. Kirillov–Noumi in [13] get it as a byproduct of a beautiful Rodriguez formula conjectured by Lapointe–Vinet in [16]. All of these proofs involve Cherednik [2] operators and extensions thereof. In contrast, the present proof as well as the Garsia–Tesler proof [12] use only identities that can be derived in a straightforward way from the contents of Chapter VI of the new edition of Macdonald's book. These proofs are also the only ones that produce explicit expressions for the $K_{\lambda\mu}(q,t)$ and show that the dependence on λ and μ splits as a symmetric polynomial depending on λ plethystically evaluated at a polynomial depending only on μ.

1. Basic identities

Differentiation of a symmetric polynomial by the power symmetric function p_1 will be denoted by ∂_{p_1}. It is well known that ∂_{p_1} is dual to multiplication by p_1 with respect to the Hall inner product. Using this fact, we can convert one of the Macdonald Pieri rules ([18] (6.24) (iv) p. 340 for $r = 1$) into the following basic identity.

Proposition 1.1. *For any $\mu \vdash n$ we have*

$$\partial_{p_1} \tilde{H}_\mu(x;q,t) \;=\; \sum_{\nu \to \mu} c_{\mu\nu}(q,t)\, \tilde{H}_\nu(x;q,t)\,, \tag{1.1}$$

where $\nu \to \mu$ denotes that ν is obtained from μ by removing a corner square, and

$$c_{\mu\nu}(q,t) \;=\; \prod_{s \in R_{\mu/\nu}} \frac{t^{l_\mu(s)} - q^{a_\mu(s)+1}}{t^{l_\nu(s)} - q^{a_\nu(s)+1}} \prod_{s \in C_{\mu/\nu}} \frac{q^{a_\mu(s)} - t^{l_\mu(s)+1}}{q^{a_\nu(s)} - t^{l_\nu(s)+1}}\,. \tag{1.2}$$

Here, $R_{\mu/\nu}$ and $C_{\mu/\nu}$ denote the cells of ν that are in the same row and respectively the same column as the cell that we must remove from μ to obtain ν.

Let Δ denote the linear operator defined by setting for any symmetric polynomial P

$$\Delta P \;=\; P \;-\; P\Big[\,X+\tfrac{(1-t)(1-q)}{z}\,\Big]\;\Omega[-zX]\,\big|_{z^0}\;, \qquad (1.3)$$

[3]

where for convenience we have set

$$\Omega[-zX] \;=\; \sum_{m\geq 0}(-z)^m\, e_m[X]\;. \qquad (1.4)$$

The polynomials $P_\mu(x;q,t)$ were characterized by Macdonald (see Th. (4.7) p. 322 of [18]) as eigenfunctions of a certain difference operator E. Translating that result in the present notation we obtain the following characterization of the symmetric polynomials $\tilde H_\mu(x;q,t)$.

Proposition 1.2. *The polynomial $\tilde H_\mu(x;q,t)$ is uniquely determined by the two identities*

$$\begin{aligned} a)\quad & \Delta\,\tilde H_\mu(x;q,t) \;=\; (1-t)(1-q)B_\mu(q,t)\,\tilde H_\mu(x;q,t)\\ b)\quad & \tilde H_\mu(x;q,t)\,|_{S_n(x)} \;=\; 1\;. \end{aligned} \qquad (1.5)$$

Proof. The Macdonald operator D_n^1 (defined by (3.4) p. 315 of [18]) may be given the plethystic form

$$D_n^1\,P(x) \;=\; \frac{1}{1-t}\;P(x)\;+\;\frac{t^n}{t-1}\;P[X+\tfrac{q-1}{tz}]\;\Omega[zX(t-1)]\,|_{z^o}\;. \qquad (1.6)$$

This is easily established using partial fraction expansions. This given 1.5 a) follows immediately from (I.2), (I.4) and the fact that

$$D_n^1\,P_\lambda \;=\; \Big(\sum_i t^{n-i}q^{\lambda_i}\Big)\;P_\lambda$$

Remark 1.1. It might be good to say a few words here of our use of the symbol Ω in representing the right hand side of (1.4). The basic idea is to set for any "alphabet" X

$$\Omega[X] \;=\; \prod_{x\in X}\frac{1}{1-x} \;=\; \exp\Big(\textstyle\sum_{k\geq 1}p_k[X]\Big)\;.$$

---

[3] Here and after the symbol "|" denotes the operation of taking a coefficient. In particular " $|_{z^k}$" denotes the operation of taking the coefficient of z^k in the preceding expression.

This given, using λ-ring notation, we can extend this definition by letting X represent any formal power series in an arbitrary number of variables. As a matter of example we see that the Cauchy, Hall–Littlewood and Macdonald kernels, for $X = x_1 + x_2 + \cdots$ and $Y = y_1 + y_2 + \cdots$ may be respectively written in the form

$$\Omega[XY] \ , \qquad \Omega[XY(1-t)] \ , \qquad \Omega[XY\tfrac{1-t}{1-q}] .$$

Another useful ingredient in our development is the following reformulation of the Macdonald "Cauchy formula" ((4.13) p. 324 of [18]).

Proposition 1.3. *For any $n \geq 1$ we have*

$$e_n[\tfrac{XY}{(1-t)(1-q)}] \ = \ \sum_{\mu \vdash n} \frac{\tilde{H}_\mu[X;q,t]\, \tilde{H}_\mu[Y;q,t]}{\tilde{h}_\mu(q,t)\tilde{h}'_\mu(q,t)} \ , \tag{1.7}$$

where for convenience we have set

$$\tilde{h}_\mu(q,t) \ = \ \prod_{s\in\mu}(q^{a_\mu(s)} - t^{l_\mu(s)+1}) \quad \textit{and} \quad \tilde{h}'_\mu(q,t) \ = \ \prod_{s\in\mu}(t^{l_\mu(s)} - q^{a_\mu(s)+1}) . \tag{1.8}$$

Proof. Equating the homogeneous components of x, y-degree $2n$ in both sides of formula (4.13) p. 324 of [18] we obtain

$$h_n\left[X\tfrac{1-t}{1-q}\right] \ = \ \sum_{\lambda\vdash n} P_\lambda(x;q,t)Q_\lambda(y;q,t) \ .$$

Rewriting this by means of (I.2), (I.3), and (I.4) gives (1.7).

In the same vein, we must replace the Macdonald scalar product by the scalar product $\langle \ , \ \rangle_*$ defined by setting for any two power basis elements

$$\langle p_{\rho^{(1)}} \ , \ p_{\rho^{(2)}} \rangle_* \ = \ \begin{cases} (-1)^{|\rho|-l(\rho)} z_\rho\, p_\rho[(1-t)(1-q)] & \text{if } \rho^{(1)} = \rho^{(2)} = \rho \\ 0 & \text{otherwise} \ , \end{cases} \tag{1.9}$$

where $l(\rho)$ denotes the number of parts of ρ and z_ρ is as in [18] p. 24.

For any given symmetric polynomial $P[X]$ it will be convenient to let P^* denote the polynomial

$$P^*[X] \ = \ P[\tfrac{X}{(1-t)(1-q)}] \ .$$

Given this we have the following corollary of (1.7).

Proposition 1.4. *For any pair of partitions λ, μ we have*

$$a)\quad \langle \tilde{H}_\lambda \,,\, \tilde{H}_\mu \rangle_* = \begin{cases} \tilde{h}_\mu \tilde{h}'_\mu & \text{if } \lambda = \mu \,, \\ 0 & \text{if } \lambda \neq \mu \,, \end{cases}$$

$$b)\quad \langle S_\mu \,,\, S^*_{\lambda'} \rangle_* = \begin{cases} 1 & \text{if } \lambda = \mu \,, \\ 0 & \text{if } \lambda \neq \mu \,. \end{cases} \tag{1.10}$$

Proof. The standard "dual" Cauchy identity yields that

$$e_n[\tfrac{XY}{(1-t)(1-q)}] = \sum_{\lambda \vdash n} S_\lambda[X]\, S_{\lambda'}[\tfrac{Y}{(1-t)(1-q)}] \,. \tag{1.11}$$

On the other hand we also have

$$e_n[\tfrac{XY}{(1-t)(1-q)}] = \sum_{\rho \vdash n} \frac{(-1)^{|\rho|-l(\rho)}}{z_\rho\, p_\rho[(1-t)(1-q)]}\, p_\rho[X]\, p_\rho[Y] \,. \tag{1.12}$$

Comparing this formula with (1.9), we see that $e_n[\tfrac{XY}{(1-t)(1-q)}]$ is precisely the reproducing kernel associated to the scalar product $\langle \,,\, \rangle_*$. Thus (1.7) and (1.10) are simply expressing the duality of the two pairs of bases

$$\{\tilde{H}_\mu\}_{\vdash n} \,,\quad \{\tilde{H}_\mu \,/\, \tilde{h}_\mu \tilde{h}_{\mu'}\}_{\vdash n} \quad \text{and} \quad \{S_\lambda\}_{\vdash n} \,,\quad \{S_{\lambda'}\}_{\vdash n}$$

with respect to this scalar product. This gives (1.10) a) and b).

The following further consequences of (1.7) will play an important role in our development:

Proposition 1.5. *For any two symmetric polynomials P and Q we have*

$$\begin{aligned} a)\quad & \langle \Delta P \,,\, Q \rangle_* = \langle P \,,\, \Delta Q \rangle_* \,, \\ b)\quad & \langle \partial_{p_1} P \,,\, Q \rangle_* = \langle P \,,\, e_1^* Q \rangle_* \,, \end{aligned} \tag{1.14}$$

Moreover,

$$\begin{aligned} a)\quad & e_n^*(x) = \sum_{\mu \vdash n} \frac{\tilde{H}_\mu(x;q,t)}{\tilde{h}_\mu(q,t)\tilde{h}'_\mu(q,t)} && \textit{(for all } n \geq 1) \\ b)\quad & \langle e_n^* \,,\, \tilde{H}_\mu \rangle_* = 1 && \textit{(for all } \mu \vdash n \geq 1) \end{aligned} \tag{1.15}$$

Proof. The identity in (1.14) a) simply states that the operator Δ is self-adjoint with respect to the $*$-scalar product. This follows from (1.5) a) and (1.7) due to the fact $\tilde{H}_\mu$ and $\tilde{H}_\mu/\tilde{h}_\mu\tilde{h}_{\mu'}$ are dual bases with respect to the $*$-scalar product. The identity in (1.14) b) is a reformulation, in terms of the $*$-scalar product, of the well known fact that the Hall scalar product adjoint of ∂_{p_1} is multiplication by e_1. The identity in (1.15) a) is obtained from (1.7) by specializing Y be a single variable and using (1.5) b). Finally (1.15) b) follows by expanding $e_n^*(x)$ according to (1.15) a) and using (1.10) a). This completes our proof.

Our algorithm for constructing the polynomials $\mathbf{k}_\gamma$ yielding (I.7) is based on the following basic result given in [12].

Theorem 1.1. *Let Γ be the linear operator on symmetric polynomials defined by setting*

$$\begin{aligned}\Gamma\, S_\lambda[X] \qquad\qquad & \qquad\qquad\qquad\qquad\qquad\qquad (1.16)\\ = e_1\, S_\lambda + \sum_{\rho\neq\lambda,\lambda/\rho\in V} & S_\rho[X]\,\Big(\frac{-1}{tq}\Big)^{|\lambda-\rho|}\ \frac{h_{|\lambda-\rho|+1}[\,(1-t)(1-q)\,X-1]}{(1-t)(1-q)}\,,\end{aligned}$$

where $\lambda/\rho\in V$ means that λ/ρ is a vertical strip. Then for any symmetric polynomial P and any partition μ we have

$$\sum_{\nu\to\mu} c_{\mu\nu}(q,t)\ P[B_\nu(q,t)] \;=\; (\Gamma P)[B_\mu(q,t)]\,. \tag{1.17}$$

Proof. To get an explicit evaluation of a sum such as in (1.17) we must be more specific as to what partition μ we are dealing with. The reason for this is that the products in (1.2) undergo massive cancellations along the horizontal and vertical portions of the diagram of μ. This given, let μ be a partition with m corners $A_1, A_2, \ldots, A_m$ labelled as we encounter them from left to right. For convenience let us set $A_i = (\alpha_i, \beta_i)$ with α_i and β_i respectively giving the coleg and coarm of A_i in μ. Similarly, for $i = 1, 2, \ldots, m-1$, let $B_i = (\alpha_{i+1}, \beta_i)$ denote the cell of μ with coleg α_{i+1} and coarm β_i. Finally, set $\alpha_{m+1} = \beta_o = -1$ and let $B_o = (\alpha_1, \beta_o)$, $B_m = (\alpha_{m+1}, \beta_m)$ denote the cells that are respectively immediately to the left of the highest row of μ and immediately below the last column of μ.

Remarkably, all the ingredients involved in both sides of (1.17) may be expressed in terms of the "weights" of the cells $A_1, A_2, \ldots, A_m; B_o, B_1, \ldots, B_m$. That is in terms of the monomials $x_i = t^{\alpha_i}q^{\beta_i}$ and $u_i = t^{\alpha_{i+1}}q^{\beta_i}$. For instance, we can easily derive from the definition of $B_\mu(q,t)$ that

$$x_1+x_2+\cdots+x_m \;-\; u_o-u_1-\cdots-u_m \;=\; (1-1/t)(1-1/q)\,B_\mu(q,t) \;-\; \frac{1}{tq}\,. \tag{1.18}$$

With a little more labor, but in a completely straighforward manner, we can carry out all possible cancellations in (1.2) and reduce it to the simple expression

$$c_{\mu\nu^{(i)}} = \frac{1}{(1-1/t)(1-1/q)} \frac{1}{x_i} \frac{\prod_{s=0}^{m}(u_s - x_i)}{\prod_{s=1\,,\,s\neq i}^{m}(x_s - x_i)} , \tag{1.19}$$

where for convenience we have let $\nu^{(i)}$ denote the partition obtained by removing A_i from μ.

This formula places us in a position to obtain an explicit evaluation for the sums

$$\Phi_\mu^{(k)}(q,t) = \sum_{\nu\to\mu} c_{\mu\nu}\,(B_\mu - B_\nu)^k \qquad \text{for arbitrary } k \geq 0 ,$$

To see how this comes about, note that when $\nu = \nu^{(i)}$ then $B_\mu - B_\nu$ is simply x_i. Thus, using (1.19), we may write

$$\Phi_\mu^{(k)}(q,t) = \frac{1}{(1-1/t)(1-1/q)} \sum_{i=1}^{m} \frac{1}{x_i} \frac{\prod_{s=0}^{m}(u_s - x_i)}{\prod_{s=1\,,\,s\neq i}^{m}(x_s - x_i)}\, x_i^k. \tag{1.20}$$

Now, standard partial fraction techniques of the theory of symmetric functions give us that

$$\sum_{i=1}^{m} \frac{1}{x_i} \frac{\prod_{s=0}^{m}(x_i - u_s)}{\prod_{\substack{s=1\\ s\neq i}}^{m}(x_i - x_s)}\, x_i^k \tag{1.21}$$

$$= \begin{cases} h_{k+1}[x_1 + \cdots + x_m - u_o - \cdots - u_m] & \text{for } k \geq 1 , \\[2ex] x_1 + \cdots + x_m - u_o - \cdots - u_m + \frac{u_o \cdots u_m}{x_1 \cdots x_m} & \text{for } k = 0 . \end{cases}$$

Thus, combining (1.20), (1.21) with (1.18) and the fact that $\frac{u_o \cdots u_m}{x_1 \cdots x_m} = 1/qt$ we finally obtain the remarkably simple evaluations

$$\sum_{\nu\to\mu} c_{\mu\nu}\,(B_\mu - B_\nu)^k = \begin{cases} \frac{1}{(1-t)(1-q)t^k q^k}\, h_{k+1}\left[(1-t)(1-q)B_\mu - 1\right] & \text{for } k \geq 1 , \\[2ex] B_\mu & \text{for } k = 0 . \end{cases} \tag{1.22}$$

Theorem 1.1 is an immediate consequence of this. In fact, by linearity, it is sufficient to verify 1.17 when $P = S_\lambda$. To this end note that the addition formula for Schur functions gives

$$S_\lambda[B_\nu] = S_\lambda[B_\mu + (B_\nu - B_\mu)] = \sum_{\rho\,:\,\lambda/\rho\in V} S_\rho[B_\mu]\,(B_\nu - B_\mu)^{|\lambda/\rho|} ,$$

and thus we can write

$$\sum_{\nu \to \mu} c_{\mu\nu}\, S_\lambda[B_\nu] \;=\; \sum_{\rho\,:\,\lambda/\rho \in V} S_\rho[B_\mu] \sum_{\nu\to\mu} c_{\mu\nu}\,(B_\nu - B_\mu)^{|\lambda/\rho|}\;.$$

Using (1.22) this becomes

$$\sum_{\nu\to\mu} c_{\mu\nu} S_\lambda[B_\nu] = B_\mu S_\lambda[B_\mu]$$

$$+ \sum_{\rho\neq\lambda,\lambda/\rho\in V} S_\rho[B_\mu]\,(\frac{-1}{tq})^{|\lambda-\rho|}\;\frac{h_{|\lambda-\rho|+1}[\,(1-t)(1-q)\,B_\mu - 1]}{(1-t)(1-q)}\,,$$

which is (1.17) for $P = S_\lambda$.

Theorem 1.1 yields us the following important auxiliary result.

Theorem 1.2. *For each $\gamma \vdash k$ we can construct a symmetric polynomial $\mathbf{E}_\gamma(x;q,t)$ whose Schur function expansion is of the form*

$$\mathbf{E}_\gamma(x;q,t) \;=\; \sum_{|\rho|\le k} S_\rho(x)\,\mathbf{E}_{\gamma,\rho}(q,t) \tag{1.23}$$

with coefficients $\mathbf{E}_{\gamma,\rho}(q,t)$ integral polynomials in $q, t, 1/q, 1/t$ giving

$$\langle e^*_{n-k}\; e_\gamma\;,\;\tilde{H}_\mu\;\rangle_* \;=\; \mathbf{E}_\gamma[B_\mu(q,t);q,t] \qquad \forall \qquad \mu\vdash n\ge k\;. \tag{1.24}$$

Proof. For $\gamma = (\gamma_1,\gamma_2,\ldots,\gamma_m)$ set $\gamma^- = (\gamma_2,\gamma_3,\ldots,\gamma_m)$. Our proof is algorithmic and will proceed by a triple induction argument which ascends in k, descends in γ_1, and ascends in the refinement order of γ^-. More precisely, for $\gamma\vdash k$ and $\gamma_1 = a$, we shall construct the polynomial $\mathbf{E}_\gamma(x;q,t)$ given that we have already constructed $\mathbf{E}_\alpha(x;q,t)$ for any $\alpha\vdash k-1$, or for any $\alpha\vdash k$ with $\alpha_1\ge a+1$ and for any $\alpha\vdash k$ with $\alpha_1 = a$ but α^- a strict refinement of γ^-.

Next note that from the Definition 1.3 we derive that, when $\gamma = (a,\gamma_2,\ldots,\gamma_m)$, we have

$$\Delta\; p_1^{a-1}\,p_{\gamma^-}\;e^*_{n-k+1} \;=\; p_1^{a-1}\,p_{\gamma^-}\;e^*_{n-k+1} \tag{1.25}$$

$$-\;\left(e^*_{n-k+1} + \tfrac{1}{z}e^*_{n-k}\right)\left(p_1 + \tfrac{(1-q)(1-t)}{z}\right)^{a-1}\prod_{i=2}^{m}\left(p_{\gamma_i} + \tfrac{(1-q^{\gamma_i})(1-t^{\gamma_i})}{z^{\gamma_i}}\right)\,\Omega[-zX]\;|_{z^o}$$

where, p_s denotes the ordinary power symmetric function. Expanding the products appearing on the right-hand side and expressing each of the power sums in terms of the elementaries we derive that the $*$-scalar product of the left-hand side of (1.25) with $\tilde{H}_\mu$ may be expressed in the form

$$\begin{aligned}
\langle \Delta p_1^{a-1} p_{\gamma^-}\ e^*_{n-k+1}\ , \tilde{H}_\mu\rangle_* = & \cdots + a_\alpha\ \langle e^*_{n-k+1} e_\alpha\ ,\ \tilde{H}_\mu\rangle_* + \cdots \quad \leftarrow \text{ with } \alpha \vdash k-1 \\
& \cdots + b_\alpha\ \langle e^*_{n-k}\ e_1\ e_\alpha\ ,\ \tilde{H}_\mu\rangle_* + \cdots \quad \leftarrow \text{ with } \alpha \vdash k-1 \\
& \cdots + c_{b,\alpha}\ \langle e^*_{n-k}\ e_b\ e_\alpha\ ,\ \tilde{H}_\mu\rangle_* + \cdots \quad \leftarrow \text{ with } b > a \text{ and } \alpha \vdash k-a \\
& \cdots + d_\alpha\ \langle e^*_{n-k}\ \ e_a e_{\alpha^-}\ ,\ \tilde{H}_\mu\rangle_* + \cdots \quad \leftarrow \text{ with } \alpha^- \text{ finer than } \gamma^- \\
& + (-1)^a\ \langle e^*_{n-k}\ \ e_\gamma\ ,\ \tilde{H}_\mu\rangle_*\,,
\end{aligned} \tag{1.26}$$

where $a_\alpha, b_\alpha, c_{b,\alpha}$, and d_α denote integer coefficients. This results by grouping terms according as: **(i)** we pick e^*_{n-k+1} in the first factor **(ii)** we pick e^*_{n-k} from the first factor and pick at least one p_1 **(iii)** we pick e^*_{n-k} , none of the p_1 and at least one $(1-q^{\gamma_i})(1-t^{\gamma_i})/z^{\gamma_i}$. **(iv)** and finally the terms obtained from the elementary expansion of the power sums in $e^*_{n-k}\ e_a\, p_{\gamma_2} p_{\gamma_3} \cdots p_{\gamma_m}$ except for the crucial last term $(-1)^a\ e^*_{n-k}\ e_\gamma$.

It is easily seen that the inductive hypothesis yields plethystic expressions for all the terms occurring in the first, third and fourth lines in (1.26). As for the terms in the second line of (1.26) we note that we may write (using (1.14) b) and (1.1))

$$\langle e^*_{n-k} e_1 e_\alpha\ ,\ \tilde{H}_\mu\ \rangle_* \quad = \quad (1-t)(1-q) \sum_{\nu \to \mu} c_{\mu\nu}\ \langle e_\alpha\ e^*_{n-k}\ ,\ \tilde{H}_\nu\ \rangle_* \quad .$$

Thus, since here $\alpha \vdash k-1$, the inductive hypothesis and Theorem 1.1 give that

$$\langle e^*_{n-k} e_1 e_\alpha\ ,\ \tilde{H}_\mu\ \rangle_* \quad = \quad (1-t)(1-q)\ (\Gamma \mathbf{E}_\alpha)[B_\mu(q,t)]\,.$$

Finally, expanding the factor $p_1^{a-1}\ p_{\gamma^-}$ in the left-hand side of (1.26) in terms of the elementaries, and using the $*$-self-adjointness of Δ together with (1.5) a) we obtain the expansion

$$\langle \Delta p_1^{a-1} p_{\gamma^-}\ \ e^*_{n-k+1}, \tilde{H}_\mu\rangle_* = \cdots + \ a_\alpha\ \ (1-t)(1-q) B_\mu(q,t) \langle e^*_{n-k+1} e_\alpha\ ,\ \tilde{H}_\mu\rangle_*\ \ . + \cdots \tag{1.27}$$

So we can use again the inductive hypothesis and derive that all terms produced by the left-hand side of (1.26) may be expressed in terms of the polynomials $(1-t)(1-q)e_1\mathbf{E}_\alpha$ (for $\alpha \vdash k-1$) plethystically evaluated at $B_\mu(q,t)$. Thus using (1.26) and (1.27) we can construct an expression for the polynomial $\mathbf{E}_\gamma$ giving (1.24), as an integral linear combination of the polynomials **(i)** $\mathbf{E}_\alpha$ for $\alpha \vdash k-1$, **(ii)** $(1-t)(1-q)e_1\Gamma\mathbf{E}_\alpha$ for $\alpha \vdash k-1$, **(iii)** $\mathbf{E}_{b,\alpha}$ for $b > a$ and $\alpha \vdash k-a$, **(iv)** $\mathbf{E}_{a,\alpha^-}$ for $\alpha^- \vdash k-a$ finer than γ^-. **(v)** $(1-t)(1-q)e_1\mathbf{E}_\alpha$ for $\alpha \vdash k-1$.

In view of the nature of the operator Γ defined in (1.17), we see that the resulting polynomial $\mathbf{E}_\gamma$ will have an expansion as in (1.23) with coefficients $\mathbf{E}_{\gamma,\rho}$ integral polynomials in $q, t, 1/q, 1/t$ as desired. This completes our inductive proof, since the identity in (1.15) b) shows that the induction can be started at $k = 0$.

Let ν be a partition with at most $r \leq n$ parts and let μ be the partition obtained by adding a column of length r to ν. This given, it follows from ([18], eq. (6.17)' p. 334) that we may write

$$P_\nu \, e_r \; = \; P_\mu \; + \sum_{\lambda/\nu \in V_r} \psi_{\lambda\nu} \, P_\lambda \, , \tag{1.28}$$

where $\lambda/\nu \in V_r$ means that λ/ν is a vertical r-strip. This formula yields a fast recursive algorithm for constructing the Macdonald polynomials. Unfortunately, the denominators introduced by the coefficients $\psi_{\lambda\nu}$ appear to make it useless in the proof of polynomiality results. Nevertheless, by a clever observation Vinet–LaPointe were able to transform (1.28) into a formula which can be used to exclude some of the potential denominator factors.

The Vinet–La Pointe Rodriguez Formula [16]. *Let* $\mathbf{B}$ *denote the operator obtained by substituting* $X = -1/qt^{n-r-1}$ *in the Macdonald operator* $D(X; q, t)$ *then if* $l(\nu) \leq r \leq n$ *and* μ *is obtained by adding a column of length* r *to* ν*, we have*

$$J_\mu(x; q, t) \; = \; \frac{1}{\prod_{i=r+1}^{n} (1 - q^{-1} t^{r-i+1})} \, \mathbf{B} \, e_r \, J_\nu(x; q, t) \, . \tag{1.29}$$

Proof. The proof given in [16] is so simple that we feel compelled to include it here. The idea is to note that with $X = -1/q\, t^{n-r-1}$, ([18], eq. (4.15) p. 324) becomes

$$\mathbf{B} P_\lambda = \prod_{i=1}^{n} (1 - q^{\lambda_i - 1} t^{r+1-i}) \, P_\lambda \ ,$$

and this yields $\mathbf{B} P_\lambda = 0$ when $\lambda_{r+1} = 1$ and

$$\mathbf{B} P_\mu = \prod_{i=r+1}^{n} (1 - \tfrac{1}{qt^{i-r-1}}) \, \tfrac{h_\mu(q,t)}{h_\nu(q,t)} \, P_\mu$$

when $l(\nu) \leq r$ and μ is obtained by adding a column of length r to ν. This given, (1.29) immediately follows from (I.2) by applying $\mathbf{B}$ to both sides of (1.28).

2. Polynomiality

We are now in possession of all the identities that are needed to obtain a relatively short proof of polynomiality for the $\tilde{K}_{\lambda\mu}(q, t)$. To begin with let us note that Theorem 1.2 has the following immediate corollary.

Theorem 2.1. *For each $\gamma \vdash k$ we have a symmetric polynomial*

$$\mathbf{\Pi}_\gamma(x;q,t) \;=\; \sum_{|\rho|\le k} S_\rho(x)\,\mathbf{\Pi}_{\gamma,\rho}(q,t) \tag{2.1}$$

such that

$$\langle e^*_{n-k} e^*_\gamma \,,\, \tilde H_\mu\rangle_* \;=\; \mathbf{\Pi}_\gamma[B_\mu(q,t);q,t] \qquad \forall\;\; \mu\vdash n\ge k+\max(\gamma)\;\;. \tag{2.2}$$

The polynomial $\mathbf{\Pi}_\gamma$ may be constructed by an algorithm which expresses the coefficients $\mathbf{\Pi}_{\gamma,\rho}(q,t)$ as integral polynomials in q,t divided by factors of the form t^i, q^j $1-t^r$ and $1-q^s$.

Proof. Expanding the factor e^*_γ in terms of the elementaries by repetitive uses of the expansion

$$e^*_m(x) \;=\; \sum_{\rho\vdash m} e_\rho(x)\, m_\rho[\tfrac{1}{(1-t)(1-q)}]\,, \tag{2.3}$$

we derive that the polynomial $\mathbf{\Pi}_\gamma$ giving (2.2) may be expressed as a linear combination of the polynomials $\mathbf{E}_\alpha$ with $\alpha\vdash k$ with coefficients integral polynomials divided by factors $1-t^r$, $1-q^s$. Thus the result follows from Theorem 1.2.

It will be convenient to extend the definition of $\mathbf{\Pi}_\gamma$ so as to allow γ to be an arbitrary integer vector. Of course we shall have to set $\mathbf{\Pi}_\gamma = 0$ if any of the components of γ are negative and set $\mathbf{\Pi}_\gamma = \mathbf{\Pi}_{\gamma^+}$ when γ is a composition that rearranges to the partition γ^+. Keeping this in mind, note that (I.5) and (1.10) b) yield

$$\tilde K_{\lambda\mu}(q,t) \;=\; \langle S^*_{\lambda'} \,,\, \tilde H_\mu\rangle_* \;\;. \tag{2.4}$$

Setting $\lambda=(n-k,\gamma)$ with $\gamma=(\gamma_1,\gamma_2,\ldots,\gamma_r)$, the Jacobi–Trudi identity gives

$$\begin{aligned}&\tilde K_{(n-k,\gamma),\mu}(q,t)\\ &= \sum_{\sigma\in S_{r+1}} \mathrm{sign}(\sigma)\,\langle e^*_{n-k+\sigma_1-1} e^*_{\gamma_1+\sigma_2-2} e^*_{\gamma_2+\sigma_3-3}\cdots e^*_{\gamma_r+\sigma_{r+1}-r-1} \,,\, \tilde H_\mu\rangle \;\;.\end{aligned}$$

Thus, from (2.2) we derive that Theorem I.1 and (I.7) hold true with

$$\mathbf{k}_\gamma \;=\; \sum_{\sigma\in S_{r+1}} \mathrm{sign}(\sigma)\,\mathbf{\Pi}_{(\gamma_1+\sigma_2-2,\gamma_2+\sigma_3-3,\ldots,\gamma_r+\sigma_{r+1}-r-1)}\;\;. \tag{2.5}$$

Corollary 2.1. *$\tilde K_{\lambda\mu}(q,t)$ as well as $K_{\lambda\mu}(q,t)$ is a polynomial with integer coefficients.*

Proof. Computing $\tilde{K}_{\lambda\mu}(q,t)$ by plethystically evaluating at $B_\mu(q,t)$ the polynomial $\mathbf{k}_\gamma$ given in (2.5), will yield (by Theorem 2.1) an expression consisting of an integral polynomial divided by denominator factors of the form

$$1-t^r \ , \ 1-q^s \ , \ t^r q^s \ . \tag{2.6}$$

On the other hand, if we compute the polynomials $J_\mu(x;q,t)$ by means of (1.29) we obtain a formula expressing $J_\mu(x;q,t)$ in terms of the monomial basis with coefficients integral polynomials divided by factors of the form $1-t^r$, $1-t^r q$, $t^r q^s$. From (I.4) and (I.5) we then derive that $\tilde{K}_{\lambda\mu}(q,t)$, in this manner, may be given an expression consisting of an integral polynomial divided by denominator factors of the form

$$1-t^r \ , \ q-t^r \ , \ t^r q^s \ . \tag{2.7}$$

Now it is an easy consequence of the duality formula ([18], (5.1) p. 327) that we must have $\tilde{K}_{\lambda\mu}(q,t) = \tilde{K}_{\lambda\mu'}(t,q)$. This yields that $\tilde{K}_{\lambda\mu}(q,t)$ may also be given an expression as an integral polynomial divided by denominator factors of the form

$$1-q^r \ , \ t-q^r \ , \ q^r t^s \ . \tag{2.8}$$

Comparing (2.6), (2.7) and (2.8), we deduce that when all these expressions for $\tilde{K}_{\lambda\mu}(q,t)$ are reduced to their normal forms the only possible denominator factors are

$$1-t \ , \ 1-q \ , \ q^r t^s \ .$$

However, these factors are easily excluded by the observation that $\tilde{K}_{\lambda\mu}(q,t)$ has been shown[4] to have a polynomial limit as $t \to 1$, $q \to 1$, $t \to 0$ and $q \to 0$. This proves the polynomiality of $\tilde{K}_{\lambda\mu}(q,t)$. Note that another consequence of duality is the identity $\tilde{K}_{\lambda'\mu}(q,t) = t^{n(\mu)} q^{n(\mu')} \tilde{K}_{\lambda\mu}(1/q,1/t)$. This shows that $\tilde{K}_{\lambda\mu}(q,t)$ is a polynomial of degree $\leq n(\mu)$ in t and $\leq n(\mu')$ in q. Thus the polynomiality of $K_{\lambda\mu}(q,t)$ follows from (I.5.) This completes our proof.

We terminate by showing that our plethystic formulas may be used to obtain positivity results. To this end, note that equating coefficients of S_ρ in both sides of (1.1) we derive that

$$\sum_{\lambda \leftarrow \rho} \tilde{K}_{\lambda,\mu}(q,t) \ = \ \sum_{\nu \to \mu} c_{\mu\nu}(q,t) \, \tilde{K}_{\rho,\nu}(q,t) \ .$$

[4] See [18] Ch. VI ex. 7. p. 364.

Setting $\rho = (n-k-1, 1^k)$ we get

$$\tilde{K}_{(n-k-1,1^{k+1}),\mu} + \tilde{K}_{(n-k-1,2,1^{k-1}),\mu} + \tilde{K}_{(n-k,1^k),\mu} = \sum_{\nu\to\mu} c_{\mu\nu}\, \tilde{K}_{(n-k-1,1^k),\nu} .$$

Using (I.6) we can rewrite this as

$$\tilde{K}_{(n-k-1,2,1^{k-1}),\mu} = \sum_{\nu\to\mu} c_{\mu\nu}\, e_k[B_\nu - 1] - e_{k+1}[B_\mu - 1] - e_k[B_\mu - 1] . \quad (2.9)$$

Theorem 2.2. *For any $k \geq 1$ and $\mu \vdash n \geq k+2$ we have*

$$\tilde{K}_{(n-k-1,2,1^{k-1}),\mu} = \frac{a}{M} e_{k+1}[B_\mu - 1] + \frac{a+b}{M} e_k[B_\mu - 1] - \frac{1}{M} e_{k+1}[aB_\mu + b] \quad 2.10$$

Where for convenience we have set $M = (1-1/t)(1-1/q)$, $a = 1/t + 1/q - 1/tq$ and $b = 1/tq - 1$.

Proof. Just make the substitution $B_\nu = B_\mu - T_{\mu/\nu}$ in (2.9) and use (1.22).

For convenience, setting $\tilde{K}_{(n-k-1,2,1^{k-1}),\mu} = A_{k,\mu}$ and replacing B_μ by $B_\nu + T_{\mu/\nu}$ in (2.10) we easily derive that

$$\begin{aligned} A_{k,\mu} = \frac{a}{M} (e_{k+1}[B_\nu - 1] &+ e_k[B_\nu - 1]\, T_{\mu/\nu} + \\ &+ \frac{a+b}{M} (e_k[B_\nu - 1] + e_{k-1}[B_\nu - 1]\, T_{\mu/\nu}) \\ &- \frac{1}{M} \sum_{s=0}^{k+1} e_{k+1-s}[aB_\nu + b]\, T^s_{\mu/\nu}\, e_s[a] . \end{aligned}$$

Comparing with (2.10) we see that we may rewrite this as

$$\begin{aligned} A_{k,\mu} = A_{k,\nu} &+ \frac{a}{M} e_k[B_\nu - 1]\, T_{\mu/\nu} + \frac{a+b}{M} e_{k-1}[B_\nu - 1]\, T_{\mu/\nu} \\ &- \frac{1}{M} \sum_{s=1}^{k+1} e_{k+1-s}[aB_\nu + b]\, T^s_{\mu/\nu}\, e_s[a] . \end{aligned} \quad (2.11)$$

Since

$$e_s[a] = e_s[1/t + 1/q - 1/tq] = \begin{cases} 1 & \text{if } s = 0 \\ 1 - M & \text{if } s = 1 \\ (-1)^s M/(tq)^{s-1} & \text{if } s \geq 2 , \end{cases}$$

we can extract $A_{k-1,\nu}\, T_{\mu/\nu}$ out of the right-hand side of (2.11) and finally obtain that for $k > 1$ we have

$$\begin{aligned} A_{k,\mu} &= A_{k,\nu} + A_{k-1,\nu}\, T_{\mu/\nu} + T_{\mu/\nu} \sum_{s=0}^{k} e_{k-s}[aB_\nu + b]\, (-T_{\mu/\nu}/tq)^s \\ &= A_{k,\nu} + A_{k-1,\nu}\, T_{\mu/\nu} + T_{\mu/\nu}\, e_k[aB_\nu + b - T_{\mu/\nu}/tq] \ . \end{aligned} \tag{2.12}$$

On the olther hand setting $k = 1$ in (2.11) gives

$$A_{1,\mu} = A_{1,\nu} + T_{\mu/\nu}\, e_1[aB_\nu + b - T_{\mu/\nu}/tq] \ . \tag{2.13}$$

Theorem 2.3. *When $\lambda \vdash n$ is an augmented hook, the polynomial $\tilde{K}_{\lambda\mu}(q,t)$ has nonnegative integer coefficients for any $\mu \vdash n$.*

Proof. Note that we may write

$$e_k \left[aB_\nu + b - T_{\mu/\nu}/tq\right] = \frac{1}{t^k q^k}\, e_k \left[(t+q-1)\, B_\nu + 1 - tq - T_{\mu/\nu}\right] \ .$$

Simple geometric considerations show that for any partition ν that is obtained from μ by removing a corner square the expression

$$(t+q-1)\, B_\nu + 1 - tq - T_{\mu/\nu}$$

evaluates to a polynomial with nonnegative integer coefficients. Given this, we see that (2.12) and (2.13) are precisely what is needed to establish the positivity of $A_{k,\nu}$ by a double induction argument which ascends in k and n.

The positivity of $\tilde{K}_{\lambda\mu}(q,t)$ has been established for a number of special choices of λ and μ. The case when μ is a hook and λ arbitrary is treated in [5]. For μ a two-row or a two-column partition $\tilde{K}_{\lambda\mu}(q,t)$ was given a combinatorial interpretation in terms of rigged configurations by S. Fishel in [3]. The case when μ is an augmented hook and λ is arbitrary follows from the work of E. Reiner [20]. Making use of plethystic formulas Garsia and Tesler [12] were able to verify the positivity of $\tilde{K}_{\lambda\mu}(q,t)$ for all partitions of $n \leq 12$. The corresponding tables may be recovered by anonymous ftp from macaulay.ucsd.edu.

References

[1] Y. M. Chen, A. M. Garsia, and J. Remmel, *Algorithms for Plethysm*, Contemporary Math. #34. Combinatorics and Algebra, Curtis Greene ed. (1984), 109–153.

[2] I. Cherednik, *Double affine Hecke algebras and Macdonald conjectures*, Annals of Math. **141** (1995), 191–216.

[3] S. Fishel, *Statistics for general q,t-Kostka polynomials*, Proc. Amer. Math. Soc **123** #10 (1995), 120–138.

[4] A. Garsia, *Orthogonality of Milne's polynomials and raising operators*, Discrete Mathematics **99** (1992), 247–264.

[5] A. M. Garsia and M. Haiman, *A graded representation module for Macdonald's polynomials*, Proc. Natl. Acad. Sci. USA V 90 (1993), 3607–3610.

[6] A. M. Garsia and M. Haiman, *Factorizations of Pieri rules for Macdonald polynomials*, Discrete Mathematics 139 (1995), 219–256.

[7] A. M. Garsia and M. Haiman, *Orbit Harmonics and Graded Representations* (Research Monograph to appear as part of the Collection Published by the Lab. de Comb. et Informatique Mathématique, edited by S. Brlek, U. du Québec à Montréal).

[8] A. Garsia and M. Haiman, *A Remarkable q,t-Catalan Sequence and q-Lagrange inversion*; to appear in the Journal of Algebraic Combinatorics.

[9] A. Garsia and M. Haiman, *Some bigraded S_n-modules and the Macdonald q,t-Kostka coefficients*, Electronic Journal of Algebraic Combinatorics, Foata Festschrift, Paper R24,
(web site http://ejc.math.gatech.edu:8080/Journal/journalhome.html).

[10] A. Garsia and M. Haiman, *A random q,t-hook walk and a Sum of Pieri Coefficients*; (submitted to the J. of Comb. Theory Series A).

[11] A. M. Garsia and C. Procesi, *On certain graded S_n-modules and the q-Kostka polynomials*, Advances in Mathematics **94** (1992), 82–138.

[12] A. M. Garsia and G. Tessler, *Plethystic Formulas for the Macdonald q,t-Kostka coefficients*, Advances in Mathematics; to appear.

[13] A. Kirillov and M. Noumi, *Raising operators for Macdonald Polynomials*, (preprint).

[14] F. Knop, *Integrality of Two Variable Kostka Functions*, (preprint).

[15] F. Knop, *Symmetric and non-symmetric Quantum Capelli Polynomials*, (preprint).

[16] L. Lapointe and L. Vinet, *Creation Operators for Macdonald and Jack Polynomials*, (preprint)

[17] I. G. Macdonald, *A new class of symmetric functions*, Actes du 20e Séminaire Lotharingien, Publ. I.R.M.A. Strasbourg, (1988) 131–171.

[18] I. G. Macdonald, *Symmetric functions and Hall polynomials*, Second Edition, Clarendon Press, Oxford (1995).

[19] I. G. Macdonald, *Affine Hecke Algebras and orthogonal Polynomials*, Séminaire Bourbaki, (1995) # 797.

[20] E. Reiner, *A Proof of the $n!$ Conjecture for Generalized Hooks*, to appear in the Journal of Combinatorial Theory, Series A.

[21] S. Sahi, *Interpolation and integrality for Macdonald's Polynomials*, (preprint).

A.M. Garsia and J. Remmel
Department of Mathematics
University of California San Diego
La Jolla, CA 92093
U.S.A.

An Alternative Evaluation of the Andrews–Burge Determinant

C. Krattenthaler[†]

Dedicated to Gian-Carlo Rota

Abstract

We give a short, self-contained evaluation of the Andrews–Burge determinant (Pacific J. Math. **158** (1994), 1–14).

1. Introduction

In [9, Theorem 1], Andrews and Burge proved a determinant evaluation equivalent to

$$\det_{0\le i,j\le n-1}\left(\binom{x+i+j}{2i-j}+\binom{y+i+j}{2i-j}\right)$$
$$=(-1)^{\chi(n\equiv 3 \mod 4)}2^{\binom{n}{2}+1}$$
$$\times\prod_{j=1}^{n-1}\frac{\left(\frac{x+y}{2}+j+1\right)_{\lfloor (j+1)/2\rfloor}\left(-\frac{x+y}{2}-3n+j+\frac{3}{2}\right)_{\lfloor j/2\rfloor}}{(j)_j},\qquad(1.1)$$

where the shifted factorial $(a)_k$ is given by $(a)_k := a(a+1)\cdots(a+k-1)$, $k\ge 1$, $(a)_0 := 1$, and where $\chi(\mathcal{A})$=1 if $\mathcal{A}$ is true and $\chi(\mathcal{A})$=0 otherwise. This determinant identity arose in connection with the enumeration of symmetry classes of plane partitions. The known proofs [9, 10] of (1.1) require that

[†]Supported in part by EC's Human Capital and Mobility Program, grant CHRX-CT93-0400 and the Austrian Science Foundation FWF, grant P10191-MAT

1991 *Mathematics Subject Classification.* Primary 15A15; Secondary 05A15, 05A17, 33C20.

Key words and phrases. determinant evaluations, hypergeometric series, enumeration of symmetry classes of plane partitions.

one knows (1.1) to hold for $x = y$. Indeed, the latter was first established by Mills, Robbins and Rumsey [15, p. 53], in turn using another determinant evaluation, due to Andrews [3], whose proof is rather complicated. In the meantime, simpler proofs of the $x = y$ special case of (1.1) were found by Andrews [7], Andrews and Stanton [8], and Petkovšek and Wilf [16]. In this note we describe a new, concise, and self-contained proof of (1.1), see Section 3.

In fact, the main purpose of this note is to popularize the method that I use to prove (1.1) (see Section 2 for a description). This method is simple but powerful. Aside from this note, evidence for this claim can be found e.g., in [11, 12, 13, 14]. Thus, the method enlarges the not at all abundant collection of methods for evaluating determinants. In fact, aside from elementary manipulations by row and column operations, we are just aware of one other method, namely Andrews' "favourite" method of evaluating determinants (cf. [1, 2, 3, 4, 5, 6, 8]), which basically consists of guessing and then proving the LU-factorization of the matrix in question (i.e., the factorization of the matrix into a product of a lower triangular times an upper triangular matrix).

We should, however, also point out a limitation of our method. Namely, in order to be able to apply our method, we need a free parameter occuring in the determinant. (In (1.1) there are even two, x and y.) Andrews' method, on the other hand, might still be applicable if there is no free parameter present. Still, it is safe to speculate that many more applications of our method are going to be found in the future.

2. The method

The method that I use to prove (1.1) is as follows. Suppose we have a matrix $\left(f_{ij}(x)\right)_{0\le i,j\le n-1}$ with entries $f_{ij}(x)$ which are polynomials in x, and we want to prove the explicit factorization of $\det\left(f_{ij}(x)\right)$ as a polynomial in x,

$$\det_{0\le i,j\le n-1}\left(f_{ij}(x)\right) = C(n)\prod_{l}(x-a_l(n))^{m_l(n)}, \tag{2.1}$$

where $C(n), a_l(n), m_l(n)$ are independent of x, and the $a_l(n)$'s are pairwise different for fixed n.

Then, in the first step, for each l we find $m_l(n)$ linearly independent linear combinations of the columns, or of the rows, which vanish for $x = a_l(n)$. In different, but equivalent terms, we find $m_l(n)$ linearly independent vectors in the kernel of our matrix evaluated at $x = a_l(n)$, i.e., in the kernel of $\left(f_{ij}(a_l(n))\right)$, or of its transpose. That this really guarantees that

$(x - a_l(n))^{m_l(n)}$ is a factor of the determinant is a fact that might not be well-known enough. Therefore, for the sake of completeness, we state it as a lemma at the end of this section and provide a proof for it.

The finding of $a_l(n)$ linearly independent linear combinations of columns or rows which vanish for $x = a_l(n)$ (equivalently, linearly independent vectors in the kernel of the matrix, or its transpose, evaluated at $x = a_l(n)$) can be done with some skill (and primarily patience) by setting $x = a_l(n)$ in the matrix $\big(f_{ij}(x)\big)$, computing tables for the coefficients of the linear combinations for $n = 1, 2, \ldots$ (by solving the respective systems of linear equations on the computer), and finally guessing what the general pattern of the coefficients could be. To prove that the guess is correct, in case of binomial determinants (such as the one in (1.1)) one has to verify certain binomial identities. But this is pure routine today, by means of Zeilberger's algorithm [18, 19].

In the second step, one checks the degrees of both sides of (2.1) as polynomials in x. If it should happen that the degree of $\det\big(f_{ij}(x)\big)$ is not larger than $\sum_l m_l(n)$, the degree of the right-hand side, then it follows by what we did in the first step that the determinant $\det\big(f_{ij}(x)\big)$ has indeed the form of the right-hand side of (2.1), where $C(n)$ is some unknown constant. This constant can then be determined in the third step by comparison of coefficients of a suitable power of x.

Finally, here is the promised lemma, and its proof.

Lemma. *Let $A = A(x)$ be a matrix whose entries are polynomials in x, and u a number. If $\dim \operatorname{Ker} A(u) \geq k$, then u is a root of $\det A(x)$ of multiplicity at least k.*

Proof. Let $v_1, v_2, \ldots, v_k$ be k linearly independent (column) vectors in the kernel of $A(u)$. Without loss of generality, we may assume that $v_1, v_2, \ldots, v_k$ are such that the matrix $[v_1, v_2, \ldots, v_k]$, formed by gluing the columns $v_1, v_2, \ldots, v_k$ to a matrix, is in column-echelon form. In addition, again without loss of generality, we may assume that for any $i = 1, 2, \ldots, k$ the vector v_i is of the form $v_i = (0, \ldots, 0, 1, \ldots)^t$, i.e., the first $i - 1$ entries are 0, the i-th entry is 1, and the remaining entries could be anything. (To justify "without loss of generality" one would possibly have to permute the columns of A.)

Now we consider the matrix $\tilde{A} = \tilde{A}(x)$, formed by replacing for $i = 1, 2, \ldots, k$ the i-th column of A by the column Av_i. It is an easy observation that $\tilde{A}$ and A are related by elementary column operations. Therefore, their determinants are the same. On the other hand, the i-th column of $\tilde{A}(u)$ is $A(u)v_i = 0$, for any $i = 1, 2, \ldots, k$. Hence, each entry of Av_i, being a polynomial in x, must be divisible by $(x - u)$. Therefore, in the determinant

$\det \tilde{A}$, we may take $(x-u)$ out of the i-th column, $i=1,2,\dots,k$, with the entries in the remaining determinant still being polynomials in x. This proves that $(x-u)^k$ divides $\det \tilde{A} = \det A$, and, thus, the lemma. ■

3. Proof of (1.1)

As announced in the previous Section, the proof consists of three steps. For convenience, let us denote the determinant in (1.1) by $AB(x,y;n)$, or sometimes just $AB(n)$ for short.

Step 1. Identification of the factors. We show that the product on the right-hand side of (1.1),

$$\prod_{j=1}^{n-1} \left(\tfrac{x+y}{2}+j+1\right)_{\lfloor (j+1)/2\rfloor} \left(-\tfrac{x+y}{2}-3n+j+\tfrac{3}{2}\right)_{\lfloor j/2\rfloor},$$

is indeed a factor of $AB(n)$, in the way that was described in Section 2. Let us first consider just one part of this product,

$$\prod_{j=1}^{n-1} ((x+y)/2+j+1)_{\lfloor (j+1)/2\rfloor}\,.$$

Let us concentrate on a typical factor $(x+y+2j+2l)$, $1\le j\le n-1$, $1\le l\le (j+1)/2$. We claim that for each such factor there is a linear combination of the columns that vanishes if the factor vanishes. More precisely, we claim that for any j,l with $1\le j\le n-1$, $1\le l\le (j+1)/2$ there holds

$$\sum_{s=2l-1}^{\lfloor (j+2l-1)/2\rfloor} \frac{(j-2l+1)}{(j-s)}\frac{(j+2l-2s)_{s-2l+1}}{(s-2l+1)!} \\ \cdot (\text{column } s \text{ of } AB(-y-2j-2l,y;n)) \\ + (\text{column } j \text{ of } AB(-y-2j-2l,y;n)) = 0. \tag{3.1}$$

To avoid confusion, for $j=2l-1$ it is understood by convention that the sum in the first line of (3.1) vanishes.

To establish the claim, we have to check

$$\sum_{s=2l-1}^{\lfloor (j+2l-1)/2\rfloor} \frac{(j-2l+1)}{(j-s)}\frac{(j+2l-2s)_{s-2l+1}}{(s-2l+1)!} \\ \cdot\left(\binom{-y-2j-2l+i+s}{2i-s}+\binom{y+i+s}{2i-s}\right) \\ +\left(\binom{-y-j-2l+i}{2i-j}+\binom{y+i+j}{2i-j}\right)=0. \tag{3.2}$$

The exceptional case $j = 2l-1$ can be treated immediately. By assumption the sum in the first line of (3.2) vanishes for $j = 2l-1$, and, by inspection, also the last line in (3.2) vanishes for $j = 2l-1$. So we are left with establishing (3.2) for $l \le j/2$. In terms of the usual hypergeometric notation

$$ {}_rF_s\left[\begin{matrix} a_1,\dots,a_r \\ b_1,\dots,b_s \end{matrix}; z\right] = \sum_{k=0}^{\infty} \frac{(a_1)_k \cdots (a_r)_k}{k!\,(b_1)_k \cdots (b_s)_k} z^k , $$

this means to check

$$ \begin{aligned} &\frac{(-1-i-2j+2l-y)_{1+2i-2l}}{(1+2i-2l)!} \\ &\quad \times {}_4F_3\left[\begin{matrix} -\frac{1}{2}-\frac{j}{2}+l, -\frac{j}{2}+l, -1-2i+2l, i-2j-y \\ -j+2l, -\frac{1}{2}-\frac{i}{2}-j+l-\frac{y}{2}, -\frac{i}{2}-j+l-\frac{y}{2} \end{matrix}; 1\right] \\ &+\frac{(-1-i+4l+y)_{1+2i-2l}}{(1+2i-2l)!}\,{}_4F_3\left[\begin{matrix} -\frac{1}{2}-\frac{j}{2}+l, -\frac{j}{2}+l, -1-2i+2l, i+2l+y \\ -j+2l, -\frac{1}{2}-\frac{i}{2}+2l+\frac{y}{2}, -\frac{i}{2}+2l+\frac{y}{2} \end{matrix}; 1\right] \\ &\qquad + \left(\binom{-y-j-2l+i}{2i-j} + \binom{y+i+j}{2i-j}\right) = 0, \qquad (3.3) \end{aligned} $$

for $1 \le j \le n-1$, $1 \le l \le j/2$. This identity can be proved routinely by means of Zeilberger's algorithm [18, 19] and Salvy and Zimmermann's Maple package GFUN [17]. However, it happens that a ${}_4F_3$-summation is already known that applies to both ${}_4F_3$-series in (3.3), namely Lemma 1 in [9]. (Here we need the assumption $l \le j/2$.) Little simplification then establishes (3.3) and hence the claim. Thus, $\prod_{j=1}^{n-1} \left((x+y)/2+j+1\right)_{\lfloor (j+1)/2 \rfloor}$ is a factor of $AB(n)$.

Now we prove that

$$ \prod_{j=2}^{n-1} \left(-(x+y)/2-3n+j+\frac{3}{2}\right)_{\lfloor j/2 \rfloor} $$

is a factor of $AB(n)$. Also here, let us concentrate on a typical factor $(x+y+6n-2j-2l-1)$, $2 \le j \le n-1$, $1 \le l \le j/2$. This time we claim that for each such factor there is a linear combination of the columns that vanishes if the factor vanishes. More precisely, we claim that for any j,l with $2 \le j \le n-1$, $1 \le l \le j/2$ there holds

$$ \begin{aligned} &\sum_{s=1}^{n-l} \frac{(2n-j-s)_{j-2l}}{(n+l-j)_{j-2l}} \frac{(s)_{n-l-s}\,(3n+l-2j-2)_{n-l-s}}{4^{n-l-s}\,(n-l-s)!\,(2n-j-\frac{1}{2})_{n-l-s}} \\ &\qquad\qquad \cdot (\text{column } s \text{ of } AB(-y-6n+2j+2l+1, y; n)) = 0. \end{aligned} $$

This means to check

$$\sum_{s=1}^{n-l} \frac{(2n-j-s)_{j-2l}}{(n+l-j)_{j-2l}} \frac{(s)_{n-l-s}\,(3n+l-2j-2)_{n-l-s}}{4^{n-l-s}\,(n-l-s)!\,(2n-j-\frac{1}{2})_{n-l-s}} \cdot \left(\binom{-y-6n+2j+2l+1+i+s}{2i-s} + \binom{y+i+s}{2i-s}\right) = 0.$$

Converting this into hypergeometric notation and cancelling some factors, we see that we have to check

$$\begin{aligned}
&(4-i+2j+2l-6n-y)_{2i-1} \\
&\quad\times {}_5F_4\left[\begin{matrix} 1-2i, \frac{5}{2}+j+l-3n, 2+j-2n, \\ 4+2j-4n, 2+2l-2n, 2-\frac{i}{2}+j+l-3n-\frac{y}{2}, \end{matrix}\right. \\
&\qquad\qquad \left.\begin{matrix} 1+l-n, 3+i+2j+2l-6n-y \\ \frac{5}{2}-\frac{i}{2}+j+l-3n-\frac{y}{2} \end{matrix}; 1\right] \\
&\quad = -(3-i+y)_{2i-1} \\
&\quad\times {}_5F_4\left[\begin{matrix} 1-2i, \frac{5}{2}+j+l-3n, 2+j-2n, 1+l-n, 2+i+y \\ 4+2j-4n, 2+2l-2n, \frac{3}{2}-\frac{i}{2}+\frac{y}{2}, 2-\frac{i}{2}+\frac{y}{2} \end{matrix}; 1\right].
\end{aligned} \tag{3.4}$$

Again, in view of Zeilberger's algorithm, this is pure routine. However, also this identity happens to be already in the literature. It is exactly identity (5.7) in [8], with $a = \frac{3}{2} - i + j + l - 3n - y$, $x = \frac{3}{2} + j + l - 3n$, $z = 1 + j - 2n$, $p = 2i - 1$. Thus, $\prod_{j=2}^{n-1} \left(-(x+y)/2 - 3n + j + \frac{3}{2}\right)_{\lfloor j/2 \rfloor}$ is a factor of $AB(n)$.

Step 2. Bounding the polynomial degrees. The degree of $AB(n)$ as a polynomial in x is obviously at most $\binom{n}{2}$. But the degree of the product on the right-hand side of (1.1) is exactly $\binom{n}{2}$. Therefore it follows that

$$\det_{0 \le i,j \le n-1} \left(\binom{x+i+j}{2i-j} + \binom{y+i+j}{2i-j}\right) = C(n) \prod_{j=1}^{n-1} \left(\tfrac{x+y}{2} + j + 1\right)_{\lfloor (j+1)/2 \rfloor} \left(-\tfrac{x+y}{2} - 3n + j + \tfrac{3}{2}\right)_{\lfloor j/2 \rfloor}, \tag{3.5}$$

with some $C(n)$ independent of x, and, by symmetry, also independent of y.

Step 3. Determining the constant. To compute $C(n)$, on both sides of (3.5) set $y = x$ and then compare coefficients of $x^{\binom{n}{2}}$. On the right-hand side the coefficient is $(-1)^{\chi(n\equiv 3 \mod 4)}C(n)$, whereas on the left-hand side the coefficient is

$$\begin{aligned}\det_{0\le i,j\le n-1}\left(2\frac{1}{(2i-j)!}\right) &= 2^n \prod_{i=0}^{n-1}\frac{1}{(2i)!}\det_{0\le i,j\le n-1}\left((2i-j+1)_j\right)\\ &= 2^n \prod_{i=0}^{n-1}\frac{1}{(2i)!}\det_{0\le i,j\le n-1}\left((2i)^j\right)\\ &= 2^n \prod_{i=0}^{n-1}\frac{1}{(2i)!}\prod_{0\le i<j\le n-1}(2j-2i)\\ &= 2^{n+\binom{n}{2}}\prod_{i=0}^{n-1}\frac{i!}{(2i)!}\\ &= 2^{1+\binom{n}{2}}\prod_{i=1}^{n-1}\frac{1}{(i)_i},\end{aligned}$$

where in the step from the first to the second line we used elementary column operations, and the subsequent step is just the Vandermonde determinant evaluation.

This completes the proof of (1.1). ■

References

[1] G. E. Andrews, *Plane partitions (II): The equivalence of the Bender–Knuth and the MacMahon conjectures*, Pacific J. Math. **72** (1977), 283–291.

[2] G. E. Andrews, *Plane partitions (I): The MacMahon conjecture*, in: Studies in Foundations and Combinatorics, G.-C. Rota (ed.), Adv. in Math. Suppl. Studies, Vol. 1, 1978, 131–150.

[3] G. E. Andrews, *Plane partitions (III): The weak Macdonald conjecture*, Inventiones Math. **53** (1979), 193–225.

[4] G. E. Andrews, *Macdonald's conjecture and descending plane partitions*, in: Combinatorics, representation theory and statistical methods in groups, Young Day Proceedings, T. V. Narayana, R. M. Mathsen, J. G. Williams (eds.), Lecture Notes in Pure Math., vol. 57, Marcel Dekker, New York, Basel, 1980, 91–106.

[5] G. E. Andrews, *Plane partitions (IV): A conjecture of Mills–Robbins–Rumsey*, Aequationes Math. **33** (1987), 230–250.

[6] G. E. Andrews, *Plane partitions V: The t.s.s.c.p.p. conjecture*, J. Combin. Theory Ser. A **66** (1994), 28–39.

[7] G. E. Andrews, *Pfaff's method (I): The Mills–Robbins–Rumsey determinant*, preprint.

[8] G. E. Andrews and D. W. Stanton, *Determinants in plane partition enumeration*, preprint.

[9] G. E. Andrews and W. H. Burge, *Determinant identities*, Pacific J. Math. **158** (1993), 1–14.

[10] I. P. Goulden and D. M. Jackson, *Further determinants with the averaging property of Andrews–Burge*, J. Combin. Theory Ser. A **73** (1996), 368–375.

[11] C. Krattenthaler, *Determinant identities and a generalization of the number of totally symmetric self-complementary plane partitions*, Elect. J. Combin. (to appear).

[12] C. Krattenthaler, *Some q-analogues of determinant identities which arose in plane partition enumeration*, Séminaire Lotharingien Combin. (to appear).

[13] C. Krattenthaler, *A new proof of the MRR-conjecture — including a generalization*, preprint.

[14] G. Kuperberg, *Another proof of the alternating sign matrix conjecture*, Math. Research Letters (1996), 139–150.

[15] W. H. Mills, D. H. Robbins and H. Rumsey, *Enumeration of a symmetry class of plane partitions*, Discrete Math. **67** (1987), 43–55.

[16] M. Petkovšek and H. Wilf, *A high-tech proof of the Mills-Robbins-Rumsey determinant formula*, Elect. J. Combin. **3** (no. 2, "The Foata Festschrift") (1996), #R19, 3 pp.

[17] B. Salvy and P. Zimmermann, *GFUN — A MAPLE package for the manipulation of generating and holonomic functions in one variable*, ACM Trans. Math. Software **20** (1994), 163–177.

[18] D. Zeilberger, *A fast algorithm for proving terminating hypergeometric identities*, Discrete Math. **80** (1990), 207–211.

[19] D. Zeilberger, *The method of creative telescoping*, J. Symbolic Comput. **11** (1991), 195–204.

Institut für Mathematik der Universität Wien,
Strudlhofgasse 4, A-1090 Wien, Austria.
e-mail: KRATT@Pap.Univie.Ac.At
WWW: `http://radon.mat.univie.ac.at/People/kratt`

The Number of Points in a Combinatorial Geometry with No 8-Point-Line Minors

Joseph E. Bonin and Joseph P. S. Kung

Abstract

We show that when n is greater than 3, the number of points in a combinatorial geometry (or simple matroid) G of rank n containing no minor isomorphic to the 8-point line is at most $\frac{1}{4}(5^n - 1)$. This bound is sharp and is attained if and only if the geometry G is the projective geometry $\mathrm{PG}(n-1, 5)$ over the field $\mathrm{GF}(5)$.

1. Introduction

Synthetic geometry, the direct study of geometrical configurations without the use of coordinates, bifurcated in the twentieth century into two areas: finite geometry and matroid theory. Finite geometry is the study of projective geometries, in particular, projective planes, and related objects, such as designs. The objects studied in finite geometry have homogeneity or regularity properties which follow or are abstracted from group actions on the objects, so that, roughly speaking, the object "looks the same" from each point. Matroid theory, on the other hand, is concerned with the geometric properties of arbitrary sets of points. In the words of Crapo and Rota, matroid theory "may be considered as a revival of projective geometry in its most synthetic form" ([3], Chapter 1).

Matroid theory aims to dispense not only with coordinates but with the ambient space altogether. One way to do this is to replace the properties of the ambient space (which one assumes are preserved under projections) by excluded-minor conditions. For example, no rank-2 contraction of a set of points in the projective space $\mathrm{PG}(n-1, q)$ over the finite field of order q can have more than $q+1$ points. Hence, one of the properties of sets of points in $\mathrm{PG}(n-1, q)$ is that they have no minors isomorphic to the $(q+2)$-point line $U_{2,q+2}$. This numerical condition, by itself, suffices to determine the expected upper bound on the number of points. More specifically, the following result holds ([4], Theorem 4.3).

Theorem 1.1 *Assume G is a rank-n geometry (or simple matroid) containing no minor isomorphic to the $(q+2)$-point line. Then the number $|G|$ of points in G is at most*

$$\frac{q^n - 1}{q - 1} = q^{n-1} + q^{n-2} + \cdots + q + 1.$$

For rank greater than 3, this bound is sharp if and only if q is a prime power; when this is the case, $|G|$ equals the upper bound if and only if G is (isomorphic to) the projective geometry $\mathrm{PG}(n-1, q)$.

A natural question arising from this result is: *what is the sharp upper bound when q is not a prime power?* In [1], it is shown that if a rank-n geometry G has no minor isomorphic to the $(q+2)$-point line and there is no projective geometry of rank n and order q, then $|G|$ is at most $q^{n-1} - 1$; if in addition q is odd, then $|G|$ is at most

$$q^{n-1} - \frac{q^{n-2} - 1}{q - 1} - 1.$$

These upper bounds, however, are not sharp and can be improved using an induction argument similar to that in the proof of Lemma 5.4. Starting the induction at $n = 3$, it can be shown that when there is no projective plane of order q,

$$q^{n-1} - q^{n-3} + \frac{q^{n-3} - 1}{q - 1}$$

is an upper bound, as is

$$q^{n-1} - 2q^{n-3} + \frac{q^{n-3} - 1}{q - 1}$$

when q is odd. Without using any information about the non-existence of projective planes of order q, we get the following weaker upper bounds by starting the induction at $n = 4$:

$$q^{n-1} - q^{n-4} + \frac{q^{n-4} - 1}{q - 1}$$

for q in general, and

$$q^{n-1} - q^{n-3} - 2q^{n-4} + \frac{q^{n-4} - 1}{q - 1}$$

for q odd. These tighter upper bounds are most probably not sharp.

In this paper, we derive the sharp upper bound when q equals 6, the first positive integer which is not a prime power.

The geometries having no minor isomorphic to the $(q+2)$-point line form a minor-closed class $\mathcal{U}(q)$. For every prime power q' not exceeding q, all geometries representable over the finite field $\mathrm{GF}(q')$ are in $\mathcal{U}(q)$. (For $q > 2$, the class $\mathcal{U}(q)$ also contains many non-representable geometries.) In particular, the rank-n projective geometry $\mathrm{PG}(n-1, 5)$ is in $\mathcal{U}(6)$. Hence, there is a rank-n geometry in $\mathcal{U}(6)$ having $\frac{1}{4}(5^n - 1)$ points. The main theorem in this paper asserts that when the rank n is greater than 3, the projective geometry $\mathrm{PG}(n-1, 5)$ is the rank-n geometry having the maximum number of points in $\mathcal{U}(6)$.

Theorem 1.2 *Let n be greater than 3 and let G be a rank-n geometry in $\mathcal{U}(6)$. Then*

$$|G| \leq \frac{5^n - 1}{5 - 1} = \tfrac{1}{4}(5^n - 1).$$

This upper bound is sharp and is attained only by the rank-n projective geometry $\mathrm{PG}(n-1, 5)$ *over the finite field* $\mathrm{GF}(5)$.

Theorem 1.2 verifies the case $q = 6$ of a conjecture made in [4], p. 35.

The proof of Theorem 1.2 is given in Section 5. Sections 2, 3, and 4 treat several lemmas used in the proof. The final section, Section 6, outlines an alternative proof of the bound in Theorem 1.2.

Although we assume some familiarity with basic matroid theory (see, for example, the classic text [3]), this paper is written so as to be reasonably accessible to any synthetic geometer. If G is a geometry on the point set S, we refer to the flats of ranks $r(S) - 1$, $r(S) - 2$, and $r(S) - 3$ respectively as copoints, colines, and coplanes. The following elementary lemma ([4], p. 42) will be used freely.

Lemma 1.3 *Let G be a geometry in $\mathcal{U}(q)$. Then a flat of G having rank k and $(q^k - 1)/(q - 1)$ points is modular. In particular, 6-point lines are modular in geometries in $\mathcal{U}(5)$ and 7-point lines are modular in geometries in $\mathcal{U}(6)$.*

2. Extensions of geometries

A lemma from [1] that is fundamental to this paper asserts that under certain conditions, a geometry in $\mathcal{U}(q)$ can be extended to a bigger geometry in $\mathcal{U}(q)$ having the same rank. Before stating this lemma, we recall some notions from Crapo's theory of single-element extensions (see [2] and Chapter 10 in [3]). A *single-element extension* $G^+(S \cup \{e\})$ of a geometry $G(S)$ is a matroid such that the restriction $G^+|S$ equals G. Let $\mathcal{M}$ be the collection of flats or closed sets A in the lattice $L(G)$ of flats of G such that the closure of A in G^+ is $A \cup \{e\}$. The set $\mathcal{M}$ is a filter in the lattice $L(G)$. In addition, it satisfies the following property (which is equivalent to the property that $\mathcal{M}$ is closed under intersections of modular pairs of flats): if A and B are flats in $\mathcal{M}$ and $r(A \cap B) = r(A) - 1 = r(B) - 1$, then their intersection $A \cap B$ is also a flat in $\mathcal{M}$. A filter $\mathcal{M}$ in $L(G)$ satisfying this property is called a *modular filter*. Crapo proved that single-element extensions of G are in one-to-one correspondence with the modular filters in $L(G)$.

If the extension G^+ is a geometry, then $\mathcal{M}$ does not contain any point of G. Therefore, if two lines in G intersect in a point, then at most one of the lines is in $\mathcal{M}$. When both G and G^+ are rank-3 geometries, this implies that if the filter $\mathcal{M}$ contains a modular line, then it contains no other line. In the rank-3 case, we also have the following useful counting result: the number of lines containing the point e in the extension G^+ is the sum of the number of lines in $\mathcal{M}$ and the number $|S - \bigcup_{\ell \in \mathcal{M}} \ell|$ of points in G not contained in any line in $\mathcal{M}$.

The next lemma is essentially Lemma 6 in [1].

Lemma 2.1 *Let $G(S)$ be a geometry in $\mathcal{U}(q)$ and let ℓ be a q-point line in G. Suppose that for each coplane X with $X \cap \ell = \emptyset$ and $r(X \cup \ell) = r(X)+2$, at least one of the following conditions holds:*

(1) *there exists a copoint Y containing X such that the rank-2 interval $[X, Y]$ contains $q+1$ colines,*

(2) *there exist at least $q^2 - 1$ colines in the rank-3 upper interval $[X, S]$, or*

(3) *every coline in the upper interval $[X, S]$ is contained in $q+1$ copoints and there are at most q^2+q+1 copoints in $[X, S]$.*

Then G has an extension $G^+(S \cup \{e\})$ of the same rank in $\mathcal{U}(q)$ in which the q-point line ℓ is extended to the $(q+1)$-point line $\ell \cup \{e\}$.

The proof of Lemma 6 in [1] can be used to prove this lemma without any changes. The basic idea in the proof is to observe that the union $\mathcal{M}$ of the two sets

$$\{A \in L(G) : A \cap \ell = \emptyset \text{ and } r(A \cup \ell) = r(A) + 1\}$$

(the set of flats which "should" intersect ℓ at a point) and

$$\{A \in L(G) : \ell \subseteq A\},$$

is a modular filter in $L(G)$. Let G^+ be the single-element extension determined by $\mathcal{M}$. To show that G^+ is in $\mathcal{U}(q)$, it suffices, by the scum theorem (see, for example, [3], Chapter 9), to check that every rank-2 upper interval in G^+ contains at most $q+1$ points. This is guaranteed by any one of the three conditions in the lemma.

We shall use the cases q equals 5 or 6 of Lemma 2.1. Note that when q equals 6,

$$q+1=7, \quad q^2-1=35, \quad \text{and} \quad q^2+q+1=43.$$

Two immediate consequences of Lemma 2.1 are the following.

Corollary 2.2 *Let $G(S)$ be a rank-3 geometry in $\mathcal{U}(6)$ with a 6-point line ℓ and a 7-point line. Then $G(S)$ has a single-element extension to a rank-3 geometry $G^+(S \cup \{e\})$ in $\mathcal{U}(6)$ in which ℓ has been extended to the 7-point line $\ell \cup \{e\}$.*

Corollary 2.3 *Let $G(S)$ be a rank-4 geometry in $\mathcal{U}(6)$ with a 6-point line ℓ. Assume that for each point x of G, there is at least one 7-point line of G not containing x. Then $G(S)$ has a single-element extension to a rank-4 geometry $G^+(S \cup \{e\})$ in $\mathcal{U}(6)$ in which ℓ has been extended to the 7-point line $\ell \cup \{e\}$.*

The next lemma is a special case of Corollary 1 in [1]. It can be proved by an easy *ad hoc* counting argument using Lemma 2.1.

Lemma 2.4 *Let G be a rank-3 geometry in $\mathcal{U}(5)$ with at least 28 points. Then G can be extended to the projective plane* PG(2, 5).

We are now ready to prove the following lemma.

Lemma 2.5 *Let $G(S)$ be a rank-3 geometry in $\mathcal{U}(5)$ with at least 28 points and let $G^+(S \cup E)$ be an extension of G to a geometry with 32 points. Then G^+ is not in $\mathcal{U}(6)$.*

Proof. Suppose G has m points, where $28 \le m \le 31$. By Lemma 2.4, we may assume that G is $\mathrm{PG}(2,5) - X$ for some set X of $31-m$ points in PG(2, 5). Let e be a point in E and let G' be the restriction $G^+|S \cup \{e\}$ of G^+. It suffices to show that either the extension G' of G is a subgeometry of PG(2, 5) or it is not in $\mathcal{U}(6)$. If the modular filter $\mathcal{M}$ for the extension G' consists of S alone, then e is on m lines in G', and so G' is not in $\mathcal{U}(6)$. If $\mathcal{M}$ contains a 6-point line, then, since 6-point lines in geometries in $\mathcal{U}(5)$ are modular and G' is a geometry, no other line is in $\mathcal{M}$. Therefore, in this case, e is on $1+(m-6)$ lines in G', and so again G' is not in $\mathcal{U}(6)$. Thus we may assume that all lines in $\mathcal{M}$ have five or fewer points (and hence, m is strictly less than 31). Let i be the number of lines in $\mathcal{M}$. Since the i lines in $\mathcal{M}$ contain at most $5i$ points in G, we have that e is on at least $i+(m-5i)$ lines in G'. If $i \leq 5$, it follows that G' is not in $\mathcal{U}(6)$. Therefore we may assume that $i \geq 6$. Since no two lines in $\mathcal{M}$ can have a point in common, there is some point a in X such that the lines in $\mathcal{M}$ are precisely the lines ℓ for which $\ell \cup \{a\}$ is a line of $\mathrm{PG}(2,5) - (X - \{a\})$. Thus, up to relabeling, the point e is a. It follows that G' is a subgeometry of $\mathrm{PG}(2,5)$. ■

Corollary 2.6 *Let G be a rank-4 geometry in $\mathcal{U}(6)$. Let P be a plane in G that has at least 28 points and is in $\mathcal{U}(5)$. Then any point x in the complement $G-P$ is on at most 31 lines in G.*

3. Some preliminary bounds

Although we prove Theorem 1.2 when the rank is greater than 3, the proof relies on the following upper bound for rank-3 geometries in $\mathcal{U}(6)$.

Lemma 3.1 *A rank-3 geometry in $\mathcal{U}(6)$ has at most 35 points.*

This lemma is a special case of Corollary 2 in [1]. The upper bound given in Lemma 3.1 is probably not sharp. It would be interesting to determine the sharp bound.

Since seven copunctual 6-point lines contain 36 points, Lemma 3.1 implies that there is no such configuration in a rank-3 geometry in $\mathcal{U}(6)$. This observation yields the following result about rank-4 geometries in $\mathcal{U}(6)$.

Corollary 3.2 *Let G be a rank-4 geometry in $\mathcal{U}(6)$, let ℓ be a line of G, and let a be a point on ℓ. Then among the seven or fewer planes in G containing ℓ, there are at most six in which there is a 6- or 7-point line not equal to ℓ through a.*

Proof. Assume the conclusion is false. Let b be a point not equal to a in ℓ. Then the rank-3 geometry G/b formed by contracting b has seven 6- or 7-point lines through x, contradicting our earlier observation. ■

The proof of Theorem 1.2 also uses a better upper bound on the number of points in planes in $\mathcal{U}(6)$ that have no 7-point lines. (Strictly speaking, we do not need the full strength of this lemma to prove Theorem 1.2.)

Lemma 3.3 *Let G be a rank-3 geometry in $\mathcal{U}(6)$. If G has no 7-point lines, then $|G| \leq 33$.*

Proof. By Lemma 3.1, $|G|$ is at most 35. Assume $|G| = 35$. Since G has no 7-point lines, we can have 35 points only by having each point incident with six 6-point lines and a single 5-point line. (Note that 35 is $5 \cdot 6 + 4 + 1$; this reflects the six 6-point lines through a given point x, each contributing five points in addition to x, the four points besides x on the 5-point line, and x itself.) This and Lemma 2.1 (using condition 2) yield a geometry in $\mathcal{U}(6)$ with 36 points, contrary to Lemma 3.1.

Now assume $|G| = 34$. Note that each point is on seven lines and there are two types of points in G: type 1 points are incident on six 6-point lines and a single 4-point line; type 2 points are incident on five 6-point lines and two 5-point lines. Let there be n_1 points of type 1 and n_2 points of type 2. To count the number of lines, we start by counting the number of 6-point lines, and to do this, we count the number of pairs consisting of a 6-point line and an incident point. By considering the number of incident 6-point lines at points of the two types, we see that the number of such pairs is $6n_1 + 5n_2$. Thus the number of 6-point lines is $(6n_1 + 5n_2)/6$. Applying these ideas to 5- and 4-point lines also, we see that the number of lines is

$$\frac{6n_1 + 5n_2}{6} + \frac{2n_2}{5} + \frac{n_1}{4} = \frac{5}{4}n_1 + \frac{37}{30}n_2.$$

This is at most $5(n_1 + n_2)/4$, or $5 \cdot 34/4$, which is less than 43. Since the number of lines is less than 43, we can apply Lemma 2.1 (using condition 3) to any 6-point line in G, yielding a geometry in $\mathcal{U}(6)$ with 35 points. Since this contradicts what we established in the last case, the lemma has been proven. ∎

4. Orthogonal Latin squares

One reason Theorem 1.2 has a relatively easy proof is that one can apply the following classical result arising from Euler's problem of 36 officers: *there exists no pair of orthogonal Latin squares of order* 6. This non-existence theorem was proved by G. Tarry [6] in 1900. For a modern proof, see, for example, [5].

A *Latin square* L of order q is a $q \times q$ array $(L_{ij})_{1 \leq i,j \leq q}$ filled with symbols from the set $\{1, 2, \ldots, q\}$ such that each row and each column contains every symbol exactly once. Two Latin squares L and M are *orthogonal* if for each ordered pair (h, k), there is a unique position (i, j) such that $L_{ij} = h$ and $M_{ij} = k$. There are several ways to associate Latin squares with configurations of lines in geometries. We shall use the following construction. Let ℓ_r, ℓ_s, ℓ_t be three $(q+1)$-point lines meeting at the common point a in a rank-3 geometry in $\mathcal{U}(q)$. Let the points other than a on ℓ_r be denoted $r_1, r_2, \ldots, r_q$, and similarly for the other two lines ℓ_s and ℓ_t. Construct a $q \times q$ array L_t by the rule: the entry in row i and column j is k if r_i, s_j, and t_k are collinear. Since any two points in a geometry are on a unique line and $(q+1)$-point lines are modular, L_t is a Latin square.

Lemma 4.1 *Let a be a point of a rank-3 geometry G in $\mathcal{U}(6)$. If a is on a 7-point line, then there are at most two other lines in G through a with more than 5 points.*

Proof. By Corollary 2.2, every 6-point line through a can be extended to a 7-point line through a in some extension of G. Therefore it suffices to show that there can be at most three 7-point lines through a. However, if $\ell_r, \ell_s, \ell_t, \ell_u$ are four 7-point lines through a, then, because a point t_h in ℓ_t and a point u_k in ℓ_u determine a unique line and this line intersects the lines ℓ_r and ℓ_s at points r_i and s_j, the 6×6 Latin squares L_t and L_u are orthogonal, contradicting Tarry's theorem. ■

Corollary 4.2 *Let a be a point of a rank-4 geometry G in $\mathcal{U}(6)$. There are at most seven 7-point lines through a.*

Proof. Let ℓ be a 7-point line through a. By contracting any point in $\ell - a$ and applying Lemma 4.1, it follows that at most three of the planes through ℓ contain 7-point lines through a. By Lemma 4.1, each of these planes contains at most two 7-point lines through a in addition to ℓ. ■

5. The proof of the main theorem

We now give the proof of Theorem 1.2, which is by induction. To establish the case $n = 4$, we use three cases: G has a 7-point line; G has no 7-point line but G has a point on seven coplanar lines; G has no point on seven coplanar lines.

Lemma 5.1 *Let G be a rank-4 geometry in $\mathcal{U}(6)$ and assume that the point a of G is on at least one 7-point line ℓ in G. Then $|G| \leq 155$.*

Proof. By Corollary 3.2, at most six of the planes through ℓ can contain 6- or 7-point lines through a. Since ℓ is a 7-point line through a, Lemma 4.1 implies that each plane through ℓ contains at most two other 6- or 7-point lines through a.

First assume that ℓ is the only 7-point line through a. There are at most twelve 6-point lines through a, namely, at most two in each of six or fewer planes through ℓ. Therefore $|G|$ is at most $7+34 \cdot 4+12$, or 155. (This counts the seven points on ℓ, the four points other than a on the other lines through a, and an additional point on twelve or fewer of the lines through a.)

Assume there is another 7-point line through a. Therefore, the contraction G/x for any point x in $\ell - \{a\}$ contains a 7-point line through a. Hence, by Lemma 4.1, at most three planes through ℓ have 6- or 7-point lines through a. Hence there are at most seven 6- or 7-point lines through a. Therefore $|G|$ is at most $35 \cdot 4 + 7 \cdot 2 + 1$, or 155. ∎

Lemma 5.2 *Let G be a rank-4 geometry in $\mathcal{U}(6)$ with no 7-point lines but with a point a on seven coplanar lines. Then $|G| \leq 153$.*

Proof. Assume the plane P contains seven lines through a. Let x be in $P - \{a\}$. Since the contraction G/a contains a 7-point line through x, Lemma 4.1 implies that 6-point lines through x can occur in at most three of the planes through the line $x \vee a$, with P being one of these planes. Since each plane contains at most six 6-point lines through a point, the number of 6-point lines through x is at most 18.

First assume there is a point x in P such that the contraction G/x contains no 7-point lines. It follows from Lemma 3.3 that x is on at most 33 lines in G. Therefore $|G|$ is at most $33 \cdot 4 + 18 + 1$, or 151.

We may now assume that each point in P is on some set of seven coplanar lines. Let z be a point in $P - \{a\}$ and let x be a point in P not on the line $a \vee z$. Let P_1 and P_2 be planes other than P through $a \vee x$ with the property that any 6-point line through x is in one of P, P_1, or P_2. Since the contraction G/z contains a 7-point line, and P_1 and P_2 are restrictions of G/z, it follows from Corollary 2.2 and Lemma 4.1 that each of P_1 and P_2 contains at most three 6-point lines through x. Thus the number of 6-point lines through x is at most $6 + 2 \cdot 3$, so $|G|$ is at most $35 \cdot 4 + 12 + 1$, or 153. ∎

Having treated rank-4 geometries in $\mathcal{U}(6)$ with either a 7-point line or some point on seven coplanar lines, the one remaining case when the rank is 4 is the following.

Lemma 5.3 *Let G be a rank-4 geometry in $\mathcal{U}(6)$ with no seven coplanar copunctual lines. Then $|G| \leq 156$. Furthermore, $|G|$ is 156 if and only if G is* $\mathrm{PG}(3,5)$.

Proof. Since G contains no seven coplanar copunctual lines, G has no 7-point lines. If there are no 6-point lines, then $|G|$ is at most $35 \cdot 4 + 1$, or 141. Thus we may assume G has at least one 6-point line, say ℓ. Since G has no 7-point lines and no seven coplanar copunctual lines, each plane in G is in $\mathcal{U}(5)$, and hence has at most 31 points. If each plane containing ℓ has at most 27 points, then $|G|$ is at most $7(27-6)+6$, or 153. Thus we may assume that ℓ is in a plane P with at least 28 points. Let x be any point in $G-P$. By Corollary 2.6, x is on at most 31 lines, so $|G|$ is at most $31 \cdot 5 + 1$, or 156.

Having established the bound, we analyze the case of equality. Since $7(27-6)+6$ is 153, we have that some plane P through ℓ has at least 28 points. By Corollary 2.6, we can have $|G| = 156$ only if for each point x in $G-P$, x is on 31 lines and all lines through x are 6-point lines. From this and the fact that all planes of G are in $\mathcal{U}(5)$, it follows that for any such point x, the plane $x \vee \ell$ has 31 points. Thus all planes through ℓ, except perhaps P, have 31 points. Since this argument can now be made using any of these 31-point planes through ℓ, we deduce that all lines in G are 6-point lines.

Since all lines in G are 6-point lines and all planes in G are in $\mathcal{U}(5)$, it follows that all planes in G have 31 points. Therefore each line of G is on six planes. It follows from this and the scum theorem that G is in $\mathcal{U}(5)$. Theorem 1.1 with $q=5$ implies that G is $\mathrm{PG}(3,5)$. ∎

Having established the rank-4 case, we now give the inductive step.

Lemma 5.4 *For $n \geq 5$, rank-n geometries in $\mathcal{U}(6)$ have $(5^n-1)/4$ or fewer points. Furthermore,* $\mathrm{PG}(n-1,5)$ *is the only rank-n geometry in $\mathcal{U}(6)$ with $(5^n-1)/4$ points.*

Proof. Because the proof involves examining the planes through a given line, we need to establish the case $n=5$ separately before we can treat the general case. Let G be a rank-5 geometry in $\mathcal{U}(6)$. Assume G has a 7-point line ℓ. Consider a point x in ℓ. Each 6- or 7-point line ℓ' not equal to ℓ through

x determines a plane $\ell \vee \ell'$ with ℓ. There are at most 35 planes through ℓ. By Lemma 4.1, each of these planes contains at most two 6- or 7-point lines through x in addition to ℓ. Therefore the maximum number of 6- or 7-point lines through x is $35 \cdot 2 + 1$, or 71. Since there are at most $(5^4 - 1)/4$ lines through x and most contain five or fewer points, $|G|$ is at most

$$\tfrac{1}{4}(5^4 - 1)4 + 71 \cdot 2 + 1 = 767.$$

This is strictly less than $(5^5 - 1)/4$, or 781. Now assume G has no 7-point lines, and let x be a point of G. Since x is on at most $(5^4 - 1)/4$ lines, $|G|$ is at most

$$\tfrac{1}{4}(5^4 - 1)5 + 1 = \tfrac{1}{4}(5^5 - 1).$$

Now assume equality holds. It follows that all lines in G are 6-point lines. Therefore each copoint of G has at least $(5^4 - 1)/4$ points. This and the upper bound in the rank-4 case imply that each copoint has $(5^4 - 1)/4$ points and is isomorphic to $\mathrm{PG}(3, 5)$. Therefore the colines of G have $(5^3 - 1)/4$ points. Thus there are

$$\frac{\frac{1}{4}(5^5 - 1) - \frac{1}{4}(5^3 - 1)}{\frac{1}{4}(5^4 - 1) - \frac{1}{4}(5^3 - 1)} = 6$$

copoints over each coline. Therefore by the scum theorem, G is in $\mathcal{U}(5)$. Theorem 1.1 with $q = 5$ implies that G is $\mathrm{PG}(4, 5)$.

Having established the case $n = 5$, we can proceed to the general case. Assume $n > 5$ and that the result holds for ranks $n - 1$ and $n - 2$. Let G be a rank-n geometry in $\mathcal{U}(6)$. Assume G has a 7-point line ℓ. Consider a point x in ℓ. There are at most $(5^{n-2} - 1)/4$ planes through ℓ, and hence at most

$$\tfrac{1}{4}(5^{n-2} - 1)2 + 1$$

6- or 7-point lines through x. There are at most $(5^{n-1} - 1)/4$ lines through x. Therefore $|G|$ is at most

$$\tfrac{1}{4}(5^{n-1} - 1)4 + \left(\tfrac{1}{4}(5^{n-2} - 1)2 + 1\right)2 + 1.$$

This is strictly less than $(5^n - 1)/4$. Now assume G has no 7-point lines, and let x be a point of G. Since x is on at most $(5^{n-1} - 1)/4$ lines, $|G|$ is at most

$$\tfrac{1}{4}(5^{n-1} - 1)5 + 1 = \tfrac{1}{4}(5^n - 1).$$

Assume equality holds. It follows that all lines in G are 6-point lines, so each copoint of G has at least $(5^{n-1} - 1)/4$ points. Therefore each copoint has

$(5^{n-1}-1)/4$ points and is isomorphic to $\mathrm{PG}(n-2,5)$. The colines of G have $(5^{n-2}-1)/4$ points. Thus there are

$$\frac{\frac{1}{4}(5^n-1)-\frac{1}{4}(5^{n-2}-1)}{\frac{1}{4}(5^{n-1}-1)-\frac{1}{4}(5^{n-2}-1)}=6$$

copoints over each coline. As in the rank-5 case, the scum theorem and Theorem 1.1 with $q=5$ imply that G is $\mathrm{PG}(n-1,5)$. ■

6. An alternative proof

A major step in the proof of Theorem 1.2 is to prove the upper bound in the rank-4 case. In this section, we sketch an alternative proof of the rank-4 upper bound, that if G is a rank-4 geometry in $\mathcal{U}(6)$, then $|G|\leq 156$.

We begin the two-part proof with the following lemma.

Lemma 6.1 *Let G be a rank-4 geometry in $\mathcal{U}(6)$ such that for each plane P of G, there is a point x in the complement $G-P$ on no 7-point lines. Then $|G|\leq 156$.*

Proof. Let a be a point in G on no 7-point lines. Since $35\cdot 4+15+1$ is 156, we may assume that there are at least sixteen 6-point lines on a. From this, one can prove that for some 6-point line ℓ on a, at least two planes through ℓ each contain three or more additional 6-point lines on a. Note that ℓ is modular in each plane P containing ℓ. (Otherwise, there exists a line in P not intersecting ℓ; contracting by a point b on that line projects a seventh point onto ℓ and so the contraction G/b has a 7-point line through a and at least three other 6- or 7-point lines through a, contradicting Lemma 4.1.)

Since ℓ is modular in P, each point c in $P-\ell$ is on precisely six lines in P. If there is a point c in $P-\ell$ on no 6-point line, then $|P|\leq 6\cdot 4+1$, and hence, $|P|\leq 25$. Therefore we may assume that each point in $P-\ell$ is on a 6-point line. To show that P is in $\mathcal{U}(5)$, it suffices to show that each point d in ℓ is on at most six lines in P. Suppose that a point d in ℓ is on seven lines in P. If there is a 6-point line ℓ' in P not containing d, then ℓ' intersects ℓ at a point not equal to d and the six points on ℓ' determine six lines through d of the form $d\vee e$, where e is a point on ℓ'. Let f be a point on the seventh line through d not equal to d. Then there are seven lines through f, contradicting the fact that ℓ is modular. Thus, we may assume that all 6-point lines in P contain d. Since all points in $P-\ell$ are on 6-point lines, all seven lines through

d are 6-point lines and there are 36 points in P, contradicting Lemma 3.1. We conclude that every point in ℓ is on at most six lines in P.

To finish the proof, observe that if each of the planes containing ℓ has at most 27 points, then $|G|$ is at most $7(27-6)+6$, or 153. On the other hand, if ℓ is in a plane P with at least 28 points, then that plane P is in $\mathcal{U}(5)$. By hypothesis, there is a point x in $G-P$ on no 7-point lines. Hence, by Corollary 2.6, x is on at most 31 lines, and so $|G|$ is at most $31 \cdot 5+1$, or 156. ∎

The second part of the proof deals with the cases not covered by Lemma 6.1.

Lemma 6.2 *Let G be a rank-4 geometry in $\mathcal{U}(6)$. Suppose that there is a plane P in G such that each point x in the complement $G-P$ is on at least one 7-point line of G. Then $|G| \leq 155$.*

Proof. Begin with the special case in which all 7-point lines not in P contain some fixed point a. If a is in P, then by Corollary 4.2, $|G-P| \leq 42$ and hence $|G| \leq 77$. If a is not in P, then each point x of $G-P$ is on a 7-point line through a and hence $|G|$ is at most $7 \cdot 6+(35-7)+1$, or 71. We may therefore assume that each point of G is disjoint from at least one 7-point line. By Corollary 2.3, we may assume that there are no 6-point lines in G. Consider the lines through any point a. By Corollary 4.2, at most seven of those lines have seven points. The rest have five or fewer points. From this, we conclude that $|G|$ is at most $35 \cdot 4+7 \cdot 2+1$, or 155. ∎

With the rank-4 upper bound just obtained, we can complete the proof of the upper bound in Theorem 1.2 using the argument in the proof of Lemma 5.4.

Acknowledgment. The second author was supported by a University of North Texas Faculty Research Grant.

References

[1] J. E. Bonin, Matroids with no $(q+2)$-point-line minors, *Adv. Appl. Math.* **17** (1996), 460–476.

[2] H. H. Crapo, Single-element extensions of matroids, *J. Res. Nat. Bur. Standards Sect. B* **69B** (1965), 55–65.

[3] H. H. Crapo and G.-C. Rota, *On the Foundations of Combinatorial Theory: Combinatorial Geometries*, Preliminary edition, M. I. T. Press, Cambridge MA, 1970.

[4] J. P. S. Kung, "Extremal Matroid Theory," in: *Graph Structure Theory*, N. Robertson and P. D. Seymour, eds., American Mathematical Society, Providence RI, 1993, 21–61.

[5] D. R. Stinson, A short proof of the nonexistence of a pair of orthogonal Latin squares of order 6, *J. Combin. Theory Ser. A* **36** (1984), 373–376.

[6] G. Tarry, Le problème des 36 officiers, *Comptes Rendus de L'Association Française pour L'Avancement de Science* **1** (1900), 122–123; **2** (1901), 170–203.

Joseph E. Bonin
The George Washington University
Washington, DC 20052

Joseph P. S. Kung
University of North Texas
Denton, TX 76203

Umbral Shifts and Symmetric Functions of Schur Type

Miguel A. Méndez

1. Introduction

Though not explicit in the Roman–Rota [12] treatment of umbral calculus, it is apparent that the underlying notion in their approach is the Hopf algebra structure of the polynomials in one variable, given by the augmentation $\langle \epsilon | p(x) \rangle = p(0)$, the comultiplication $E^y p(x) = p(x+y)$, and the antipode $\theta p(x) = p(-x)$. The algebraic dual endowed with the multiplication $\langle L.M | p(x) \rangle = \langle L^x M^y | p(x+y) \rangle$, and with a suitable topology, becomes a topological algebra. This topological algebra, isomorphic to the algebra of formal power series, is called the *umbral algebra.*

The Hopf algebra structure of the symmetric functions $\Lambda(X)$ over an infinite set of variables X is known since the work of L. Geissenger [5]. $\Lambda(X)$ is isomorphic to its graded dual. To make the analogy with the one variable case work, we have to get rid of the notion of graded dual and work instead with the full dual $\Lambda(X)^*$, which is isomorphic to the algebra of symmetric series. The coalgebra automorphisms of $\Lambda(X)$(umbral maps) are classified in terms of families of symmetric functions of binomial type (images of the power symmetric functions p_λ by an umbral map) and the dual notion: the delta systems of functionals. Chen's compositional calculus [3] is criptomorphic to the present calculus by identifying x_n with the power sum $\sum_{r=1}^{\infty} x_r^n$.

In the Roman–Rota paper [12], the classical recursive formula (Rodrigues formula) for the generation of sequences of binomial type is generalized by the use of umbral shifts. The umbral shifts are the adjoint operators of the continuous derivations in the umbral algebra. This procedure was later extended by Roman [11] to polynomials in a finite number of variables. We extend this duality to the present situation by associating to each delta system of linear functionals an infinite system of derivations in the umbral algebra. Their adjoint operators form an infinite system of umbral shifts. We obtain in this way a recursive formula for the connecting constants between two basic families of symmetric functions of binomial type.

We go one step forward in this direction. A family of symmetric functions of Schur type is defined as the set of images by an umbral map of the classical Schur functions. The dual notion of a family of functions of Schur type is that of a family of functionals of Schur type. A family of symmetric functions of Schur type satisfies all the coalgebraic properties of the Schur functions. The functionals of Schur type preserve all the algebraic properties of the Schur functions (in particular, analogs of the Jacobi–Trudi identity). Taking exterior powers of umbral shifts we generalize the recursive formulas for the connecting constants between symmetric functions of binomial type. When the umbral map is also an algebra map, that generalization is the bridge between the Jacobi–Trudi formulas and the quotient of alternant formulas: a general formula for the symmetric functions of Schur type as a quotient of alternants is proved.

Finally, we define a generalized Schur function in the variables $x_1, x_2, \ldots, x_n$ as the quotient of determinants

$$\frac{|t_{\lambda_r+r-1}(x_s)|_{r,s=1}^n}{|x_s^{r-1}|_{r,s=1}^n},$$

where $t_n(x)$ is any polynomial sequence. They form a basis of $\Lambda(x_1, \ldots, x_n)$. By using the exterior power of umbral shifts, we obtain a generalization of the Jacobi–Trudi formula for the dual pseudobasis. The factorial symmetric functions [1, 2, 4] and Macdonald's $6th$ variation in [6] are special cases of the generalized Schur functions. We obtain a necessary and sufficient condition over the polynomials $t_n(x)$, for the existence of the inverse limit of the generalized Schur functions.

2. Preliminaries

A *partition* of a positive integer n is a collection of positive integers whose sum is equal to n. We represent a partition as a nondecreasing sequence $\lambda = \{\lambda_1 \leq \lambda_2 \leq \lambda_3 \ldots \leq \lambda_l\}$. We also use the multiset notation $\lambda = (1^{\alpha_1}2^{\alpha_2}3^{\alpha_3}\ldots)$. It means that exactly α_i parts in the partition are equal to i. We identify the partition λ with the vector $\alpha(\lambda) = \vec{\alpha} = (\alpha_1, \alpha_2, \alpha_3, \ldots)$. The expression $|\lambda|$(resp. $|\vec{\alpha}|$) denotes the sum of the parts of the partition λ(resp. $\vec{\alpha}$), and $l(\lambda)$ (resp. $l(\vec{\alpha})$) the number of parts of λ. We denote by $z_\lambda = z_{\vec{\alpha}}$ the number $z_{\vec{\alpha}} = \prod_i i^{\alpha_i}\alpha_i!$. The symbol $\mathfrak{P}$ will stand for the set of all partitions.

For three partitions $\vec{\alpha}$, $\vec{\beta}$, $\vec{\gamma}$ satisfying $\vec{\beta} + \vec{\gamma} = \vec{\alpha}$, we define the binomial coefficient

$$\binom{\vec{\alpha}}{\vec{\beta}, \vec{\gamma}} = \frac{z_{\vec{\alpha}}}{z_{\vec{\beta}} z_{\vec{\gamma}}} = \frac{\vec{\alpha}!}{\vec{\beta}!\vec{\gamma}!},$$

where $\vec{\alpha}! = \prod_i \alpha_i!$.

Let $X = \{x_1, x_2, x_3, \ldots\}$ be an infinite set of variables. We denote by $\Lambda(X)$ the $\mathbb{C}$-algebra freely generated by the symmetric functions $p_n(X) = \sum_{r=1}^{\infty} x_r^n$. The family of all the functions $p_{\vec{\alpha}}(X) = \prod_i p_i(X)^{\alpha_i}$ is a basis of $\Lambda(X)$. We say that $\mu \sqsubseteq \lambda$ if $\mu_i \leq \lambda_i$ for every i. For $\mu \sqsubseteq \lambda$ we denote by $S_{\lambda/\mu}(X)$ the skew Schur function. We will follow Macdonald's [7] notation for the rest of the classical bases of $\Lambda(X)$. Consider an alphabet $Y = \{y_1, y_2, y_3, \ldots\}$. Denote by $X + Y$ the disjoint union of X and Y, $\{x_1, x_2, x_3, \ldots, y_1, y_2, y_3, \ldots\}$, and by $\Lambda(X, Y)$ the $\mathbb{C}$-algebra of polynomials symmetric in X and in Y. $\Lambda(X, Y)$ is isomorphic to $\Lambda(X) \otimes \Lambda(Y)$. $\Lambda(X)$ is a Hopf algebra with comultiplication $E^Y : \Lambda(X) \to \Lambda(X, Y)$, given by $E^Y R(X) = R(X + Y)$, counit $\epsilon : \Lambda(X) \to \mathbb{C}$ which is evaluation at zero, and antipode $\theta(p_{\vec{\alpha}}(X)) = (-1)^{l(\vec{\alpha})} p_{\vec{\alpha}}(X)$.

For a $\mathbb{C}$-algebra $\mathfrak{A}$, the vector space of linear homomorphisms $\mathrm{Hom}(\Lambda(X), \mathfrak{A})$ is an algebra. If we consider the discrete topology in $\mathfrak{A}$ and in $\Lambda(X)$, all the elements of $\mathrm{Hom}(\Lambda(X), \mathfrak{A})$ are continuous. The discrete topology over $\mathfrak{A}$ induces a topology over $\mathrm{Hom}(\Lambda(X), \mathfrak{A})$, described as follows. A sequence $\{L_j\}_{j \in \mathbb{P}}$, where $\mathbb{P}$ denotes the set of positive integers, converges to a functional L if for any symmetric function $R(X)$ there exists some j_0 such that $\langle L_j | R(X) \rangle = \langle L | R(X) \rangle$ for $j \geq j_0$. The algebraic dual $\Lambda(X)^* = \mathrm{Hom}(\Lambda(X), \mathbb{C})$ is then a topological algebra that we call the *umbral algebra*. In this topology a series $\sum_{\vec{\alpha}} L_{\vec{\alpha}}$ is convergent iff $L_{\vec{\alpha}} \to 0$ when $|\vec{\alpha}| \to \infty$.

3. The umbral algebra and sequences of binomial type

Since $p_n(X + Y) = p_n(X) + p_n(Y)$, it is easy to see that $p_{\vec{\alpha}}(X)$ satisfies the identity

$$p_{\vec{\alpha}}(X + Y) = \sum_{\vec{\beta} + \vec{\gamma} = \vec{\alpha}} \binom{\vec{\alpha}}{\vec{\beta}, \vec{\gamma}} p_{\vec{\beta}}(X) p_{\vec{\gamma}}(Y).$$

Definition 1 A basic family $\{q_{\vec{\alpha}}(X)\}$ of symmetric functions is said to be of binomial type if $q_{\vec{0}}(X) \equiv 1$, $q_{\vec{\alpha}}(0) = 0$, for every $\vec{\alpha} \neq \vec{0}$, and

$$q_{\vec{\alpha}}(X + Y) = \sum_{\vec{\beta} + \vec{\gamma} = \vec{\alpha}} \binom{\vec{\alpha}}{\vec{\beta}, \vec{\gamma}} q_{\vec{\beta}}(X) q_{\vec{\gamma}}(Y).$$

If $\{q_{\vec{\alpha}}(X)\}_{\vec{\alpha} \in \mathfrak{P}}$ is of binomial type, for a finite sequence of alphabets $X^{(1)}, X^{(2)}, \ldots, X^{(n)}$, we have

$$E^{X^{(2)} + \cdots + X^{(n)}} q_{\vec{\alpha}}(X) = \sum_{\vec{\alpha}^{(1)} + \cdots + \vec{\alpha}^{(n)} = \vec{\alpha}} \binom{\vec{\alpha}}{\vec{\alpha}^{(1)}, \vec{\alpha}^{(2)}, \ldots, \vec{\alpha}^{(n)}} \prod_{i=1}^{n} q_{\vec{\alpha}^{(i)}}(X^{(i)}), \quad (1)$$

where $X^{(1)} = X$, and

$$\binom{\vec{\alpha}}{\vec{\alpha}^{(1)}, \vec{\alpha}^{(2)}, \ldots, \vec{\alpha}^{(n)}} = \frac{z_{\vec{\alpha}}}{\prod_{i=1}^{n} z_{\vec{\alpha}^{(i)}}}.$$

Let M and N be two functionals in $\Lambda(X)^*$. The product $M.N$ is explicitly given by $\langle M.N|R(X)\rangle = \langle M^X N^Y|R(X+Y)\rangle$. For a basic family of binomial type $\{q_{\vec{\alpha}}(X)\}$, the product $\langle M.N|q_{\vec{\alpha}}(X)\rangle$ is given by the formula

$$\langle M.N|q_{\vec{\alpha}}(X)\rangle = \sum_{\vec{\beta}+\vec{\gamma}=\vec{\alpha}} \binom{\vec{\alpha}}{\vec{\beta}, \vec{\gamma}} \langle M|q_{\vec{\beta}}(X)\rangle\langle L|q_{\vec{\gamma}}(X)\rangle.$$

Example 1 The family of functionals A_n is defined by the relations $\langle A_n|p_{\vec{\alpha}}(X)\rangle = n\delta(\vec{\alpha}, \vec{e}_n)$. It is clear that $A_n \to 0$ when $n \to \infty$. Define $A^{\vec{\alpha}} = \prod_i A_i^{\alpha_i}$. Using equation (1) it is not difficult to see that $\langle A^{\vec{\alpha}}|p_{\vec{\beta}}(X)\rangle = z_{\vec{\beta}}\delta_{\vec{\alpha},\vec{\beta}}$. The sequence $A^{\vec{\alpha}}$ converges to zero when $|\vec{\alpha}| \to \infty$.

For a multiset of complex numbers $\mathfrak{M}$, the functional $\epsilon_{\mathfrak{M}}$, described by $\langle \epsilon_{\mathfrak{M}}|R(X)\rangle = R(\mathfrak{M})$, substitutes a finite number of the variables of X in $R(X)$ by the elements of $\mathfrak{M}$ and sets the rest of the variables to zero. The functional $\epsilon_{\mathfrak{M}}^{(n)}$ is defined by $\langle \epsilon_{\mathfrak{M}}^{(n)}|R(X)\rangle = R^{(n)}(\mathfrak{M})$, where $R^{(n)}(X)$ is the homogeneous component of degree n of $R(X)$.

Proposition 1 *Let M be an element of $\Lambda(X)^*$ and let $R(X)$ be any symmetric function. Then we have the expansions*

$$M = \sum_{\vec{\alpha}\in\mathfrak{P}} \langle M|p_{\vec{\alpha}}(X)\rangle \frac{A^{\vec{\alpha}}}{z_{\vec{\alpha}}}. \tag{2}$$

and

$$R(X) = \sum_{\vec{\alpha}\in\mathfrak{P}} \langle A^{\vec{\alpha}}|R(X)\rangle \frac{p_{\vec{\alpha}}(X)}{z_{\vec{\alpha}}}. \tag{3}$$

Proof. Since $A^{\vec{\alpha}} \to 0$, the series in equation (2) converges to some functional L. We have $\langle L|p_{\vec{\beta}}\rangle = \sum_{\vec{\alpha}} \langle M|p_{\vec{\alpha}}(X)\rangle\delta_{\vec{\alpha},\vec{\beta}} = \langle M|p_{\vec{\beta}}\rangle$. Identity (3) is proved in a similar way. ∎

The indicator of a functional M is defined as the symmetric series

$$\mathfrak{I}_M(X) = \sum_{\vec{\alpha}} \langle M|p_{\vec{\alpha}}(X)\rangle \frac{p_{\vec{\alpha}}(X)}{z_{\vec{\alpha}}}$$

For example, the indicators of $\epsilon_{\mathfrak{M}}^{(n)}$ and $\epsilon_{\mathfrak{M}}$ are respectively

$$\mathfrak{I}_{\epsilon^{(n)}_{\mathfrak{M}}}(X) = \sum_{|\vec{\alpha}|=n} p_{\vec{\alpha}}(\mathfrak{M}) \frac{p_{\vec{\alpha}}(X)}{z_{\vec{\alpha}}} = h_n(\mathfrak{M}X),$$

and

$$\mathfrak{I}_{\epsilon_{\mathfrak{M}}}(X) = \sum_{\vec{\alpha}} p_{\vec{\alpha}}(\mathfrak{M}) \frac{p_{\vec{\alpha}}(X)}{z_{\vec{\alpha}}} = \sum_{n \geq 0} h_n(\mathfrak{M}X) := h(\mathfrak{M}X),$$

where $h_n(X)$ is the n-th homogeneous symmetric function.

The umbral algebra is isomorphic to the algebra of symmetric series $\Lambda((X))$ via the isomorphism $M \mapsto \mathfrak{I}_M(X)$. For a symmetric series $R(X)$, we denote by $R(A)$ the corresponding functional in $\Lambda(X)^*$. For example $h_n(A)$ denotes the functional $\epsilon_1^{(n)}$, and $h(X)$ the functional ϵ_1.

Let V be an arbitrary linear space, and V^* its linear dual. Using a procedure analogous to that used to define a topology on $\Lambda(X)^*$ we define a topology on V^*. We have the following proposition.

Proposition 2 *Let $U : V \to \Lambda(X)$ be a linear operator. The adjoint $T = U^* : \Lambda(X)^* \to V^*$ is a continuous linear operator. Conversely, if $T : \Lambda(X) \to V^*$ is a continuous linear operator, then there exists a linear operator $U : V \to \Lambda(X)$ such that $T = U^*$.*

Proof. The proof of the first part of the proposition is straightforward. Assume that T is continuous. Choose a basis $\{v_i\}_{i \in I}$ of V. Since $L_{\vec{\alpha}} = T(A^{\vec{\alpha}}) \to 0$, the symmetric function

$$q_i(X) = \sum_{\vec{\beta}} \langle TA^{\vec{\beta}} | v_i \rangle \frac{p_{\vec{\beta}}(X)}{z_{\vec{\beta}}(X)}$$

is well defined. Clearly $\langle TA^{\vec{\beta}} | v_i \rangle = \langle A^{\vec{\beta}} | q_i(X) \rangle$ for every $\vec{\beta}$ and i. Define the operator $U : V \to \Lambda(X)$ by $Uv_i = q_i(X)$. Then we have $\langle U^* A^{\vec{\beta}} | v_i \rangle = \langle A^{\vec{\beta}} | q_i(X) \rangle$. Since T is continuous we have $T = U^*$. ∎

The rest of the propositions of this section are consequences of Proposition 2. Detailed proofs of them can be seen in [9].

Corollary 1 *A linear operator $T : \Lambda(X)^* \to V^*$ is continuous if and only if $L_{\vec{\alpha}} = TA^{\vec{\alpha}}$ converges to zero when $|\vec{\alpha}| \to \infty$.*

Definition 2 A family of functionals $\{L_{\vec{\alpha}}\}_{\vec{\alpha} \in \mathfrak{P}}$ in $\Lambda(X)^*$ is called a pseudobasis if

(1) $L_{\vec{\alpha}} \to 0$ when $|\vec{\alpha}| \to \infty$, and

(2) Every functional M can be uniquely expanded in a series of the form $M = \sum_{\vec{\alpha}} a_{\vec{\alpha}} L_{\vec{\alpha}}$.

For every basis $\{R_{\vec{\alpha}}(X)\}_{\vec{\alpha}\in\mathfrak{P}}$ the family of functionals $\{L_{\vec{\alpha}}\}_{\vec{\alpha}\in\mathfrak{P}}$ defined by $\langle L_{\vec{\alpha}}, R_{\vec{\beta}}\rangle = \delta_{\vec{\alpha},\vec{\beta}}$, $\vec{\alpha}, \vec{\beta} \in \mathfrak{P}$ is easily seen to be a pseudobasis. The converse is also true.

Proposition 3 *Given a pseudobasis $\{L_{\vec{\alpha}}\}_{\vec{\alpha}\in\mathfrak{P}}$ there exists a sequence of symmetric functions $\{R_{\vec{\alpha}}(X)\}_{\vec{\alpha}\in\mathfrak{P}}$ satisfying $\langle L_{\vec{\alpha}}, R_{\vec{\beta}}\rangle = \delta_{\vec{\alpha},\vec{\beta}}$. Moreover, we have the expansions*

$$M = \sum_{\vec{\alpha}} \langle M | R_{\vec{\alpha}}\rangle L_{\vec{\alpha}}, \qquad M \in \Lambda(X)^*, \tag{4}$$

$$R(X) = \sum_{\vec{\alpha}} \langle L_{\vec{\alpha}} | R(X)\rangle R_{\vec{\alpha}}, \qquad R(X) \in \Lambda(X), \tag{5}$$

and hence $\{R_{\vec{\alpha}}\}_{\vec{\alpha}\in\mathfrak{P}}$ is a basis.

Proposition 4 (1) *Let $U : \Lambda(X) \to \Lambda(X)$ be a linear map. Then U is a coalgebra automorphism if and only if $q_{\vec{\alpha}}(X) = Up_{\vec{\alpha}}(X)$ is a basic family of binomial type.*

(2) *If $U : \Lambda(X) \to \Lambda(X)$ is a coalgebra automorphism then its adjoint $U^* : \Lambda(X)^* \to \Lambda(X)^*$ is a continuous algebra automorphism. Conversely, if $T : \Lambda(X)^* \to \Lambda(X)^*$ is a continuous algebra automorphism, then there exists a coalgebra automorphism $U : \Lambda(X) \to \Lambda(X)$ such that $U^* = T$.*

A coalgebra automorphism U is called an *umbral map*. Let U be an umbral map. Any system $\vec{L}$ of the form $\vec{L} = (U^* A_n)_{n=1}^{\infty}$ is called a *delta system*. Since U^* is continuous, $L_n = U^* A_n$ converges to zero.

Definition 3 Let U be an umbral map. Define the delta systems $\vec{M} = (U^* A_n)_{n=1}^{\infty}$, and $\vec{L} = ((U^{-1})^* A_n)_{n=1}^{\infty}$, where U^{-1} is the inverse of U. The binomial family $q_{\vec{\alpha}}(X) = Up_{\vec{\alpha}}(X)$ is called the *conjugate family* of $\vec{M}$ and the *associated* family of $\vec{L}$. It is easy to see that $\langle L_n | q_{\vec{\alpha}}(X)\rangle = n\delta(\vec{\alpha}, \vec{e}_n)$.

For a delta system $\vec{L}$ we denote by $\vec{L}^{\vec{\alpha}}$ the product $\prod_i L_i^{\alpha_i}$. By example (1), and since $(U^{-1})^* A^{\vec{\alpha}} = L^{\vec{\alpha}}$, we obtain $\langle L^{\vec{\alpha}} | q_{\vec{\beta}}(X)\rangle = z_{\vec{\alpha}} \delta_{\vec{\alpha},\vec{\beta}}$. Therefore $L_{\vec{\alpha}}$ is the dual pseudobasis of $\{q_{\vec{\alpha}}(X)/z_{\vec{\alpha}}\}$, and we have:

Proposition 5 *Let $q_{\vec{\alpha}}(X)$ and $\vec{L}$ be as above. Any functional M in $\Lambda(X)^*$ can be expanded as a convergent series*

$$M = \sum_{\vec{\alpha}} \langle M | q_{\vec{\alpha}} \rangle \frac{\vec{L}^{\vec{\alpha}}}{z_{\vec{\alpha}}}. \tag{6}$$

Dually, any symmetric function $R(X)$ has the following expansion

$$R(X) = \sum_{\vec{\alpha}} \langle L^{\vec{\alpha}} | R(X) \rangle \frac{q_{\vec{\alpha}}(X)}{z_{\vec{\alpha}}}. \tag{7}$$

Example 2 Let $m_\lambda(X) = m_{\alpha(\lambda)}(X) = m_{\vec{\alpha}}(X)$ denote the elements of the basic family of monomial symmetric functions. Assume that $\alpha_i = 0$ for $i > m$. Then $m_{\vec{\alpha}}(X)$ can be written in the form

$$m_{\vec{\alpha}}(X) = \sum_{S_1, S_2, \ldots, S_m} \prod_{i=1}^{m} \prod_{r \in S_i} x_r^i, \tag{8}$$

where the sum is over all the m-tuples $(S_1, S_2, \ldots, S_m)$ of pairwise disjoint sets of positive integers satisfying $|S_i| = \alpha_i$. From equation (8) we easily get that

$$m_{\vec{\alpha}}(X+Y) = \sum_{\vec{\beta}+\vec{\gamma}=\vec{\alpha}} m_{\vec{\beta}}(X) m_{\vec{\gamma}}(Y).$$

Then, the family $\{\hat{m}_{\vec{\alpha}}(X)\} = \{\vec{\alpha}! m_{\vec{\alpha}}(X)\}$ is of binomial type. Since $\langle \epsilon_1^{(n)} | m_{\vec{\alpha}}(X) \rangle = m_{\vec{\alpha}}(1) = \delta(\vec{e}_n, \vec{\alpha})$, $\{\hat{m}_{\vec{\alpha}}(X)\}$ is the associated family of the delta system $\vec{K} = (n\epsilon_1^{(n)})_{n=1}^{\infty}$.

Given a delta system $\vec{L}$ we denote by $M(\vec{L})$ the functional obtained by substituting $\vec{A}^{\vec{\alpha}}$ by $\vec{L}^{\vec{\alpha}}$ in the expansion of M in terms of $\vec{A}$. If $\vec{M}$ is another delta system, $\vec{M}(\vec{L})$ will denote the system of functionals $(M_n(\vec{L}))_{n=1}^{\infty}$. A delta system $\vec{L}$ is said to be *invertible* if there exists another delta system $\vec{M}$ such that $\vec{L}(\vec{M}) = \vec{M}(\vec{L}) = \vec{A}$. The inverse of $\vec{L}$ is denoted $\vec{L}^{\langle -1 \rangle}$.

Note that $\vec{M}(\vec{L})$ is also a delta system. To see that, denote by $U_{\vec{L}}$ the umbral map satisfying $U^* A_n = L_n$. It is easy to see that $(U_{\vec{M}} \circ U_{\vec{L}})^* A_n = M_n(\vec{L})$. Our claim follows, since the composition of umbral maps is an umbral map. Furthermore we have

$$U_{\vec{M}} \circ U_{\vec{L}} = U_{\vec{M}(\vec{L})}. \tag{9}$$

From equation (9), every delta system is invertible, and conversely. We have $U_{\vec{L}}^{-1} = U_{\vec{L}^{\langle -1 \rangle}}$.

4. Hammond derivatives and umbral shifts

Definition 4 Consider a delta system of functionals $\vec{L}$. The Hammond derivative δ_{L_n} is the unique continuous derivation of the umbral algebra satisfying $\delta_{L_n} L_m = n\delta(m,n)\epsilon$. Clearly, $\delta_{L_n} = n\frac{\partial}{\partial L_n}$. For any system $\vec{T}$, such that $\lim_{n\to\infty} T_n = 0$, we define the infinite matrix

$$\delta_{\vec{L}}\vec{T} := \left(\frac{\delta_{L_i} T_j}{j}\right)_{i,j=1}^{\infty}.$$

Since δ_{L_i} is continuous, $lim_{j\to\infty}\delta_{L_i}T_j = 0$ for every i. We call $\delta_{\vec{A}}\vec{T}$ the Jacobian matrix of $\vec{T}$.

Denote by $\mathcal{R}_f$ the set of infinite matrices of the form $B = (b_{i,j})_{i,j=1}^{\infty}$, with complex entries, such that in every row there are only a finite number of non-zero entries. With the usual operations $\mathcal{R}_f$ is a $\mathbb{C}$-algebra. The elements of the topological algebra $\mathrm{Hom}(\Lambda(X), \mathcal{R}_f)$ can be identified with infinite matrices $\mathcal{M} = (M_{i,j})_{i,j=1}^{\infty}$, where the $M_{i,j}$ are in $\Lambda(X)^*$, satisfying for every i, $\lim_{j\to\infty} M_{i,j} = 0$. We say that $\mathcal{M}$ is invertible if it is invertible as an element of $\mathrm{Hom}(\Lambda(X), \mathcal{R}_f)$.

Proposition 6 (Chain rule) *Let W be any element of the umbral algebra. Let $\vec{L}$ and $\vec{M}$ be two delta systems. Then*

$$\delta_{L_n} W = \sum_{j=1}^{\infty} \frac{\delta_{L_n} M_j}{j} \delta_{M_j} W. \tag{10}$$

Proof. Since $\vec{M}$ is a delta system, there exists a functional R such that $W = R(\vec{M})$. It is easy to verify the chain rule

$$\frac{\partial W}{\partial L_n} = \sum_{j=1}^{\infty} \frac{\partial R}{\partial A_j}(\vec{M}) \frac{\partial M_j}{\partial L_n}.$$

Multiplying both sides by n we get

$$\delta_{L_n} W = \sum_{j=1}^{\infty} \frac{\delta_{L_n} M_j}{j} (\delta_{A_j} R)(\vec{M}).$$

The result follows since $(\delta_{A_j} R)(\vec{M}) = \delta_{M_j} W$. ∎

In operator notation (10) may be written as

$$\delta_{L_n} = \sum_{j=1}^{\infty} \frac{1}{j} \mu(\delta_{L_n} M_j).\delta_{M_j}, \tag{11}$$

where $\mu(L) : \Lambda(X)^* \to \Lambda(X)^*$ represents the operator of multiplication by L, this is $\mu(L)M = L.M$. Writing it in matrix notation we obtain

$$\delta_{\vec{L}} = \mu(\delta_{\vec{L}}\vec{M}).\delta_{\vec{M}}, \tag{12}$$

where $\delta_{\vec{L}}$ symbolizes the vector of operators $(\delta_{L_1}, \delta_{L_2}, \delta_{L_3}, \ldots)^t$ and $\mu(\delta_{\vec{L}}\vec{M})$ the matrix $\left(\mu(\frac{\delta_{L_i}M_j}{j})\right)_{i,j=1}^{\infty}$.

Proposition 7 *Let $\vec{M}$ and $\vec{L}$ be delta systems. Then the matrix $\delta_{\vec{L}}\vec{M}$ is invertible, and we have*

$$(\delta_{\vec{L}}\vec{M})^{-1} = \delta_{\vec{M}}\vec{L}. \tag{13}$$

Proof. By equation (10) we have

$$k\delta(n,k)\epsilon = \delta_{L_n}L_k = \sum_{j=1}^{\infty} \frac{\delta_{L_n}M_j}{j}.\delta_{M_j}L_k.$$

Therefore

$$\sum_{j=1}^{\infty} \frac{\delta_{L_n}M_j}{j}.\frac{\delta_{M_j}L_k}{k} = \delta(n,k)\epsilon.$$

In a similar way we prove that $(\delta_{\vec{M}}\vec{L}).(\delta_{\vec{L}}\vec{M}) = Id$. ∎

From a classical result about Hopf algebras, the adjoint $\mu(L)^*$ of $\mu(L)$ is given by the formula $\mu(L)^*R(X) = \langle L^Y | R(X+Y) \rangle$.

Denote by D_n the operator $n\frac{\partial}{\partial p_n}$ and by $D^{\vec{\alpha}}$ the product $\prod_i D_i^{\alpha_i}$. We easily get that $\mu(A^{\vec{\alpha}}) = D^{\vec{\alpha}}$. The operator $\mu(L)^*$ can be written as the formal sum

$$\mu(L)^* = \sum_{\vec{\alpha}} \langle L | p_{\vec{\alpha}}(X) \rangle \frac{D^{\vec{\alpha}}}{z_{\vec{\alpha}}} = L(D).$$

Definition 5 Let $\{q_{\vec{\alpha}}(X)\}$ be the basic family of symmetric functions of binomial type associated with $\vec{L}$. The linear operator $\vartheta_{L_n} : \Lambda(X) \to \Lambda(X)$, defined by $\vartheta_{L_n} q_{\vec{\alpha}}(X) = q_{\vec{\alpha}+\vec{e}_n}(X)$, is called an umbral shift.

It is easy to verify that $\vartheta^*_{L_n} = \delta_{L_n}$. Hence, taking adjoints in equation (12) we get

$$\vartheta_{\vec{L}} = \vartheta_{\vec{M}}(\mu(\delta_{\vec{L}}\vec{M}))^{*t}, \tag{14}$$

where $\vartheta_{\vec{L}}$ symbolizes the vector of operators $\vartheta_{\vec{L}} = (\vartheta_{L_1}, \vartheta_{L_2}, \vartheta_{L_3}, \ldots)$, and $(\mu(\delta_{\vec{L}}\vec{M}))^{*t}$ is the infinite matrix of operators $\left(\frac{\delta_{L_i}M_j(D)}{j}\right)_{j,i=1}^{\infty}$. Then, we obtain

Proposition 8 *Let $\{q_{\vec{\alpha}}\}$ be the binomial family associated with $\vec{L}$. We have the following recursive formula*

$$q_{\vec{\alpha}+\vec{e}_n}(X) = \sum_j \vartheta_{M_j} \frac{\delta_{L_i} M_j(D)}{j} q_{\vec{\alpha}}(X). \tag{15}$$

Example 3 Let $\vec{K}$ be as in example (2). We have $\delta_{A_i} K_j = \delta_{A_i} j\epsilon_1^{(j)} = j\epsilon_1^{(j-i)}$, where $\epsilon_1^{(j-i)}$ is assumed to be zero when $j < i$. The Jacobian matrix $\delta_{\vec{A}}\vec{K}$ is equal to $(\epsilon_1^{(j-i)})_{i,j=1}^{\infty} = (h_{j-i}(A))_{i,j=1}^{\infty}$. By Newton's identities, $(\delta_{\vec{A}}\vec{K})^{-1} = ((-1)^{j-i}e_{j-i}(A))_{i,j=1}^{\infty}$, where the e_j are the elementary symmetric functions. Therefore, we obtain

$$\vartheta_{K_i} \hat{m}_{\vec{\alpha}}(X) = \sum_{j \geq i} \underline{p_j(X)}(-1)^{j-i} e_{j-i}(D) \hat{m}_{\vec{\alpha}}(X). \tag{16}$$

Equivalently

$$(\alpha_i + 1) m_{\vec{\alpha}+\vec{e}_i}(X) = \sum_{j \geq i} \underline{p_j(X)}(-1)^{j-i} e_{j-i}(D) m_{\vec{\alpha}}(X). \tag{17}$$

For example, to expand $m_{(n,1)}(X)$ in terms of the $p_{\vec{\alpha}}$ we have $m_{(n,1)}(X) = \vartheta_{K_2} m_{(n)}(X) = \vartheta_{K_2} e_n(X)$. Then

$$m_{(n,1)}(X) = \sum_{j \geq 2} (-1)^{j-2} \underline{p_j(X)} e_{j-2}(D) e_n(X) = \sum_{j=2}^{n+2} (-1)^j p_j(X) e_{n-j+2}(X)$$

$$= \sum_{j=2}^{n+2} \sum_{|\vec{\alpha}|=n-j+2} (-1)^j (-1)^{\alpha_2+\alpha_4+\ldots} \frac{p_{\vec{\alpha}+\vec{e}_j}(X)}{z_{\vec{\alpha}}}.$$

Denote by $\mathbb{P}^{[n]}$ the set of n-subsets of the positive integers. For $I \in \mathbb{P}^{[n]}$, let $\|I\|$ denote the sum $\sum_{i \in I} i$. For I and J in $\mathbb{P}^{[n]}$ we say that $I \preceq J$ if $|I| < |J|$ or if $|I| = |J|$ and I precedes J in the reverse lexicographic order. $\mathbb{P}^{[n]} = \{I_1 \prec I_2 \prec I_3 \ldots\}$ is a totally ordered set. For example:

$$\mathbb{P}^{[2]} = \{\{1,2\} \prec \{1,3\} \prec \{2,3\} \prec \{1,4\} \prec \ldots\}$$

Let $\mathcal{M} = (M_{i,j})_{i,j=1}^{\infty}$ be an infinite matrix in $\mathrm{Hom}(\Lambda(X), \mathcal{R}_f)$. For I and J in $\mathbb{P}^{[n]}$, denote by $\mathcal{M}(I,J)$ the determinant $|M_{i,j}|_{i \in I, j \in J}$. We denote by $\mathcal{M}^{[n]}$ the compound matrix of the minors $(\mathcal{M}(I,J))_{I,J \in \mathbb{P}^{[n]}}$. $\mathcal{M}^{[n]}$ can be thought of as an element of $\mathrm{Hom}(\Lambda(X), \mathcal{R}_f)$ by defining the entry $\mathcal{M}_{r,s}^{[n]}$ to be the minor $\mathcal{M}(I_r, I_s)$, I_r and I_s being the r-th and s-th elements of $\mathbb{P}^{[n]}$ respectively. It is not difficult to verify that $\mathcal{M}_{r,s} \to 0$ when $s \to \infty$.

For a set $I = \{i_1, i_2, \ldots, i_n\}$ in $\mathbb{P}^{[n]}$ let $\delta^I_{\vec{L}}$ be the continuous map

$$\delta^I_{\vec{L}} : \Lambda(X^{(1)}, X^{(2)}, \ldots, X^{(n)})^* \to \Lambda(X)^*$$

that assigns to a multivariate decomposed functional of the form $\prod_{s=1}^n T_s^{X^{(s)}}$, the determinant $|\delta_{L_{i_r}} T_s|_{r,s=1}^n$ in $\Lambda(X)^*$.

Proposition 9 *The transformation $\vartheta^I_{\vec{L}} : \Lambda(X) \to \Lambda(X^{(1)}, X^{(2)}, \ldots, X^{(n)})$ whose adjoint is $\delta^I_{\vec{L}}$ is given by*

$$\vartheta^I_{\vec{L}} R(X) = \left|\vartheta^{X^{(s)}}_{L_{i_r}}\right|^n_{r,s=1} R(X^{(1)} + X^{(2)} + \cdots + X^{(n)}). \tag{18}$$

Proof. The adjoint of $E^{X^{(2)}+\cdots+X^{(n)}}$ is the n-multiplication $\mu^{(n-1)}$: $\Lambda(X^{(1)}, X^{(2)}, \ldots, X^{(n)})^* \to \Lambda(X)^*$, where $\mu^{(n-1)} \prod_{s=1}^n T_s^{X^{(s)}} = \prod_{s=1}^n T_s$. Using this notation we have

$$\vartheta^I_{\vec{L}} = \sum_{\sigma \in \mathfrak{S}_n} \operatorname{sign}(\sigma) \prod_{r=1}^n \vartheta^{X^{(\sigma(r))}}_{L_{i_r}} \circ E^{X^{(2)}+X^{(3)}+\cdots+X^{(n)}}$$

Taking adjoints we get

$$(\vartheta^I_{\vec{L}})^* = \mu^{(n-1)} \circ \sum_{\sigma \in \mathfrak{S}_n} \operatorname{sign}(\sigma) \prod_{r=1}^n \delta^{X^{(\sigma(r))}}_{L_{i_r}} = \delta^I_{\vec{L}}.$$

■

We denote by $\vartheta^{(n)}_{\vec{L}}$ the infinite vector $(\vartheta^{I_1}_{\vec{L}}, \vartheta^{I_2}_{\vec{L}}, \ldots)$ and by $\delta^{(n)}_{\vec{L}}$, the vector $(\delta^{I_1}_{\vec{L}}, \delta^{I_2}_{\vec{L}}, \ldots)^t$.

Theorem 1 *Let $\vec{M}$ and $\vec{N}$ be two delta systems. We have the identities*

$$\delta^{(n)}_{\vec{L}} = \mu((\delta_{\vec{L}} \vec{M})^{(n)}) \delta^{(n)}_{\vec{M}}, \tag{19}$$

and

$$\vartheta^{(n)}_{\vec{L}} = \vartheta^{(n)}_{\vec{M}} (\delta_{\vec{L}} \vec{M})^{(n)t}(D). \tag{20}$$

Proof. Equation (20) is the dual form of (19). To prove (19) we only have to show that for every multivariate functional of the form $T_1^{X^{(1)}} T_2^{X^{(2)}} \cdots T_n^{X^{(n)}}$ and any set $I \in \mathbb{P}^{[n]}$,

$$\delta^I_{\vec{L}} \prod_{s=1}^n T_s^{X^{(s)}} = \sum_{J \in \mathbb{P}^{[n]}} (\delta_{\vec{L}} \vec{M})^{[n]}(I, J) \delta^J_{\vec{M}} \prod_{s=1}^n T_n^{X^{(s)}}. \tag{21}$$

Consider the system $\vec{T} = (T_1, T_2, \ldots, T_n, 0, 0, \ldots)$. By equation (12) we have

$$\delta_{\vec{L}}\vec{T} = (\delta_{\vec{L}}\vec{M})(\delta_{\vec{M}}\vec{T}).$$

Taking compound matrices and using the Cauchy–Binet identity we get

$$(\delta_{\vec{L}}\vec{T})^{(n)} = (\delta_{\vec{L}}\vec{M})^{(n)}(\delta_{\vec{M}}\vec{T})^{(n)}. \tag{22}$$

We obtain equation (21) by computing the entry $(I, \{1, 2, \ldots, n\})$ in both sides of equation (22). ∎

5. Symmetric functions of Schur type

Definition 6 A family of symmetric functions of the form $\{t_{\lambda/\mu}(X)\}_{\mu\sqsubseteq\lambda}$ is called of Schur type if they are the images of the skew Schur functions $S_{\lambda/\mu}(X)$ by an umbral operator. Similarly, a family of functionals $\{L_{\lambda/\mu}\}_{\lambda\sqsubseteq\mu}$ is called of Schur type if they are the images of the Schur functionals $S_{\lambda/\mu}(A)$ by an algebra automorphism of $\Lambda(X)^*$.

If U is an umbral operator, we denote by $S^U_{\lambda/\mu}(X)$ the Schur type symmetric function $US_{\lambda/\mu}(X)$. Similarly, we denote by $H^U_{\lambda/\mu}$ the Schur type functional $(U^{-1})^* S_{\lambda/\mu}(A)$. From now on, L^U will denote the functional $(U^{-1})^*L$.

It is clear that $\langle H^U_\lambda | S^U_\tau(X)\rangle = \delta_{\lambda,\mu}$. In other words, $\{H^U_\lambda\}_\lambda$ is the dual pseudobasis of $\{S^U_\lambda(X)\}_\lambda$.

Any family $\{S^U_{\lambda/\mu}(X)\}$ satisfies the coalgebra properties of the Schur functions,

$$S^U_{\lambda/\mu}(0) = 0, \text{ when } \mu \neq \lambda,$$

$$S^U_{\lambda/\mu}(X+Y) = \sum_{\mu\sqsubseteq\tau\sqsubseteq\lambda} S^U_{\tau/\mu}(X) S^U_{\lambda/\tau}(Y).$$

Given a set $I = \{i_1, i_2, \ldots, i_n\}$ in $\mathbb{P}^{[n]}$, the nonzero elements of the sequence $i_1 - 1, i_2 - 2, \ldots, i_n - n$ form a partition of length less than or equal to n. We denote it by $\pi_n(I)$.

Denote by $\mathfrak{P}_n$ the set of all the partitions whose length is less than or equal to n. A partition λ in $\mathfrak{P}_n$ can be represented as a vector of integers $(\lambda_1, \lambda_2, \ldots, \lambda_n)$, by placing zeroes in the $n-k$ first positions when $l(\lambda) = k < n$. We denote by $\varsigma_n(\lambda)$ the set $\{\lambda_1 + 1, \lambda_2 + 2, \ldots, \lambda_n + n\}$ in $\mathbb{P}^{[n]}$, ς_n is a bijection from $\mathfrak{P}_n$ to $\mathbb{P}^{[n]}$ with inverse π_n.

Given two partitions $\mu \sqsubseteq \lambda$ in $\mathfrak{P}_n$, the entry $(\varsigma_n(\mu), \varsigma_n(\lambda))$ of the matrix $(\delta_{\vec{A}}\vec{K})^{(n)}$ is the determinant $|\epsilon_1^{(\lambda_j+j-\mu_i-i)}|_{i,j=1}^n = |h_{\lambda_j-\mu_i+j-i}(A)|_{i,j=1}^n$. By the Jacoby-Trudi identity, $S_{\lambda/\mu}(A) = (\delta_{\vec{A}}\vec{K})^{(n)}(\varsigma_n(\mu), \varsigma_n(\lambda))$. Since the algebra properties are preserved by the operator $(U^{-1})^*$, $H^U_{\lambda/\mu}$ also satisfies Jacobi–Trudi:

$$H^U_{\lambda/\mu} = (\delta_{\vec{A}^U}\vec{K}^U)^{(n)}(\varsigma_n(\mu), \varsigma_n(\lambda)) = |\epsilon_1^{(\lambda_j-\mu_i+j-i),U}|_{i,j=1}^n.$$

Assume now that U is a bialgebra map. Since $Up_n(X+Y) = Up_n(X) + Up_n(Y)$ we have that $q_n(X) = Up_n(X)$ is of the form

$$q_n(X) = \sum_{k>0} c_{n,k}p_k(X) = \sum_{s=1}^{\infty}\sum_{k>0} c_{n,k}x_s = \sum_{r=1}^{\infty} q_n(x_r),$$

where $\{q_n(x)\}$ is a basis of $\mathbb{C}[x]$. Since U is an algebra map, $Up_{\vec{\alpha}}(X) = q_{\vec{\alpha}}(X) = \prod_n q_n^{\alpha_n}(X)$. Our goal is to generalize the quotient of alternants formula for the Schur function $S_\lambda(x_1, x_2, \ldots, x_n)$ to $S^U_\lambda(x_1, x_2, \ldots, x_n)$, U being a bialgebra map.

Theorem 2 *Let U be a bialgebra map as above. For any sequence of alphabets $X^{(1)}, X^{(2)}, \ldots, X^{(n)}$, $S^U_\lambda(X^{(1)} + \ldots + X^{(n)})$ satisfies the identity*

$$S^U_\lambda(X^{(1)} + \ldots + X^{(n)})|q_r(X^{(s)})|_{r,s=1}^n = \sum_{\mu\in\mathfrak{P}_n}\sum |\vartheta_{K^U_{\mu_r+r}} S^U_{\lambda^{(s)}/\lambda^{(s-1)}}(X^{(s)})|_{r,s=1}^n, \tag{23}$$

where the second sum is over the sequences of partitions

$$\mu = \lambda^{(0)} \sqsubseteq \lambda^{(1)} \ldots \sqsubseteq \lambda^{(n)} = \lambda.$$

Proof.
By equation (20), for $I \in \mathbb{P}^{[n]}$ we have $\vartheta^I_{\vec{A}^U} = \sum_{J\in\mathbb{P}^{[n]}} \vartheta^J_{\vec{K}^U}(\delta_{\vec{A}^U}K^U)^{(n)}(I,J)(D)$. Since $(\delta_{\vec{A}^U}K^U)^{(n)}(I,J)(D) = H_{\mu/\tau}(D)$, $\mu = \pi_n(J)$, $\tau = \pi_n(I)$, we have

$$\vartheta^{\varsigma_n(\tau)}_{\vec{A}^U} = \sum_{\mu\in\mathfrak{P}_n} \vartheta^{\varsigma_n(\mu)}_{\vec{K}^U} H^U_{\mu/\tau}(D). \tag{24}$$

Applying both sides of equation (24), with $\tau = 0$, to $S^U_\lambda(X)$ we obtain

$$S^U_\lambda(X^{(1)} + X^{(2)} + \ldots + X^{(n)})\left|q_r(X^{(s)})\right|^n_{r,s=1} =$$

$$\sum_{\mu\in\mathfrak{P}_n}\left|\vartheta^{X^{(s)}}_{K^U_{\mu_r+r}}\right|^n_{r,s=1} S^U_{\lambda/\mu}(X^{(1)} + \ldots + X^{(n)}). \tag{25}$$

The theorem follows, since the $S^U_{\lambda/\mu}(X)$ have the same coalgebraic properties as the $S_{\lambda/\mu}(X)$. ∎

Corollary 2 *If $q_n(x)$ is a polynomial sequence* $(\deg(q_n(x)) = n)$, *then*

$$S_\lambda^U(x_1, x_2, \ldots, x_n) = \sum_{\mu \in \mathfrak{P}_n} \frac{1}{|q_r(x_s)|_{r,s=1}^n} \sum \left|(\vartheta_{K^U_{\mu_r+r}} S^U_{\lambda^{(s)}/\lambda^{(s-1)}})(x_s)\right|_{r,s=1}^n, \quad (26)$$

where the second sum ranges over all chains of partitions

$$\mu = \lambda^{(0)} \sqsubseteq \lambda^{(1)} \ldots \sqsubseteq \lambda^{(n)} = \lambda.$$

Proof. Taking one element alphabets $X^{(s)} = \{x_s\}$ in equation (23) we obtain

$$S_\lambda^U(x_1, x_2, \ldots, x_n)\, |q_r(x_s)|_{r,s=1}^n = \sum_{\mu \in \mathfrak{P}_n} \sum \left|(\vartheta_{K^U_{\mu_r+r}} S^U_{\lambda^{(s)}/\lambda^{(s-1)}})(x_s)\right|_{r,s=1}^n.$$

If $q_n(x)$ is a polynomial sequence, the determinant $|q_r(x_s)|_{r,s=1}^n$ is of the form $c_n |x_r^s|_{r,s=1}^n$, for some sequence of complex numbers c_n. Since the Vandermonde determinant divides any alternant polynomial, it divides $(\vartheta_{K^U}^{\varsigma_n(\mu)} S^U_{\lambda/\mu})(x_1, x_2, \ldots, x_n)$ for every pair of partitions λ, μ. Then, we obtain the corollary. ■

Observe that, if the umbral shifts $\vartheta_{K_n^U}$ have the following property,

$$(\vartheta_{K_n^U} R)(x) = 0, \text{ if } R(0) = 0, \quad (27)$$

then, $(\vartheta_{K^U_{\mu_r+r}} S^U_{\lambda^{(s)}/\lambda^{(s-1)}})(x) = 0$ unless $\lambda^{(s)} = \lambda^{(s-1)}$. Hence, equation (26) would reduce to

$$S_\lambda^U(x_1, x_2, \ldots, x_n) = \frac{1}{|q_s(x_r)|_{r,s=1}^n} \left|(U m_{\vec{e}_{\lambda_r+r}})(x_s)\right|_{r,s=1}^n.$$

Equation (27) is satisfied when U is the identity (and we obtain the classical formula for the Schur functions). We shall construct in the next section a general class of umbral operators U that satisfy (27).

6. Generalized Schur functions

Let $q_0 \equiv 1$ and $q_n(x) = x t_{n-1}(x)$, where $\{t_n(x) : n \geq 0\}$ is a polynomial sequence $(\deg(t_n(x)) = n)$. Define the map

$$U_q m_{\vec{\alpha}}(X) = \sum_{S_1, S_2, \ldots, S_m} \prod_{i=1}^m \prod_{s \in S_i} q_i(x_s),$$

where the sum is as in example (2). The symmetric functions $m_{\vec{\alpha}}^{U_q}(X)$ can be written in the form $c_{\vec{\alpha}} m_{\vec{\alpha}}(X) +$ terms of lower degree. Then $m_{\vec{\alpha}}^{U_q}(X)$ is a basis. Moreover, since $\hat{m}^{U_q}_{\vec{\alpha}}(X) = \vec{\alpha}! m^{U_q}_{\vec{\alpha}}(X)$ is of binomial type, U_q is an umbral operator. Equation (27) is satisfied by U_q since this is equivalent to saying that $\hat{m}^{U_q}_{\vec{\alpha}}(x) = 0$ when $\vec{\alpha}$ has more than one non-zero component.

Let ρ_n be the projection $\Lambda(X) \to \Lambda(x_1, x_2, \ldots, x_n)$ and $\rho_n^* : \Lambda(x_1, x_2, \ldots, x_n)^* \to \Lambda(X)^*$ its adjoint. For a sequence of functionals $\{\tilde{L}_i\}_{i=1}^n$ of $\mathbb{C}[x]^* = \Lambda(x)^*$, define the functional $\tilde{L}$ in $\Lambda(x_1, x_2, \ldots, x_n)$ by

$$\langle \tilde{L} | m_{\vec{\alpha}}(x_1, x_2, \ldots, x_n)\rangle = \sum_{\vec{\alpha}^{(1)}+\ldots+\vec{\alpha}^{(n)}=\vec{\alpha}} \prod_{i=1}^{n} \langle \tilde{L}_i | m_{\vec{\alpha}^{(i)}}(x)\rangle.$$

It is easy to check that $\rho_n^* \tilde{L} = \prod_{i=1}^n \rho_1^* \tilde{L}_i$.

Define the evaluation $\tilde{\epsilon}^n_a \in \mathbb{C}[x]^*$ by $\langle \tilde{\epsilon}^{(n)}_a | x^m \rangle = \delta_{m,n} a^m$. Let $c_{n,k}$ be the coefficients connecting $q_k(X)$ with x^n, this is $x^n = \sum_{k=1}^n c_{n,k} q_k(x)$. If $\tilde{L}_j := \sum_{n \geq j} c_{n,j} \tilde{\epsilon}^{(n)}_1$, then $\langle \tilde{L}_j | q_i(x)\rangle = \sum_{n \geq j} c_{n,j} \hat{c}_{i,n} = \delta_{i,j}$. Let $L_j = \rho_1^* \tilde{L}_j$. If $|\vec{\alpha}| = n$,

$$\langle L_j | \hat{m}^{U_q}_{\vec{\alpha}}(X)\rangle = \langle \tilde{L}_j | \hat{m}^{U_q}_{\vec{\alpha}}(x)\rangle = \langle \tilde{L}_j | \delta_{\vec{\alpha},\vec{e}_n} q_n(x)\rangle = \delta_{\vec{\alpha},\vec{e}_j}.$$

Then $jL_j = K^{U_q}_j = j\epsilon_1^{(j),U_q}$ is the delta-system associated to $\hat{m}^{U_q}_{\vec{\alpha}}(X)$. A simple computation will give us

$$\frac{\delta_{A_i} K^{U_q}_j}{j} = \sum_{k \geq j-i} c_{k+i,j} \epsilon_1^{(k)}.$$

Definition 7 Let $L_{\lambda,n}$, for $\lambda \in \mathfrak{P}_n$, be an element of $\Lambda(X)^*$, and $\tilde{L}_{\lambda,n}$ be in $\Lambda(x_1, x_2, \ldots, x_n)^*$ defined as the functionals

$$L_{\lambda,n} = (\delta_{\vec{A}} K^{U_q})(\varsigma_n(0), \varsigma_n(\lambda)) = \left| \sum_{k \geq \lambda_r + r - s} c_{k+s,\lambda_r+r} \epsilon_1^{(k)} \right|_{r,s=1}^{n},$$

and

$$\tilde{L}_{\lambda,n} = \rho_n^* L_{\lambda,n} = \left| \sum_{k \geq \lambda_r + r - s} c_{k+s,\lambda_r+r} \tilde{\epsilon}_1^{(k)} \right|_{r,s=1}^{n}.$$

We define the *generalized Schur functions* as

$$R^q_\lambda(x_1, x_2, x_3, \ldots, x_n) = \frac{|t_{\lambda_r+r-1}(x_s)|_{r,s=1}^n}{|x_s^{r-1}|_{r,s=1}^n}.$$

Theorem 3 *The symmetric functions $R^q_\lambda(x_1, x_2, \ldots, x_n)$ form a basis of $\Lambda(x_1, x_2, \ldots, x_n)$. For every $Q(x_1, x_2, \ldots, x_n)$ in $\Lambda(x_1, x_2, \ldots, x_n)$ we have the expansion*

$$Q(x_1, x_2, \ldots, x_n) = \sum_{\lambda \in \mathfrak{P}_n} \langle \tilde{L}_{\lambda,n} | Q(x_1, x_2, \ldots, x_n)\rangle R^q_\lambda(x_1, x_2, \ldots, x_n). \quad (28)$$

Proof. By equation (20)

$$\vartheta_{\vec{A}}^{\varsigma_n(0)} = \sum_{\lambda\in\mathfrak{P}_n} \vartheta_{\vec{K}^{U_q}}^{\varsigma_n(\lambda)} L_{\lambda,n}(D). \tag{29}$$

Applying both sides of (29) to $p_{\vec{\alpha}}(X)$ we get

$$p_{\vec{\alpha}}(X^{(1)} + \ldots + X^{(n)})|p_r(X^{(s)})|_{r,s=1}^{n} =$$

$$\sum_{\lambda\in\mathfrak{P}_n} \sum_{\vec{\beta}+\vec{\gamma}=\vec{\alpha}} \binom{\vec{\alpha}}{\vec{\beta},\vec{\gamma}} \langle L_\lambda | p_{\vec{\beta}}(X)\rangle |\vartheta_{K_{\lambda_r+r}^{U_q}}^{X^{(s)}}|_{r,s=1}^{n} p_{\vec{\gamma}}(X^{(1)} + \cdots + X^{(n)}). \tag{30}$$

Since U_q satisfies equation (27), for $\vec{\gamma} \neq 0$ we have

$$\left(\left|\vartheta_{K_{\lambda_r+r}^{U_q}}^{X^{(s)}}\right|_{r,s=1}^{n} p_{\vec{\gamma}}\right)(x_1, \ldots, x_n) =$$

$$\sum_{\vec{\gamma}^{(1)}+\ldots+\vec{\gamma}^{(n)}=\vec{\gamma}} \binom{\vec{\gamma}}{\vec{\gamma}^{(1)}, \ldots, \vec{\gamma}^{(n)}} \left|(\vartheta_{K_{\lambda_r+r}^{U_q}}^{X^{(s)}} p_{\vec{\gamma}^{(s)}})(x_s)\right|_{r,s=1}^{n} = 0.$$

Then, taking one element alphabets in (30) we obtain

$$\begin{aligned} p_{\vec{\alpha}}(x_1, x_2, \ldots, x_n) &= \sum_{\lambda\in\mathfrak{P}_n} \langle L_{\lambda,n} | p_{\vec{\alpha}}(X)\rangle \frac{|\hat{m}_{\vec{e}_{\lambda_r+r}}^{U_q}(x_s)|_{r,s=1}^{n}}{|x_s^r|_{r,s=1}^{n}} \\ &= \sum_{\lambda\in\mathfrak{P}_n} \langle L_{\lambda,n} | p_{\vec{\alpha}}(X)\rangle \frac{|x_s t_{\lambda_r+r-1}(x_s)|_{r,s=1}^{n}}{|x_s^r|_{r,s=1}^{n}} \end{aligned}$$

$$= \sum_{\lambda\in\mathfrak{P}_n} \langle L_{\lambda,n} | p_{\vec{\alpha}}(X)\rangle R_\lambda^q(x_1, x_2, \ldots, x_n). \tag{31}$$

By linearity, for any symmetric function $Q(X)$

$$Q(x_1, x_2, \ldots, x_n) = \sum_{\lambda\in\mathfrak{P}_n} \langle L_{\lambda,n} | Q(X)\rangle R_\lambda^q(x_1, x_2, \ldots, x_n)$$

$$= \sum_{\lambda\in\mathfrak{P}_n} \langle \tilde{L}_{\lambda,n} | Q(x_1, x_2, \ldots, x_n)\rangle R_\lambda^q(x_1, x_2, \ldots, x_n). \tag{32}$$

The result follows since the map $\rho_n Q(X) = Q(x_1, x_2, \ldots, x_n)$ is surjective. ∎

Proposition 10 *Let C be the infinite lower triangular constant matrix $C = (c_{i,j})_{i,j=1}^{\infty}$ of the coefficients connecting the family $q_n(x)$ with the powers x^n, and $\tilde{C}$ its inverse. Then we have the expansions*

$$L_{\lambda,n} = \sum_{\mu \sqsupseteq \lambda} C^{n}_{\mu,\lambda} H_{\mu}, \tag{33}$$

$$H_{\lambda} = \sum_{\mu \sqsupseteq \lambda} \tilde{C}^{n}_{\mu,\lambda} L_{\lambda,n}, \tag{34}$$

and

$$R^{q}_{\mu}(x_1, \ldots, x_n) = \sum_{\lambda \sqsubseteq \mu} \tilde{C}^{n}_{\mu,\lambda} S_{\lambda}(x_1, \ldots, x_n), \tag{35}$$

where $C^{n}_{\mu,\lambda} = C^{(n)}(\varsigma_n(\mu), \varsigma_n(\lambda)) = |c_{\mu_r+r,\lambda_s+s}|^{n}_{r,s=1}$.

Proof. By the chain rule we have $\delta_{\vec{A}}\vec{K}^{U_q} = (\delta_{\vec{A}}\vec{K}).(\delta_{\vec{K}}\vec{K}^{U_q}) = (\delta_{\vec{A}}\vec{K})C$. We also have $\delta_{\vec{A}}\vec{K} = (\delta_{\vec{A}}\vec{K}^{U_q})\tilde{C}$. By Cauchy -Binet we obtain (33) and (34). Equation (35) is the dual form of (34). ∎

Example 4 (Factorial symmetric function) For $q_n(x) = x(x)_{n-1}$, the corresponding generalized Schur function

$$R^{q}_{\lambda}(x_1, x_2, \ldots, x_n) = \frac{|(x_s)_{\lambda_r+r-1}|^{n}_{r,s=1}}{|x_s^{r-1}|^{n}_{r,s=1}}$$

is the factorial symmetric function [1, 2, 4, 6]. Since $x^n = \sum_{k=1}^{n} S(n,k)(x)_k = \sum_{k=1}^{n} S(n-1,k-1)q_k(x)$, where $S(n,k)$ is the Stirling number of the second kind, the dual pseudobasis is given by

$$\tilde{L}_{\lambda,n} = \left| \sum_{k \geq \lambda_r + r - s} S(k+s-1, \lambda_r + r - 1)\tilde{\epsilon}_1^{(k)} \right|^{n}_{r,s=1}.$$

Example 5 (Macdonald's $6th$ variation) *Let* $(a_n)_{n=1}^{\infty}$ *be any sequence of integers. Defining* $(x|a)^r = \prod_{i=1}^{r}(x + a_i)$, *and* $q_n(x) = x(x|a)^{n-1} = \sum_{k=1}^{n} e_{n-k}(a_1, a_2, \ldots, a_{n-1})x^k$ *we obtain*

$$R^{q}_{\lambda}(x_1, x_2, \ldots, x_n) = \frac{|(x_s|a)^{\lambda_r+r-1}|^{n}_{r,s=1}}{|x_s^{r-1}|^{n}_{r,s=1}} = S_{\lambda}(x_1, \ldots, x_n|a),$$

which is Macdonald's $6th$ *variation of Schur functions [6]. It generalizes the factorial Schur functions and the* α*-paired Schur functions [4]. The functionals* L_j *are obtained by computing the inverse of the triangular matrix* $(e_{n-k}(a_1, a_2, \ldots, a_{n-1}))_{n,k=1}^{\infty}$.

The next proposition is easy to prove, and is left as an exercise for the reader.

Proposition 11 *The symmetric functions $R^q_\lambda(x_1, x_2, \ldots, x_n)$ satisfy the condition*

$$R^q_\lambda(x_1, x_2, \ldots, x_n, 0) = R^q_\lambda(x_1, x_2, \ldots, x_n),$$

for every λ and n, if and only if the polynomials $t_n(x)$ are monic, and satisfy the divided difference equation

$$\frac{t_n(x) - t_n(0)}{x} = t_{n-1}(x), \qquad n \geq 1. \tag{36}$$

Example 6 The polynomial sequences satisfying (36) are easily classified. They are determined by the sequence $(a_n)_{n\geq 1}$, $a_n = t_n(0)$;

$$t_n(x) = x^n + a_1 x^{n-1} + \ldots + a_n.$$

The inverse limit $R^q_\lambda(X)$ exists for every λ. By theorem (3), they form a basis of $\Lambda(X)$. Since the matrix C is Toeplitz, by equation (33) the functionals $L_{\lambda,n}$ do not depend on n, and $L_\lambda = L_{\lambda,n}$ is the dual pseudobasis of $R^q_\lambda(X)$.

References

[1] L. Biederharn and J. Louck, A new class of symmetric polynomials defined in terms of tableaux, *Adv. Appl. Math.* **10** (1989), 396–438.

[2] L. Biederharn and J. Louck, Inhomogeneous basis set of symmetric polynomials defined by tableaux, *Proc. Nat. Acad. Sci. U.S.A.* **87**(1990) 1441–1445.

[3] W. Chen, Compositional calculus, *J. Combin. Theory, Ser.* A **64** (1993), 149–188.

[4] W. Chen and J. Louck, The factorial Schur function, *J. Math. Phys.* **34**(1993), 4144–4160.

[5] L. Geissenger, Hopf algebras of symmetric functions and class functions, in Combinatoire et représentations du groupe symétrique, Springer Lecture Notes in Mathematics **579** (1977), 168–181.

[6] I.G. Macdonald, "Symmetric functions and Hall polynomials," Oxford University Press, 1979.

[7] I.G. Macdonald, Schur functions: theme and variations, Publ. I.R.M.A. Strasbourg, 1992, 498/S-27, Actes 28e Séminaire Lotharingien, pp. 5–39.

[8] M.A. Méndez, Plethystic exponential polynomials and plethystic Stirling numbers, *Stud. Appl. Math. 96* (1996), 1–8.

[9] M.A. Méndez, The umbral calculus of symmetric functions, *Adv. in Math.* **124** (1996), 207–271.

[10] A. Lascoux, Wronski's factorization of polynomials, Topics in Algebra, Banach Center Publications, **26**, part 2, PWN-Polish Scientific Publishers, Warsaw 1990.

[11] S. Roman, The algebra of formal series III: several variables, *J. Approx. Theory* **26** (1979), 340–381.

[12] S. Roman and G.-C. Rota, The umbral calculus, *Adv. in Math.* **27** (1978), 95–188.

Venezuelan Institute for Scientific Research (IVIC)
Departamento de Matemáticas
Apartado 21827, Caracas 1020-A
Venezuela

An Axiomization for Cubic Algebras

Colin Bailey and Joseph Oliveira

Abstract

We provide a list of universal axioms for cubic algebras inside the variety of implication algebras. This list is shown to axiomatize the subvariety generated by cubic algebras.

1 Introduction

The algebraic characterization of the lattice of closed intervals of a Boolean algebra has been well-studied. The first results were due to Metropolis and Rota, who were studying the lattice of faces of n-cubes and showed that (see [7]):

Theorem 1.1 *Let $\mathcal{L}$ be a finite lattice with minimum 0 and maximum 1. For every $x \neq 0$, let Δ_x be a function defined on the segment $[0, x]$ and taking values in $[0, x]$. Assume*

(i) If $a \leq b \leq x$ then $\Delta_x(a) \leq \Delta_x(b)$;

(ii) $\Delta_x^2 = id$ (the identity map);

(iii) Let $a < x$ and $b < x$. Then the following two conditions are equivalent:

$$\Delta_x(a) \vee b < x \text{ and } a \wedge b = 0.$$

Then $\mathcal{L}$ is isomorphic to the lattice of faces of an n-cube, for some n.

Conversely, if $\mathcal{L}$ is the lattice of faces of an n-cube, and $\Delta_x(y)$ is the antipodal face of y within the face x, then $\mathcal{L}$ satisfies conditions (i) through (iii).

The proof of this result depends heavily on the assumption of finiteness. Finiteness implies that the lattice is atomic, co-atomic and complete. Using these extra assumptions, Oliveira (see [8] or [6]) has extended this theorem to arbitrary cardinals.

One can take an alternative approach to extending the Metropolis-Rota theorem by noting that the face lattice of the n-cube is exactly the lattice of closed intervals of the Boolean algebra 2^n, and the Δ operator is induced by local complementation. One can then ask for an algebraic characterization of such lattices. This was provided by Bailey and Oliveira in [5] where we showed that such lattices are characterized (essentially) by the MR axioms and atomicity together with the fact that every interval $[x, y]$ where $x > 0$ is a Boolean algebra.

In studying such lattices one quickly discovers that 0 plays a rather ambiguous role. It makes the structure into a lattice, but it prevents the class of such structures from being closed under cartesian products. It also makes the natural homomorphisms trivial (they are forced to be embeddings).

The obvious next step is to eliminate the 0. The resulting structures are upper semi-lattices, and (as is clear from the Bailey-Oliveira theorem) they have a natural structure as implication algebras (this was first pointed out and studied by Oliveira, see [8]). Furthermore the class of such structures is closed under cartesian products, and images of the (now interesting) homomorphisms, but not subalgebras. We will call the elements of this class *MR-algebras*. They are all isomorphic to structures $\mathcal{I}(B) = \{[x, y] \mid x \leq y,\ x, y \in B\}$ for some Boolean algebra B – see [5] for a more detailed exposition of MR-algebras and an axiomatization.

The last axiom

$$\Delta_x(a) \vee b < x \text{ iff } a \wedge b \text{ exists}$$

will be referred to as the *MR axiom*. It is not equational and the class of MR-algebras does not form a variety.

There are several ways to make this class into a variety. One way is to close under subobjects. Another way is to add a new constant naming an atom. This method is highly unsatisfactory, as the resulting category is naturally isomorphic to the category of Boolean algebras.

In this paper we are considering the first method, closing under subobjects, and showing that there is a relatively simple axiomatic description of the resulting variety. Although it has not been made explicit, and will not be used explicitly, the Δ operators can be treated as a single binary operation by letting $\Delta(x, y) = \Delta_{x \vee y}(y)$.

Definition 1.2 *A* cubic algebra *is any isomorphic copy of a subalgebra of an MR-algebra that contains one and is closed under join and Δ.*

It is easy to see that the class of cubic algebras is a variety and hence must have an axiomatization via equations.

In this paper we provide one such set of equations and prove that they do in fact axiomatize this variety. There is an implication operation definable by an equation in join and delta, and so we obtain a subvariety of the variety of implication algebras.

Suppose that we let an *A-algebra* be an algebra as described by the hypotheses of theorem 2.1. We will show that every A-algebra is a cubic algebra. There are two steps to this proof. First we obtain a structural theorem for our A-algebras showing that every A-algebra is a union of MR-algebras – this is theorem 2.1. Then we explicitly embed those A-algebras that are a finite union of MR-algebras into an MR-algebra (theorem 5.1), and lastly use the compactness theorem (from first-order logic) to obtain the full result – theorem 6.1.

As is typical with using the compactness theorem we get very little information about the resulting embedding. A stronger embedding theorem is possible but obtaining it is beyond the scope of this paper–see [4] for more details.

2 The Axioms

In this section we will provide the equational axioms and show that the structures satisfying these axioms are unions of MR-algebras.

Theorem 2.1 *Let $\mathcal{L}$ be an upper-semi-lattice with* 1 *and a binary operator Δ satisfying the following axioms:*

a. $x \leq y$ *implies* $\Delta(y, x) \vee x = y$;

b. $x \leq y \leq z$ *implies* $\Delta(z, \Delta(y, x)) = \Delta(\Delta(z, y), \Delta(z, x))$;

c. $x \leq y$ *implies* $\Delta(y, \Delta(y, x)) = x$;

d. $x \leq y \leq z$ *implies* $\Delta(z, x) \leq \Delta(z, y)$;

Let $xy = \Delta(1, \Delta(x \vee y, y)) \vee y$ *for any* x, y *in* $\mathcal{L}$. *Then:*

e. $(xy)y = x \vee y$;

f. $x(yz) = y(xz)$;

Then for any $x \in \mathcal{L}$ *the set* $\mathcal{L}_x = \bigcup\{[\Delta(y, x), 1] \mid x \leq y\}$ *is an MR-algebra with* $\Delta_x = \Delta(x, -)$ *which contains* x.

Remark 2.1 (a) It is easy to show (using the axiomatization of MR-algebras from [5], or directly from the interval-algebra representation given above) that every cubic algebra satisfies these axioms.

(b) It is not known if this list of axioms is an independent set. In our view it is rather unlikely that these are all independent.

Proof We proceed by a series of lemmas. The first few establish some of the basic facts about these algebras that will be used in deducing the required properties of $\mathcal{L}_x$.

Lemma 2.2 *For all $x \in \mathcal{L}$ we have $\Delta(x, x) = x$.*

Proof $x = \Delta(x, \Delta(x, x)) \leq \Delta(x, x) \leq x$. ■

Lemma 2.3 $\Delta(x, -): [\leftarrow, x] \to [\leftarrow, x]$ *is an order preserving bijection.*

Proof That it preserves order follows from part (d), and it has a two-sided inverse, namely itself by part (c). ■

We note that this lemma implies that $\Delta(x, -)$ preserves joins of elements below x and also any meets that may exist.

Lemma 2.4 *If $w \leq x, y \leq z$ then $y \vee \Delta(z, x) = z$.*

Proof

$$\begin{aligned} z \geq y \vee \Delta(z, x) &\geq w \vee \Delta(z, w) && \text{by axiom (d)} \\ &= z && \text{by axiom (a).} \end{aligned}$$

■

Lemma 2.5 *$\mathcal{L}$ with the binary operation $\langle x, y \rangle \mapsto xy$ is an implication algebra.*

Proof We need only prove that $(xy)x = x$ for any x and y since (e) and (f) provide the other two axioms.

$$\begin{aligned} (xy)x &= \Delta(1, \Delta(xy \vee x, x)) \vee x && \text{by definition} \\ xy \vee x &= \Delta(1, \Delta(x \vee y, y)) \vee y \vee x && \text{by definition} \\ &= 1 && \text{by lemma 2.4. Hence} \\ (xy)x &= \Delta(1, \Delta(1, x)) \vee x \\ &= x \vee x \\ &= x. \end{aligned}$$

■

It therefore follows that for any x the interval $[x, 1]$ is a Boolean algebra with the complement of $y \geq x$ being given by yx. Furthermore we know that $y, z \geq x$ implies the existence of a greatest lower bound to z, y which, as usual, we will denote by $z \wedge y$. We are now able to define the very important relation $\preccurlyeq$ similar to the commutation relation in orthomodular lattices.

Definition 2.6

$$x \preccurlyeq y \iff y = (x \vee y) \wedge (\Delta(1, x) \vee y).$$

Lemma 2.7 *Let x, y and z be arbitrary elements of $\mathcal{L}$. Then*

a. For any $w \geq x$ we have $x \preccurlyeq \Delta(w,x)$ and $\Delta(w,x) \preccurlyeq x$;

b. $x \leq y$ implies $x \preccurlyeq y$;

c. $x \preccurlyeq y$ and $y \preccurlyeq z$ implies $x \preccurlyeq z$;

d. $\Delta(x \vee y, x) \leq \Delta(1,x) \vee y$;

e. $x \preccurlyeq y$ implies $\Delta(x \vee y, x) \leq y$.

f. $x < y$ implies $y \not\preccurlyeq x$.

Proof

a. It clearly suffices to show that $\Delta(w,x) \preccurlyeq x$ (since then we also have $\Delta(w, \Delta(w,x)) = x \preccurlyeq \Delta(w,x)$).

$$\begin{aligned} x \vee \Delta(w,x) &= w && \text{by axiom (a)} \\ x \vee \Delta(1, \Delta(w,x)) &= wx && \text{by definition.} \end{aligned}$$

Hence

$$\begin{aligned} (x \vee \Delta(w,x)) \wedge (x \vee \Delta(1,\Delta(w,x))) &= w \wedge wx \\ &= x && \text{by lemma 2.5.} \end{aligned}$$

b.

$$\begin{aligned} x \vee y &= y \\ \Delta(1,x) \vee y &= 1 && \text{by lemma 2.4.} \\ \text{Hence } (x \vee y) \wedge (\Delta(1,x) \vee y) &= y \wedge 1 \\ &= y. \end{aligned}$$

c.

$$\begin{aligned} & x \vee z \\ &\quad = x \vee ((y \vee z) \wedge (\Delta(1,y) \vee z)) && \text{as } y \preccurlyeq z \\ &\quad = (x \vee z) \vee ((y \vee z) \wedge (\Delta(1,y) \vee z)) \\ &\qquad\qquad \text{as } z \leq (y \vee z) \wedge (\Delta(1,y) \vee z) \\ &\quad = (x \vee y \vee z) \wedge (x \vee \Delta(1,y) \vee z) && \text{as } [z,1] \text{ is distributive} \\ & \Delta(1,x) \vee z \\ &\quad = (\Delta(1,x) \vee y \vee z) \wedge (\Delta(1,x) \vee \Delta(1,y) \vee z) && \text{similarly.} \end{aligned}$$

Putting these together and using the associativity of meet and join we get

$$\begin{aligned}
&(x \vee z) \wedge (\Delta(1,x) \vee z) \\
&\quad = (x \vee y \vee z) \wedge (x \vee \Delta(1,y) \vee z) \wedge \\
&\qquad \wedge (\Delta(1,x) \vee y \vee z) \wedge (\Delta(1,x) \vee \Delta(1,y) \vee z) \\
&\quad = [(x \vee y \vee z) \wedge (\Delta(1,x) \vee y \vee z)] \wedge \\
&\qquad \wedge [(x \vee \Delta(1,y) \vee z) \wedge (\Delta(1,x) \vee \Delta(1,y) \vee z)] \\
&\quad = [z \vee ((x \vee y) \wedge (\Delta(1,x) \vee y))] \wedge \\
&\qquad \wedge [z \vee ((x \vee \Delta(1,y) \wedge (\Delta(1,x) \wedge \Delta(1,y)))] \\
&\quad = (z \vee y) \wedge (z \vee \Delta(1,y)) && \text{as } x \preccurlyeq y \\
&\quad = z && \text{as } y \preccurlyeq z.
\end{aligned}$$

d. Since we have $x \leq x \vee \Delta(1,y)$ and $(x \vee y) \vee (x \vee \Delta(1,y)) = 1$ we see that $x \vee \Delta(1,y)$ is greater than the complement of $x \vee y$ in $[x,1]$. Thus

$$x \vee \Delta(1,y) \geq (x \vee y)x = \Delta(1, \Delta(x \vee y, x)) \vee x \geq \Delta(1, \Delta(x \vee y, x)).$$

Taking $\Delta(1,-)$ of both sides, we get

$$\Delta(x \vee y, x) \leq \Delta(1,x) \vee y.$$

e. $$\begin{aligned}
\Delta(x \vee y, x) &\leq (x \vee y) \wedge (\Delta(1,x) \vee y) && \text{by (d)} \\
&= y && \text{as } x \preccurlyeq y.
\end{aligned}$$

f. $y \preccurlyeq x$ and $x < y$ implies $y = \Delta(y,y) = \Delta(x \vee y, y) \leq x$ (by (e)), a contradiction.

■

Corollary 2.8 *Let $x, y, z \in \mathcal{L}$.*

(a) If $x \leq y$ then $\Delta(y,x) = y \wedge \Delta(1, yx)$;

(b) if $x \leq y \leq z$ then $z \wedge \Delta(1, yx)$ exists;

(c) if $y \preccurlyeq z$ and $z \preccurlyeq y$ then $\Delta(y \vee z, y) = z$.

Proof

(a) We have $\Delta(1, \Delta(y,x)) \leq \Delta(1, \Delta(y,x)) \vee x = yx$ and so $\Delta(y,x) \leq \Delta(1, yx)$. Thus the meet exists.

From lemma 2.7 (a) we have $x \preccurlyeq \Delta(y,x)$ and so

$$\begin{aligned}
\Delta(y,x) &= (\Delta(y,x) \vee x) \wedge (\Delta(y,x) \vee \Delta(1,x)) \\
&= y \wedge \Delta(1, x \vee \Delta(1, \Delta(y,x))) \\
&= y \wedge \Delta(1, yx).
\end{aligned}$$

(b) $z \wedge \Delta(1, yx) \geq y \wedge \Delta(1, yx) = \Delta(y, x)$ from (a).

(c) As $y \preccurlyeq z$, by lemma 2.7 (e) we have $\Delta(y \vee z, y) \leq z$. Likewise, as $z \preccurlyeq y$ we have $\Delta(y \vee z, z) \leq y$ and taking $\Delta(z \vee y, -)$ of both sides of this last inequality, we get $z \leq \Delta(z \vee y, y)$. Together these give the result.

■

Lemma 2.9 *Let $y \geq x$. Then x and $\Delta(y, x)$ have a lower bound iff $x = y$.*

Proof The right to left proof follows from lemma 2.2, so we will only do the left to right proof. Suppose that a lower bound exists, and then let $w = x \wedge \Delta(y, x)$. Then

$$\begin{aligned} \Delta(y, x) &= \Delta(y, x) \vee w && \text{as } w \leq \Delta(y, x) \\ &= y && \text{by lemma 2.4 and as } w \leq x \\ \text{Hence } x &= \Delta(y, \Delta(y, x)) && \\ &= \Delta(y, y) && \\ &= y && \text{by lemma 2.2.} \end{aligned}$$

■

This ends the elementary properties of the algebra that we need. We now turn to looking at some of the structure theory we use in establishing our result.

Fix $x \in \mathcal{L}$ and let

$$\mathcal{L}_x = \bigcup_{y \geq x} [\Delta(y, x), 1]$$
$$\mathcal{V}_x = \{\Delta(y, x) \mid y \geq x\}.$$

The next few lemmas are to establish relationships between $\mathcal{L}_x$ and $\mathcal{V}_x$ that, although elementary, are important in understanding the underlying symmetry of $\mathcal{L}_x$. This symmetry is a key part of our proof that $\mathcal{L}_x$ is an MR-algebra, and is also used in the proof of the embedding theorem.

Lemma 2.10 *$\mathcal{V}_x$ is the set of atoms for $\mathcal{L}_x$.*

Proof By definition, if $z \in \mathcal{L}_x$ then $z \geq \Delta(y, x)$ for some $y \geq x$. But $\Delta(y, x) \in \mathcal{V}_x$.

Also, if w is any atom of $\mathcal{L}_x$ then $w \geq \Delta(y, x)$ for some $\Delta(y, x) \in \mathcal{V}_x \subseteq \mathcal{L}_x$ and so $w = \Delta(y, x)$ - as w is an atom. ■

Lemma 2.11 *Suppose that $v, w \in \mathcal{V}_x$ and $w \leq z$. Then*

(a) $\Delta(v \vee w, v) = w$;

(b) if $v \neq w$ then $v \wedge w$ does not exist;

(c) $\Delta(z, w) \in \mathcal{V}_x$.

Proof

(a) We know that $v \preccurlyeq x \preccurlyeq w$ and so $v \preccurlyeq w$. Likewise $w \preccurlyeq v$. By corollary 2.8 (c) we have the result.

(b) As $\Delta(v \vee w, v) = w$, by lemma 2.9 we see that $v \wedge w$ exists iff $v = w$.

(c) As $x \preccurlyeq w \preccurlyeq \Delta(z, w)$ we get $x \preccurlyeq \Delta(z, w)$. Likewise $\Delta(z, w) \preccurlyeq w \preccurlyeq x$ and so $\Delta(z, w) \preccurlyeq x$. By corollary 2.8 (c) this gives $\Delta(\Delta(z, w) \vee x, x) = \Delta(z, w)$ is in $\mathcal{V}_x$.

■

Lemma 2.12 *Suppose that $y \in \mathcal{L}$ and $v \in \mathcal{V}_x$. Then*

(a) $y \in \mathcal{L}_x$ *iff* $v \preccurlyeq y$;

(b) $y \in \mathcal{L}_x$ *iff* $\Delta(x \vee y, x) \leq y$;

(c) $\mathcal{L}_x = \mathcal{L}_v$.

Proof

(a) If $y \in \mathcal{L}_x$ then we have $y \geq \Delta(w, x)$ for some $w \geq x$. Also $v = \Delta(u, x) \preccurlyeq x \preccurlyeq \Delta(w, x) \preccurlyeq y$ and so $v \preccurlyeq y$.

Conversely, if $v \preccurlyeq y$ then $x \preccurlyeq v \preccurlyeq y$ gives us $x \preccurlyeq y$ and so $\Delta(y \vee x, x) \leq y$ showing that $y \in \mathcal{L}_x$.

(b) $y \in \mathcal{L}_x$ iff $x \preccurlyeq y$ by (a)
iff $\Delta(x \vee y, x) \leq y$ by lemma 2.7(a) and (e).

(c) As $v \in \mathcal{V}_x$ implies $x \preccurlyeq v \preccurlyeq x$ and so, for any y we have $x \preccurlyeq y$ iff $v \preccurlyeq y$.

■

Putting some of these results together we see that $\mathcal{L}_x$ is an upwards-closed subalgebra of $\mathcal{L}$.

Proposition 2.13 *If $y \in \mathcal{L}_x$ and $y \leq z$ then $z \in \mathcal{L}_x$ and $\Delta(z, y) \in \mathcal{L}_x$.*

Proof That $z \in \mathcal{L}_x$ is clear.
Also $x \preccurlyeq y \preccurlyeq \Delta(z, y)$ shows us that $\Delta(z, y)$ is in $\mathcal{L}_x$. ■

Now we can move on and show that $\mathcal{L}_x$ is an MR-algebra. We will do this by showing that, for the Boolean algebra $[x, 1]$ there is a natural isomorphism between $\mathcal{L}_x$ and $\mathcal{I}([x, 1])$. The next lemma defines the key components of this isomorphism.

Definition 2.14 *Let $y \in \mathcal{L}_x$. Then*

$$\begin{aligned} \ell_x(z) &= z \vee x && (1) \\ k_x(z) &= (\Delta(1, z) \vee x)x. && (2) \end{aligned}$$

Lemma 2.15 *Let $w, z \in \mathcal{L}_x$. Then*

a. $k_x(z) \leq \ell_x(z)$;

b. If $\ell_x(z) = \ell_x(w)$ *and* $k_x(z) = k_x(w)$ *then* $z = w$.

Proof We will suppress subscripts for ease of reading.

a.

$$\begin{aligned} k(z) &= \Delta(1, \Delta(\Delta(1, z) \vee x, x)) \vee x && \text{by definition} \\ &= \Delta(z \vee \Delta(1, x), \Delta(1, x)) \vee x && \text{by axiom (b)} \end{aligned}$$

From lemma 2.7(e) and $\Delta(1, x) \preccurlyeq z$ it follows that

$$\begin{aligned} \Delta(z \vee \Delta(1, x), \Delta(1, x)) &\leq z \\ \text{and so} \qquad k(z) &\leq z \vee x \\ &= \ell(z). \end{aligned}$$

b.

$$\begin{aligned} k(z) = k(w) \Rightarrow \Delta(1, z) \vee x &= k(z)x \\ &= k(w)x \\ &= \Delta(1, w) \vee x && \text{by axiom (e)} \\ \Rightarrow z \vee \Delta(1, x) &= w \vee \Delta(1, x) \end{aligned}$$

and so we have

$$\begin{aligned} z &= (z \vee x) \wedge (z \vee \Delta(1, x)) && \text{as } x \preccurlyeq z \\ &= (w \vee x) \wedge (w \vee \Delta(1, x)) && \text{by above} \\ &= w && \text{as } x \preccurlyeq w. \end{aligned}$$

■

Now we define our (intended) isomorphism:

$$f\colon \mathcal{L}_x \to \mathcal{I}([x,1]) \quad \text{by} \tag{3}$$
$$f(z) = [k_x(z), \ell_x(z)]. \tag{4}$$

By the last lemma f is well-defined and injective. The work that remains is to show that f is onto, order-preserving and preserves Δ. We split this into the last two lemmas.

Lemma 2.16 *Let f be the map defined in equation* (4).

a. Let $x \leq a \leq b$ and $z = b \wedge \Delta(1, ax)$. Then $z \in \mathcal{L}_x$ and $f(z) = [a, b]$.

b. f preserves order.

Proof Again we will suppress subscripts for ease of reading.

a. By corollary 2.8 (b) the meet $b \wedge \Delta(1, ax)$ exists. As $b \geq a$ this meet is greater than $\Delta(a, x)$ and so $z \in \mathcal{L}_x$.

$$\begin{aligned}
b \geq z \vee x &= (b \wedge \Delta(1, ax)) \vee x \\
&\geq (b \wedge \Delta(1, bx)) \vee x && \text{as } a \leq b \text{ implies } bx \leq ax \\
&= \Delta(b, x) \vee x && \text{by corollary 2.8(a)} \\
&= b. \\
ax \geq \Delta(1, z) \vee x &= (\Delta(1, b) \wedge ax) \vee x \\
&\geq (\Delta(1, a) \wedge ax) \vee x \\
&= \Delta(1, a \wedge \Delta(1, ax)) \vee x \\
&\geq \Delta(1, \Delta(a, x)) \vee x && \text{by corollary 2.8(a)} \\
&= ax.
\end{aligned}$$

Thus we have $b = z \vee x = \ell(z)$ and $ax = \Delta(1, z) \vee x$ and so $k(z) = (\Delta(1, z) \vee x)x = (ax)x = a \vee x = a$.

b. Let $w \leq z$ in $\mathcal{L}_x$. Then $\ell(w) = w \vee x \leq z \vee x = \ell(z)$ and $wx \geq zx$ which implies $k(w) = \Delta(1, wx) \geq \Delta(1, zx) = k(z)$. Thus $f(w) \leq f(z)$. ■

Lemma 2.17 *f preserves the Δ operation.*

Proof Since we know that

$$v \leq u \text{ implies } \Delta(u, v) = u \wedge \Delta(1, uv) \qquad \text{by corollary 2.8(a)}$$

and that f is an order-preserving bijection, we see that f must preserve meets and joins, and hence preserves the implication operation (by the uniqueness of complements in Boolean algebras). Thus we need only show that f preserves $\Delta(1,-)$. However since $\mathcal{I}([x,1])$ has a Δ operator satisfying the MR axiom

$$\text{if } a,b < y \text{ then } a \wedge b \text{ exists} \iff a \vee \Delta(y,b) = y$$

we know that f induces another Δ operator on $\mathcal{L}_x$ (call it Δ' say) that satisfies the same MR-axiom. We just need to show that $\Delta(1,-) = \Delta'(1,-)$.

First we do this for $v \in \mathcal{V}_x$:
Since $\mathcal{V}_x$ is the set of atoms for $\mathcal{L}_x$ and we know that $\Delta'(1,-)$ preserves atoms we have that $\Delta(1,v)$ and $\Delta'(1,v)$ are both in $\mathcal{V}_x$. If they are not equal then we know by lemma 2.11 (b) that $\Delta(1,v)$ and $\Delta'(1,v)$ have no lower bound. But then we apply the MR-axiom for Δ' to get $\Delta(1,v) \vee v < 1$ which we know to be false. Thus for all $v \in \mathcal{V}_x$ $\Delta(1,v) = \Delta'(1,v)$.

For arbitrary $y \in \mathcal{L}_x$:
Since we know that $\mathcal{I}([x,1])$ is atomic, it follows that $\mathcal{L}_x$ is also, so that if $\Delta(1,y) \neq \Delta'(1,y)$ then there is an atom below one that is not below the other. Call this atom v. Then $v \in \mathcal{V}_x$ and $\Delta(1,v) = \Delta'(1,v) \leq y$. But this is impossible as it implies $v \leq \Delta(1,y) \wedge \Delta'(1,y)$ which we assumed it was not. Thus we must have $\Delta(1,y) = \Delta'(1,y)$. ■

■

Remark 2.2 The proof of this theorem almost shows that any atomic algebra satisfying the axioms of theorem 2.1 is isomorphic to $\mathcal{I}(B)$ for some Boolean algebra B. The only additional fact needed is showing that for any atom x the set $\mathcal{V}_x$ is the set of **all** atoms. The MR axiom is used to show this fact.

Having obtained this theorem we may use it to find interesting examples of cubic algebras. Here is one such construction.

Definition 2.18 *Let $\mathcal{L}$ be a cubic algebra, and let $\mathcal{C}$ be any chain in $\mathcal{L}$. Define*

$$\mathcal{L}[\mathcal{C}] = \bigcup_{c \in \mathcal{C}} \mathcal{L}_c.$$

Theorem 2.19 *Let $\mathcal{L}$ be an upper-semi-lattice with* 1 *and* Δ *such that for every x the subalgebra $\mathcal{L}_x$ is an MR-algebra. Then $\mathcal{L}$ satisfies the axioms of theorem 2.1.*

Proof This is clear as the axioms are about elements above some x and so are true iff they are true in some $\mathcal{L}_x$. ■

Corollary 2.20 *Let $\mathcal{L}$ be a cubic algebra, and let $\mathcal{C}$ be any chain in $\mathcal{L}$. Then $\mathcal{L}[\mathcal{C}]$ satisfies the axioms of theorem 2.1.*

Proof This is immediate. ∎

We observe from this theorem that there exist atomless cubic algebras satisfying the MR axiom–for instance let B be the countable atomless Boolean algebra, and let $\mathcal{L} = \mathcal{I}(B)$. Then let $\mathcal{C}$ be a maximal chain in $[v,1]$ for some atom v in $\mathcal{L}$. As $[v,1] \simeq B$, we know that $\mathcal{C}$ is order-isomorphic to $\{r \mid r \in \mathbb{Q} \text{ and } 0 \leq r \leq 1\}$. Take $\mathcal{C}/2$ to be the top half of $\mathcal{C}$ i.e. that part corresponding to $]1/2, 1]$, and consider $\mathcal{L}[\mathcal{C}/2]$. Then by the theorem, this is a cubic algebra, and also it satisfies the MR axiom, as if $x, y, z \in \mathcal{L}[\mathcal{C}/2]$, then there is some $c \in \mathcal{C}/2$ such that $x, y, z \in \mathcal{L}_c$. It is also clear that $\mathcal{L}[\mathcal{C}/2]$ is atomless.

It would be nice to conjecture that every cubic algebra is isomorphic to some $\mathcal{L}[\mathcal{C}]$ for some MR-algebra and chain C. But this is false, since the face lattice of a line with four points is a cubic algebra which does not satisfy the MR axiom. Although it is somewhat harder to show, it is also false for cubic algebras satisfying the MR axiom.

3 Cubic Embeddings

In this section we will do a construction generalizing the interval algebra of a Boolean algebra. This construction motivates our approach to the later embedding results, and will provide part of the technical apparatus we use to obtain the embedding theorem.

Definition 3.1 *Let $\mathcal{L}$ and $\mathcal{M}$ be cubic algebras. Then*

(a) $f: \mathcal{L} \to \mathcal{M}$ is a cubic homomorphism *iff f preserves one, joins and delta.*

(b) $f: \mathcal{L} \to \mathcal{M}$ is a cubic embedding *iff f is a one-one cubic homomorphism.*

We recall that if B is any Boolean algebra, then $\mathcal{I}(B) = \{[a,b] \mid a, b \in B \text{ and } a \leq b\}$ ordered by inclusion is an MR-algebra. As in the previous section, for any cubic algebra $\mathcal{L}$ and any $x \in \mathcal{L}$ we have $\mathcal{L}_x = \bigcup\{[\Delta(y,x), 1] \mid x \leq y\}$ is an MR-algebra and a subalgebra of $\mathcal{L}$.

Our first step is to construct a simple cubic embedding based on product Boolean algebras that we will later generalize.

Lemma 3.2 *Let $\mathcal{B} = \mathcal{B}_1 \times \mathcal{B}_2$ be a product of Boolean algebras. Then the mapping $\Phi: \mathcal{I}(\mathcal{B}_1) \to \mathcal{I}(\mathcal{B})$ defined by*

$$\Phi([a,b]) = [\langle a, 0_2\rangle, \langle b, 1_2\rangle]$$

is a cubic embedding, whose range is upwards closed (in fact the range is $\mathcal{I}(\mathcal{B})_{\Phi([0,0])}$).

Proof We first check that Φ preserves 1, joins and deltas. Let $x = [a, b]$ and $y = [c, d]$ be any elements of $\mathcal{I}(\mathcal{B}_1)$

$$\begin{aligned}
\Phi(1) &= \Phi([0_1, 1_1]) \\
&= [\langle 0_1, 0_2\rangle, \langle 1_1, 1_2\rangle] \\
&= 1. \\
\Phi(x \vee y) &= \Phi([a, b] \vee [c, d]) \\
&= \Phi([a \wedge c, b \vee d]) \\
&= [\langle a \wedge c, 0_2\rangle, \langle b \vee d, 1_2\rangle] \\
&= [\langle a, 0_2\rangle, \langle b, 1_2\rangle] \vee [\langle c, 0_2\rangle, \langle d, 1_2\rangle] \\
&= \Phi(x) \vee \Phi(y).
\end{aligned}$$

$$\begin{aligned}
&\Phi(\Delta(y, x)) \\
&\quad = \Phi(\Delta([c, d], [a, b])) \\
&\quad = \Phi([c \vee (d \wedge \bar{b}), c \vee (d \wedge \bar{a})]) \\
&\quad = \left[\left\langle c \vee (d \wedge \bar{b}), 0_2\right\rangle, \langle c \vee (d \wedge \bar{a}), 1_2\rangle\right]. \\
&\Delta(\Phi(y), \Phi(x)) \\
&\quad = \Delta([\langle c, 0_2\rangle, \langle d, 1_2\rangle], [\langle a, 0_2\rangle, \langle b, 1_2\rangle]) \\
&\quad = \left[\left\langle c \vee (d \wedge \bar{b}), 0_2 \vee (1_2 \wedge \bar{1}_2)\right\rangle, \left\langle c \vee (d \wedge \bar{a}), 0_2 \vee (1_2 \wedge \bar{0}_2)\right\rangle\right] \\
&\quad = \left[\left\langle c \vee (d \wedge \bar{b}), 0_2\right\rangle, \langle c \vee (d \wedge \bar{a}), 1_2\rangle\right] \\
&\quad = \Phi(\Delta(y, x)).
\end{aligned}$$

Now if $\Phi(x) = \Phi(y)$ then we have $[\langle a, 0_2\rangle, \langle b, 1_2\rangle] = [\langle c, 0_2\rangle, \langle d, 1_2\rangle]$ from which it is clear that $x = y$.

Now suppose that $[\langle p, q\rangle, \langle s, t\rangle] \geq \Phi([0, 0]) = [\langle 0_1, 0_2\rangle, \langle 0_1, 1_2\rangle]$. Then we have

$$p \leq 0_1, \quad q \leq 0_2, \quad s \geq 0_1 \text{ and } \quad t \geq 1_2$$

and so we get

$$p = 0_1, \quad q = 0_2 \text{ and } \quad t = 1_2.$$

Now then we see that $\Phi([0_1, s]) = [\langle p, q\rangle, \langle s, t\rangle]$. Hence Φ is onto $[\Phi([0, 0]), 1]$, and so $\mathcal{I}(\mathcal{B})_{\Phi([0,0])}$ must be a subset of the range of Φ. Furthermore since every element of $\mathcal{I}(\mathcal{B}_1)$ is the join of two delta images of $[0, 0]$, we see that the range of Φ is a subset of $\mathcal{I}(\mathcal{B})_{\Phi([0,0])}$, and so this set is exactly the range of Φ. ∎

Lemma 3.3 *Let $\mathcal{B}_1$ be isomorphic to $\mathcal{B}_2$ via f. Then the mapping $\widehat{f}: \mathcal{I}(\mathcal{B}_1) \to \mathcal{I}(\mathcal{B}_2)$ defined by*

$$\widehat{f}([a,b]) = [f(a), f(b)]$$

is a cubic isomorphism.

Proof This is clear. ∎

Now we will generalize this embedding construction to the case where two Boolean algebras have some "overlap" – by which we mean a common factor in some representation as a product.

Theorem 3.4 *Let $\mathcal{B}_1$, $\mathcal{B}_2$ be two Boolean algebras, and suppose that there is some $b_i \in \mathcal{B}_i$ with $[b_1, 1] \simeq [b_2, 1]$, via e say. Then there is an MR-algebra $\mathcal{L}$ and a_1, a_2 in $\mathcal{L}$ such that*

(a) there exist isomorphisms $\Phi_i: \mathcal{B}_i \to [a_i, 1]$;

(b) $a_1 \wedge a_2$ exists, and $\mathcal{L} = \mathcal{L}_{a_1 \wedge a_2}$;

(c) there is an isomorphism $\Phi: [b_1, 1] \to [a_1 \vee a_2, 1]$ making the following diagram commute:

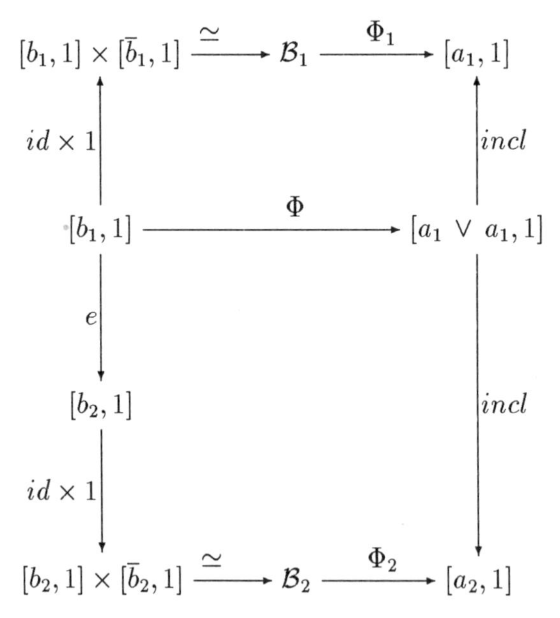

(d) $\mathcal{L}_{a_1} \cap \mathcal{L}_{a_2} = \mathcal{L}_{a_1 \vee a_2}$.

Proof For the sake of easing notation we will identify b_1 and b_2 with b, and suppose that $\mathcal{B}_i = [b, 1] \times [c_i, 1]$. Then we let $\mathcal{L} = \mathcal{I}([b, 1] \times [c_1, 1] \times [c_2, 1])$. Mappings $\widehat{\Phi}_i: \mathcal{I}(\mathcal{B}_i) \to \mathcal{L}$ are defined by

$$\begin{aligned}\widehat{\Phi}_1([\langle x_1, y_1\rangle, \langle x_2, y_2\rangle]) &= [\langle x_1, y_1, c_2\rangle, \langle x_2, y_2, 1\rangle] \\ \widehat{\Phi}_2([\langle u_1, v_1\rangle, \langle u_2, v_2\rangle]) &= [\langle u_1, c_1, v_1\rangle, \langle u_2, 1, v_2\rangle].\end{aligned}$$

By lemma 3.2, we know that both of these mappings are cubic embeddings whose ranges are upwards closed. Now let $a_i = \widehat{\Phi}_i([\langle b, c_i\rangle, \langle b, c_i\rangle])$. Again by the lemma we have $\mathcal{L}_{a_i} \simeq \mathcal{I}(\mathcal{B}_i)$ via $\widehat{\Phi}_i^{-1}$, and

$$\begin{aligned} a_1 \wedge a_2 &= [\langle b, c_1, c_2\rangle, \langle b, c_1, 1\rangle] \wedge [\langle b, c_1, c_2\rangle, \langle b, 1, c_2\rangle] \\ &= [\langle b, c_1, c_2\rangle, \langle b, c_1, c_2\rangle] \end{aligned}$$

is an atom of $\mathcal{L}$, and so $\mathcal{L} = \mathcal{L}_{a_1 \wedge a_2}$.

Also we have

$$a_1 \vee a_2 = [\langle b, c_1, c_2\rangle, \langle b, 1, 1\rangle].$$

We now define the mappings Φ_i. We note that the mapping $\jmath_i \colon \mathcal{B}_i \to \mathcal{I}(\mathcal{B}_i)$ defined by

$$\jmath_i(x) = [\langle b, c_i\rangle, x]$$

is a mapping which preserves meet, join and one. So we define $\Phi_i = \widehat{\Phi}_i \circ \jmath_i$.

To show that the diagram commutes, we see that for $x \in [b, 1]$

$$\begin{aligned} \Phi_1 \circ (id \times 1)(x) &= \Phi_1(\langle x, 1\rangle) \\ &= \widehat{\Phi}_1([\langle b, c_1\rangle, \langle x, 1\rangle]) \\ &= [\langle b, c_1, c_2\rangle, \langle x, 1, 1\rangle] \\ \text{and } \Phi_2 \circ (id \times 1)(x) &= \Phi_2(\langle x, 1\rangle) \\ &= \widehat{\Phi}_2([\langle b, c_2\rangle, \langle x, 1\rangle]) \\ &= [\langle b, c_1, c_2\rangle, \langle x, 1, 1\rangle] \\ &= \Phi_1 \circ (id \times 1)(x) \\ &\geq a_1 \vee a_2. \end{aligned}$$

So we define the mapping Φ as $\Phi_1 \circ (id \times 1)$.

It is clear that $\mathcal{L}_{a_1} \cap \mathcal{L}_{a_2} \supseteq \mathcal{L}_{a_1 \vee a_2}$. Now suppose that $x \in \mathcal{L}_{a_1} \cap \mathcal{L}_{a_2}$. Then we have $x = \widehat{\Phi}_i(u_i)$ for some $u_i \in \mathcal{I}(\mathcal{B}_i)$. But then we see that

$$\begin{aligned} x &= [\langle v_1, v_2, c_2\rangle, \langle w_1, w_2, 1\rangle] \\ &= [\langle s_1, c_1, s_2\rangle, \langle t_1, 1, t_2\rangle] \end{aligned}$$

as it is in the range of both mappings. Hence $v_1 = s_1 = p$, $w_1 = t_1 = q$, $v_2 = c_1$, $s_2 = c_2$, $w_2 = 1$ and $t_2 = 1$ so that

$$\begin{aligned} x &= [\langle p, c_1, c_2\rangle, \langle q, 1, 1\rangle] \\ &\in \mathcal{L}_{a_1 \vee a_2} \end{aligned}$$

as $b \leq p \leq q$. ∎

From this theorem we see that the cubic algebra $\mathcal{L}_{a_1} \cup \mathcal{L}_{a_2}$ sits in a nice way inside $\mathcal{L}$, and that $\mathcal{L}_{a_1} \cap \mathcal{L}_{a_2} = \mathcal{L}_{a_1 \vee a_2}$. We will show that this is essentially the only way to get a cubic algebra of the form $\mathcal{L}_a \cup \mathcal{L}_b$.

4 Preliminaries to Embeddings

Let us fix an algebra $\mathcal{L}$ satisfying the axioms of theorem 2.1. We want to show that this is a cubic algebra, i.e. can be embedded into an MR-algebra. In order to find this embedding we need to make the MR axiom true, that is for any a, b and x with $a, b < x$ we have

$$\Delta(x, a) \vee b < x \iff a \wedge b \text{ does not exist.}$$

Alternatively

$$a \wedge b \text{ exists} \iff \Delta(x, a) \vee b = x.$$

Since the existence of $a \wedge b$ implies $\Delta(x, a) \vee b = x$ (by lemma 2.4), we need only arrange the converse. The major concern for us now is the possibility that there is some a, b and x_1, x_2 strictly above a, b for which $\Delta(x_1, a) \vee b = x_1$ but $\Delta(x_2, a) \vee b < x_2$. In this section we will show that this cannot happen, and from this deduce an important representation for $\mathcal{L}_a \cap \mathcal{L}_b$.

Lemma 4.1 *Let a, b be such that $a, b < x_1, x_2$ and*

$$\Delta(x_1, a) \vee b = x_1, \qquad \Delta(x_2, a) \vee b < x_2.$$

Then $x_1 \not< x_2$.

Proof Suppose that $x_1 < x_2$. Then we can work below x_2, and so without loss of generality, we have $x_2 = 1$.

By lemma 2.7 (d) we have $\Delta(b \vee \Delta(1, a), \Delta(1, a)) \leq b \vee \Delta(1, \Delta(1, a)) = b \vee a$. Now both $b \vee a$ and $b \vee \Delta(1, a)$ are in $\mathcal{L}_a$ and are above $\Delta(b \vee \Delta(1, a), \Delta(1, a))$ (also in $\mathcal{L}_a$) so that the meet

$$g = (b \vee a) \wedge (b \vee \Delta(1, a)) \leq b \vee a < x_1$$

exists in $\mathcal{L}_a \cap \mathcal{L}_b$ (the meet exists in $\mathcal{L}_b$ as both terms are above b).

Noting that $b \leq g \leq b \vee a \leq x_1$ we have $x_1 \geq \Delta(x_1, g) \vee a \geq \Delta(x_1, b) \vee a = x_1$, and so (looking in $\mathcal{L}_a$ now) $g \wedge a > 0$.

Let $c = g \wedge a$. Then we have $c \leq g \leq b \vee \Delta(1, a)$; and $\Delta(1, c) \leq \Delta(1, a) \leq b \vee \Delta(1, a)$.

This implies $1 = c \vee \Delta(1, c) \leq b \vee \Delta(1, a)$, a contradiction. ∎

Lemma 4.2 *Let a, b be such that $a, b < x_1, x_2$ and*

$$\Delta(x_1, a) \vee b = x_1, \qquad \Delta(x_2, a) \vee b < x_2.$$

Then $x_2 \not< x_1$.

Proof Suppose that $x_2 < x_1 = 1$ (wolog), $\Delta(x_1, a) \vee b = x_1$ and $\Delta(x_2, a) \vee b < x_2$. Then we must have $\Delta(b \vee a, a) \vee b < b \vee a$ and a, b are incomparable.

- If a, b are comparable, then by lemma 2.4 we have $\Delta(x_2, a) \vee b = x_2$, contradicting our hypothesis.

- If $b \vee a < x_2$, as $a, b < b \vee a < x_2$, by the last lemma we cannot have $\Delta(b \vee a, a) \vee b = b \vee a$ and so we have $\Delta(b \vee a, a) \vee b < b \vee a$.

So we have incomparable a, b such that $\Delta(b \vee a, a) \vee b < b \vee a$ and $\Delta(1, a) \vee b = 1$.

Now $b \vee \Delta(b \vee a, a) \geq b$ and so we have $\Delta(1, b \vee \Delta(b \vee a, a)) \vee a \geq \Delta(1, b) \vee a = 1$. As $b \vee \Delta(b \vee a, a) \in \mathcal{L}_a$ which models the MR-axiom, we have $g = a \wedge (b \vee \Delta(b \vee a, a))$ exists in $\mathcal{L}_a$.

Clearly $g \leq a$ and so $\Delta(b \vee a, g) \leq \Delta(b \vee a, a)$. Now we have

$$\begin{aligned} b \vee a &= g \vee \Delta(b \vee a, g) \\ &\leq (b \vee \Delta(b \vee a, a)) \vee \Delta(b \vee a, a) = b \vee \Delta(b \vee a, a) \\ &< b \vee a \end{aligned}$$

– a contradiction. ■

Theorem 4.3 *Let a, b be in $\mathcal{L}$. There is no pair of elements x_1, x_2 such that $a, b < x_1, x_2$ and*

$$\Delta(x_1, a) \vee b = x_1, \qquad \Delta(x_2, a) \vee b < x_2.$$

Proof Suppose there were such a pair x_1, x_2. Then a and b must be incomparable and so we have $a, b < x_1 \wedge x_2 < x_1, x_2$. Now it either happens that

$$\Delta(x_1 \wedge x_2, b) \vee a < x_1 \wedge x_2 \quad \text{or} \quad \Delta(x_1 \wedge x_2, b) \vee a = x_1 \wedge x_2.$$

In either case we have a contradiction to one of the above lemmas. ■

The next step is to consider how to embed an A-algebra of the form $\mathcal{L} = \mathcal{L}_a \cup \mathcal{L}_b$ into an MR-algebra. The real problem here is determining what $\mathcal{L}_a \cap \mathcal{L}_b$ looks like. The next lemma shows us that for some a, b the intersection has the best possible form.

Lemma 4.4 *Suppose that* $\Delta(b \vee a, b) \vee a = b \vee a$. *Then* $\mathcal{L}_a \cap \mathcal{L}_b = \mathcal{L}_{b \vee a}$.

Proof It is clear that $\mathcal{L}_{b \vee a} \subseteq \mathcal{L}_a \cap \mathcal{L}_b$.

Now pick $x \in \mathcal{L}_a \cap \mathcal{L}_b$. Then

$$\begin{aligned} \Delta(x \vee a, a) &\leq x \quad \text{by lemma 2.7 (e)} \\ \Delta(x \vee b, b) &\leq x \quad \text{by lemma 2.7 (e).} \end{aligned}$$

Since $x \vee b \in \mathcal{L}_a$ we also have

$$\Delta(x \vee a \vee b, a) \leq x \vee b.$$

But also $x \vee a \vee b \geq b \vee a$ and so by lemma 4.1, we must have $\Delta(x \vee a \vee b, a) \vee b = x \vee a \vee b$. Now we get

$$\begin{aligned} x \vee a \vee b &= \Delta(x \vee a \vee b, a) \vee b \\ &\leq (x \vee b) \vee b \\ &= x \vee b. \end{aligned}$$

Therefore $x \vee a \vee b = x \vee b$. Likewise we have $x \vee a \vee b = x \vee a$.

Now we see that

$$\begin{aligned} \Delta(x \vee a \vee b, b \vee a) &= \Delta(x \vee a \vee b, b) \vee \Delta(x \vee a \vee b, a) \\ &= \Delta(x \vee b, b) \vee \Delta(x \vee a, a) \\ &\leq x \vee x \\ &= x. \end{aligned}$$

By lemma 2.12 (b), this entails $x \in \mathcal{L}_{b \vee a}$. ∎

Remark 4.1 The proof of the lemma shows us a little more. In fact we have $x \in \mathcal{L}_a \cap \mathcal{L}_b$ iff $x \vee a = x \vee b$.

Now we need to show that there are such nice pairs a' and b' satisfying the hypothesis of the lemma with $\mathcal{L}_a = \mathcal{L}_{a'}$ and $\mathcal{L}_b = \mathcal{L}_{b'}$. In fact there are many such pairs.

Lemma 4.5 *Let* $a, b \in \mathcal{L}$. *Let* $b' = \Delta(a \vee b, b)$. *Then*

$$\Delta(a \vee b', b') \vee a = b' \vee a.$$

Proof If $a \leq b'$ this reduces to $\Delta(b', b') \vee a = b' \vee a$.

If $a \leq b$ then $b' = b$ and we are done.

If $b' \leq a$ then $\Delta(a \vee b', b') \vee a = \Delta(a, b') \vee a = a = b' \vee a$.

If $b \leq a$ then $b' = \Delta(a, b) \leq a$ and we are done.

So we may assume that a is incomparable to both b and b'. Suppose that $\Delta(a \vee b', b') \vee a < b' \vee a$. Then we have $a, b' < b' \vee a$ and $a, b' < b \vee a$ and so by theorem 4.3 above we must have $\Delta(b \vee a, b') \vee a < b \vee a$ but this is absurd as $\Delta(b \vee a, b') = b$. ∎

Theorem 4.6 *Let $a, b \in \mathcal{L}$ be arbitrary. Then*

$$\mathcal{L}_a \cap \mathcal{L}_b = \mathcal{L}_{a \vee \Delta(a \vee b, b)}.$$

Proof This follows from the last two lemmas. ∎

We note that if $a \preccurlyeq b$ (respectively $b \preccurlyeq a$) this reduces to $\mathcal{L}_{\Delta(a \vee b, b)} = \mathcal{L}_b$ (resp. $\mathcal{L}_a$).

5 The Finite Embedding

In this section we will take an algebra $\mathcal{L}$ satisfying theorem 2.1 which is a finite union of subalgebras of the form $\mathcal{L}_x$ and show that it may be embedded into an MR-algebra.

The last result (theorem 4.6) suggests how we might do this if there are only two components $\mathcal{L}_a$ and $\mathcal{L}_b$ with intersection $\mathcal{L}_{a \vee b}$. We just note that $\mathcal{L}_a \simeq \mathcal{I}([a, 1])$, $\mathcal{L}_b \simeq \mathcal{I}([b, 1])$, $[a, 1] \simeq [b \vee a, 1] \times [(b \vee a)a, 1]$ and $[b, 1] \simeq [b \vee a, 1] \times [(b \vee a)b, 1]$. Then we can use theorem 3.4 to embed both of the MR-algebras $\mathcal{I}([a, 1])$ and $\mathcal{I}([b, 1])$ into $\mathcal{M} = \mathcal{I}([b \vee a, 1] \times [(b \vee a)a, 1] \times [(b \vee a)b, 1])$ in a way that agrees on $[b \vee a, 1]$. Then we have mappings from $\mathcal{L}_a$ into $\mathcal{M}$ and from $\mathcal{L}_b$ into $\mathcal{M}$. The only thing left to verify is that these last two mappings agree on the intersection $\mathcal{L}_a \cap \mathcal{L}_b = \mathcal{L}_{b \vee a}$. They do, and so we can naturally define an embedding from $\mathcal{L}_a \cup \mathcal{L}_b$ into the MR-algebra $\mathcal{M}$.

This deals with two components easily enough. To get any finite number of components we proceed by induction. For $\mathcal{L} = \left(\bigcup_{i=1}^{k} \mathcal{L}_{a_i}\right) \cup \mathcal{L}_c$ where $c \notin \bigcup_{i=1}^{k} \mathcal{L}_{a_i}$ and $\bigcup_{i=1}^{k} \mathcal{L}_{a_i}$ embedded into an MR-algebra $\mathcal{M} = \mathcal{L}_a$ we want to extend $\mathcal{M}$ and the embedding to include $\mathcal{L}_c$. The proof basically does this by gluing $\mathcal{M}$ and $\mathcal{L}_c$ together and using the case of two components to get the result.

The "gluing together" proceeds by looking at $\bigcup_{i=1}^{k}(\mathcal{L}_{a_i} \cap \mathcal{L}_c)$ inside both $\mathcal{M}$ and $\mathcal{L}_c$. In $\mathcal{M}$ they are shown to sit inside some $\mathcal{L}_b$ and in $\mathcal{L}_c$ they are shown to sit inside some $\mathcal{L}_{b'}$ for which $[b, 1] \simeq [b', 1]$.

The gluing and the two component case will be done simultaneously using theorem 3.4 as discussed above and then showing the the composite mapping from $\mathcal{L}$ into the new MR-algebra is a well-defined embedding.

Since the gluing is only possible if the range of the embedding into $\mathcal{M}$ is upwards-closed, we shall also have to arrange this for each of our embeddings. It is this point that makes the general case so difficult.

Theorem 5.1 *Let $\mathcal{L}$ be a cubic algebra such that there exists a finite set $\{a_1, \ldots, a_k\} \subseteq \mathcal{L}$ for which $\mathcal{L} = \bigcup_{i=1}^k \mathcal{L}_{a_i}$. Then there is an MR-algebra $\mathcal{L}^*$ into which $\mathcal{L}$ embeds as an upper segment.*

Proof The proof proceeds by induction on k. If $k = 1$ there is nothing to prove, we just take $\mathcal{L}^* = \mathcal{L}$. So now suppose that

$$\mathcal{L} = \left(\bigcup_{i=1}^{k} \mathcal{L}_{a_i} \right) \cup \mathcal{L}_c,$$

with $c \notin \bigcup_{i=1}^k \mathcal{L}_{a_i}$, and that we can embed $\bigcup_{i=1}^k \mathcal{L}_{a_i}$ into $\mathcal{L}_a$ where $\bigwedge_{i=1}^k a_k = a$. (This extra assumption will be arranged along the way.) We will identify a_i with its image in $\mathcal{L}_a$ for notational convenience, and we will suppress all reference to the embedding from $\bigcup_{i=1}^k \mathcal{L}_{a_i}$ into $\mathcal{L}_a$.

We still need to be careful where operations are performed, and this will usually be made clear by subscripts – for example $(x \wedge y)_{\mathcal{L}_a}$ means that the meet is computed in $\mathcal{L}_a$.

For each i define

$$c_i = \Delta(a_i \vee c, c).$$

We know that $\mathcal{L}_{a_i} \cap \mathcal{L}_c = \mathcal{L}_{a_i \vee c_i}$ (by theorem 4.6). Now define

$$b = \left(\bigwedge_{i=1}^{k} a_i \vee c_i \right)_{\mathcal{L}_a}.$$

This meet exists as each $a_i \vee c_i \geq a_i \geq a$.

$\mathcal{L}_b$ is the natural MR-algebra such that $\bigcup_{i=1}^k (\mathcal{L}_{a_i} \cap \mathcal{L}_c)$ embeds into $\mathcal{L}_b$. We want to show that there is an equivalent inside $\mathcal{L}_c$.

Lemma 5.2

(a) The meet

$$b' = \bigwedge_{i=1}^{k} a_i \vee c_i$$

exists in $\mathcal{L}_c$.

There is an isomorphism $\phi\colon [b, 1] \to [b', 1]$ such that

(b) for any $x \in [b, 1] \cap \mathcal{L}_{a_i}$ we have $\phi(x) = x$; and

(c) If $x \in \mathcal{L}_{a_i}$ then $\phi(x \vee b) = x \vee b'$.

(d) If $x \in \mathcal{L}_{a_i}$ then $x \vee b = x \vee b'$.

Proof We proceed by induction on $l \leq k$. For $l = 1$ the result is clear. In general let

$$\begin{aligned} b_l &= \left(\bigwedge_{i=1}^{l} a_i \vee c_i\right)_{\mathcal{L}_a}, \\ b'_l &= \left(\bigwedge_{i=1}^{l} a_i \vee c_i\right)_{\mathcal{L}_c} \end{aligned}$$

and let

$$\phi_l : [b_l, 1] \rightarrow [b'_l, 1]$$

be the isomorphism with the given properties. Then $b_l \geq \left(\bigwedge_{i=1}^{l} a_i\right)_{\mathcal{L}_a} = a'$. As $a' \wedge a_{l+1}$ exists in $\mathcal{L}_a$, we have $a' \vee \Delta(1, a_{l+1}) = 1$. This implies $b_l \vee \Delta(1, a_{l+1} \vee c_{l+1}) = 1$ and so

$$\begin{aligned} 1 &= \phi_l(1) \\ &= \phi_l(b_l \vee \Delta(1, a_{l+1} \vee c_{l+1})) \\ &= b'_l \vee \Delta(1, a_{l+1} \vee c_{l+1}). \end{aligned}$$

From this we see that $b'_{l+1} = b'_l \wedge (a_{l+1} \vee c_{l+1})$ exists in $\mathcal{L}_c$ – as $\mathcal{L}_c$ is an MR-algebra.

Now to construct the isomorphism $\phi_{l+1} : [b_{l+1}, 1] \rightarrow [b'_{l+1}, 1]$.

Let

$$\alpha_{l+1} = b_l \vee a_{l+1} \vee c_{l+1}; \qquad \alpha'_{l+1} = b'_l \vee a_{l+1} \vee c_{l+1}.$$

By the induction hypothesis, we have $\alpha_{l+1} = \alpha'_{l+1}$, and also b_l is the complement of $a_{l+1} \vee c_{l+1}$ in $[b_{l+1}, \alpha_{l+1}]$. Likewise b'_l is the complement of $a_{l+1} \vee c_{l+1}$ in $[b'_{l+1}, \alpha_{l+1}]$. Hence

$$\begin{aligned} [b_{l+1}, \alpha_{l+1}] &\simeq [b_l, \alpha_{l+1}] \times [a_{l+1} \vee c_{l+1}, \alpha_{l+1}] \qquad \text{and} \\ [b'_{l+1}, \alpha_{l+1}] &\simeq [b'_l, \alpha_{l+1}] \times [a_{l+1} \vee c_{l+1}, \alpha_{l+1}]. \end{aligned}$$

Since $\phi_l(\alpha_{l+1}) = \alpha'_{l+1} = \alpha_{l+1}$ we have $\phi_l \restriction [b_l, \alpha_{l+1}]$ is an isomorphism between $[b_l, \alpha_{l+1}]$ and $[b'_l, \alpha_{l+1}]$. We now define ϕ_{l+1} as the composite of the following isomorphisms:

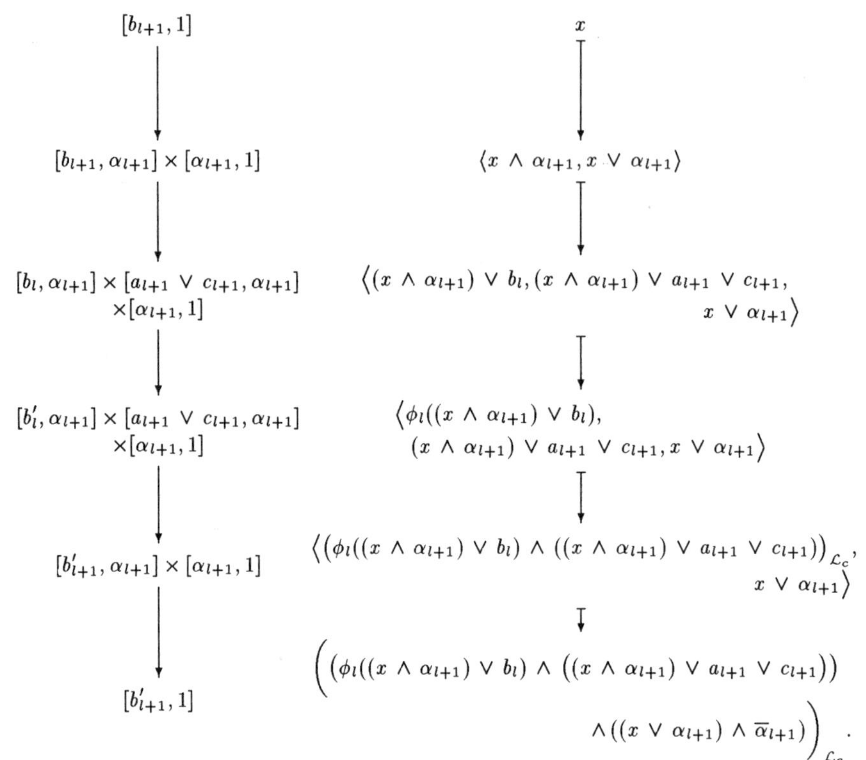

Although this looks rather complex, it is really just composing ϕ_l with standard Boolean isomorphisms on products and factors.

Now suppose that $x \in \mathcal{L}_{a_j}$ for some $1 \leq j \leq k$ and $x \geq b_{l+1}$. We need to show that $\phi_{l+1}(x) = x$. First we have

$$
\begin{aligned}
x = x \vee b_{l+1} &= x \vee \bigwedge_{i=1}^{l+1}(a_i \vee c_i) \\
&= \bigwedge_{i=1}^{l+1}(x \vee a_i \vee c_i) \qquad \text{in } [b_{l+1}, 1]. \\
x \vee b_l &= \bigwedge_{i=1}^{l+1}(x \vee a_i \vee c_i) \vee \bigwedge_{s=1}^{l}(a_s \vee c_s) \\
&= \bigwedge_{i=1}^{l+1}\bigwedge_{s=1}^{l}(x \vee a_i \vee c_i \vee a_s \vee c_s) \qquad \text{in } [b_{l+1}, 1].
\end{aligned}
$$

Now this is an $\mathcal{L}_a$ meet, and so must be an $\mathcal{L}_{a_j}$ meet (as everything is above x and so must be in $\mathcal{L}_{a_j}$) and an $\mathcal{L}_c$ meet as everything is greater than b'_{l+1}. Therefore we have

$$\begin{aligned} x \vee b_l &= \bigwedge_{i=1}^{l+1}(x \vee a_i \vee c_i) \vee \left(\bigwedge_{s=1}^{l} a_s \vee c_s\right)_{\mathcal{L}_c} \\ &= x \vee b'_l. \end{aligned}$$

Also because $x = \bigwedge_{i=1}^{l+1} x \vee a_i \vee c_i$ is an $\mathcal{L}_a$-meet, it must agree with the $\mathcal{L}_{a_j}$-meet, which must agree with the $\mathcal{L}_c$-meet (which exists as each term is greater than $b'_{l+1} \in \mathcal{L}_c$) and so we have $x \geq b'_{l+1}$.

Therefore we have (in $\mathcal{L}_c$) that

$$\begin{aligned} x &= (((x \wedge \alpha_{l+1}) \vee b'_l) \wedge ((x \wedge \alpha_{l+1}) \vee a_{l+1} \vee c_{l+1})) \wedge \\ &\quad ((x \vee \alpha_{l+1}) \wedge \overline{\alpha}_{l+1}). \end{aligned}$$

But

$$\begin{aligned} \phi_l((x \wedge \alpha_{l+1}) \vee b_l) &= \phi_l((x \vee b_l) \wedge \alpha_{l+1}) \\ &= \phi_l(x \vee b_l) \wedge \phi_l(\alpha_{l+1}) \\ &= (x \vee b_l) \wedge \alpha_{l+1} \\ &= (x \vee b'_l) \wedge \alpha_{l+1} \\ &= (x \wedge \alpha_{l+1}) \vee b'_l. \end{aligned}$$

It follows then that $\phi_{l+1}(x) = x$.

Now for any $x \in \mathcal{L}_{a_j}$ we have $\phi_{l+1}(x \vee b_{l+1}) = x \vee b_{l+1} \geq \phi_{l+1}(b_{l+1}) = b'_{l+1}$ in $\mathcal{L}$. Also $\phi_{l+1}(x \vee b_{l+1}) \geq x$ and so $\phi_{l+1}(x \vee b_{l+1}) = x \vee b_{l+1} \geq x \vee b'_{l+1}$. Using ϕ_{l+1}^{-1} we also have $x \vee b'_{l+1} \geq x \vee b_{l+1}$, and so they are equal. This takes care of the last two parts. ∎

The lemma sets us up to do a construction similar to the one in theorem 3.4. The feature lacking is a c' with $c' \leq b'$ and $\mathcal{L}_c = \mathcal{L}_{c'}$.

So we will do this next and show that the c' we find is in some sense the right choice.

Define

$$c' = \Delta(b' \vee c, c).$$

Then $c' \leq b'$ as $b' \in \mathcal{L}_c$ (using lemma 2.12 (b)).

As we are using the isomorphisms we defined in definition 2.14 we need to know how the isomorphisms $\mathcal{L}_a \simeq \mathcal{I}([a,1])$ and $\mathcal{L}_b \simeq \mathcal{I}([b,1])$ fit together when $a \leq b$.

Lemma 5.3 *If $y \in \mathcal{L}_x$ and $w \leq x$ then*

(a) $y \vee w = y \vee x$;

(b) $k_w(y) \vee x = k_x(y)$.

Proof

(a) We have $y \leq y \vee w \in \mathcal{L}_x$. Hence $\Delta(y \vee w \vee x, x) \leq y \vee w$. As $w \leq x$ we also have $\Delta(y \vee w \vee x, w) \leq \Delta(y \vee w \vee x, x)$ and so $\Delta(y \vee w \vee x, w) \leq y \vee w$.

Now we have $y \vee x \leq y \vee w \vee x = \Delta(y \vee w \vee x, w) \vee w \leq (y \vee w) \vee w = y \vee w \leq y \vee x$.

(b) We recall that in a Boolean algebra, if $r \leq s \leq t$ then the complement of t in $[s, 1]$ is equal to the complement of t in $[r, 1]$ joined with s.

Now as $y \in \mathcal{L}_x$ we also have $\Delta(1, y) \in \mathcal{L}_x$ and so by (a) we have $\Delta(1, y) \vee w = \Delta(1, y) \vee x$. Then we have

$$\begin{aligned} k_w(y) \vee x &= (\Delta(1,y) \vee w)w \vee x \\ &= (\Delta(1,y) \vee x)w \vee x \\ &= (\Delta(1,y) \vee x)x \\ &\qquad\qquad \text{by the remark about complements} \\ &= k_x(y). \end{aligned}$$

■

Now we define $\mathcal{L}^*$ as $\mathcal{I}([b', 1] \times [a, b] \times [c', b'])$. By theorem 3.4, we know that the mappings

$$\begin{aligned} \Phi_a \colon \ & \mathcal{L}_a \to \mathcal{L}^* \qquad \text{defined by} \\ \Phi_a(x) = \ & [\langle \phi(k_a(x) \vee b), k_a(x) \wedge b, c' \rangle, \langle \phi(\ell_a(x) \vee b), \ell_a(x) \wedge b, b' \rangle] \\ \text{and } \Phi_c \colon \ & \mathcal{L}_c \to \mathcal{L}^* \qquad \text{defined by} \\ \Phi_c(x) = \ & [\langle k_{c'}(x) \vee b', a, k_{c'}(x) \wedge b' \rangle, \langle \ell_{c'}(x) \vee b', b, \ell_{c'}(x) \wedge b' \rangle]. \end{aligned}$$

are cubic embeddings. This uses the fact that $\mathcal{L}_a \simeq \mathcal{I}([a, 1])$ via $x \mapsto [k_a(x), \ell_a(x)]$ – see lemmas 2.15-2.17.

The "overlap" of the domains of these two mappings is essentially $\mathcal{L}_b$ or its isomorphic copy $\mathcal{L}_{b'}$. We first need to show that these two algebras are in fact isomorphic, and then show, that modulo this isomorphism, the two mappings Φ_a and Φ_c agree on the "overlap".

There is a natural mapping, $\widehat{\phi}$, from $\mathcal{L}_b$ to $\mathcal{L}_{b'}$ induced by ϕ, namely $x \mapsto \phi(x \vee b) \wedge \Delta(1, \phi(k_b(x))b')$. The next two lemmas will show that this mapping is both defined for all x and is an isomorphism.

Lemma 5.4 *Let $\mathcal{L}$ be an MR-algebra, and suppose that $x \geq b$ and $y \geq \Delta(1, b)$ and $x \wedge y$ exists. Then $(x \wedge y) \vee b = x$ and $(x \wedge y) \vee \Delta(1, b) = y$.*

Proof We will prove the first result as the second follows from the first by interchanging b with $\Delta(1, b)$ and x with y.

If either x or y is 1, or if $x = y$ the result is clear. Otherwise we have

$$\begin{array}{lll} (x \wedge y) \vee b < x & & \\ \iff & \Delta(x,b) \wedge (x \wedge y) \text{ does not exist} & \text{by the MR axiom} \\ \iff & \Delta(x,b) \wedge y \text{ does not exist} & \text{as } \Delta(x,b) \leq x \\ \iff & \Delta(x,b) \vee \Delta(1,y) < 1 & \text{by the MR axiom.} \end{array}$$

But $x, \Delta(1,y) \geq b$ and so $1 = \Delta(x,b) \vee \Delta(1,y)$ by lemma 2.4. Hence $(x \wedge y) \vee b = x$. ■

Lemma 5.5 *For all* $x \in \mathcal{L}_b$

(a) $\widehat{\phi}(x)$ *is defined;*

(b) $\widehat{\phi}(x) \vee b' = \phi(x \vee b)$*;*

(c) $\widehat{\phi}(x) \vee \Delta(1,b') = \Delta(1, \phi(k_b(x))b')$*;*

(d) $k_{b'}(\widehat{\phi}(x)) = \phi(k_b(x))$*;*

(e) $\widehat{\phi}\colon \mathcal{L}_b \to \mathcal{L}_{b'}$ *is an isomorphism.*

Proof We know that $\phi\colon [b,1] \to [b',1]$ is a Boolean isomorphism. Let $x \in \mathcal{L}_b$.

(a) We have $k_b(x) \leq \ell_b(x)$ by lemma 2.15. As ϕ is an isomorphism, we have $b' \leq \phi(k_b(x)) \leq \phi(x \vee b)$ and hence by corollary 2.8 (b) the meet $\phi(x \vee b) \wedge \Delta(1, \phi(k_b(x))b')$ exists.

(b, c) $\phi(x \vee b) \geq b'$ and $\phi(k_b(x))b' \geq b'$ and so $\Delta(1, \phi(k_b(x))b') \geq \Delta(1, b')$. By the lemma above we have

(b) $\widehat{\phi}(x) \vee b' = \phi(x \vee b)$; and

(c) $\widehat{\phi}(x) \vee \Delta(1,b') = \Delta(1, \phi(k_b(x))b')$.

(d)
$$\begin{aligned} k_{b'}(\widehat{\phi}(x)) &= (\Delta(1, \widehat{\phi}(x)) \vee b')b' \\ &= \Delta(1, \widehat{\phi}(x) \vee \Delta(1,b'))b' \\ &= \Delta(1, \Delta(1, \phi(k_b(x))b'))b' \\ &= (\phi(k_b(x))b')b' \\ &= \phi(k_b(x)). \end{aligned}$$

(e) $\widehat{\phi}$ **preserves order** as $x \leq y$ implies $b \leq k_b(y) \leq k_b(x) \leq \ell_b(x) \leq \ell_b(y)$ by lemma 2.15 (a). Then we have $\phi(k_b(y)) \leq \phi(k_b(x)) \leq \phi(\ell_b(x)) \leq \phi(\ell_b(y))$ as ϕ preserves order.

This gives $\phi(k_b(x))b' \leq \phi(k_b(y))b'$ and so $\Delta(1, \phi(k_b(x))b') \leq \Delta(1, \phi(k_b(y))b')$.

Putting all this together gives $\widehat{\phi}(x) \leq \widehat{\phi}(y)$.

$\widehat{\phi}$ **is one-one** If $\widehat{\phi}(x) = \widehat{\phi}(y)$ then by (a) and (c) we have $\phi(x \vee b) = \phi(y \vee b)$ and $\phi(k_b(x)) = \phi(k_b(y))$. As ϕ is one-one this gives us $x \vee b = y \vee b$ and $k_b(x) = k_b(y)$. By lemma 2.15 (b) we have $x = y$.

$\widehat{\phi}$ **is onto** as if $y \in \mathcal{L}_{b'}$ then $y \vee b' \geq k_{b'}(y) \geq b'$. As ϕ is an isomorphism, there exist $p \geq q \geq b$ with $\phi(p) = y \vee b'$ and $\phi(q) = k_{b'}(y)$.

Now we apply lemma 2.16 (a) to find $z \in \mathcal{L}_b$ such that $z \vee b = p$ and $k_b(z) = q$. Then we have

$$\begin{aligned}
\widehat{\phi}(z) &= \phi(z \vee b) \wedge \Delta(1, \phi(k_b(z))b') \\
&= \phi(p) \wedge \Delta(1, \phi(q)b') \\
&= (y \vee b') \wedge \Delta(1, k_{b'}(y)b') \\
&= (y \vee b') \wedge \Delta(1, \Delta(1, y) \vee b') \\
&= (y \vee b') \wedge (y \vee \Delta(1, b')) \\
&= y \qquad\qquad \text{as } b' \preccurlyeq y.
\end{aligned}$$

■

We want to show that the mappings agree on the "overlap", i.e. for all $x \in \mathcal{L}_b$ we have $\Phi_a(x) = \Phi_c(\widehat{\phi}(x))$.

$$\begin{aligned}
\Phi_c(\widehat{\phi}(x)) &= \Big[\Big\langle k_{c'}(\widehat{\phi}(x)) \vee b', a, k_{c'}(\widehat{\phi}(x)) \wedge b'\Big\rangle, \\
&\qquad \Big\langle \ell_{c'}(\widehat{\phi}(x)) \vee b', b, \ell_{c'}(\widehat{\phi}(x)) \wedge b'\Big\rangle\Big] \\
\Phi_a(x) &= [\langle \phi(k_a(x) \vee b), k_a(x) \wedge b, c'\rangle, \\
&\qquad \langle \phi(\ell_a(x) \vee b), \ell_a(x) \wedge b, b'\rangle].
\end{aligned}$$

We will show that each of the components are equal:

-

$$\begin{aligned}
k_{c'}(\widehat{\phi}(x)) \vee b' &= k_{b'}(\widehat{\phi}(x)) && \text{by lemma 5.3 (b)} \\
&= \phi(k_b(x)) && \text{by lemma 5.5 (d)} \\
&= \phi(k_a(x) \vee b) && \text{by lemma 5.3 (b).}
\end{aligned}$$

- $k_a(x)$ is the $[a, 1]$-complement of $\Delta(1, x) \vee a = \Delta(1, x) \vee b$ – this equality is lemma 5.3 (a). As $\Delta(1, x) \vee b \geq b$ we have $k_a(x) \leq ba$ and so $a \leq k_a(x) \wedge b \leq (ba) \wedge b = a$.

- In similar fashion we have $k_{c'}(\widehat{\phi}(x)) \wedge b' = c'$.

-
$$\begin{aligned}
\phi(\ell_a(x) \vee b) &= \phi(x \vee a \vee b) \\
&= \phi(x \vee b) && \text{as } a \leq b \\
&= \widehat{\phi}(x) \vee b' && \text{by lemma 5.5 (b)} \\
&= (\widehat{\phi}(x) \vee c') \vee b' && \text{as } c' \leq b' \\
&= \ell_{c'}(\widehat{\phi}(x)) \vee b'.
\end{aligned}$$

-
$$\begin{aligned}
\ell_a(x) \wedge b &= (x \vee a) \wedge b \\
&= (x \vee b) \wedge b && \text{by lemma 5.3 (a)} \\
&= b.
\end{aligned}$$

- In similar fashion we have $\ell_{c'}(\widehat{\phi}(x)) \wedge b' = b'$.

Noting that if $x \in \mathcal{L}_{a_i} \cap \mathcal{L}_b$ then

$$\begin{aligned}
\widehat{\phi}(x) &= \phi(x \vee b) \wedge \Delta(1, \phi(k_b(x))b') && \text{by lemma 5.2 (c)} \\
&= (x \vee b') \wedge \Delta(1, \phi((\Delta(1,x) \vee b)b)b') \\
&= (x \vee b') \wedge \Delta(1, (\phi(\Delta(1,x) \vee b)\phi(b))b') && \text{as } \phi \text{ is an isomorphism} \\
&= (x \vee b') \wedge \Delta(1, ((\Delta(1,x) \vee b')b')b') && \text{by lemma 5.2 (c)} \\
&= (x \vee b') \wedge \Delta(1, \Delta(1,x) \vee b') \\
&= (x \vee b') \wedge (x \vee \Delta(1,b')) \\
&= (x \vee b) \wedge (x \vee \Delta(1,b)) && \text{by lemma 5.2 (d)} \\
&= x && \text{as } b \preccurlyeq x.
\end{aligned}$$

which shows us that $\mathcal{L}_{a_i}$ is correctly embedded.

The last thing we need to verify is that one, join and delta are preserved. Now one and delta must be preserved as they are local operations. The join of two things both in $\mathcal{L}_c$ or both in $\mathcal{L}_a$ must be preserved also, as the mappings Φ preserve join. It remains therefore to deal with mixed joins.

The mixed join only makes sense where we have some $x \in \mathcal{L}_{a_i}$ and $y \in \mathcal{L}_c$, in which case there is a join $\Phi_a(x) \vee \Phi_c(y)$ defined in $\mathcal{L}^*$, and a join $x \vee y$ defined in $\mathcal{L}$. Since $x \vee y \in \mathcal{L}_{a_i} \cap \mathcal{L}_c$ it has image in $\mathcal{L}^*$ given by $\Phi_a(x \vee y) = \Phi_c(x \vee y)$. We need to show that these are the same.

Now

$$\begin{aligned}
&\Phi_a(x \vee y) \\
&\quad = [\langle k_{c'}(x \vee y) \vee b', a, c' \rangle, \langle \ell_{c'}(x \vee y) \vee b', b, b' \rangle]
\end{aligned}$$

and $\Phi_a(x) \vee \Phi_c(y)$

$$\begin{aligned}
&= [\langle \phi(k_a(x) \vee b), k_a(x) \wedge b, c' \rangle, \langle \phi(x \vee b), (x \vee a) \wedge b, b' \rangle] \vee \\
&\qquad [\langle k_{c'}(y) \vee b', a, k_{c'}(y) \wedge b' \rangle, \langle \ell_{c'}(y) \vee b', b, \ell_{c'}(y) \wedge b' \rangle] \\
&= [\langle \phi(k_a(x) \vee b) \wedge (k_{c'}(y) \vee b'), (k_a(x) \wedge b) \wedge a, c' \wedge (k_{c'}(y) \wedge b') \rangle, \\
&\qquad \langle \phi(x \vee b) \vee (\ell_{c'}(y) \vee b'), ((x \vee a) \wedge b) \vee b, b' \vee (\ell_{c'}(y) \wedge b') \rangle] \\
&= [\langle \phi(k_a(x) \vee b) \wedge (k_{c'}(y) \vee b'), a, c' \rangle, \langle \phi(x \vee b) \vee (\ell_{c'}(y) \vee b'), b, b' \rangle].
\end{aligned}$$

We will show that these are componentwise the same –

- We know that $x \in \mathcal{L}_{a_i}$ implies $\phi(x \vee b) = x \vee b'$ (by lemma 5.2 (c)) and so
  $\phi(x \vee b) \vee (\ell_{c'}(y) \vee b') = (x \vee b') \vee ((y \vee c') \vee b') = (x \vee y) \vee b'$.
  Therefore

$$\begin{aligned}
&\phi(k_a(x) \vee b) \wedge (k_{c'}(y) \vee b') && \\
&\quad = \phi(k_b(x)) \wedge k_{b'}(y) && \text{by lemma 5.3 (b)} \\
&\quad = \phi((\Delta(1,x) \vee b)b) \wedge (\Delta(1,y) \vee b')b' && \\
&\quad = \phi(\Delta(1,x) \vee b)\phi(b) \wedge (\Delta(1,y) \vee b')b' && \text{as } \phi \text{ is an isomorphism} \\
&\quad = (\Delta(1,x) \vee b')b' \wedge (\Delta(1,y) \vee b')b' && \text{by lemma 5.2(c)} \\
&\quad = (\Delta(1,x) \vee \Delta(1,y) \vee b')b' && \text{De Morgan's law} \\
&\quad = (\Delta(1,x \vee y) \vee b')b' && \text{by lemma 2.3} \\
&\quad = k_{b'}(x \vee y) && \\
&\quad = k_{c'}(x \vee y) \vee b' && \text{by lemma 5.3 (b).}
\end{aligned}$$

-

$$\begin{aligned}
&\phi(x \vee b) \vee (\ell_{c'}(y) \vee b') && \\
&\quad = (x \vee b') \vee ((y \vee c') \vee b') && \text{by lemma 5.2 (c)} \\
&\quad = x \vee y \vee b' && \text{as } c' \leq b' \\
&\quad = \ell_{c'}(x \vee y) \vee b'. &&
\end{aligned}$$

The last thing to note is that

$$\begin{aligned}
&\Phi_a(a) \wedge \Phi_c(c') \\
&\quad = [\langle\phi(k_a(a) \vee b), k_a(a) \wedge b, c'\rangle, \langle\phi(\ell_a(a) \vee b), \ell_a(a) \wedge b, b'\rangle] \wedge \\
&\qquad\qquad [\langle k_{c'}(c') \vee b', a, k_{c'}(c') \wedge b'\rangle, \langle\ell_{c'}(c') \vee b', b, \ell_{c'}(c') \wedge b'\rangle] \\
&\quad = [\langle\phi(b), a, c'\rangle, \langle\phi(b), a, b'\rangle] \wedge [\langle b', a, c'\rangle, \langle b', b, c'\rangle] \\
&\quad = [\langle b', a, c'\rangle, \langle b', a, c'\rangle]
\end{aligned}$$

is an atom of $\mathcal{L}^*$ and so we have

$$\mathcal{L}^* = \mathcal{L}_{\left(\bigwedge_{i=1}^{k} \Phi_a(a_i) \wedge \Phi_c(c')\right)}.$$

■

6 The General Embedding

For the general case we apply the compactness theorem. Let $\mathcal{L}$ be any cubic algebra. Let $\mathbb{L}$ be the language $\{1, \rightarrow, \vee, \Delta, \leq, =\} \cup \{\underline{l} \mid l \in \mathcal{L}\}$. Let $\mathbb{L}^+ = \mathbb{L} \cup \{c\}$ where c is a new constant symbol. Let T be the set of all atomic sentences of $\mathbb{L}$ which are true in $\mathcal{L}$ together with

$\{\neg(\underline{a} = \underline{b}) \mid a \neq b \text{ in } \mathcal{L}\}$ and the axioms of theorem 2.1. Let $T' = T \cup \{\forall x\, \Delta(x \vee c, c) \leq x\}$. We wish to show that T' has a model.

Let T_0 be a finite subset of T', and suppose $a_1, \ldots, a_k$ are all the elements l of $\mathcal{L}$ such that $\underline{l}$ appears in some sentence in T_0. We first note that
$\left\langle \bigcup_{i=1}^k \mathcal{L}_{a_i}, a_1, \ldots, a_k \right\rangle \models T_0 \setminus \{\forall x\, \Delta(x \vee c, c) \leq x\}$ as $\bigcup_{i=1}^k \mathcal{L}_{a_i}$ is a subalgebra of $\mathcal{L}$.

By theorem 5.1 above we can embed $\mathcal{L}_{a_1} \cup \ldots \cup \mathcal{L}_{a_k}$ into an MR-algebra $\mathcal{L}^0 = \mathcal{L}_a$ say. If we interpret c in $\mathcal{L}^0$ as a, then all of the sentences in T_0 are true in $\langle \mathcal{L}^0, a \rangle$.

So now let $\mathcal{L}^*$ be a model for T'. Then since $\mathcal{L}^*$ models all the axioms of theorem 2.1, we know that $\mathcal{L}_{c^{\mathcal{L}^*}}$ is an MR-algebra. But we also know, as $\left\langle \mathcal{L}^*, c^{\mathcal{L}^*} \right\rangle \models \forall x\, \Delta(x \vee c, c) \leq x$ that every element of $\mathcal{L}^*$ is in $\mathcal{L}_{c^{\mathcal{L}^*}}$. Hence $\mathcal{L}^*$ is an MR-algebra.

Furthermore the mapping $e\colon \mathcal{L} \to \mathcal{L}^*$ defined by $e(x) = \underline{x}^{\mathcal{L}^*}$ is an embedding (this is a general fact about theories of this type).

We now have

Theorem 6.1 *Let $\mathcal{L}$ satisfy the axioms of theorem 2.1. Then $\mathcal{L}$ is isomorphic to a subalgebra of an MR-algebra.*

References

[1] Abbott,J.C. *Sets, Lattices, and Boolean Algebras,* Allyn and Bacon, Boston, MA, 1969.

[2] Bennet,M.K. *The face lattice of an n-dimensional cube,* Algebra Universalis 14 (1982), 82-86.

[3] Birkhoff,G. *Lattice Theory,* AMS Colloquium Publications, Vol.25, Providence RI, 1979.

[4] Bailey,C.G. *Free Cubic Algebras,* in preparation.

[5] Bailey,C.G. and Oliveira,J.S. *Cubic lattices,* preprint.

[6] Chen,W.Y.C. and Oliveira,J.S. *Implication Algebras and the Metropolis-Rota Axioms for Cubic Lattices,* preprint.

[7] Metropolis,N. and Rota,G.-C. *Combinatorial Structure of the Faces of the n-cube,* SIAM J.Appl.Math 35 (1978), 689-694.

[8] Oliveira,J.S. *The Theory of Cubic Lattices,* Ph.D. thesis, MIT, 1992.

School of Mathematical and Computing Sciences, Victoria University of Wellington, Wellington, New Zealand
UCSF Medical School, San Francisco, U.S.A.

An Elementary Proof of Roichman's Rule for Irreducible Characters of Iwahori–Hecke Algebras of Type A

*Arun Ram**

Dedicated to Gian-Carlo Rota

Introduction

The purpose of this note is to give an elementary proof of a recent formula of Y. Roichman [Ro] which describes the irreducible characters of the Iwahori-Hecke algebras of type A. Roichman's original proof is via a detailed analysis of the action of certain elements in the Iwahori-Hecke algebra in the terms of the Kazhdan–Lusztig basis of each irreducible representation. Here we shall show that the Robinson–Schensted–Knuth insertion algorithm can be used to show that Roichman's rule is equivalent to the "Frobenius formula" for the characters of the Iwahori-Hecke algebras [Ra].

Let $\chi_q^\lambda(\mu)$ be the irreducible character of the Iwahori-Hecke algebra of type A associated to the partition λ evaluated at the element T_{γ_μ} where $\gamma_\mu = \gamma_{\mu_1} \times \cdots \times \gamma_{\mu_\ell} \in S_{\mu_1} \times \cdots \times S_{\mu_\ell}$ and $\gamma_r = (1, 2, \ldots, r) \in S_r$ (in cycle notation). Then, Roichman's formula is

$$\chi_q^\lambda(\mu) = \sum_Q \mathrm{rw}_q^\mu(Q), \tag{0.1}$$

where the sum is over all *standard tableaux* Q of shape λ and the μ-weight of a standard tableau is given by

$$\mathrm{rw}_q^\mu(Q) = \prod_{\substack{1 \le j \le k \\ j \notin B(\mu)}} f_\mu(j, Q), \quad \text{where } B(\mu) = \{\mu_1 + \cdots + \mu_r \mid 1 \le r \le \ell\}, \text{ and} \tag{0.2}$$

* Research supported by an Australian Research Council Fellowship.

$$f_\mu(j,Q) = \begin{cases} -1, & \text{if } j+1 \text{ is southwest of } j \text{ in } Q, \\ 0, & \text{if } j+1 \text{ is northeast of } j \text{ in } Q,\ j+1 \notin B(\mu), \\ & \qquad \text{and } j+2 \text{ is southwest of } j+1 \text{ in } Q, \\ q, & \text{otherwise.} \end{cases} \tag{0.3}$$

The notations and directions for partitions and their Ferrers diagrams are as in [Mac], "northeast" means weakly north and strictly east and "southwest" means strictly south and weakly west.

A consequence of this new proof of Roichman's rule is that we obtain a new insertion scheme proof of the formula, [Mac] Ch. I (7.8),

$$p_\mu = \sum_{\lambda \vdash k} \chi^\lambda(\mu) s_\lambda, \tag{0.4}$$

where p_μ is the power sum symmetric function, s_λ is the Schur function and $\chi^\lambda(\mu)$ is the irreducible character of the symmetric group S_k indexed by the partition λ and evaluated at a permutation of cycle type $\mu = (\mu_1, \ldots, \mu_\ell)$. We also obtain new insertion scheme proofs of the "Frobenius formula" which was proved in [Ra] and a new RSK-insertion scheme approach to the irreducible characters of the symmetric group.

One should note that Roichman's formula is actually more general than what I have stated above because I am considering only the Type A case. D. White [Wh] has also analyzed the characters of the symmetric group by an analogue of the RSK insertion scheme. His methods are different from those used in this paper.

Symmetric functions

Fix a positive integer n, let $x_1, \ldots, x_n$ be commuting variables and let q be an indeterminate. If λ is a partition let $s_\lambda = s_\lambda(x_1, \ldots, x_n)$ denote the Schur function associated to λ [Mac]. Define

$$\begin{aligned} q_r &= q_r(x_1, x_2, \ldots, x_n; q) \\ &= \sum_{1 \le i_1 \le \cdots \le i_r \le n} x_{i_1} \cdots x_{i_r} q^{(\#\text{ of } i_j = i_{j+1})}(q-1)^{(\#\text{ of } i_j < i_{j+1})}, \end{aligned} \tag{1.1}$$

and for a partition $\mu = (\mu_1, \ldots, \mu_\ell)$ define $q_\mu = q_\mu(x_1, \ldots, x_n; q) = q_{\mu_1} q_{\mu_2} \cdots q_{\mu_\ell}$. The following are well known facts:

If $\mu \vdash k$ then

$$p_\mu = \sum_{\lambda \vdash k} \chi^\lambda(\mu) s_\lambda, \tag{1.2}$$

where $\chi^\lambda(\mu)$ is the character of the symmetric group S_k associated to the partition λ evaluated at an element of cycle type μ.

[Ra] If $\mu \vdash k$ then

$$q_\mu = \sum_{\lambda \vdash k} \chi_q^\lambda(\mu) s_\lambda, \tag{1.3}$$

where $\chi_q^\lambda(\mu)$ is the character of the Iwahori-Hecke algebra of type A associated to the partition λ evaluated at the element T_{γ_μ} where $\gamma_\mu = \gamma_{\mu_1} \times \cdots \times \gamma_{\mu_\ell} \in S_{\mu_1} \times \cdots \times S_{\mu_\ell}$ and $\gamma_r = (1, 2, \ldots, r) \in S_r$ (in cycle notation).

$\chi^\lambda(\mu) = \chi_q^\lambda(\mu)\big|_{q=1}$ and $q_\mu(x_1, \ldots, x_n; 1) = p_\mu(x_1, \ldots, x_n)$, where p_μ is the power sum symmetric function. (1.4)

Lemma 1.5. *If $(i_1, \ldots, i_r)$ is a sequence $1 \le i_1, \ldots, i_r \le n$, define*

$$\mathrm{wt}_q(i_1, \ldots, i_r) = \begin{cases} (-1)^s q^{r-1-s}, & \text{if there exists an } s,\ 0 \le s < r, \text{ such} \\ & \text{that } i_1 < \cdots < i_s < i_{s+1} \ge \cdots \ge i_r, \\ 0, & \text{otherwise.} \end{cases}$$

Then $$q_r(x_1, \ldots, x_n; q) = \sum_{1 \le i_1, \ldots, i_r \le n} x_{i_1} \cdots x_{i_r} \mathrm{wt}_q(i_1, \ldots, i_r).$$

Proof. Let $1 \le j_1 \le j_2 \le \cdots \le j_r \le n$ be an increasing sequence and let us show that

$$\sum_{1 \le i_1, \ldots, i_r \le n} x_{i_1} \cdots x_{i_r} \mathrm{wt}_q(i_1, \ldots, i_r)\big|_{x_{j_1} \cdots x_{j_r}} = q^{r-1-\ell}(q-1)^\ell,$$

where ℓ is the number of $j_k < j_{k+1}$ in the sequence $j_1 \le j_2 \le \ldots \le j_r$.

Let D be the set of distinct elements in the sequence $j_1 \le \cdots \le j_r$, let m be the maximal element of D and let $D' = D \backslash \{m\}$. For each subset S of D' let

$s = \mathrm{Card}(S)$,

$i_1 < i_2 < \cdots < i_s$ be the elements of S in increasing order,

$i_{s+1} = m$, and let

$i_{s+2} \geq i_{s+3} \geq \cdots \geq i_r$ be the remainder of the elements of the sequence $j_1, \ldots, j_r$ arranged in decreasing order.

Then $x_{i_1} \cdots x_{i_r} = x_{j_1} \cdots x_{j_r}$ and $\mathrm{wt}_q(i_1, \ldots, i_r) = (-1)^s q^{r-1-s}$.

We have $x_{i_1} \cdots x_{i_r} = x_{j_1} \cdots x_{j_r}$ and $\mathrm{wt}_q(i_1, \ldots, i_r) \neq 0$ if and only if

(1) there exists a permutation $\sigma \in S_r$ such that $(i_1, \ldots, i_r) = (j_{\sigma(1)}, \ldots, j_{\sigma(r)})$ and

(2) there exists an s such that $0 \leq s < r$ and

$$j_{\sigma(1)} < \cdots < j_{\sigma(2)} < \cdots < j_{\sigma(s)} < j_{\sigma(s+1)} \geq j_{\sigma(s+2)} \geq \cdots \geq j_{\sigma(r)}.$$

It follows from this that every sequence $1 \leq i_1, \ldots, i_r \leq n$ such that $x_{i_1} \cdots x_{i_r} = x_{j_1} \cdots x_{j_r}$ and $\mathrm{wt}_q(i_1, \ldots, i_r) \neq 0$ is of the form given in the previous paragraph for a unique subset S of D'. Thus

$$\sum_{1 \leq i_1, \ldots, i_r \leq n} x_{i_1} \cdots x_{i_r} \mathrm{wt}_q(i_1, \ldots, i_r)\big|_{x_{j_1} \cdots x_{j_r}} = \sum_{S \subseteq D'} q^{r-|S|-1}(-1)^{|S|}$$
$$= \sum_{s=0}^{\ell} \binom{\ell}{s} (-1)^s q^{r-s-1} = q^{r-1-\ell}(q-1)^{\ell},$$

where $\ell = \mathrm{Card}(D) = (\#\text{ of } j_k < j_{k+1} \text{ in } j_1 \leq \cdots \leq j_r)$. ∎

Corollary 1.6. *Let $\mu = (\mu_1, \ldots, \mu_\ell)$ be a partition of k and let $B(\mu) = \{\mu_1 + \cdots + \mu_r \mid 1 \leq r \leq \ell\}$. Then*

$$q_\mu = \sum_{i_1, \ldots, i_k} x_{i_1} \cdots x_{i_k} \mathrm{wt}_q^{\mu}(i_1, \ldots, i_k),$$

where the sum is over all $1 \leq i_1, \ldots i_k \leq n$,

$$\mathrm{wt}_q^{\mu}(i_1, \ldots, i_k) = \prod_{\substack{1 \leq j \leq k \\ j \notin B(\mu)}} \phi_\mu(j; i_1, \ldots, i_k), \qquad \textit{and}$$

$$\phi_\mu(j; i_1, \ldots, i_k) = \begin{cases} -1, & \textit{if } i_j < i_{j+1}, \\ 0, & \textit{if } i_j \geq i_{j+1} < i_{j+2} \textit{ and } j+1 \notin B(\mu), \\ q, & \textit{otherwise.} \end{cases}$$

Proof. The result follows immediately from the previous lemma once we note that with $\mathrm{wt}_q(i_1, \ldots, i_r)$ defined as in Lemma 1.5 we have

$$\mathrm{wt}_q(i_1, \ldots, i_r) = \prod_{j=1}^{r} \phi(j; i_1, \ldots, i_r), \quad \text{where}$$

$$\phi(j; i_1, \ldots, i_r) = \begin{cases} -1, & \text{if } i_j < i_{j+1}, \\ 0, & \text{if } i_j \geq i_{j+1} < i_{j+2} \text{ and } j+1 \neq r. \\ q, & \text{otherwise.} \end{cases} \quad \blacksquare$$

The following corollary follows from the last one by setting $q = 1$.

Corollary 1.7.

$$p_\mu = \sum_{i_1, \ldots, i_k} x_{i_1} \cdots x_{i_k} \mathrm{wt}_1^\mu(i_1, \ldots, i_k),$$

where the sum is over all $1 \leq i_1, \ldots, i_k \leq n$, *and* wt_1^μ *is as given in Corollary 1.6 except with* $q = 1$.

Column insertion

Let us assume that the variables $x_1, \ldots, x_n$ are ordered so that $x_1 < x_2 < \cdots < x_n$. A *column strict tableau of shape* λ is a filling of the boxes of the Ferrers diagram of λ with entries from $\{x_1, \ldots, x_n\}$ such that

(1) the entries in the rows are weakly increasing, left to right, and

(2) the entries in the columns are increasing, top to bottom.

If P is a column strict tableau let x^P be the product of the entries in P. One has $s_\lambda = \sum_P x^P$, where the sum is over all column strict tableaux P of shape λ.

The Robinson–Schensted–Knuth (RSK) insertion scheme gives a bijection between sequences $(x_{i_1}, \ldots, x_{i_k})$ and pairs (P, Q) where P is a column strict tableau and Q is a standard tableau and P and Q are the same shape. (The original references for the RSK insertion scheme are [Sz], [Sch] and [Kn]. For an expository treatment see [Sag]). The pair of tableaux obtained by RSK insertion of the sequence $x_{i_1}, \ldots, x_{i_k}$ is the pair $(P, Q) = (P_k, Q)$ determined recursively by setting

$P_0 = \emptyset$, $P_j = (P_{j-1} \leftarrow x_{i_j})$, and

the box of Q containing k is the box created upon the insertion of x_{i_j} into P_{j-1}.

Here $P_{j-1} \leftarrow x_{i_j}$ denotes the column strict tableau obtained by column insertion of the letter x_{i_j} into the column strict tableau P_{j-1}. The following proposition is a well known fact about RSK-insertion.

Proposition 2.1. *Let P_{j-1} be a column strict tableau and consider the insertions $(P_{j-1} \leftarrow x_{i_j}) \leftarrow x_{i_{j+1}}$.*

(1) *If $x_{i_j} < x_{i_{j+1}}$ then the box created upon insertion of $x_{i_{j+1}}$ into $P_j = (P_{j-1} \leftarrow x_{i_j})$ appears southwest of the box created upon insertion of x_{i_j} into P_{j-1}.*

(2) *If $x_{i_j} \geq x_{i_{j+1}}$ then the box created upon insertion of $x_{i_{j+1}}$ into $P_j = (P_{j-1} \leftarrow x_{i_j})$ appears northeast of the box created upon insertion of x_{i_j} into P_{j-1}.*

Our goal now is to show that Roichman's formula is equivalent to (1.3). For each pair of partitions $\lambda, \mu \vdash k$ define a polynomial $\eta_q^\lambda(\mu) \in \mathbb{Z}[q]$ by the right-hand side of Roichman's formula (0.1), i.e., define

$$\eta_q^\lambda(\mu) = \sum_Q \mathrm{rw}_q^\mu(Q), \tag{2.1}$$

where the sum is over all standard tableaux Q of shape λ and the μ-weight of a standard tableau is as given in (0.2).

Theorem 2.3. *If $\mu \vdash k$, then* $q_\mu = \sum_{\lambda \vdash k} \eta_q^\lambda(\mu) s_\lambda.$

Proof. RSK-insertion of a sequence $x_{i_1}, \ldots, x_{i_k}$ produces a pair (P, Q) consisting of a column-strict tableau P and a standard tableau Q where the shape λ of the tableau P is a partition with k boxes. Thus RSK-insertion combined with Proposition 2.1 implies the following identity

$$q_\mu = \sum_{i_1,\ldots,i_k} x_{i_1}\cdots x_{i_k}\mathrm{wt}_q^\mu(i_1,\ldots,i_k) = \sum_{(P,Q)} x^P \mathrm{rw}_q^\mu(Q)$$

$$= \sum_{\lambda\vdash k}\left(\sum_Q \mathrm{rw}_q^\mu(Q)\right)\left(\sum_P x^P\right) = \sum_{\lambda\vdash k}\eta_q^\lambda(\mu)s_\lambda. \qquad \blacksquare$$

It follows that (1.3) is equivalent to Roichman's theorem. We get the following corollary of Theorem 2.3 by setting $q = 1$.

Corollary 2.4.
(a) *If* $\mu \vdash k$ *then* $p_\mu = \sum_{\lambda\vdash k}\eta_1^\lambda(\mu)s_\lambda$.

(b) *For* $\lambda, \mu \vdash k$ *the character of the symmetric group* S_k *associated to the partition* λ *evaluated at an element of cycle type* μ *is given by*

$$\chi^\lambda(\mu) = \sum_Q \mathrm{rw}_1^\mu(Q),$$

where the sum is over all standard tableaux Q *of shape* λ *and* $\mathrm{rw}_1^\mu(Q)$ *is as given in (0.2) except with* $q = 1$.

References

[Kn] D.E. Knuth, Permutations, matrices and generalized Young tableaux, *Pacific J. Math.* **34**, No. 3 (1970).

[Mac] I.G. Macdonald, *Symmetric Functions and Hall Polynomials*, Second Edition, Oxford Univ. Press, Oxford, 1995.

[Ra] A. Ram, A Frobenius formula for the characters of the Hecke algebras, *Invent. Math.* **106** (1991), 461–488.

[Ro] Y. Roichman, A recursive rule for Kazhdan–Lusztig characters, *Adv. Math.* **129** (1997), No. 1, 25–29.

[Sag] B. Sagan, *The Symmetric Group*, Wadsworth and Brooks, Pacific Grove, California, 1991.

[Sch] C. Schensted, Longest increasing and decreasing subsequences, *Canad. J. Math.* **13** (1961), 179–191.

[Sz] M.P. Schützenberger, La correspondance de Robinson, in *Combinatoire et Représentation du Groupe Symétrique*, 1976 (D. Foata, Ed.), 59-113. Lecture Notes in Math. **579** Springer-Verlag, Berlin–New York, 1977.

[Wh] D. White, A bijection proving the orthogonality of the characters of S_n, *Adv. in Math.* **50** (1983), 160–186.

Department of Mathematics
Princeton University
Princeton, NJ 08544-1000 U.S.A.

Universal Constructions in Umbral Calculus

Nigel Ray

1. Introduction

Modern umbral calculus is steadily approaching maturity, as applications develop in several areas of mathematics. To maximize this utility it is important to work in the most general (as opposed to the most abstract) setting.

The origins of the 19th century theory lie in analysis. In a beautiful recent article [13] G.-C. Rota and B.D. Taylor have returned to these roots, and their bibliography details many of the appropriate works. Given this context the subject naturally developed over the real and complex numbers, and versions with a more algebraic flavour often maintained the requirement of working over a field. Even in S.Roman's book [12] of 1984, for example, the author invites us to suppose that the coefficients of the generic delta operator are invertible, although the hypothesis is never fully used.

Once we accept the need for a general theory, we may jettison the field of scalars and work over a commutative ring R (which we assume to have an identity). It is then a short step to describing generic examples in the categorical language of universality, and insisting on invertibility only when necessary. This viewpoint offers many new insights and challenges, and informs our belief that the most flexible basis for the study of umbral calculus lies in the category $\mathcal{C}_R$ of coassociative coalgebras over R together with the category $\mathcal{A}_R$ of dual algebras. Additional features such as gradings and Hopf algebra structures may naturally be present in certain circumstances. This viewpoint was conceived by S.A. Joni and G.-C. Rota [5] and developed by W. Nichols and M. Sweedler [8], although both works continued to suggest that R should usually be a field.

Our purpose here is to popularize elements of umbral calculus which have already been translated into the language of universal algebra, and to introduce new and related constructions which are motivated by emerging applications. For readers who are unfamiliar with the standard definitions and notations of coalgebra theory, we refer to [8] as a convenient source.

2. Basics

We define an *umbral calculus* (C, r) to consist of a coalgebra C in $\mathcal{C}_{\mathcal{R}}$ and an *evaluation* functional r in C^* (the dual $\mathrm{Hom}_R(C, R)$). For added convenience we insist that C be *supplemented*, in the sense that it is equipped with a summand of *scalars* whose projection is a counit $\varepsilon: C \to R$, and that r acts on this summand as the identity. Since C is a left C-comodule under the coproduct δ, it is also a right C^*-module by duality; in this guise, r acts on C as a right-invariant (or *shift-invariant*) R-linear endomorphism, whose value on c in C is usually denoted by $c \prec r$, or simply cr. In cocommutative situations we write this action on the left as rc. We define positive powers $r^{\star m}$ of r by convolution, and may then identify the action of $r^{\star m}$ with the m-fold iterate of the action of r.

A fundamental example is provided by taking C to be the polynomial algebra $R[x]$, with binomial coproduct $\delta(x^n) = \sum_k \binom{n}{k} x^k \otimes x^{n-k}$ for all $n \geq 0$, and counit $\varepsilon(p(x)) = p(0)$. Such an r is determined by its *umbra*, or sequence of values $r(x^n) = r_n$, and acts on the left of $R[x]$ by cocommutativity. Of course $R[x]$ is actually a Hopf algebra with respect to the polynomial product and the antipode $x \mapsto -x$, but we do not yet need this additional structure. The dual algebra consists of the divided power algebra (or *Hurwitz algebra* [1]) $R\{\{D\}\}$, where the functional D satisfies $D(x^n) = \delta_{1,n}$ and therefore acts on $R[x]$ as d/dx. In this context r may be expressed as the formal differential operator

$$r_1 D + r_2 D^2/2! + \cdots + r_n D^n/n! + \cdots, \tag{2.1.}$$

and composition of operators is given by the Cauchy product of formal power series.

The original umbral calculus acquired its shady reputation because of seemingly arbitrary manipulations with symbolic notation. Following Rota's lead, these manipulations are demythologized by the use of the functional r, expressing its value on an arbitrary polynomial $p(x)$ by the substitution

$$r(p(x)) = p(r) \qquad r^n \equiv r_n.$$

Similarly, we may rewrite the expression (2.1.) for r as e^{rD} in $R\{\{D\}\}$, so long as we insist that r^n denotes $r(x^n)$ whenever we formally expand in powers of rD; we may again describe the appropriate substitution in the form $r^n \equiv r_n$. With this convention, the action of r as an endomorphism on $R[x]$ may be abbreviated to

$$e^{rD} p(x) = p(x + r) \qquad r^n \equiv r_n,$$

thereby revealing e^{rD} as the *umbral shift*. If we have two functionals r and s then their convolution product $r \star s$ acts as $e^{rD} e^{sD}$, which we may contract to $e^{(r+s)D}$ so long as we insist that $(r+s)^n$ denotes $r \star s(x^n)$ whenever we expand. Since

$$r \star s(x^n) = \mu(r \otimes s(x \otimes 1 + 1 \otimes x)^n) = \sum_k \binom{n}{k} r_k s_{n-k} \tag{2.2.}$$

(where μ is the product map in $R[x]$), we may write $r \star s(x^n)$ as

$$(r+s)^n \qquad r^k \equiv r_k, \quad s^k \equiv s_k \tag{2.3.}$$

by applying the binomial theorem and the appropriate substitutions.

If r_1 is 1 we refer to the endomorphism $e^{rD} - 1$ as a *delta operator*, and label it Δ^r; it corresponds to the functional $r - \varepsilon$. An alternative pseudobasis for $R\{\{D\}\}$ is then given by the divided powers $\Delta^r_{(n)}$, and dualizing back (continuously, to recover $R[x]$) yields a new basis $B^r_n(x)$ of monic polynomials, whose elements form the *associated sequence* for Δ^r. This sequence is characterized by the binomial property

$$\delta(B^r_n(x)) = \sum_k \binom{n}{k} B^r_k(x) \otimes B^r_{n-k}(x),$$

together with the fact that $\varepsilon(B^r_n(x)) = 0$ for all $n > 0$. When working with Δ^r it is extremely convenient to reindex the umbra by $r(x^n) = r_{n-1}$ (so that $r_0 = 1$), and we adopt this convention henceforth.

We construct the universal delta operator Δ^ϕ by taking R to be the polynomial algebra $\mathbb{Z}[\phi_1, \phi_2, \ldots]$, abbreviated to Φ, and r to be the functional ϕ satisfying $\phi(x^n) = \phi_{n-1}$ for $n \geq 1$. Thus the endomorphism ϕ is the universal umbral shift. The associated sequence $B^\phi_n(x)$ consists of the *conjugate Bell polynomials*, for which (as we shall explain below) there exists an alternative description in terms of posets of partitions. The first three such polynomials are

$$x, \qquad x^2 - \phi_1 x, \qquad \text{and} \qquad x^3 - 3\phi_1 x^2 + (3\phi_1^2 - \phi_2)x. \tag{2.4.}$$

We often refer to the m-fold convolution $\phi^{\star m}$ as the mth *umbral integer*, and note that elementary computation along the lines of (2.2.) reveals

$$\begin{gathered} \phi^{\star m}(x) = m, \quad \phi^{\star m}(x^2) = m^2 + m(\phi_1 - 1), \\ \phi^{\star m}(x^3) = m^3 + 3m^2(\phi_1 - 1) + m(\phi_2 - 3\phi_1 + 2), \ldots \end{gathered} \tag{2.5.}$$

Thus we may interpret $(\phi^{\star x})^n$ as a polynomial of degree n in $\Phi[x]$, for all $n \geq 1$. In fact the formulae (2.5.) are equivalent to

$$\phi^{\star m}(B^\phi_n(x)) = [m]_n \tag{2.6.}$$

(where $[m]_n$ denotes the falling factorial $m(m-1)\cdots(m-n+1)$ for all $m,n \geq 0$), which follows directly from the definitions.

Our arbitrary delta operator Δ^r defines a homomorphism $\Phi \to R$ by $\phi_n \mapsto r_n$, through which its properties may be studied in terms of the universal example. The conjugate Bell polynomials are the universal binomial sequence, and the same homomorphism maps them to the associated sequence for Δ^r; again, the properties of the $B_n^r(x)$ may be investigated in terms of those of the universal example. One particular instance motivates much of our terminology, namely when R is $\mathbb{Z}$ and each r_n is 1. Then the functional $r^{\star m}$ is the substitution $x = m$ and the endomorphism r is the forward shift e^D, whilst Δ^r is the forward difference operator Δ and the associated sequence consists of the falling factorial polynomials $[x]_n$.

An important variation is provided by the subalgebra $R[[\Delta^r]]$ of the Hurwitz algebra. When we form its continuous dual we obtain an R-coalgebra, freely generated by elements b_n^r (where $n \geq 0$ and $b_0^r = 1$) which are characterized by the *divided power* property

$$\delta(b_n^r) = \sum_k b_k^r \otimes b_{n-k}^r,$$

together with the fact that $\varepsilon(b_n^r) = 0$ for all $n > 0$. We label this coalgebra as the *penumbral coalgebra* $\Pi(\Delta^r)$, and remark that the homomorphism $R[x] \to \Pi(\Delta^r)$ (dual to the inclusion) acts such that $B_n^r(x)$ maps to $n!\, b_n^r$. We therefore refer to the b_n^r as the *divided sequence* of Δ^r, although they may only be expressed as rational polynomials in x when R is free of additive torsion. The functional r, its associated umbral shift, and the delta operator Δ^r all extend naturally to the penumbral coalgebra; for example, Δ^r maps b_n^r to b_{n-1}^r for each $n \geq 1$, and so is an isomorphism modulo scalar. In order to ensure that $\Pi(\Delta^r)$ is closed under multiplication (and is therefore a Hopf algebra), we may have to extend the ring of scalars. By duality, this is tantamount to insisting that the differential operator Δ^r admits a Leibniz formula for its action on products, so we label the appropriate extension as R^L and the resulting Hopf algebra as $R^L\langle b_n^r\rangle$. Determining the explicit structure of R^L in any given case remains an intriguing unsolved problem.

In the universal example Φ is torsion free, so the elements of the universal divided sequence b_n^ϕ may be written as normalized conjugate Bell polynomials $B_n^\phi(x)/n!$ in the rationalization $\Phi\mathbb{Q}[x]$. In this case Φ^L is a genuine extension of Φ, satisfying $\Phi < \Phi^L < \Phi\mathbb{Q}$; it is known as the *Lazard ring*, and as explained in [3] it features prominently in the theories of formal group laws and stable homotopy. The Hopf algebra $\Phi^L\langle b_n^\phi\rangle$ is the covariant bialgebra of the universal formal group law, and masquerades as the complex bordism module of infinite dimensional complex projective space in algebraic topology.

The homomorphism $\Phi \to R$ extends naturally to $\Phi^L \to R^L$, and again allows the properties of an arbitrary case to be studied in terms of the universal example.

3. Sheffer Sequences

Associated sequences are special cases of Sheffer sequences, which arise when we invest our umbral calculus (C, r) with a right-invariant R-linear isomorphism u of C, which fixes the scalars. Thus u is a map of right C-comodules with respect to δ.

When C is $R[x]$ we may describe the isomorphism by a power series

$$u(\Delta^r) = 1 + u_1\Delta^r + u_2(\Delta^r)^2 + \cdots + u_n(\Delta^r)^n + \cdots, \tag{3.1.}$$

where the elements r_n and u_n of R may (or may not) be algebraically independent. Thus u is a unit in $R[[\Delta^r]]$, although it is more usually expressed in terms of D (as in [12], for example); however, the form (3.1.) is equivalent whenever $\mathbb{Q}$ is a subring of R, and is imperative for the universal viewpoint. The corresponding *Sheffer sequence* $S_n^{u,r}(x)$ consists of the monic, degree n polynomials $u(\Delta^r)B_n^r(x)$ in $R[x]$, where $n \geq 0$. Thus the $S_n^{u,r}(x)$ are given by $\sum_j [n]_j u_j B_{n-j}^r(x)$, and are characterized by the properties

$$\delta(S_n^{u,r}(x)) = \sum_k \binom{n}{k} S_k^{u,r}(x) \otimes B_{n-k}^r(x) \tag{3.2.}$$

and $\varepsilon(S_n^{u,r}(x)) = n!\, u_n$. Alternatively, the $S_n^{u,r}(x)$ may be defined to satisfy the orthogonality relations

$$\langle (\Delta^r)^m u^{-1}, S_n^{u,r}(x) \rangle = m!\, \delta_{m,n} \quad \text{for all} \quad m, n \geq 0,$$

with respect to the sequence of operators $(\Delta^r)^m u^{-1}$.

The universal Sheffer sequence lies over $\Psi \otimes \Phi$, where Ψ denotes the polynomial algebra $\mathbb{Z}[\psi_1, \psi_2, \ldots]$. Writing ψ_n for $\psi_n \otimes 1$ and ϕ_n for $1 \otimes \phi_n$, we may describe the universal isomorphism by means of the power series

$$\psi(\Delta^\phi) = 1 + \psi_1\Delta^\phi + \psi_2(\Delta^\phi)^2 + \cdots + \psi_n(\Delta^\phi)^n + \cdots.$$

The first three universal polynomials $S_n^{\psi,\phi}(x)$ are

$$x + \psi_1, \qquad x^2 + (2\psi_1 - \phi_1)x + 2\psi_2, \qquad \text{and}$$
$$x^3 + 3(\psi_1 - \phi_1)x^2 + (6\psi_2 - 3\psi_1\phi_1 + 3\phi_1^2 - \phi_2)x + 6\psi_3,$$

although there are other, equivalent forms.

Our arbitrary unit u defines a homomorphism $\Psi \otimes \Phi \to R$ by means of $\psi_n \mapsto u_n$ and $\phi_n \mapsto r_n$, which acts on polynomials by reducing each $S_n^{\psi,\phi}(x)$ to $S_n^{u,r}(x)$. The properties of the latter may therefore be investigated in terms of the universal example.

By way of illustration, we again consider the forward difference operator Δ over $\mathbb{Z}$; if we then let u be $\Delta/\log(1+\Delta)$, we obtain the Bernoulli polynomials of the second kind as the corresponding Sheffer sequence over $\mathbb{Q}$ (see [12] for further details).

Returning to the penumbral coalgebra $\Pi(\Delta^r)$ and the divided sequence b_n^r, we define the *divided Sheffer sequence* $s_n^{u,r}$ for the pair (u, Δ^r) by $u(\Delta^r)b_n^r$ for each $n \geq 0$. Thus $s_n^{u,r} = \sum_j u_j b_{n-j}^r$, and the sequence is characterized by the property

$$\delta(s_n^{u,r}) = \sum_k s_k^{u,r} \otimes b_{n-k}^r,$$

together with the values $\varepsilon(s_n^{u,r}) = u_n$ for all $n \geq 0$. Alternatively, the basis $s_n^{u,r}$ is dual to the pseudobasis $(\Delta^r)^n u^{-1}$ for $R[[\Delta^r]]$. In the universal example $\Psi \otimes \Phi$ is torsion free, so the elements $s_n^{\psi,\phi}$ may be written as $S_n^{\psi,\phi}(x)/n!$ in $\Psi \otimes \Phi\mathbb{Q}[x]$.

There is more structure to the universal example than meets the eye! This involves the polynomial algebra $\Psi \otimes \Psi$, in which we write ψ_n for $\psi_n \otimes 1$ and ψ'_n for $1 \otimes \psi_n$, as convenient.

We define a second delta operator $\psi_+(\Delta^\phi) = \Delta^\phi \psi(\Delta^\phi)$, and employ its compositional inverse to construct a second unit $\overline{\psi}(\Delta^\phi)$ satisfying $\psi_+(\Delta^\phi \overline{\psi}(\Delta^\phi)) = \Delta^\phi$ over $\Psi \otimes \Phi$. Thus the units $\psi'(\psi_+(\Delta^\phi))$ and $\pi(\Delta^\phi) = \psi(\Delta^\phi)\psi'(\psi_+(\Delta^\phi))$ over $\Psi \otimes \Psi \otimes \Phi$ are such that

$$\pi_+ = \psi'_+ \circ \psi_+ \qquad \text{and} \qquad \psi_+ \circ \overline{\psi}_+ = 1 \ ,$$

respectively. Simple computation (and Lagrange inversion) reveals that

$$\pi_n = \sum_{k=1}^{n} \psi'_k \psi_{n-k}^{k+1} \qquad \text{and} \qquad \overline{\psi}_n = \psi_n^{-(n+1)}/(n+1),$$

where $\psi = 1 + \psi_1 + \psi_2 + \cdots$ and ψ_{n-k}^k denotes the component of degree $n-k$ in the formal expansion ψ^k (assuming that each ψ_m has degree m).

We consider the homomorphism $\delta \colon \Psi \to \Psi \otimes \Psi$ induced by $\psi_n \mapsto \pi_n$ for all $n \geq 0$; this defines a coproduct map

$$\delta(\psi_n) = \sum_{k=0}^{n} \psi_{n-k}^{k+1} \otimes \psi_k,$$

for which there is an alternative construction in terms of the *Sheffer operator* $s_n^{\psi,\phi} \mapsto s_n^{\pi,\phi}$ (in the sense of [12]). We may verify directly that δ is coassociative, respects the product map μ, and has counit the projection onto $\mathbb{Z}$. The endomorphism χ of Ψ, induced similarly by $\psi_n \mapsto \overline{\psi}_n$ for all $n \geq 0$, is defined so that the composition $\mu \circ (1 \otimes \chi) \circ \delta$ coincides with the counit, and we therefore conclude that Ψ is a Hopf algebra with coproduct δ and antipode χ.

This Hopf algebra is better known to topologists as the dual of the Landweber–Novikov algebra, and to algebraic geometers as representing the affine group scheme which assigns to R the group of formal power series $t + t^2R[[t]]$ under composition. We conclude by outlining one of its more important properties.

Proposition 3.3. *The Hopf algebra Ψ coacts on the integral subcoalgebras of $\Phi[x]$ and $\Pi(\Delta^\phi)$ spanned by the $B_n^\phi(x)$ and b_n^ϕ respectively; furthermore, the coactions are compatible with respect to inclusion.*

Proof. For convenience we work rationally and consider the units $b(\Delta^\phi) = \sum_{n\geq 0} b_n^\phi(\Delta^\phi)^n$ over $\Phi\mathbb{Q}[x]$, and $b(\psi_+(\Delta^\phi))$ over $\Psi \otimes \Phi\mathbb{Q}[w]$. We consider the Φ-linear homomorphism $c\colon \Phi\mathbb{Q}[w] \to \Psi \otimes \Phi\mathbb{Q}[w]$ induced by $b_n^\phi \mapsto b(\psi_+)_n = \sum_{k=1}^n \psi_{n-k}^k \otimes b_k^\phi$, for each $n \geq 1$. This clearly restricts compatibly to both $\Phi[x]$ and $\Pi(\Delta^\phi)$, and it remains only to confirm that it satisfies the coaction condition $(1 \otimes c) \circ c = (\delta \otimes 1) \circ c$ as homomorphisms $\Phi\mathbb{Q}[w] \to \Psi \otimes \Psi \otimes \Phi\mathbb{Q}[w]$. Since both possibilities are induced by the same map $b_n^\phi \mapsto b(\psi'_+(\psi_+))_n$, the result follows. ∎

We remark that $b_n^\phi \mapsto \psi_{n-1}$ defines a homomorphism $\sigma\colon \mathbb{Z}\langle b_n^\phi\rangle \to \Psi$ of left Ψ-comodules. Moreover, we may extend the structures above so that $\Phi \otimes \Psi$ becomes a Hopf algebroid, which coacts on the Φ-modules $\Phi[w]$ and $\Pi(\Delta^\phi)$ in their entirety. Correspondingly, σ extends to a comodule homomorphism $\Pi(\Delta^\phi) \to \Phi \otimes \Psi$; but thereby hangs another tale.

4. Number theory

To define the classical Bernoulli numbers B_n, we work in the dual of the coalgebra $\mathbb{Q}[x]$ and consider the right-invariant operator

$$D/(e^D - 1) = \sum_{n\geq 0} B_n D^n/n! \tag{4.1.}$$

(abbreviated to e^{BD}) in $\mathbb{Q}[[D]]$; of course $e^D - 1$ is the forward difference operator Δ. Thus e^{BD} arises from the functional B on $\mathbb{Q}[x]$, specified by $B(x^n) = B_n$ for all $n \geq 0$. The B_n are rational numbers, and are clearly zero when n is odd and > 1.

A convenient method of computation (which is recursive, and therefore subject to the usual limitations) may be succinctly described by the symbolic manipulations of (2.3.). We rearrange (4.1.) as $D+e^{BD}=e^{(B+1)D}$ in $\mathbb{Q}[[D]]$, and deduce that

$$B_n=(B+1)^n \qquad B^k\equiv B_k, \qquad \text{for all } n>1,$$

where 1 denotes the functional defined by $1(x^n)=1$ for all $n\geq 0$. Observing that $B_1=-\frac{1}{2}$, we then calculate the next three Bernoulli numbers to be

$$B_2=\frac{1}{6}, \qquad B_3=0, \quad \text{and} \quad B_4=-\frac{1}{30}.$$

A famous theorem of von Staudt (from 1840) asserts that

$$B_n\equiv -\sum\frac{1}{p} \bmod \mathbb{Z} \tag{4.2.}$$

for even n, where the summation ranges over all primes p such that $p-1$ divides n.

The operator e^{BD} is an isomorphism on $\mathbb{Q}[x]$, and its inverse J is the operator $\int_x^{x+1}$. Since J is Δ/D by (4.1.), we may write $1=\Delta e^{BD}/D$ in $\mathbb{Q}[[D]]$ and immediately obtain the classical Euler–MacLaurin summation formula

$$p(x)=Jp(x)+\sum_{n\geq 1}\frac{B_n}{n!}\Delta p^{(n-1)}(x)$$

for any polynomial $p(x)$ in $\mathbb{Q}[x]$.

To define the *universal Bernoulli numbers* B_n^ϕ, we work in the dual of the coalgebra $\Phi\mathbb{Q}[x]$ and consider the right-invariant operator

$$D/\Delta^\phi=\sum_{n\geq 0}B_n^\phi D^n/n!$$

(abbreviated to $e^{B^\phi D}$) in $\Phi\mathbb{Q}[[D]]$. The B_n^ϕ are therefore homogeneous rational polynomials in the ϕ_n. By analogy with the classical case, we may compute them recursively after noting that $B_1^\phi=-\frac{1}{2}\phi_1$ and applying

$$B_n^\phi=(B^\phi+\phi)^n \qquad (B^\phi)^k\equiv B_k^\phi, \quad \phi^k\equiv\phi_{k-1}$$

for all $n>1$. We obtain

$$B_2^\phi=\tfrac{1}{6}(3\phi_1^2-2\phi_2), \qquad B_3^\phi=\tfrac{1}{4}(-3\phi_1^3+4\phi_1\phi_2-\phi_3),$$

$$\text{and} \qquad B_4^\phi=\tfrac{1}{30}(45\phi_1^4-90\phi_1^2\phi_2+30\phi_1\phi_3+20\phi_2^2-6\phi_4).$$

These computations may readily be automated using standard symbolic algebra packages, and the first ten numbers are displayed in [2]. We remark that

B_n^ϕ reduces to B_n under the homomorphism $\Phi \to \mathbb{Z}$ representing the forward difference operator; this sets each ϕ_n to 1.

The universal Bernoulli numbers were first defined explicitly by H. Miller in [6], but special cases had already appeared in the literature many years earlier. For a useful bibliography, see [2].

It is often more convenient to express the B_n^ϕ in terms of the power series which is compositionally inverse to Δ^ϕ. This is traditionally written as

$$D + c_1 D^2/2 + \cdots + c_n D^n/n + \cdots$$

in $\Phi\mathbb{Q}[[D]]$ by analogy with the standard exponential and logarithmic series, so that each c_n is a homogeneous rational polynomial in the ϕ_k, where $k \leq n$. Moreover, each c_n maps to $(-1)^n$ whenever we reduce Φ to $\mathbb{Z}$, and lies in the Lazard ring $\Phi^L < \Phi\mathbb{Q}$ of § 2. We may now state the *universal von Staudt Theorem* as

$$B_n^\phi \equiv -\sum \frac{1}{p} c_{p-1}^{n/(p-1)} \bmod \Phi^L$$

for even n, where the summation is over all primes p such that $p-1$ divides n, and as $B_n^\phi \equiv 0 \bmod \Phi^L$ for odd $n > 1$. Different proofs may be found in [2] and [11], neither of which appeals to (4.2.) and both of which therefore reprove the classical result.

We may provide a *universal Euler–Maclaurin formula*, by writing J^ϕ for the symbolic operator $\int_x^{x+\phi}$, which is inverse to $e^{B^\phi D}$ on $\Phi\mathbb{Q}[x]$. Since $1 = \Delta^\phi e^{B^\phi D}/D$ in $\Phi\mathbb{Q}[[D]]$, we immediately obtain

$$p(x) = J^\phi p(x) + \sum_{n \geq 1} \frac{B_n^\phi}{n!} \Delta^\phi p^{(n-1)}(x)$$

for any polynomial $p(x)$ in $\Phi\mathbb{Q}[x]$.

As in [11], we may also construct universal Stirling numbers $S^\phi(n,k)$ and $s^\phi(n,k)$ from the Bell polynomials and their conjugates. These numbers lie in Φ, and have interesting properties which generalize their classical counterparts; in particular, they are closely related to the B_n^ϕ.

5. Negative integers

We have utilized the convolution product $\phi^{\star m}$ to define the positive umbral integers, thereby raising the question of finding a consistent interpretation when m is negative. Since the umbral shift is an isomorphism on $\Phi[x]$ we may consider its inverse, which is known as the *backward umbral shift.* We write the associated functional as ϕ^{-1}, and its m-fold convolution as $\phi^{\star(-m)}$ for all positive integers m; these functionals are our candidates for the negative umbral integers. We note immediately that $\phi^{\star(-m)}$ acts on $\Phi[x]$ as $(1+\Delta^{\phi})^{-m}$, and so deduce that

$$\phi^{\star(-m)}(B_n^{\phi}(x)) = (-1)^n[m]^n \tag{5.1.}$$

(where $[m]^n$ is the rising factorial) directly from the definitions.

A comparison with (2.6.) demonstrates that the formulae (2.5.) are equally valid for negative values of m, confirming that our candidates pass the simplest available test. By way of corroboration, we shall now consider a more subtle application in the theory of chromatic polynomials.

We recall that any partition π of an n-element set V has *type* $\tau(\pi)$, namely the monomial $\prod \phi_j$ in Φ to which each block of cardinality $j+1$ contributes a factor ϕ_j. Thus any function $f: V \to [m]$ also has a type $\tau(f)$, defined by its kernel. As described in [10], we may utilize τ to enrich the standard theory of zeta and Möbius functions of posets $\mathcal{P}$ of partitions, and in particular to define the characteristic *type polynomial* $c^{\phi}(\mathcal{P}; x)$ in $\Phi[x]$. For example, we obtain the conjugate Bell polynomial $B_n^{\phi}(x)$ when $\mathcal{P}$ is the complete partition lattice on V, and x^n when it consists solely of the partition of V into singletons. If we apply the homomorphism $\Phi \to \mathbb{Z}$ which sets each ϕ_n to 1, then $c^{\phi}(\mathcal{P}; x)$ reduces to the classical characteristic polynomial.

For any simple graph G there is an *umbral chromatic polynomial* $\chi^{\phi}(G; x)$ in $\Phi[x]$. This may either be defined as the characteristic type polynomial of a certain poset $\mathcal{A}$ of admissible partitions of the vertices V of G, or else in interpolated form as $\sum_{\pi} \tau(\pi) B_{|\pi|}^{\phi}(x)$, where the summation ranges over all proper colour partitions of V. In either event the evaluation $\phi^{\star m}(\chi^{\phi}(G; x))$ enumerates the colourings f of G by type, as a sum of monomials in Φ. If we apply the homomorphism $\Phi \to \mathbb{Z}$, then $\chi^{\phi}(G; x)$ reduces to the classical chromatic polynomial $\chi(G; x)$, and $\phi^{\star m}$ reduces to the substitution $x = m$. We remark that $\chi^{\phi}(G; x)$ encodes the same information as Stanley's symmetric function $X(G)$ [15]. In the case of the complete graph K_n the poset $\mathcal{A}$ consists of the complete partition lattice on $[n]$, so that $\chi^{\phi}(K_n; x)$ is $B_n^{\phi}(x)$ (as also follows from the interpolated form); for the null graph N_n the poset $\mathcal{A}$ is trivial, so that $\chi^{\phi}(N_n; x)$ is x^n.

We stress that the study of umbral chromatic polynomials and partition types is best construed as a combinatorial realization of the theory of the universal formal group law, or alternatively of the universal delta operator Δ^{ϕ}.

Motivated by the fact that the classical chromatic polynomial yields important combinatorial information when evaluated at negative integers, we aim for a generalization in terms of the elements $\phi^{\star(-m)}(\chi^{\phi}(G; x))$. To this end we introduce the set $D(m)$ of proper colourings $g: V \to [s+m-1]$ whose image has cardinality $s = s(g)$ for some $1 \leq s \leq n$; we declare that such colourings have *deficiency* m, and refer to $s(g)$ as the *span* of g.

Proposition 5.2. *For all positive integers m, we have that*

$$\phi^{\star(-m)}(\chi^{\phi}(G;x)) = \sum_{D(m)} (-1)^{s(g)}\tau(g)$$

in Φ.

Proof. We apply $\phi^{\star(-m)}$ to the interpolated form of $\chi^{\phi}(G;x)$, and immediately obtain from (5.1.) that

$$\phi^{\star(-m)}(\chi^{\phi}(G;x)) = \sum(-1)^{|\pi|}\tau(\pi)[m]^{|\pi|},$$

summed over all proper colour partitions. Since the rising factorial $[m]^{|\pi|}$ enumerates those colourings with kernel π and deficiency m, the formula follows at once. ■

Even when reduced to the classical case by the homomorphism $\Phi \to \mathbb{Z}$, this result does not appear to be well documented in the literature.

Every colouring g in $D(m)$ gives rise to an acyclic orientation $\mathcal{O}$ of G by insisting that each edge be oriented in the direction of increasing colour. Furthermore, we may decompose the image of g into maximal subintervals $I(n_1)$, $I(n_2)$, ... within $[s+m-1]$; there are at most m of these, and the indices n_j are chosen to exceed by 1 the number of omitted elements less than the smallest member of $I(n_j)$. We may then define a function $q\colon V \to [m]$ by requiring that $g(v)$ lie in $I(q(v))$, and observe that q is compatible with $\mathcal{O}$ insofar as $q(u) \leq q(v)$ whenever $u < v$. We thereby obtain a *Stanley m-pair* $(\mathcal{O}, q)$, consisting of an acyclic orientation $\mathcal{O}$ and a compatible function q with codomain $[m]$. We say that g *covers* the pair $(\mathcal{O}, q)$, and write $C(\mathcal{O}, q)$ for the subset of $D(m)$ consisting of all such colourings.

Proposition 5.3. *For any Stanley m-pair $(\mathcal{O}, q)$, we have that*

$$\sum(-1)^{s(g)} = (-1)^n,$$

where the summation ranges over all $g \in C(\mathcal{O}, q)$.

Proof. Suppose that the result holds for all graphs with n vertices and $> q$ edges, and with $< n$ vertices. For induction, choose G with n vertices and q edges, and let $(\mathcal{O}, q)$ be a Stanley m-pair. If G is K_n (including the possibility that $n = 1$) then there is a unique g which covers $(\mathcal{O}, q)$, and q dictates precisely which n of the $n+m-1$ colours are used. Thus $\sum(-1)^{s(g)} = (-1)^n$ immediately, and our induction may begin.

We select a non-edge $d = uv$ of G, and suppose either $q(u) \neq q(v)$, or that there exists a directed path between u and v. Then $\mathcal{O}$ extends uniquely (and compatibly with q) to $\mathcal{O}'$ on $G \cup d$; moreover, there is a bijection of $(\mathcal{O}, q)$ with $(\mathcal{O}', q)$ which preserves spans. So

$$\sum_{(\mathcal{O},q)} (-1)^{s(g)} = \sum_{(\mathcal{O}',q)} (-1)^{s(g)} = (-1)^n,$$

where the second equality follows by induction.

On the other hand, if $q(u) = q(v)$ and there is no directed path between u and v, then $\mathcal{O}$ extends to $\mathcal{O}_1$ on $G \cup (u, v)$, to $\mathcal{O}_2$ on $G \cup (v, u)$, and to $\mathcal{O}_3$ on G/d. Each of the first two extensions is compatible with q, and the third is compatible with q', induced on G/d. So we may partition $C(\mathcal{O}, q)$ into three blocks, namely $C(\mathcal{O}_1, q)$, $C(\mathcal{O}_2, q)$, and a block which corresponds bijectively to $C(\mathcal{O}_3, q')$, preserving spans. Thus

$$\sum_{(\mathcal{O},q)} (-1)^{s(g)} = \sum_{(\mathcal{O}_1,q)} (-1)^{s(g)} + \sum_{(\mathcal{O}_2,q)} (-1)^{s(g)} + \sum_{(\mathcal{O}_3,q')} (-1)^{s(g')},$$

yielding $(-1)^n + (-1)^n + (-1)^{n-1}$ by induction and hence completing the proof. ■

We may now recover Stanley's original result [14].

Corollary 5.4. *For any simple graph G and positive integer m, the Stanley m-pairs are enumerated by the expression $(-1)^{|V|}\chi(G; -m)$.*

Proof. We set each ϕ_n to 1 in 5.2., and apply 5.3.. Since each $\tau(g)$ reduces to 1, the result follows. ■

Our candidates have passed another test!

6. Hopf rings

A ring object in the category C_R is known as a Hopf ring (although the epithet *coalgebraic ring* would be strictly more appropriate). Such objects arise naturally in the study of ring schemes as described by Mumford [7], but were first fully exploited in algebraic topology, where they have served to organize a mass of complicated algebraic and geometric information into a coherent framework. The definitive treatment was given by D.C. Ravenel and W.S. Wilson in 1977 [9].

After discussion with Gian-Carlo, we suggest that Hopf rings will eventually find similar employment in combinatorics. We therefore take the opportunity to abstract from [4] certain applications of umbral calculus to computations in algebraic topology, and place them in the more sytematic context of the previous sections. The appropriate Hopf rings are actually $\mathbb{Z}$-graded, and therefore belong to an enriched category $\mathcal{GC}_R$. We could easily incorporate these into our description, and so pay additional tribute to Gian-Carlo; we resist this temptation, and reserve such discussions for the future.

The most basic example of a Hopf ring may be constructed from two commutative rings R and T with identity. We first restrict attention to the additive group structure $+$ of T, and form the classical group ring $R[T]$ of finite linear combinations $\sum_\alpha r_\alpha[s_\alpha]$, generated over R by the elements of T. The multiplicative structure (which we write by juxtaposition) arises from $+$. Once we take account of the multiplication $\cdot$ in T, we induce a second binary operation $\circ$ by means of $[t_1] \circ [t_2] = [t_1 \cdot t_2]$; in this sense we refer to $R[T]$ as the *ring ring* of T over R. In fact $R[T]$ is a Hopf algebra with respect to juxtaposition, having diagonal defined by $\delta[t] = [t] \otimes [t]$ and antipode induced by $t \mapsto -t$ for all t in T. The distributive law in T ensures that multiplication and $\circ$ are linked via the formula

$$x \circ (yz) = \sum (x' \circ y)(x'' \circ z) \tag{6.1.}$$

for arbitrary x, y, and z in $R[T]$, where $\delta(x) = \sum x' \otimes x''$. There are other interrelations between the structure maps, all of which are encapsulated in the fact that $R[T]$ is a ring object in the category $\mathcal{C}_R$.

By analogy with polynomials in a single variable, we may form the *free Hopf ring* $R[T]\langle x\rangle$ over $R[T]$ on the single primitive generator x (which we assume augments to zero). In Ravenel and Wilson's terminology, this is the free $R[T]$-Hopf ring on the binomial coalgebra $\mathbb{Z}[x]$, and consists of all possible products and $\circ$ combinations of x with itself and elements of $R[T]$, subject to two types of relation. The first equates the multiplicative identity $x^0 = 1$ with the element $[0]$ (where 0 is the additive identity in T), whilst the second consists of all relations imposed by the requirements of being a ring object; rather than make the latter explicit, we refer readers to [9] for a comprehensive account. By way of clarification, we note that the monomials incorporating only x have the form $x^{\circ k_1} \cdots x^{\circ k_n}$, since any expression $x^{k_1} \circ \cdots \circ x^{k_n}$ may be simplified by repeated application of the distributivity law (6.1.). Primitivity demands that $\delta(x)$ be $x \otimes 1 + 1 \otimes x$, so there is a canonical inclusion $R[x] \to R[T]\langle x\rangle$ of R-coalgebras; this identifies each scalar r with $r[0]$. Since the relation $x \circ [0] = 0$ follows from the ring structure, every element $x^{\circ k}$ is also primitive. Readers should not confuse $R[T]\langle x\rangle$ with the Hopf ring $R[T[x]]$.

We now consider the $R[T]$-linear functional ∂ on $R[T]\langle x\rangle$, defined by $\partial(x^n) = \delta_{1,n}$ and $\partial(y \circ z) = 0$ for all y and z; this restricts to D on the terms rx^n. Straightforward computation shows that the corresponding right-invariant endomorphism ∂ acts as $\partial/\partial x$ (with respect to juxtaposition), and restricts to D on $R[x]$. Given a delta operator Δ^r on $R[x]$ it therefore extends over $R[T]\langle x\rangle$ to an operator ∂^r, defined by $\sum_{n\geq 1} r_{n-1}\partial^n$.

We refer to the R-coalgebra map $R[x] \to R[T]\langle x\rangle$ as *internal differentiation* D_*, and define $D_*^{\circ n}(x^m)$ to be $(x^m)^{\circ n}$; it is an instructive exercise to deduce from the axioms that $D_*^{\circ n}(x^m)$ is zero unless $m = nd$, in which case it yields $m!(x^{\circ n})^d/d!$. We then define the *internal delta operator* Δ_*^t to be the R-linear extension of

$$x^n \mapsto \sum \binom{n}{q_1, 2q_2, \ldots, nq_n} \prod_{i=1}^{n} \frac{(iq_i)!}{i!q_i!}[t_{i-1}] \circ (x^{\circ i})^{q_i}$$

as a map $R[x] \to R[T]\langle x\rangle$. This is tantamount to interpreting Δ_*^t as $\prod_{n\geq 1} \frac{1}{n!}[t_{n-1}] \circ D_*^{\circ n}$. Because Δ_*^t respects coproducts (it is certainly not multiplicative, in general), the polynomials $\Delta_*^t B^r(x)$ continue to exhibit the binomial property (3.2.). We write them as $B_n^{r,t}(x)$, and label them the *mixed associated sequence* for the pair (Δ^r, Δ_*^t). We record that $\partial^r B_n^{r,t}(x) = nB_{n-1}^{r,t}(x)$ for all $n \geq 1$, by construction.

The universal example is provided by selecting both r and t to be ϕ over Φ. In this case, we compute the first three *mixed conjugate Bell polynomials* $B^{\phi,\phi}(x)$ to be

$$x, \qquad x^2 - \phi_1 x + [\phi_1] \circ x^{\circ 2}, \qquad \text{and}$$

$$x^3 - 3\phi_1 x^2 + (3\phi_1^2 - \phi_2)x + 3x([\phi_1] \circ x^{\circ 2}) - 3\phi_1[\phi_1] \circ x^{\circ 2} + [\phi_2] \circ x^{\circ 3}$$

in $\Phi[\Phi]\langle x\rangle$. These reduce to (2.4.) on setting each $[\phi_n]$ to zero; it is again an instructive exercise to check from the axioms that the binomial property holds, and that $\partial^\phi B_n^{\phi,\phi}(x) = nB_{n-1}^{\phi,\phi}(x)$ for all $n \geq 1$.

Our delta operators Δ^r and Δ^t define a homomorphism $\Phi[\Phi] \to R[T]$, permitting investigation of the polynomials $B^{r,t}(x)$ in terms of the universal example. There are also analogues for the divided polynomials and penumbral coalgebra, which are closer to the topological applications.

Acknowledgements. Over the last decade the author has received a great deal of encouragement from Gian-Carlo in developing this material, and offers his grateful thanks; currently he is engaged with Brian Taylor in reconciling his approach with that of the classics. It is a pleasure to thank Cristian Lenart for many hours of fruitful discussion, and also Francis Clarke and both referees for improving the accuracy of the final draft.

References

[1] L. Carlitz, *Some properties of Hurwitz series*, Duke Mathematical Journal **16** (1949), 285–295

[2] F. Clarke, *The Universal von Staudt theorems*, Transactions of the American Mathematical Society **315** (1989), 591–603

[3] M. Hazewinkel, *Formal Groups and Applications*, Academic Press, London (1978)

[4] J. Hunton and N. Ray, *A rational approach to Hopf rings*, Journal of Pure and Applied Algebra **101** (1995), 313–333

[5] S.A. Joni and G.-C. Rota, *Coalgebras and bialgebras in combinatorics*, Studies in Applied Mathematics **61** (1979), 93–139

[6] H. Miller, *Universal Bernoulli numbers and the S^1-transfer*, in: *Current Trends in Algebraic Topology, part 2 (London, Ontario 1981)*, CMS Conference Proceedings **2**, American Mathematical Society (1982), 437–449

[7] D. Mumford, *Lectures on Curves on an Algebraic Surface*, Annals of Mathematics Studies **59** (1966), Princeton University Press.

[8] W. Nichols and M. Sweedler, *Hopf Algebras and combinatorics*, in: Umbral Calculus and Hopf Algebras, Contemporary Mathematics **6**, American Mathematical Society (1982), 49–84

[9] D.C. Ravenel and W.S. Wilson, *The Hopf ring for complex cobordism*, Journal of Pure and Applied Algebra **9** (1977), 241–280

[10] N. Ray, *Umbral calculus, binomial enumeration and chromatic polynomials*, Transactions of the American Mathematical Society **309** (1988), 191–213

[11] N. Ray, *Stirling and Bernoulli numbers for complex oriented homology theories*, in *Algebraic Topology: Proceedings, Arcata 1986*, Lecture Notes in Mathematics **1370**, Springer-Verlag (1989), 362–373

[12] S. Roman, *The Umbral Calculus*, Academic Press, Orlando FL (1984)

[13] G.-C. Rota and B.D. Taylor, *The classical umbral calculus*, SIAM Journal of Mathematical Analysis **25** (1994), 694–711

[14] R.P. Stanley, *Acyclic orientations of graphs*, Discrete Mathematics **5** (1973), 171–178

[15] R.P. Stanley, *A symmetric function generalization of the chromatic polynomial of a graph*, Advances in Mathematics **111** (1995), 166–194

University of Manchester, Oxford Road, Manchester M13 9PL, United Kingdom

Hyperplane Arrangements, Parking Functions and Tree Inversions

Richard P. Stanley

Dedicated to Gian-Carlo Rota on the occasion of his sixty-fourth birthday, for twenty-nine years of encouragement, support, and friendship

1. Introduction

A *(real) hyperplane arrangement* is a discrete set of hyperplanes in $\mathbb{R}^n$. We will be concerned with hyperplane arrangements that "interpolate" between two well-known arrangements: (1) the set $\mathcal{B}_n$ of hyperplanes $x_i = x_j$, for $1 \leq i < j \leq n$, and (2) the set $\tilde{\mathcal{B}}_n$ of hyperplanes $x_i - x_j = m$, for $1 \leq i < j \leq n$ and $m \in \mathbb{Z}$. The arrangement $\mathcal{B}_n$ is known as the *braid arrangement* or the *reflection arrangement of type* A_{n-1} (i.e., the set of reflecting hyperplanes of the symmetric group $\mathfrak{S}_n$, which is the Coxeter group of type A_{n-1}). Similarly, $\tilde{\mathcal{B}}_n$ is the *affine braid arrangement* or *reflection arrangement of type* $\tilde{A}_n$, i.e., the set of reflecting hyperplanes of the affine Weyl group $\tilde{\mathfrak{S}}_n$ of type $\tilde{A}_n$.

The class of arrangements we will discuss is the following. For $k \geq 1$ define the *extended Shi arrangement* $\mathcal{S}_n^k$ to be the collection of hyperplanes

$$x_i - x_j = -k+1, -k+2, \ldots, k, \text{ for } 1 \leq i < j \leq n.$$

The arrangement $\mathcal{S}_n^1 = \mathcal{S}_n$ is known as the *Shi arrangement* or *sandwich arrangement*, and was first considered by Shi [15, Ch. 7][16] and later by Headley [9, Ch. VI][10, §5]. Some properties of $\mathcal{S}_n$ are stated without proof in [20, §5]. In this paper we extend these results to $\mathcal{S}_n^k$ and provide the proofs. For some additional arrangements related to $\mathcal{B}_n$, see [20] and [14].

The main property of $\mathcal{S}_n^k$ to concern us here will be the number of regions R separated from a "natural" base region R_0 by a given number r of hyperplanes in the arrangement. Let us make this notion more precise. If we remove the union of the hyperplanes of an arrangement $\mathcal{A}$ from $\mathbb{R}^n$, then we obtain a disjoint union of open cells, called the *regions* of $\mathcal{A}$. Fix a region R_0 of $\mathcal{A}$, called the *base region*. Given a region R of $\mathcal{A}$, let $d(R)$ denote the number of hyperplanes H of $\mathcal{A}$ which separate R_0 from R, i.e., R_0 and R lie on different sides of H. (This number will always be finite since $\mathcal{A}$ is discrete.) For instance, $d(R_0) = 0$. Think of $d(R)$ as the "distance" of R from R_0. Define the *distance enumerator* of $\mathcal{A}$ (with respect to R_0) to be the generating function

$$D_{\mathcal{A}}(q) = \sum_R q^{d(R)},$$

where R ranges over all regions of $\mathcal{A}$. Thus $D_{\mathcal{A}}$ is a formal power series, which becomes a polynomial if $\mathcal{A}$ is finite.

Let us first consider the braid arrangement $\mathcal{B}_n$. It is most natural for us to let R_0 be defined by the conditions $x_1 > x_2 > \cdots > x_n$. There is a canonical way to label the regions R by the elements w of $\mathfrak{S}_n$, namely, $\mathfrak{S}_n$ acts on $\mathbb{R}^n$ as a group generated by reflections in the hyperplanes of $\mathcal{B}_n$. This action permutes the regions, and for any region R there is a unique $w \in \mathfrak{S}_n$ for which $w(R_0) = R$. Label by w this region $w(R_0)$. (The transitivity of $\mathfrak{S}_n$ on the regions shows that $D_{\mathcal{B}_n}$ is independent of the choice of R_0.) Equivalently, the label of region R is the unique permutation w such that for $i < j$ we have $w(i) > w(j)$ if and only if the hyperplane $x_i = x_j$ separates R_0 from R. It follows that $d(R)$ is the number $\ell(w)$ of inversions of w, i.e., the number of pairs $i < j$ for which $w(i) > w(j)$. This number $\ell(w)$ is also the *length* of w in the Coxeter group sense, i.e., the minimum number p such that w can be written as a product of p adjacent transpositions. It is then well-known, either from combinatorics [19, Cor. 1.3.10] or Coxeter group theory [1, Cor. 4.7][3, Exercise 10(a), pp. 230–231], that

$$D_{\mathcal{B}_n}(q) = (1+q)(1+q+q^2)\cdots(1+q+\cdots+q^{n-1}), \qquad (1)$$

the standard q-analogue of $n!$.

There is another way of labeling the regions and obtaining the formula (1). Let $\mathbb{N} = \{0, 1, 2, \ldots\}$. We will label each region with an n-tuple $\lambda(R) = (a_1, \ldots, a_n) \in \mathbb{N}^n$ as follows. Let $e_i \in \mathbb{N}^n$ denote the vector with a 1 in the ith coordinate and 0's elsewhere. First label the region R_0 by $\lambda(R_0) = (0, 0, \ldots, 0)$. Suppose now that R has been labelled, and that R' is an unlabelled region which is separated from R by a unique hyperplane $x_i = x_j$, where $i < j$. Then define $\lambda(R') = \lambda(R) + e_i$. It is easy to see that this labeling is independent of the order in which the regions are labelled. In fact, if $R = w(R_0)$ (i.e., R corresponds to the permutation $w \in \mathfrak{S}_n$) and $\lambda(R) = (a_1, \ldots, a_n)$, then

$$a_i = \#\{j \mid j > i \text{ and } w(j) < w(i)\}.$$

Thus $\lambda(R)$ is essentially the *inversion table* or *code* of w, as defined in [19, p. 21]. Moreover,

$$d(R) = a_1 + a_2 + \cdots + a_n.$$

The codes of permutations $w \in \mathfrak{S}_n$ are precisely those sequences $(a_1, \ldots, a_n) \in \mathbb{N}^n$ satisfying $a_i \leq n - i$. These observations make equation (1) obvious.

Similar results hold for $\tilde{\mathcal{B}}_n$. Define R_0 to be the region given by $x_1 > x_2 > \cdots > x_n > x_1 - 1$. For any region R, there is a unique element $w \in \tilde{\mathcal{S}}_n$ such that $w(R_0) = R$, and $d(R) = \ell(w)$, the length of w as an element of the Coxeter group $\tilde{\mathfrak{S}}_n$. By e.g. [3, Exercise 10(b), p. 231] we have

$$D_{\tilde{\mathcal{B}}_n}(q) = \sum_{w \in \tilde{\mathfrak{S}}_n} q^{\ell(w)} = \frac{1 + q + \cdots + q^{n-1}}{(1-q)^{n-1}}. \tag{2}$$

We can also ask if there is a labeling of the regions R by n-tuples $\lambda(R) \in \mathbb{N}^n$, similar to what was done for $\mathcal{B}_n$. Later we will describe such a labeling as a limiting case of a labeling of the regions of $\mathcal{S}_n^k$.

2. Labeling the extended Shi arrangements

We will define a labeling $\lambda(R) \in \mathbb{N}^n$ of the regions R of the extended Shi arrangement $\mathcal{S}_n^k$ similar to what is described above for the braid arrangement $\mathcal{B}_n$. For the Shi arrangement itself ($k = 1$), this method of labeling was suggested by I. Pak and is described in [20, §5]. Some similarities between the Shi arrangement and the extended Shi arrangements were pointed out by A. Postnikov[1] after which it was straightforward to extend Pak's method of labeling. (However, there remained the problem of actually proving that Pak's labeling and its extension to $\mathcal{S}_n^k$ had the desired properties.)

Define the base region R_0 of $\mathcal{S}_n^k$ by

$$R_0: \quad x_1 > x_2 > \cdots > x_n > x_1 - 1,$$

the same as for $\tilde{\mathcal{B}}_n$. First label the region R_0 by $\lambda(R_0) = (0, 0, \ldots, 0) \in \mathbb{N}^n$. Suppose now that R has been labelled, and that R' is an unlabelled region which is separated from R by a unique hyperplane $x_i - x_j = m$, where $i < j$ and $m \leq 0$. Then define $\lambda(R') = \lambda(R) + e_j$. On the other hand, if instead $m > 0$, then define $\lambda(R') = \lambda(R) + e_i$. It is easy to see that this labeling is well-defined (i.e., is independent of the order in which the regions are labelled), since $\lambda(R)$ depends only on the set of hyperplanes separating R from R_0.

From the definition of λ we see immediately that if $\lambda(R) = (a_1, \ldots, a_n)$, then

$$d(R) = a_1 + \cdots + a_n. \tag{3}$$

[1]For instance, the characteristic polynomial (as defined e.g. in [20]) of $\mathcal{S}_n^k$ is equal to $q(q - kn)^{n-1}$.

In order to describe the labels that occur, we define a *k-parking function of length n* to be a sequence $\alpha = (a_1, \ldots, a_n) \in \mathbb{N}^n$ satisfying the following condition: if $b_1 \leq b_2 \leq \cdots \leq b_n$ is the monotonic rearrangement of the terms of α, then $b_i \leq k(i-1)$. A 1-parking function is called simply a *parking function.* Parking functions (defined slightly differently, but equivalent to our definition) were first considered by A. G. Konheim and B. Weiss [11]. Other references include [4], [5], and [12]. (In [12] a sequence $(n+1-a_1, \ldots, n+1-a_n)$, where $(a_1, \ldots, a_n)$ is a parking function, is called a "suite majeure.") See for example [11], [4, p. 10], and [20, p. 2625] for the reason for the terminology "parking function."

The main theorem on the arrangements $\mathcal{S}_n^k$ is the following.

2.1 Theorem. *The labels $\lambda(R)$ of the extended Shi arrangement $\mathcal{S}_n^k$ are just the k-parking functions of length n, each occuring exactly once.*

Proof. For simplicity we will assume here that $k = 1$. A region R of the Shi arrangement $\mathcal{S}_n$ may be thought of as a pair (w, I), where $w \in \mathfrak{S}_n$ and I is a collection of sets $[w(i), w(j)] := \{w(i), w(i+1), \ldots, w(j)\}$ with the following properties: (1) if $[w(i), w(j)] \in I$ then $1 \leq i < j \leq n$ and $w(i) < w(j)$, and (2) the elements of I, ordered by inclusion, form an *antichain*, i.e., no element of I is a subset of another element of I. We regard such a pair (w, I) as defining the region

$$\begin{aligned} & x_{w(1)} > x_{w(2)} > \cdots > x_{w(n)}, \\ & x_{w(r)} - x_{w(s)} < 1 \text{ if } [w(r), w(s)] \in I, \\ & x_{w(r)} - x_{w(s)} > 1 \text{ if } r < s, w(r) < w(s), \text{ and no set} \\ & \qquad [w(i), w(j)] \in I \text{ satisfies } i \leq r < s \leq j. \end{aligned}$$

In general, define a *valid pair* or *valid t-pair* to be an ordered pair (v, J) where $v = v(1), \ldots, v(t)$ is a permutation of some t-element subset of $\{1, 2, \ldots, n\}$ and J is an antichain of subsets of the form $\{v(i), v(i+1), \ldots, v(j)\}$, where $i < j$. We call the elements ι of J *intervals*, and say that ι is an *interval of J*. If $i < j$, $v(i) < v(j)$, and no interval of J contains both $v(i)$ and $v(j)$, then we say that the pair $(v(i), v(j))$ is *separated.* Similarly if $i < j$ and $v(i) > v(j)$, then we say that the pair $(v(i), v(j))$ is an *inversion.* If (v, J) is a valid t-pair and $1 \leq i \leq t$, then define

$$\begin{aligned} F(v, J, i) &= \{j : (i, j) \text{ is an inversion}\} \cup \{j : (i, j) \text{ is separated}\} \\ f(v, J, i) &= \#F(v, J, i). \end{aligned}$$

If (w, I) corresponds to the region R, then $(w(i), w(j))$ is an inversion if and only if the hyperplane $x_{w(j)} - x_{w(i)} = 0$ separates R from R_0, while $(w(i), w(j))$

is separated if and only if $x_{w(i)} - x_{w(j)} = 1$ separates R from R_0. There follows

$$\lambda(R) = (f(w, I, 1), f(w, I, 2), \ldots, f(w, I, n)). \tag{4}$$

It is easy to see that $\lambda(R)$ is a parking function. Indeed, $f(w, I, w(i))$ cannot exceed $n - i$, the number of elements in w to the right of $w(i)$.

The essence of the proof of the theorem is to show that for every k-parking function α there is a unique region R for which $\lambda(R) = \alpha$. The following lemma on the structure of valid pairs will be of crucial importance.

Lemma. Let (v, J) be a valid pair. Suppose that $i < j$, and that either $(v(i), v(j))$ is an inversion or $(v(i), v(j))$ is separated. Then $f(v, J, v(i)) > f(v, J, v(j))$.

Proof of lemma. Suppose $(v(i), v(j))$ is an inversion. If $h > j$ then $(v(i), v(h))$ is an inversion whenever $(v(j), v(h))$ is an inversion (since $(v(i) > v(j))$. Suppose now that $(v(j), v(h))$ is separated. If $v(h) < v(i)$ then $(v(i), v(h))$ is an inversion. On the other hand, if $v(h) > v(i)$ then $(v(i), v(h))$ is separated (since any interval containing $v(i)$ and $v(h)$ would also contain $v(j)$). Hence $f(v, J, v(i)) \geq f(v, J, v(j))$. But since $(v(i), v(j))$ is an extra inversion not yet taken into account, we have strict inequality.

A similar argument works when $(v(i), v(j))$ is separated. If $h > j$ then $(v(i), v(h))$ is separated whenever $(v(j), v(h))$ is separated. Suppose now that $(v(j), v(h))$ is an inversion. If $v(i) > v(h)$ then $(v(i), v(h))$ is an inversion, while if $v(i) < v(h)$ then $(v(i), v(h))$ is separated. Thus $f(v, J, v(i)) \geq f(v, J, v(j))$, and we get strict inequality since $(v(i), v(j))$ is separated. This completes the proof of the lemma.

Now consider a parking function $\alpha = (a_1, \ldots, a_n)$, such as

$$\alpha = (2, 3, 0, 0, 7, 2, 3, 0, 3). \tag{5}$$

We will build up the pair (w, I) corresponding to the region R satisfying $\lambda(R) = \alpha$ one step at a time. After the mth step we will have a valid m-pair (w^m, I^m). Let $b_1, b_2, \ldots, b_n$ be the permutation of $1, 2, \ldots, n$ obtained by listing the indices (coordinates) of the smallest terms of α from right-to-left, then the indices of the next smallest terms from right-to-left, etc. For α given by (5) we have $b_1, \ldots, b_9 = 8, 4, 3, 6, 1, 9, 7, 2, 5$. Then w^m will be a permutation of $b_1, \ldots, b_m$, obtained by inserting b_m into a certain position of w^{m-1}, while I^m will be obtained from I^{m-1} by adjoining a certain interval (possibly empty) $[b_m, c_m]$ and removing any interval properly contained in another (so that I^m remains an antichain).

If $\lambda(R) = \alpha$, then by (4) we need that $f(w, I, i) = a_i$ for all i. It follows from the lemma that we must insert b_m into w^{m-1} so that $f(w^{m-1}, I^{m-1}, h) = f(w^m, I^m, h)$ for all terms h of w^{m-1}. This means that b_m cannot be inserted to the right of a larger element, and cannot be inserted to the right of a smaller element c unless there is some $d > c$ to the right of b_m such that (c, d) is not separated. Moreover, the interval $[b_m, c_m]$ cannot contain two terms that are separated in (w^{m-1}, I^{m-1}). (I.e., separated pairs stay separated.) We claim that there is exactly one way to insert b_m and to choose I^m according to these rules, so that $f(w^m, I^m, b_m) = a_{b_m}$.

First note that once we decide where to insert b_m, say after $w^{m-1}(p)$ (or at the beginning, in which case we set $p = 0$), then the interval $[b_m, c_m]$ is uniquely determined (if it exists at all) by the condition $f(w^m, I^m, b_m) = a_{b_m}$. Thus if there are two ways to insert b_m, then we must insert b_m into different places of w^{m-1}, say after $w^{m-1}(p)$ and $w^{m-1}(j)$, where $p < j$, to get permutations w^m and $\bar{w}^m$, respectively. Let $[b_m, c_m]$ be the interval corresponding to the insertion of b_m after $w^{m-1}(p)$, and similarly $[b_m, d_m]$ for $w^{m-1}(j)$. By the lemma, we have that $w^{m-1}(j-1) < b_m$. Thus $c_m < d_m$, so (b_m, d_m) is separated in w^m. Therefore $(w(j-1), d_m)$ is separated in w^m, and hence also in w^{m-1}. But by the lemma $(w(j-1), d_m)$ must remain separated in $\bar{w}^m$, a contradiction. Hence there is at most one choice of w^m and I^m for each m, and in particular at most one choice of the pair $(w, I) = (w^n, I^n)$.

The above argument shows that the map $R \mapsto \lambda(R)$ from regions to parking functions is injective. Since the number of regions of $\mathcal{S}_n$ is known to equal $(n+1)^{n-1}$ [15, Thm. 7.3.1], and similarly for the number of parking functions of length n [4] and [12], the proof follows for the case $k = 1$.

For general k, the proof is analogous but more complicated. The regions of $\mathcal{S}_n^k$ are specified by a $(k+1)$-tuple $(w, I_1, \ldots, I_k)$, where $w \in \mathfrak{S}_n$ and $I_1, \ldots, I_k$ are antichains of subsets of $\{1, 2, \ldots, n\}$ of the form $\{w(i), w(i+1), \ldots, w(j)\}$. The permutation w specifies the order of the coordinates (as in the case $k = 1$), and the antichains I_m specify which coordinates are within distance m of each other. There are certain compatibility conditions which w and the I_m's must satisfy. Given a k-parking function $\alpha = (a_1, \ldots, a_n)$, we build up $(w, I_1, \ldots, I_k)$ one step at a time as before, inserting elements in the same order as for $k = 1$, i.e, first the coordinates in descending order of the smallest terms of α, then the coordinates in descending order of the next smallest terms of α, etc. There will always be a unique choice with the necessary properties. The details are tedious and will be omitted. ■

Example. For the example (5), the successive valid pairs (w^m, I^m) are as follows (beginning with $m = 3$):

$$\begin{array}{rl} 348, & \{[3,8]\} \\ 6348, & \{[6,8]\} \\ 16348, & \{[1,3],[6,8]\} \\ 169348, & \{[1,3],[6,8]\} \\ 1769348, & \{[1,3],[7,8]\} \\ 21769348, & \{[2,3],[7,8]\} \\ 521769348, & \{[5,7],[2,3],[7,8]\}. \end{array}$$

Combining equation (3) and Theorem 2.1, we obtain the following corollary.

2.2 Corollary. *The distance enumerator of the extended Shi arrangement $\mathcal{S}_n^k$ is given by*

$$D_{\mathcal{S}_k^n}(q) = \sum_{(a_1,\ldots,a_n)} q^{a_1+\cdots+a_n},$$

where $(a_1, \ldots, a_n)$ ranges over all k-parking functions of length n.

If we let $k \to \infty$ in our labeling λ of the regions of $\mathcal{S}_n^k$, then we obtain a labeling of the regions of the affine braid arrangement $\tilde{\mathcal{B}}_n$ by vectors $(a_1, \ldots, a_n) \in \mathbb{N}^n$ such that at least one $a_i = 0$. Hence, letting $\mathbb{P} = \{1, 2, \ldots\}$, we get

$$\begin{aligned} D_{\tilde{\mathcal{B}}_n}(q) &= \sum_{\substack{(a_1,\ldots,a_n)\in\mathbb{N}^n \\ \text{not all } a_i>0}} q^{a_1+\cdots+a_n} \\ &= \sum_{(a_1,\ldots,a_n)\in\mathbb{N}^n q^{a_1+\cdots+a_n} - \sum_{(a_1,\ldots,a_n)\in\mathbb{P}^n} q^{a_1+\cdots+a_n}} \\ &= \frac{1}{(1-q)^n} - \frac{q^n}{(1-q)^n} \\ &= \frac{1+q+\cdots+q^{n-1}}{(1-q)^{n-1}}, \end{aligned}$$

agreeing with (2). A. Björner and F. Brenti [1, §4] describe a labeling of elements of $\tilde{S}_n$ by sequences in $\mathbb{N}^n \backslash \mathbb{P}^n$; presumably this labeling is equivalent to ours.

3. Enumeration of k-parking functions

Corollary 2.2 is not an entirely satisfactory "determination" of $D_{\mathcal{S}_n^k}(q)$ since it does not lead immediately to any explicit formulas, generating functions, recurrences, etc. We need a better understanding of k-parking functions. First let us recall the well-known situation for the case $k = 1$. A *rooted forest on* $[n]$ is a graph on the vertex set $[n] = \{1, 2, \ldots, n\}$ for which every connected component is a rooted tree. An *inversion* of a rooted forest F is a pair (i, j) for which $i < j$, and j lies on the unique path connecting k to i, where k is the root of the tree to which i belongs. Let $\mathrm{inv}(F)$ denote the number of inversions of F. The *inversion enumerator* $I_n(q)$ for labelled forests on $[n]$ is defined to be the polynomial

$$I_n(q) = \sum_F q^{\mathrm{inv}(F)},$$

where F ranges over all labelled forests on $[n]$. (Often $I_n(q)$ is called the inversion enumerator of trees on $n+1$ (labelled) vertices. A tree T can be obtained from the rooted forest F by adjoining a new vertex 0 and connecting it to the roots of F.) Since it is well-known that there are $(n+1)^{n-1}$ rooted forests on $[n]$, we have $I_n(1) = (n+1)^{n-1}$. Some values of $I_n(q)$ for small n are

$$\begin{aligned}
I_1(q) &= 1\\
I_2(q) &= q+2\\
I_3(q) &= q^3+3q^2+6q+6\\
I_4(q) &= q^6+4q^5+10q^4+20q^3+30q^2+36q+24\\
I_5(q) &= q^{10}+5q^9+15q^8+35q^7+70q^6+120q^5+180q^4\\
&\qquad +240q^3+270q^2+240q+120.
\end{aligned}$$

The next result summarizes the fundamental properties of $I_n(q)$. Property (a) is implicit in C. L. Mallows and J. Riordan [13], and appears more explicitly in [12]. An elegant bijective proof was given by I. Gessel and D.-L. Wang [7]. Property (b) is equivalent to [13, equation (2)], and appears more explicitly in [6, equation (14.6)]. Finally, property (c) is due to G. Kreweras [12].

3.1 Theorem. (a) *We have*

$$I_n(1+q) = \sum_G q^{e(G)-n},$$

where G ranges over all connected graphs (without loops or multiple edges) on $n+1$ labelled vertices, and where $e(G)$ denotes the number of edges of G.
(b) *We have the generating function identity*

$$\sum_{n\geq 0} I_n(q)(q-1)^n \frac{x^n}{n!} = \frac{\sum_{n\geq 0} q^{\binom{n+1}{2}} \frac{x^n}{n!}}{\sum_{n\geq 0} q^{\binom{n}{2}} \frac{x^n}{n!}}.$$

(c) *We have*

$$q^{\binom{n}{2}} I_n(1/q) = \sum_{(a_1,\dots,a_n)} q^{a_1+\cdots+a_n},$$

where $(a_1, \dots, a_n)$ *ranges over all parking functions of length* n. *Hence from equation (3) and Corollary 2.2 there follows*

$$D_{\mathcal{S}_n^1}(q) = q^{\binom{n}{2}} I_n(1/q).$$

We want to extend Theorem 3.1 to $\mathcal{S}_n^k$. First we need to generalize the notion of an inversion of a forest. Define a *rooted* k*-forest* to be a rooted forest on vertices $1, 2, \dots, n$ with edges colored with the colors $0, 1, ..., k-1$. There is no additional restriction on the possible colors of the edges. Denote the color of an edge e by $\kappa(e)$. Define the *length* $\ell(F)$ of a rooted k-forest F by

$$\ell(F) = \text{inv}(F) + \sum_{(v,e)} \kappa(e), \tag{6}$$

where $\text{inv}(F)$ denotes the number of inversions of F (ignoring the edge colors), and where (v, e) ranges over all vertices v and edges e such that e lies on the unique path from v to the root of the component of F to which v belongs. Define the inversion enumerator $I_n^k(q)$ by

$$I_n^k(q) = \sum_F q^{\ell(F)},$$

where F ranges over all rooted k-forests on $[n]$.

It is easy to see by standard enumerative arguments that there are $(kn + 1)^{n-1}$ rooted k-forests on $[n]$, so

$$I_n^k(1) = (kn + 1)^{n-1}. \tag{7}$$

Some values of $I_n^k(q)$ for small n and $k > 1$ are as follows:

$$\begin{aligned}
I_1^k(q) &= 1 \\
I_2^2(q) &= q^2 + 2q + 2 \\
I_3^2(q) &= q^6 + 3q^5 + 6q^4 + 9q^3 + 12q^2 + 12q + 6 \\
I_4^2(q) &= q^{12} + 4q^{11} + 10q^{10} + 20q^9 + 34q^8 + 52q^7 + 74q^6 + 96q^5 \\
&\quad +114q^4 + 120q^3 + 108q^2 + 72q + 24 \\
I_2^3(q) &= q^3 + 2q^2 + 2q + 2 \\
I_3^3(q) &= q^9 + 3q^8 + 6q^7 + 9q^6 + 12q^5 + 15q^4 + 18q^3 + 18q^2 + 12q + 6.
\end{aligned}$$

There is a formula for $I_n^k(q)$ in terms of *unlabelled* rooted forests (though in Theorem 3.3(b) we will give a more explicit formula in terms of generating functions). Let φ be an unlabelled rooted forest, with vertex set $V(\varphi)$. Regard φ as a poset whose maximal elements are the roots. Given a vertex $v \in V(\varphi)$, let $h_v = \#\{u \in V(\varphi) : u \leq v\}$. Let $e(\varphi)$ denote the number of linear extensions of φ (as defined e.g in [19, p. 110]), and let $[j] = 1+q+\cdots+q^{j-1}$, the q-analogue of the nonnegative integer j. If $\#V(\varphi) = n$, then it is well-known [18, §22] that

$$e(\varphi) = \frac{n!}{\prod_{v\in V(\varphi)} h_v}.$$

It was observed by A. Björner and M. L. Wachs [2, Thm. 1.3] that

$$\sum_F q^{\mathrm{inv}(F)} = e(\varphi) \prod_{v\in V(\varphi)} [h_v],$$

where F ranges over all $n!$ labelings of φ. If $a(\varphi)$ denotes the order of the automorphism group of φ, then the $n!$ labelings of φ include $a(\varphi)$ copies of each nonisomorphic labelled rooted forest whose underlying unlabelled rooted forest is φ. Hence

$$\sum_F q^{\mathrm{inv}(F)} = \frac{e(\varphi)}{a(\varphi)} \prod_{v\in V(\varphi)} [h_v], \tag{8}$$

where now F ranges over all nonisomorphic labelled rooted forests whose underlying unlabelled rooted forest is φ.

3.2 Theorem. *We have*

$$I_n^k(q) = \sum_\varphi \frac{e(\varphi)}{a(\varphi)} \prod_{\substack{v\in V(\varphi)\\ v \text{ not a root of } \varphi}} [kh_v] \cdot \prod_{\substack{v\in V(\varphi)\\ v \text{ a root of } \varphi}} [h_v],$$

where φ ranges over all nonisomorphic (unlabelled) rooted forests with n vertices.

Proof. By the definition (6) of $\ell(F)$ for a labelled rooted forest F, there is a contribution to $\ell(F)$ from the vertex labeling, and a completely independent contribution from the edge coloring. Given F, denote by $\Upsilon(F)$ the underlying unlabelled rooted forest, i.e., erase the vertex labels and edge colors. It follows that for fixed φ we have

$$\sum_{F:\Upsilon(F)=\varphi} q^{\ell(F)} = \left(\sum_{F'} q^{\mathrm{inv}(F')}\right)\left(\sum_\kappa q^{\sum_{(v,e)} \kappa(e)}\right), \tag{9}$$

where (a) F' ranges over all nonisomorphic vertex labelings of φ, (b) κ ranges over all edge k-colorings of φ, and (c) (v,e) is as in equation (6). By (8), the sum over F' is equal to $\frac{e(\varphi)}{a(\varphi)}\prod_{v\in V(\varphi)}[h_v]$. Let e be an edge of φ, and let t be the vertex of e farthest from the root of its component. If e is colored $\kappa(e)$ in the labeling F of φ, then $\kappa(e)$ is counted h_t times in the sum on the right-hand side of (6). Since all edges are colored independently, we get that the sum over κ in (9) is equal to

$$\prod_{\substack{v\in V(\varphi) \\ v \text{ not a root of } \varphi}} \left(1+q^{h_v}+q^{2h_v}+\cdots+q^{(k-1)h_v}\right).$$

Since $[h_v]\cdot\left(1+q^{h_v}+q^{2h_v}+\cdots+q^{(k-1)h_v}\right)=[kh_v]$, the proof follows by summing (9) over all φ. ■

Next we define an extension of the notion of a connected graph (in order to generalize Theorem 3.1(a)). A *multirooted k-graph* is a graph G on the vertex set $\{1,2,\ldots,n\}$ such that (a) a subset S of the vertices is chosen as a set of "roots," with the restriction that every connected component of G contains at least one root, and (b) the edges are colored from a set of k colors. We do not allow loops (edges from a vertex to itself) and multiple edges of the same color. However, it is permissible to have several edges between two distinct vertices as long as they all have different colors. Denote by $e(G)$ the number of edges of G and by $r(G)$ the number of roots. The concept of a multirooted 2-graph is due to C. H. Yan [21], who proved Theorem 3.3(a) below in the case $k=2$. It was then routine to extend this result to arbitrary k.

There is a simple bijection between multirooted 1-graphs G on $[n]$ and connected graphs G' on $[n+1]$, as follows. Since $k=1$ the color of the edges of G are all the same and can be ignored. Adjoin a new vertex $n+1$ to G, and connect it to all the roots of G, yielding a connected graph G' on $[n+1]$. This gives the desired bijection. Note that $e(G')=e(G)+r(G)$. Thus multirooted k-graphs are indeed a generalization of connected graphs.

We can now give our extension of Theorem 3.1.

3.3 Theorem. (a) *We have*

$$I_n^k(1+q)=\sum_G q^{e(G)+r(G)-n}, \tag{10}$$

where G ranges over all multirooted k-graphs on $[n]$.

(b) *We have the generating function identity*

$$\sum_{n\geq 0} I_n^k(q)(q-1)^n\frac{x^n}{n!}=\frac{\sum_{n\geq 0} q^{k\binom{n}{2}+n}\frac{x^n}{n!}}{\sum_{n\geq 0} q^{k\binom{n}{2}}\frac{x^n}{n!}}.$$

(c) *We have*

$$q^{k\binom{n}{2}} I_n^k(1/q) = \sum_{(a_1,\ldots,a_n)} q^{a_1+\cdots+a_n},$$

where $(a_1, \ldots, a_n)$ *ranges over all* k*-parking functions of length* n. *Hence from equation (3) and Theorem 3.1 there follows*

$$D_{\mathcal{S}_n^k}(q) = q^{k\binom{n}{2}} I_n^k(1/q).$$

Proof. (a) We follow the proof of the $k = 1$ case in [12]. For $k = 2$ the argument is due to C. H. Yan [21], and our proof is a straightforward extension of hers. We first claim that $I_n^k(q)$ satisfies the recurrence

$$I_{n+1}^k(q) = \sum_{a_1+a_2+\cdots+a_{k+1}=n} \binom{n}{a_1, a_2, \ldots, a_{k+1}} q^{a_2+2a_3+3a_4+\cdots+(k-1)a_k}$$
$$\cdot \left(1 + q + \cdots + q^{a_1+a_2+\cdots+a_k}\right) I_{a_1}^k(q) I_{a_2}^k(q) \cdots I_{a_{k+1}}^k(q), \quad (11)$$

where $a_1, a_2, \ldots, a_{k+1}$ are nonnegative integers, and where $\binom{n}{a_1,a_2,\ldots,a_{k+1}}$ denotes a multinomial coefficient. To prove (11), let $F_1, \ldots, F_{k+1}$ be rooted k-forests on disjoint vertex sets whose union is $[n]$. Let a_i be the number of vertices of F_i. We will "merge" these rooted k-forests into a single rooted k-forest F on $[n+1]$ as follows. The components of F_{k+1} remain components of F. Let the vertices of $F_1, \ldots, F_k$ be $u_1 < u_2 < \cdots < u_r$, where $r = a_1 + \cdots + a_k$. Define $u_{r+1} = n+1$. Choose an integer $1 \leq j \leq r+1$, and for all $m \geq j$ replace vertex u_m in whatever forest it appears by u_{m+1}. (If $j = r+1$ then there is nothing to replace.) We have replaced $F_1, \ldots, F_k$ with isomorphic rooted k-forests $F_1', \ldots, F_k'$ whose vertices are $u_1, \ldots, u_{r+1}$ with u_j omitted. Now let the components of $F_1', \ldots, F_k'$ be the subtrees of a root u_j, and put color $i-1$ on the edges connecting u_j with all the roots of F_i'. Putting together this tree T with the forest F_{k+1} gives a rooted k-forest F on $[n+1]$.

For each solution to $a_1 + \cdots + a_{k+1} = n$ in nonnegative integers, there are $\binom{n}{a_1,\ldots,a_{k+1}}$ choices for the vertex sets of $F_1, \ldots, F_{k+1}$. There are also $r+1$ choices for the integer j. We then get an additional $j-1$ ordinary inversions of F (each involving the root vertex u_j of T), in addition to the inversions already appearing in $F_1, \ldots, F_{k+1}$. Moreover, for each $1 \leq i \leq k$, we get an additional a_i pairs (v, e), where v is a vertex of T and e is an edge colored $i-1$ on the path from v to the root u_j. Namely, v is any vertex of F_i', and e is the last edge on the path from v to u_j. Thus the length enumerator for those rooted k-forests F obtained by fixing $a_1, \ldots, a_{k+1}$ and j is given by

$$q^{a_2+2a_3+3a_4+\cdots+(k-1)a_k} \cdot q^{j-1} I_{a_1}^k(q) I_{a_2}^k(q) \cdots I_{a_{k+1}}^k(q).$$

Summing over all $a_1 + \cdots + a_{k+1} = n$ and all $1 \le j \le r$ yields equation (11).

Let $J_n^k(q)$ denote the right-hand side of equation (10). It is clear that $J_0^k(q) = I_0^k(q+1) = 1$. Hence it suffices to show that $J_n^k(q)$ satisfies the same recurrence as $I_n^k(1+q)$, viz.,

$$J_{n+1}^k(q) = \sum_{a_1+a_2+\cdots+a_{k+1}=n} \binom{n}{a_1, a_2, \ldots, a_{k+1}} (1+q)^{a_2+2a_3+3a_4+\cdots+(k-1)a_k}$$

$$\cdot \frac{(1+q)^{a_1+a_2+\cdots+a_k+1}-1}{q} J_{a_1}^k(q) J_{a_2}^k(q) \cdots J_{a_{k+1}}^k(q). \qquad (12)$$

To prove (12), let $G_1, \ldots, G_{k+1}$ be multirooted k-graphs on disjoint vertex sets whose union is $[n]$. Let the colors of the edges of G be $1, 2, \ldots, k$. Let a_i be the number of vertices of G_i. We will "merge" these multirooted k-graphs into a single multirooted k-graph G on $[n+1]$, as follows. Adjoin a new vertex $n+1$, and for each $1 \le i \le k$, draw an edge colored i from $n+1$ to the roots of G_i. Also draw any number of edges with colors less than i from $n+1$ to the vertices of G_i (as long as there are no multiple edges of the same color). We now have a connected graph H with colored edges. "Erase" the roots of H, and choose any nonempty subset of the vertices of H to be a new set of roots. Taking the disjoint union of H with G_{k+1} gives a multirooted k-graph G on $[n+1]$.

The above procedure yields a bijection between multirooted k-graphs G on $[n+1]$ and sequences $\Gamma = (G_1, \ldots, G_{k+1}, E_1, \ldots, E_k, S)$, where the G_i's are multirooted k-graphs on disjoint vertex sets whose union is $[n]$, where E_i is a set of edges colored $1, 2, \ldots, i-1$ connecting $n+1$ with vertices in G_i, and where S is a nonempty subset of the union of the vertices of $G_1, \ldots, G_k$, together with the vertex $n+1$. Write $\nu(K)$ for the number of vertices of the graph K. We then have

$$\begin{aligned}
\sum_G q^{e(G)+r(G)-n} &= \sum_{a_1+\cdots+a_{k+1}=n} \sum_{\substack{(G_1,\ldots,G_{k+1},E_1,\ldots,E_k,S) \\ \nu(G_i)=a_i}} q^{\#E_1} \cdots q^{\#E_k} q^{\#S} \\
&\quad \cdot q^{e(G_1)+r(G_1)-a_1} \cdots q^{e(G_{k+1})+r(G_{k+1})-a_{k+1}} \\
&= \sum_{a_1+\ldots+a_{k+1}=n} (1+q)^{a_2}(1+q)^{2a_3} \cdots (1+q)^{(k-1)a_k} \\
&\quad \cdot \left((1+q)^{a_1+\ldots+a_k} - 1\right) J_{a_1}^k(q) \cdots J_{a_{k+1}}^k(q).
\end{aligned}$$

Now divide both sides by q. The left-hand side becomes $J_{n+1}^k(q+1)$, while the right-hand side agrees with the right-hand side of equation (12). This completes the proof of (a).

(b) Let

$$C_n^k(q) = \sum_G q^{e(G)},$$

where G ranges over all connected graphs on $[n]$ with k-colored edges, with no loops and with no multiple edges of the same color. (We do not choose a set of roots of G.) Without the condition that G is connected, the corresponding generating function is clearly $(1+q)^{k\binom{n}{2}}$. Hence by the exponential formula (e.g., [17, Cor. 6.2]), we have

$$\begin{aligned} F^k(q) &:= \sum_{n\geq 1} C_n^k(q)\frac{x^n}{n!} \\ &= \log \sum_{n\geq 0} (1+q)^{k\binom{n}{2}} \frac{x^n}{n!}. \end{aligned}$$

We get a multirooted k-graph on $[n]$ by choosing a partition $\pi = \{B_1, \ldots, B_j\}$ of the set $[n]$, placing a graph enumerated by $C_n^k(q)$ on each block B_i, and choosing a nonempty subset of B_i. Hence

$$q^n J_n^k(q) = \sum_{\pi=\{B_1,\ldots,B_j\}} C_{b_1}^k(q)\cdots C_{b_j}^k(q)[(1+q)^{b_1}-1]\cdots[(1+q)^{b_j}-1],$$

where π ranges over all partitions of $[n]$ and $b_i = \#B_i$. Again by the exponential formula we get

$$\begin{aligned} \sum_{n\geq 0} q^n J_n^k(q)\frac{x^n}{n!} &= \exp\left(F^k((1+q)x) - F^k(x)\right) \\ &= \exp\left(\log\sum_{n\geq 0}(1+q)^{k\binom{n}{2}}\frac{(1+q)^n x^n}{n!} - \log\sum_{n\geq 0}(1+q)^{k\binom{n}{2}}\frac{x^n}{n!}\right) \\ &= \frac{\sum_{n\geq 0}(1+q)^{k\binom{n}{2}+n}\frac{x^n}{n!}}{\sum_{n\geq 0}(1+q)^{k\binom{n}{2}}\frac{x^n}{n!}}. \end{aligned}$$

Now substitute $q-1$ for q and use (a) to get the desired formula.

(c) A proof was given by C. H. Yan and appears in [21]. (Another proof was later found by I. Pak.) Yan's proof is based on the following. Let

$$P_n^k(q) = \sum_{(a_1,\ldots,a_n)} q^{k\binom{n}{2}-a_1-\cdots-a_n},$$

where $(a_1, \ldots, a_n)$ ranges over all k-parking functions of length n. Yan then gives a combinatorial proof of the recurrence

$$P_{n+1}^k(q) = \sum_{i=0}^n \binom{n}{i}\left(1+q+\cdots+q^{k-1}\right)^{n-i}\left(1+q+\cdots+q^{ki}\right)P_i^k(q)P_{n-i}^1(q^k).$$

She then shows combinatorially that

$$J^k_{n+1}(q) = \sum_{i=0}^{n} \binom{n}{i} \left(1+(1+q)+\cdots+(1+q)^{k-1}\right)^{n-i} \left(1+(1+q)+\cdots+(1+q)^{ki}\right) J^k_i(q) J^1_{n-i}((1+q)^k-1),$$

and the proof follows from (a). ■

Note. It follows from Theorem 3.3(c) and equation (7) that there are $(kn+1)^{n-1}$ k-parking functions of length n. A direct way to see this, generalizing an argument due essentially to H. Pollak [4, §2] for the case $k=1$, is as follows. Let H be the subgroup of $(\mathbb{Z}/(kn+1)\mathbb{Z})^n$ generated by $(1,1,\ldots,1)$. Then it is not difficult to show that every coset of H contains exactly one k-parking function, and the result follows. This argument shows that the set of k-parking functions of length n has the natural structure of an abelian group isomorphic to $(\mathbb{Z}/(kn+1)\mathbb{Z})^{n-1}$. It might be interesting to see if this group structure can be exploited in some way in the study of the extended Shi arrangements and rooted k-trees.

NOTE. There is a natural two variable polynomial $D^k_n(q,t)$ that refines the distance enumerator $D_{\mathcal{S}^k_n}(q)$. Namely, define

$$D^k_n(q,t) = \sum_R q^{a(R)} t^{b(R)},$$

where (a) R ranges over all regions of $\mathcal{S}^k_n$, (b) $a(R)$ is the number of hyperplanes $x_i - x_j = m$, where $1 \le i < j \le n$ and $m > 0$, which separate R from R_0, and (c) $b(R)$ is the number of hyperplanes $x_i - x_j = m$, where $1 \le i < j \le n$ and $m \le 0$, which separate R from R_0. Thus $D^k_n(q,q) = D_{\mathcal{S}^k_n}(q)$. The coefficients of $D^1_n(q,t)$ for $2 \le n \le 4$ are given by the following tables:

$t\backslash q$	0	1
0	1	1
1	1	

$t\backslash q$	0	1	2	3
0	1	1	2	1
1	2	2	2	
2	2	2		
3	1			

$t\backslash q$	0	1	2	3	4	5	6
0	1	1	2	3	3	3	1
1	3	3	6	7	6	3	
2	5	5	8	9	5		
3	6	7	9	6			
4	5	6	5				
5	3	3					
6	1						

We do not know a direct interpretation of $D^k_n(q,t)$ in terms of rooted k-forests or k-parking functions, nor do we know of any simple recurrences or generating functions for $D_n^k(q,t)$ (though it is easy to describe $D_n^k(q,0)$, $D_n^k(0,t)$, and the coefficients of the terms of total degree $k\binom{n}{2}$). We also don't know a generalization of Theorem 3.3(a) involving $D_n^k(q,t)$. Let us note that M. Haiman [8] has a (conjectured) two variable refinement of $(n+1)^{n-1}$ which is completely different from our $D_n^1(q,t)$. We don't know of any direct connection between our work and Haiman's, and such a connection remains an intriguing area of investigation.

Acknowledgment. Research partially supported by NSF grant DMS-9500714. I am grateful to Donald Knuth for pointing out several inaccuracies in the original version of this paper.

References

[1] A. Björner and F. Brenti, Affine permutations of type A, *Elec. J. Combinatorics* **3**(2) (1996). Available at the URL http://ejc.math.gatech.edu:8080/Journal/Volume_3/foatatoc.html.

[2] A. Björner and M. L. Wachs, q-hook length formulas for forests, *J. Combinatorial Theory (A)* **52** (1989), 165–187.

[3] N. Bourbaki, *Groupes et algèbres de Lie*, Ch. 4, 5, et 6, Éléments de Mathématique, fasc. XXXIV, Hermann, Paris, 1968.

[4] D. Foata and J. Riordan, Mappings of acyclic and parking functions, *aequationes math.* **10** (1974), 10–22.

[5] J. Françon, Acyclic and parking functions, *J. Combinatorial Theory (A)* **18** (1975), 27–35.

[6] I. Gessel, A noncommutative generalization and q-analog of the Lagrange inversion formula, *Trans. Amer. Math. Soc.* **257** (1980), 455–482.

[7] I. Gessel and D.-L. Wang, Depth-first search as a combinatorial correspondence, *J. Combinatorial Theory (A)* **26** (1979), 308–313.

[8] M. Haiman, Conjectures on the quotient ring by diagonal invariants, *J. Algebraic Combinatorics* **3** (1994), 17–76.

[9] P. Headley, Reduced expressions in infinite Coxeter groups, Ph.D. thesis, University of Michigan, Ann Arbor, 1994.

[10] P. Headley, On reduced expressions in affine Weyl groups, in *Formal Power Series and Algebraic Combinatorics, FPSAC '94, May 23–27, 1994*, DIMACS preprint, pp. 225–232.

[11] A. G. Konheim and B. Weiss, An occupancy discipline and applications, *SIAM J. Applied Math.* **14** (1966), 1266–1274.

[12] G. Kreweras, Une famille de polynômes ayant plusieurs propriétés énumeratives, *Periodica Math. Hung.* **11** (1980), 309–320.

[13] C. L. Mallows and J. Riordan, The inversion enumerator for labeled trees, *Bull Amer. Math. Soc.* **74** (1968), 92–94.

[14] A. Postnikov and R. Stanley, Deformations of Coxeter hyperplane arrangements, preprint, available at the URL http://front.math.ucdavis.edu/math.CO.

[15] J.-Y. Shi, The Kazhdan-Lusztig cells in certain affine Weyl groups, *Lecture Note in Mathematics*, no. 1179, Springer, Berlin/Heidelberg/New York, 1986.

[16] J.-Y. Shi, Sign types corresponding to an affine Weyl group, *J. London Math. Soc.* **35** (1987), 56–74.

[17] R. Stanley, Generating functions, in *Studies in Combinatorics* (G.-C. Rota, ed.), Mathematical Association of America, Washington, DC, 1978, pp. 100–141.

[18] R. Stanley, Ordered structures and partitions, *Mem. Amer. Math. Soc.*, no. 119, 1972.

[19] R. Stanley, *Enumerative Combinatorics*, vol. 1, Wadsworth and Brooks/Cole, Pacific Grove, CA, 1986; second printing, Cambridge University Press, Cambridge/New York, 1996.

[20] R. Stanley, Hyperplane arrangements, interval orders, and trees, *Proc. Nat. Acad. Sci.* **93** (1996), 2620–2625.

[21] C. H. Yan, Generalized tree inversions and k-parking functions, *J. Combinatorial Theory (A)* **79** (1997), 268–280.

Department of Mathematics 2-375
Massachusetts Institute of Technology
Cambridge, MA 02139

More Orthogonal Polynomials as Moments

Mourad E. H. Ismail[1] *and Dennis Stanton*[2]

To Gian-Carlo Rota with thanks, gratitude and admiration

Abstract

Classical orthogonal polynomials as moments for other classical orthogonal polynomials are obtained via linear functionals. The combinatorics of the Al-Salam–Chihara polynomials are given, and three classification theorems for generalized moments as orthogonal polynomials are proven. Some combinatorial explanations and open problems are discussed.

1. Introduction

The symbolic method consists of manipulating power series in x, and mapping x^n to α_n, where $\{\alpha_n\}$ is a sequence of combinatorial numbers. This was used by I. Kaplansky, N. Mendelsohn and J. Riordan [K, KR, M] to treat a variety of combinatorial problems. In a beautiful series of papers [RKO, RR, JR], Rota's ideas put the umbral and symbolic calculus on solid foundations and his techniques were applied to study several combinatorial and analytic problems. The purpose of this paper is to use these ideas to consider moments of orthogonal polynomials as other orthogonal polynomials. We thank Gian-Carlo for his insight into these problems and for being the driving force behind the modern theory of the umbral calculus.

In [K2] and [IS2] several families of orthogonal polynomials are shown to be the moment sequences for other orthogonal polynomials. The proofs in [IS2] are by brute force, using the explicit form of the measures. In this paper we motivate and generalize some of these results (Theorems 1, 2, 3 and 4), by evaluating linear functionals on appropriate bases of the vector

[1]This work was supported by NSF grant DMS-9625459 and a research fellowship from the Leverhulme Foundation.

[2]This work was supported by NSF grant DMS-9400510.

space of real polynomials. We also combinatorially study the Al-Salam–Chihara polynomials in Section 5–6. Three characterizations of generalized moment sequences as specialized Al-Salam–Chihara polynomials are given in Section 7. Some open problems are discussed throughout this work.

The Rotafest, which resulted in these Proceedings, had two components; one on enumeration and a workshop on the umbral calculus. We are pleased that this work overlaps with both components since on one hand our study of functionals is umbral in nature but on the other hand our results on Hermite, Meixner and Al-Salam–Chihara polynomials are combinatorial in nature and use enumerative techniques.

We set some notation. If $\{p_n(x)\}$ is a sequence of monic orthogonal polynomials with real coefficients, it is known [Ch] that they satisfy a recursion relation

$$p_{n+1}(x) = (x - b_n)p_n(x) - \lambda_n p_{n-1}(x), \quad n \geq 0, \tag{1.1}$$

for some real b_n and λ_n, with $p_0(x) := 1$ and $\lambda_0 p_{-1}(x) := 0$. We refer to (1.1) as the three term recurrence relation for $p_n(x)$. We let L denote the linear functional on the vector space of real polynomials for which orthogonality holds,

$$L(p_n p_m) = 0 \quad \text{if } n \neq m. \tag{1.2}$$

The moments μ_n are defined by

$$\mu_n = L(x^n).$$

We note that if $p_n(x)$ satisfies (1.1), and

$$L(p_n) = 0 \text{ for } n > 0, \tag{1.3}$$

then (1.2) holds.

We shall also find the value of L at polynomials of degree n, other than x^n and $p_n(x)$. We shall consider

$$L((x+a)^n) = \sum_{k=0}^{n} \binom{n}{k} \mu_k a^{n-k}, \tag{1.4}$$

$$L((x;q)_n), \tag{1.5}$$

where

$$(A;q)_n := \prod_{i=0}^{n-1} (1 - Aq^i),$$

and

$$L(\phi_n(x;a)), \tag{1.6}$$

where

$$\phi_n(x;a) = (ae^{i\theta};q)_n(ae^{-i\theta};q)_n, \quad x = \cos\theta.$$

For many of the cases considered one can find an explicit expansion

$$e_n(x,y) = \sum_{k=0}^{n} c_k p_k(x) s_{n-k}(y), \tag{1.7}$$

where $e_n(x,y)$ is some elementary homogeneous polynomial in x, y of degree n (for instance $(x+y)^n$), the c_k are explicit constants, the p_k form a class of orthogonal polynomials and the s_k form another class of polynomial special functions, often expressible in terms of some class of orthogonal polynomials. The assumption $L(p_0) = 1$ and the expansion (1.7) imply

$$c_n s_n(y) = L(e_n(.,y)). \tag{1.8}$$

This also shows that occurrence of orthogonal polynomials as moments is only a special case of occurrence of orthogonal polynomials as expansion coefficients. Indeed the set up in (1.4), which is used in this paper is just one instance of the more general set up in (1.7) and (1.8).

In some cases, formula (1.7) can be obtained by multiplication of a generating function for $p_k(x)$ with a generating function for $s_l(y)$. It may be possible to obtain (1.7) from (1.8) by substitution of a Rodrigues type formula combined with integration or summation or q-summation by parts. Sometimes one can recognize (1.7) as a degenerate addition formula.

For instance, the Hermite case considered in Section 5 can be obtained by multiplication of the two generating functions

$$e^{2xz-z^2} = \sum_{k=0}^{\infty} \frac{H_k(x)\, z^k}{k!}, \quad e^{2iyz+z^2} = \sum_{l=0}^{\infty} \frac{i^l\, H_l(y)\, z^l}{l!}.$$

The result is

$$(x+iy)^n = \sum_{k+l=n} \binom{n}{l} i^l\, H_k(x)\, H_l(y). \tag{1.9}$$

Motivated by identities such as (1.9), W. Al-Salam and T. Chihara [AC] characterized all triples $\{p_k(x)\}$, $\{s_n(y)\}$, $\{e_n(x,y)\}$ satisfying (1.7) such that $\{p_k(x)\}$ and $\{s_n(y)\}$ are orthogonal polynomials and $\{e_n(x,y)\}$ are orthogonal polynomials in x for infinitely many values of y. In addition to some classical polynomials, Al-Salam and Chihara [AC] identified what has become known as the Al-Salam–Chihara polynomials and their weight function was found recently, see [AI].

Polynomials depending on parameters are orthogonal when the parameters lie in a certain domain. If these polynomials are represented as moments, the integral representation of the functional with respect to a positive measure restricts the parameters to outside this domain. The reason is that an orthogonal polynomial of degree n has n real and simple zeros. One must use other techniques to extend the validity of the results to the domain of orthogonality.

We use the standard notation for hypergeometric and basic hypergeometric series in [GR]. We also use the notion of basic numbers

$$[n]_q = \frac{1-q^n}{1-q}$$

and the q-binomial coefficients

$$\begin{bmatrix} n \\ k \end{bmatrix}_q = \frac{(q;q)_n}{(q;q)_k (q;q)_{n-k}}.$$

2. Meixner polynomials as moments

Here we obtain the Meixner polynomials as moments of the translated beta measure. We will see that the moments can be found directly from the orthogonal polynomials via (1.3), without knowledge of a representing measure.

First consider the normalized beta integral on $[0,1]$, and define the associated linear functional L by

$$L(p(x)) = \frac{\Gamma(\alpha+\beta+2)}{\Gamma(\alpha+1)\Gamma(\beta+1)} \int_0^1 p(x) x^\alpha (1-x)^\beta dx. \tag{2.1}$$

The monic orthogonal polynomials for L are constant multiples of the Jacobi polynomials,

$$P_n^{(\alpha,\beta)}(1-2x) = \frac{(\alpha+1)_n}{n!} \, {}_2F_1(-n, n+\alpha+\beta+1; \alpha+1; x).$$

Clearly from (2.1) and the beta function evaluation we have

$$\mu_k = \frac{(\alpha+1)_k}{(\alpha+\beta+2)_k}. \tag{2.2}$$

Thus (1.4) implies

$$L((x+a)^n) = \sum_{k=0}^{n} \binom{n}{k} \frac{(\alpha+1)_k}{(\alpha+\beta+2)_k} a^{n-k}, \tag{2.3}$$

which is a Meixner polynomial under an appropriate choice of α and β. This says that the measure for which the Meixner polynomials are moments is a translate of the orthogonality measure, for Jacobi polynomials, which is stated in [IS2].

Note that (2.2) implies that

$$L(P_n^{(\alpha,\beta)}(1-2x)) = \frac{(\alpha+1)_n}{n!} \,{}_2F_1(-n, n+\alpha+\beta+1; \alpha+\beta+2; 1) = 0 \text{ if } n > 0,$$

from the Chu–Vandermonde evaluation of a terminating ${}_2F_1$ at $x = 1$. So we could obtain (2.2) from the explicit formula for $P_n^{(\alpha,\beta)}(1-2x)$ without knowledge of an explicit measure. We shall use this method again in the next section.

3. Three q-versions

In this section we consider three different q-versions of the functional L of Section 2. These three functionals will be denoted by L_1, L_2 and L_3. They act nicely on x^n, $(x;q)_n$, and $\phi_n(x;a)$, respectively (see Theorems 1, 2, and 3). The corresponding three sets of orthogonal polynomials are the little q-Jacobi, big q-Jacobi, and the Askey-Wilson polynomials. We use the explicit formula for these polynomials to find the value of the linear functional L, in order for (1.3) to hold. Then we change the bases to find orthogonal polynomials as generalized moments.

The little q-Jacobi polynomials are defined by [GR, (7.3.1)]

$$p_n(x;a,b;q) = {}_2\phi_1(q^{-n}, abq^{n+1}; aq; q, xq).$$

For (1.3) to hold, we should try

$$L_1(x^k) = \frac{(aq;q)_k}{(abq^2;q)_k}, \tag{3.1}$$

analogous to Section 2. In this case the q-analogue of the Chu–Vandermonde evaluation [GR, (II.6)] does imply (1.3). Thus we have found the moments without explicitly knowing any representing measure.

We next obtain the analog of translating the measure by a constant.

Theorem 1. *For the little q-Jacobi functional L_1 we have*

$$L_1((cx;q)_n) = {}_2\phi_1(q^{-n}, aq; abq^2; q, cq^n).$$

Proof. Apply the q-binomial theorem in the form

$$(cx;q)_n = \sum_{k=0}^{n} \frac{(q^{-n};q)_k}{(q;q)_k} (cq^n x)^k$$

to (3.1). ■

The big q-Jacobi polynomials of Andrews and Askey are defined by [GR, (7.3.10)]

$$P_n(x;a,b,c;q) = {}_3\phi_2(q^{-n}, abq^{n+1}, x; aq, cq; q, q).$$

As for the little q-Jacobi polynomials again if we put

$$L_2((x;q)_k) = \frac{(aq;q)_k (cq;q)_k}{(abq^2;q)_k},$$

then the q-analogue of the Chu–Vandermonde sum [GR, (II.6)] implies (1.3). To find the moments we expand x^n in terms of $(x;q)_k$, by a limiting case of the above mentioned ${}_2\phi_1$ evaluation

$$x^n = \sum_{k=0}^{n} \frac{(q^{-n};q)_k}{(q;q)_k} (x;q)_k q^k.$$

Theorem 2. *For the big q-Jacobi functional L_2 we have*

$$L_2(x^n) = {}_3\phi_2(q^{-n}, aq, cq; abq^2, 0; q, q).$$

By appropriately choosing the parameters, the moments in Theorem 2 are Al-Salam–Chihara polynomials. Theorem 2 is proven from the explicit big q-Jacobi measure in [IS2, Theorem 3.1].

Finally we consider the Askey–Wilson polynomials, [GR, (7.5.2)]

$$p_n(x;a,b,c,d|q) = {}_4\phi_3(q^{-n}, abcdq^{n-1}, ae^{i\theta}, ae^{-i\theta}; ab, ac, ad; q, q).$$

This time

$$L_3(\phi_k(x,a)) = \frac{(ab;q)_k(ac;q)_k(ad;q)_k}{(abcd;q)_k}$$

works. By expanding $\phi_n(x;f)$ in terms of $\phi_n(x;a)$ [I, (2.2)]

$$\phi_n(x;f) = (af, f/a;q)_n \sum_{k=0}^{n} \frac{(q^{-n};q)_k q^k}{(q, af, aq^{1-n}/f;q)_k} \phi_k(x;a)$$

we obtain the following theorem.

Theorem 3. *For the Askey–Wilson functional L_3 we have*

$$L_3(\phi_n(x;f)) = (af, f/a;q)_n \ {}_4\phi_3(q^{-n}, ab, ac, ad; abcd, af, aq^{1-n}/f;q,q).$$

Note that the explicit form of $p_n(x)$ was crucial to determine the appropriate polynomial of degree n, $R_n(x)$, and the value of $L(R_n(x))$ which factored. In Section 4 we show that this idea can applied even if the explicit form of $p_n(x)$ is not known, but the measure is known.

4. Al-Salam–Chihara polynomials revisited

Theorem 2 gives the Al-Salam–Chihara polynomials as the moments of the measure with respect to which the big q-Jacobi polynomials are orthogonal. In this section we give another measure whose moments are multiples of the Al-Salam–Chihara polynomials. As before we find a polynomial $R_n(x)$ of degree n such that $L(R_n(x))$ factors. However, we do not know an explicit formula for the orthogonal polynomials $\{p_n(x)\}$ with respect to L, nor do we explicitly know the recurrence coefficients given by (1.1).

We consider a measure which is purely discrete with two infinite sequences of jumps,

$$\begin{aligned} L(p(x)) =& \frac{(q/A, q/B)_\infty}{(q, q/D)_\infty} \sum_{n=0}^{\infty} \frac{(A,B;q)_n}{(q,D;q)_n} (Dq/AB)^n p(uq^n) + \\ & \frac{(D/B, D/A)_\infty}{(q, D/q)_\infty} \sum_{n=0}^{\infty} \frac{(Aq/D, Bq/D;q)_n}{(q, q^2/D;q)_n} (Dq/AB)^n p(uq^{n+1}/D). \end{aligned} \tag{4.1}$$

If we let $c = ut$, $e = cq$, in [GR, (III.33)], and consider $L(1/(1-xt))$, we have a sum of two ${}_3\phi_2$'s which is a single infinite product. The result is

$$\frac{(qut/D)_\infty}{(qut/A)_\infty} L(1/(1-xt)) = \frac{(qut/B)_\infty}{(ut)_\infty}. \tag{4.2}$$

Clearly (4.2) is equivalent to a generating function which implies

$$L(R_n(x)) = u^n \frac{(q/B;q)_n}{(q;q)_n}$$

if

$$R_n(x) = \sum_{l=0}^{n} \frac{(A/D;q)_l}{(q;q)_l} (qu/A)^l x^{n-l}. \tag{4.3}$$

We also easily obtain from (4.2) the following theorem, first obtained by S. Suslov [S].

Theorem 4. *The moments for the linear functional given by (4.1) are*

$$L(x^n) = (qu/D)^n \frac{(D/A)_n}{(q;q)_n} {}_2\phi_1(q^{-n}, q/B; Aq^{1-n}/D; q, A).$$

Clearly we could rescale and put $u = 1$.
Note that [GR, (III.6)] implies

$$L(x^n) = (Bu/D)^n \frac{(Dq/AB;q)_n}{(q;q)_n} {}_3\phi_2(q^{-n}, q/B, D/B; Dq/AB, 0; q, q),$$

which is multiple of the result in Theorem 2. Thus Theorems 2 and 4 give two possible interpretations for the Al-Salam–Chihara polynomials as moments. There should also be a companion theorem for Theorem 3, but we do not know such a result.

5. Combinatorial applications

In Section 2–4 we found that moments of classical orthogonal polynomials may be other classical orthogonal polynomials. There has been much work on combinatorial models for both orthogonal polynomials [FO,FS] and their moments [V]. So if a given orthogonal polynomial is also a moment, these two possibly different combinatorial points of views should be reconciled. In this section we make some remarks in this direction.

The Hermite polynomials, $H_n(x)$, are the simplest limiting case of any classical polynomial. In [IS2], (or from a limiting case of (3.2)) it is shown that a rescaled version, $\tilde{H}_n(a)$ are the moments for a translate of the Hermite measure by a. Thus the Hermite polynomials are the moments for any translate of their own measure.

We give the combinatorial reason for this phenomenon. Consider the set $S = \{1, 2, \cdots, n\}$. A matching m of S is an involution on S. We refer to the

2-cycles of m as edges, and the 1-cycles (fixed points) of m as unmatched vertices.

It is well known [Fo] that, with the proper rescaling, the Hermite polynomials

$$\tilde{H}_n(x) := 2^{-n/2} H_n(x/\sqrt{2}) \tag{5.1}$$

have the representation

$$\tilde{H}_n(x) = \sum_m (-1)^{\#\text{edges in } m} x^{\#\text{fixed points of } m}. \tag{5.2}$$

They are the generating function for all matchings m on a set $\{1, 2, \cdots, n\}$, with edges weighted by -1, and unmatched vertices by x. It is also known that the moments μ_n are the number of complete matchings on $\{1, 2, \cdots, n\}$. Thus

$$L((x+a)^n) = \sum_{k=0}^{n/2} \binom{n}{2k} \mu_{2k} a^{n-2k} \tag{5.3}$$

is the generating function for all matchings of $\{1, 2, \cdots, n\}$, with edges weighted by 1, and unmatched vertices by a. The right-hand side of (5.3) is $i^{-n}\tilde{H}_n(ia)$, hence is just the rescaled Hermite polynomials rescaled again.

Although there is an a priori combinatorial interpretation for Meixner polynomials [V], and another interpretation for moments of general orthogonal polynomials [V], for the Meixner polynomials we do not have a combinatorial reconciliation, as we gave for the Hermite polynomials.

Another example is the Laguerre polynomials, a limiting case of the Meixner, for which there is well-studied combinatorial model [FS]. There are two possible interpretations as moments, corresponding to the limiting cases of Theorems 2 and 4. This would lead to two new models.

6. Combinatorics of Al-Salam–Chihara polynomials

The Al-Salam–Chihara polynomials are a special case of the Askey–Wilson polynomials. Theorems 2 and 4 give linear functionals whose moments are these polynomials. In this section we give the combinatorial interpretations for these polynomials and their moments.

The monic form of the Al-Salam–Chihara polynomials [AI, (3.2)] have the three term recurrence relation

$$p_{n+1}(x) = (x - aq^n)p_n(x) - (c + bq^{n-1})[n]_q p_{n-1}(x). \tag{6.1}$$

An explicit representation for the p_n's as multiples of a ${}_3\phi_2$ function is in Chapter 3 of [AI].

To combinatorially understand these polynomials and their moments, we consider matchings m of $\{1, 2, \cdots, n\}$. A *2-bicoloring* C of a matching m is a 2-coloring of the edges of the matching (say with colors b and c), and an independent 2-coloring of the unmatched vertices (say with colors x and a). We let $b(C)$, $c(C)$, $x(C)$, and $a(C)$ denote the number of these colored edges and unmatched vertices.

If only the edges are 2-colored, and not the unmatched vertices, we call such a coloring D an *edge 2-coloring* of m. We denote by $a(D)$ the number of unmatched vertices, and by $b(D)$ and $c(D)$ the number of edges colored b and c respectively.

Theorem 5. *The Al-Salam–Chihara polynomial $p_n(x)$ is the generating function of all 2-bicolorings C of all matchings m of $\{1, 2, \cdots, n\}$ with weight $w(C)$*

$$p_n(x) = \sum_C w(C),$$

where

$$w(C) = x^{x(C)}(-a)^{a(C)}(-b)^{b(C)}(-c)^{c(C)}q^{s(C)},$$

$$s(C) = s_1(C) + s_2(C) + 2s_3(C),$$

$$s_1(C) = \sum_{a\text{-vertices } i} |\{z : z < i, m(z) < i\}|,$$

$$s_2(C) = \sum_{\text{all edges } i<j} |\{z : i < z < j, m(z) < j\}|,$$

$$s_3(C) = \sum_{b\text{-edges } i<j} |\{z : z < i, m(z) < j\}|.$$

Proof. We verify (6.1) by considering $n + 1$ in a 2-bicoloring C on $\{1, 2, \cdots, n+1\}$. First ignore the power of q. If $n+1$ is unmatched, then we have an arbitrary 2-bicoloring on $\{1, 2, \cdots, n\}$, with $n+1$ colored either x or a. These are the two terms multiplying $p_n(x)$ in (6.1). If $n + 1$ is matched, there are n choices for $m(n + 1)$, what remains is an arbitrary 2-bicoloring of $\{1, 2, \cdots, n\} - \{m(n + 1)\}$. The colors for the $\{(n + 1), m(n + 1)\}$ edge are b or c, agreeing with (6.1). So it remains to check that the power of q given by $s(C)$ agrees with (6.1). If $n + 1$ is unmatched and colored x, then $n + 1$ does not contribute to $s(C)$. If $n + 1$ is unmatched and colored a, then (6.1) contributes n to $s(C)$, and n is the number of vertices i to the left of $n + 1$ such that $m(i) < n + 1$. Any $i < n + 1$ with $m(i) > n + 1$ is inserted after $n + 1$. This gives the term $s_1(C)$. If $n + 1$ is matched to

$m(n+1) < n+1$, we choose a monomial q^{j-1}, $1 \le j \le n$, from $[n]_q$ to weight the edge. If the edge is colored b we additionally weight the edge by q^{n-1}. We can choose j from left-to-right or right-to-left. For a c-edge $\{n-j+1, n+1\}$, choose q^{j-1}, for the b edge $\{j, n+1\}$ choose $q^{n-1+j-1}$. The term q^{j-1} contributes to $s_2(C)$ for the c-edges, and to $s_3(C)$ for the b-edges. The term q^{n-1} contributes to $s_2(C)+s_3(C)$ for the b-edges. ■

It is clear from the proof that several other versions of Theorem 5 could be given, with slight modifications of $s(C)$. For example, if the b-edges are read in the opposite direction, $s_2(C)$ and $2s_3(C)$ would be replaced by

$$\begin{aligned} \tilde{s}_2(C) &= \sum_{\text{all edges } i<j} |\{z : i < z < j, m(z) < j\}|, \\ \tilde{s}_3(C) &= \sum_{b\text{-edges} i<j} |\{z : z < j, m(z) < j\}|. \end{aligned} \tag{6.2}$$

Note that by taking $a = b = 0$, and $c = 1$, we obtain the continuous q-Hermite polynomials $\tilde{H}_n(x|q)$, which are defined by (1.1) with

$$b_n = 0, \quad \lambda_n = [n]_q.$$

In this case we have only matchings, and Theorem 5 (with (6.2)) becomes Proposition 3.3 in [ISV].

The moments of the continuous q-Hermite polynomials are the generating functions of the crossing numbers of complete matchings [ISV, (3.6)],

$$\text{cross}(m) = |\{\text{edges } i<j, k<l : i<k<j<l\}|.$$

or also the generating functions of the nesting numbers of complete matchings [ISV, (3.9)],

$$\text{nest}(m) = |\{\text{edges } i<j, k<l : i<k<l<j\}|.$$

For the Al-Salam–Chihara polynomials, we need a q-statistic on edge 2-colorings generalizing either of these two statistics.

Theorem 6. *The nth moment for the Al-Salam–Chihara polynomials (6.1) is the generating function for all edge 2-colorings D of matchings m of $\{1, 2, \cdots, n\}$ with weight $w(D)$*

$$\mu_n = \sum_D w(D),$$

where

$$w(D) = a^{a(D)}b^{b(D)}c^{c(D)}q^{t(D)},$$

$$t(D) = c_1(m) + c_2(D) + c_3(m),$$

$$c_1(m) = \sum_{a\text{-vertices}} |\{\text{edges } i < j : i < a < j\}|,$$

$$c_2(D) = \sum_{b\text{-edges } i<j} |\{\text{edges } k < l : k < j < l\}|,$$

and $c_3(m)$ *is either the crossing number cross*(m) *or the nesting number nest*(m).

Proof. We follow the proof of [ISV, (3.6)]. If $a = b = q = 1$ and $c = 0$, the bijection from Motzkin paths of length n gives matchings on $\{1, 2, \cdots, n\}$. We must weight the unmatched vertices by a, the edges by either b or c, and also an appropriate power of q. This gives Theorem 6, up to the power $t(D)$ of q. An unmatched vertex a has weight aq^n if there are n uncompleted edges preceding a, this contributes the term $c_1(m)$ in Theorem 6. A similar argument applies for the b edges of weight bq^{n-1}, yielding $c_2(D)$. The remaining term $c_3(m)$ appears from the term q^j, $0 \leq j \leq n-1$, chosen from $[n]_q$ for any edge, b or c. This contributes either cross(m) or nest(m). ∎

Again by reading the inserted edges in the opposite order we may find other versions of Theorem 6.

We note that the L^2-norm can be considered as the generating function for the length in Weyl groups of type B_n.

Proposition 1. *Let L be the linear functional for the Al-Salam–Chihara polynomials. Then*

$$L(p_n p_m) = \delta_{n,m} n!_q \prod_{i=0}^{n-1} (c + bq^i).$$

Proof. Since $L(1) = 1$, the L^2-norm is always given by $\lambda_n \cdots \lambda_1$, so (6.1) gives the stated constant. Another method is to use the general theory of Viennot [V], giving an involution which proves orthogonality. In this case the fixed points will be all edge 2-colorings of complete matchings of $\{1, 2, \cdots, n\}$ to $\{n+1, n+2, \cdots, 2n\}$. There are no a-vertices in this case, and Theorem 6 also gives the stated constant. The edge $(m^{-1}(2n-i), 2n-i)$ contributes c or bq^i, $0 \leq i \leq n-1$. The crossing number contributes $n!_q$, independent of the coloring. ■

It is of interest to consider the q-analog of the Hermite polynomials, which were moments of their own translated measure. If we put $c = 0$, $b = -1$ in (6.1) the Al-Salam–Chihara polynomials become Al-Salam-Carlitz (I) polynomials [KS, p. 87]. Then Theorem 6 implies that the moments are the continuous q-Hermite polynomials [ISV, (2.10)].

Corollary 1. *If L is given by $b_n = aq^n$, $\lambda_n = -q^{n-1}[n]_q$, then*

$$L(x^n) = \tilde{H}_n(a|q).$$

Proof. If we apply Theorem 6 with $c = 0$, the edges are colored only $b = -1$, while the unmatched vertices are weighted by a. Thus the moments are some q-version of the Hermite polynomials in a. In Theorem 6, $c_2(D) = \text{cross}(m) + \text{nest}(m)$. If we choose $c_3(m) = \text{nest}(m)$, then the q-statistic is $t(m) = c_1(m) + c_2(D) + c_3(m) = c_1(m) + \text{cross}(m) + 2\text{nest}(m)$. If we apply Theorem 5 to $\tilde{H}_n(a|q)$, ($a = 0$, $b = 0$, $c = 1$, $x = a$), again we have just matchings m, with edges weighted by -1. The power of q is $s(m) = \tilde{s}_2(m) = t(m)$. ■

Another q-analog is given by the discrete q-Hermite, [GR, p. 193] $\tilde{H}_n(x;q)$, which have

$$b_n = 0, \quad \lambda_n = q^{n-1}[n]_q.$$

The next corollary says that the discrete q-Hermite are the "shifted moments" for the discrete q^{-1}-Hermite.

Corollary 2. *If L is given by $b_n = 0$, $\lambda_n = -q^{-n}[n]_{1/q}$, then*

$$L(d^n(-x/d;q)_n) = \tilde{H}_n(d;q).$$

Proof.

$$L(d^n(-x/d;q)_n) = \sum_{k=0}^{n} \begin{bmatrix} n \\ k \end{bmatrix}_q q^{\binom{k}{2}} d^{n-k} L(x^k). \tag{6.3}$$

We appeal to Theorem 6 to find $L(x^k)$. The choices given for b_n and λ_n correspond to $a = c = 0$, $b = -q$, and then q replaced by $1/q$ in Theorem 6. Since $a = 0$ the matchings must be complete and k is even. As in the proof of Corollary 1, the q-statistic can be taken to be $t(m) = 2\mathrm{cross}(m)+\mathrm{nest}(m)$. Moreover the generating function for complete matchings is [SS, (5.4)]

$$\sum_m q^{2\mathrm{cross}(m)+\mathrm{nest}(m)} = [1]_q[3]_q\cdots[k-1]_q,$$

so that [GR, p. 193]

$$L(d^n(-x/d;q)_n) = \sum_{k=0}^{n} \begin{bmatrix} n \\ 2k \end{bmatrix}_q q^{k^2-k}(-1)^k d^{n-2k}[1]_q[3]_q\cdots[2k-1]_q = \tilde{H}_n(d;q).$$

■

Corollaries 1 and 2 are special cases of Corollary 3 ($A = -a$, $B = 1$, $q \to 1/q$, and $A = 0$, $B = 1/q$, respectively). It says that the shifted moment of an Al-Salam-Carlitz (I) polynomial is an Al-Salam-Carlitz (II) polynomial [KS, p. 87].

Corollary 3. *If L is given by $b_n = -A/q^n$, $\lambda_n = -Bq^{1-n}[n]_{1/q}$, then $L(d^n(-x/d;q)_n)$ is an Al-Salam–Chihara polynomial in d of degree n with $a = A$, $c = 0$, and $b = Bq$.*

Proof. The choices of b_n and λ_n imply $C = 0$ in (6.1), thus force no C-colored edge in Theorem 6. We apply Theorem 6 to (6.3) (with the matching m replacing the edge 2-coloring D) to obtain

$$L(d^n(-x/d;q)_n) = \sum_{k=0}^{n} \begin{bmatrix} n \\ k \end{bmatrix}_q q^{\binom{k}{2}} d^{n-k} \sum_{m \text{ on } \{1,\cdots,k\}} (-A)^{A(m)}(-B)^{B(m)}q^{-t(m)}. \tag{6.4}$$

The desired conclusion of Corollary 3 also forces $c = 0$ in Theorem 5, so we must show

$$L(d^n(-x/d;q)_n) = \sum_{\tilde{m} \text{ on } \{1,\cdots,n\}} (-A)^{A(\tilde{m})}(-Bq)^{B(\tilde{m})}d^{d(\tilde{m})}q^{s(\tilde{m})}. \tag{6.5}$$

Given a subset $S = \{l_1 < \cdots < l_k\}$ of $\{1,\cdots,n\}$, and a matching m on $\{1,\cdots,k\}$, define a matching $\tilde{m}$ on $\{1,\cdots,n\}$ by letting the d unmatched vertices be $\{1,\cdots,n\} - S$, and $\tilde{m}(l_i) = l_j$ if $m(i) = j$. If we show that

$$s(\tilde{m}) + t(m) + \#\text{edges in } m = (l_1 - 1) + \cdots + (l_k - 1), \tag{6.6}$$

then Corollary 3 is established, because

$$l_1 + \cdots + l_k$$

is an integer partition into k distinct parts, whose largest part is $\leq n$. It is well-known that the generating function for these partitions is

$$\begin{bmatrix} n \\ k \end{bmatrix}_q q^{\binom{k+1}{2}}.$$

The following observations verify (6.6).

(1) If $l_i \in S$ is an A-vertex, then each point $z < l_i$ appears exactly once in the l_i contribution to $c_1(m) + s_1(\tilde{m})$.

If $\{l_i < l_j\}$ is an edge, we show that all points $z < l_i$ are counted twice, all points $l_i \leq z < l_j$ are counted once, in the $\{l_i < l_j\}$ contribution to the left side of (6.6). This gives a total of $(l_j - 1) + (l_i - 1)$.

(1) If $z < l_i$ and $\tilde{m}(z) < l_i$, then $2s_3(\tilde{m})$ counts z twice.
(2) If $z < l_i$ and $\tilde{m}(z) > l_i$, then z is counted once in $c_2(m)$ and once in $c_3(m) = nest(m)$. (We count the nesting when $\{l_i < l_j\}$ is inside the other edge.)
(3) If $z = l_i$, then the edge $\{l_i < l_j\}$ in (6.6) counts z exactly once.
(4) If $l_i < z < l_j$ and $\tilde{m}(z) < l_i$, then z is counted once in $s_2(\tilde{m})$.
(5) If $l_i < z < l_j$ and $l_i < \tilde{m}(z) < l_j$, then z is counted once in $s_2(\tilde{m})$ and not in nest(m). (We do not count the nesting when $\{l_i < l_j\}$ is outside the other edge.) ■

An interested referee has pointed out that one can prove Corollary 3 from the following pair of generating functions for the Al-Salam-Carlitz polynomials, $\{U_n^{(a)}(x;q)\}$ and $\{V_n^{(a)}(x;q)\}$, [Ch, KS]

$$\frac{(z;q)_\infty \, (az;q)_\infty}{(xz;q)_\infty} = \sum_{k=0}^{\infty} \frac{U_k^{(a)}(x;q)\, z^k}{(q;q)_k} \tag{6.7}$$

and

$$\frac{(yz;q)_\infty}{(z;q)_\infty \, (az;q)_\infty} = \sum_{l=0}^{\infty} \frac{(-1)^l \, q^{l(l-1)/2} \, V_l^{(a)}(y;q)\, z^l}{(q;q)_l}. \tag{6.8}$$

Clearly multiplying (6.7) and (6.8) and using the q-binomial theorem implies

$$(y/x;q)_n \, x^n = \sum_{k+l=n} \begin{bmatrix} n \\ l \end{bmatrix}_q (-1)^l \, q^{l(l-1)/2} \, U_k^{(a)}(x;q) \, V_l^{(a)}(y;q). \tag{6.9}$$

Then (1.8), (6.9) and rescaling imply Corollary 3.

It is perhaps worth noting that $|q| < 1$ is necessary for the generating function expansions (6.7) and (6.8). This restriction is removed when applied to Corollary 3. Combinatorially the two Al-Salam-Carlitz polynomials $\{U_n^{(a)}(x;q)\}$ and $\{V_n^{(a)}(x;q)\}$ are identical, because

$$U_n^{(a)}(x;q^{-1}) = V_n^{(a)}(x;q),$$

The combinatorial proof of Corollary 3 requires no assumption on q.

7. Remarks

In [IS2] several applications of Theorem 2 are given to generating functions. All of the techniques given there apply, in particular new generating functions for Al-Salam–Chihara polynomials may be given via Theorem 4. A more elementary example is given by applying the linear functional given by Corollary 1 (the Al-Salam-Carlitz measure [Ch, p. 197]), to the generating function for the continuous q-Hermite polynomials. The result is the q-analog of Mehler's formula, [IS1, (2.2)].

One may ask if it is possible to characterize which orthogonal polynomials are moments. Since any sequence is a moment sequence [Ch, p. 74] (possibly not of a positive definite measure), we must put a restriction on the types of functionals which are available. We give two such results, below, motivated by Corollaries 1 and 3.

Proposition 2. *If $b_n = aq^n$, and λ_n is independent of a, then $L(x^n)$ is an orthogonal polynomial in a of degree n only when $\lambda_n = q^{n-1}[n]_q\lambda_1$.*

Proof. From Corollary 1 the stated choice of λ_n works. It is easy to see from [V] that $L(x^n)$ is an even function of a for n even, and an odd function of a for n odd. The remainder, upon division of $L(x^{2n}) - aL(x^{2n-1})$ by $L(x^{2n-2})$, is a linear polynomial in λ_n, so λ_n is uniquely determined for $n > 1$. ■

This raises the question of characterizing orthogonal polynomials of the form $L(d^n(-x/d;q)_n)$.

Proposition 3. *If λ_n and b_n are independent of d and $|q| \neq 0, 1$, then $L(d^n(-x/d;q)_n)$ is an orthogonal polynomial in d of degree n only when*

$$\lambda_n = q^{2-2n}[n]_q\lambda_1, \quad b_n = q^{-n}b_0.$$

Proof. The proof is similar to the proof of Proposition 2. From Corollary 3 the stated choices of b_n and λ_n work. The two leading terms of $L(d^n(-x/d;q)_n)$ are

$$L(d^n(-x/d;q)_n) = d^n + \begin{bmatrix} n \\ 1 \end{bmatrix}_q b_0 d^{n-1} + \cdots. \tag{7.1}$$

The possible three term recurrence relation for $p_n(d) = L(d^n(-x/d;q)_n)$ is

$$p_{n+1}(d) = (d + \tilde{b}_n)p_n(d) + \tilde{\lambda}_n p_{n-1}(d). \tag{7.2}$$

Clearly (7.1) implies that $\tilde{b}_n = b_0 q^n$. The remainder when $p_{n-1}(d)$ divides $p_{n+1}(d) - (d + q^n b_0)p_n(d)$ as a polynomial in d must be 0. Note that [V] implies that $p_{2m}(d)$ has a unique monomial containing λ_m,

$$\lambda_m \lambda_{m-1} \cdots \lambda_1 q^{\binom{2m}{2}},$$

while $p_{2m-1}(d)$ and $p_{2m-2}(d)$ do not contain λ_m. So λ_m will appear in the remainder for $n = 2m-1$ if $\lambda_{m-1} \cdots \lambda_1 \neq 0$, which is the case since $|q| \neq 0, 1$. This uniquely determines λ_m from $\{\lambda_{m-1}, \cdots, \lambda_1, b_{m-1}, \cdots, b_0\}$. An analogous argument on b_m and $p_{2m+1}(d)$, with monomial

$$b_m \lambda_m \lambda_{m-1} \cdots \lambda_1 q^{\binom{2m+1}{2}},$$

shows that b_m is uniquely determined. For $n = 0, 1$ there is no remainder, so b_0 and λ_1 are arbitrary. ■

Theorem 7. *Suppose λ_n and b_n are independent of d, $b_1 \neq 0$, $a_n \neq 0$, $a_0 + \cdots + a_n \neq 0$ for all n. $L(\prod_{i=0}^{n-1}(d + a_i x))$ is an orthogonal polynomial in d of degree n only when*

$$a_i = a_0 q^i, \quad b_i = b_0/q^i, \quad \lambda_i = q^{2-2n}[n]_q \lambda_1,$$

where $q = b_0/b_1$.

Proof. Let $p_n(d) = L(\prod_{i=0}^{n-1}(d + a_i x))$, so that

$$p_n(d) = \sum_{i=0}^{n} e_i(a_0, \cdots, a_{n-1}) d^{n-i} \mu_i, \tag{7.3}$$

where e_i is the elementary symmetric function of degree i. By equating the coefficients of d^{n+1-i} in (7.2) we have

$$\begin{aligned}\mu_i e_i(a_0,\cdots,a_n) =&\mu_i e_i(a_0,\cdots,a_{n-1}) + \tilde{b}_n e_{i-1}(a_0,\cdots,a_{n-1})\\ &+ \tilde{\lambda}_n e_{i-2}(a_0,\cdots,a_{n-2}).\end{aligned} \tag{7.4}$$

If $i = 1$ in (7.4) we have

$$\tilde{b}_n = \mu_1 a_n = b_0 a_n.$$

If $i = 2$ in (7.4) we have

$$\begin{aligned}\tilde{\lambda}_n =&\mu_2(e_2(a_0,\cdots,a_n) - e_2(a_0,\cdots,a_{n-1}) - \mu_1\tilde{b}_n e_1(a_0,\cdots,a_{n-1})\\ =&a_n\lambda_1(a_0+\cdots+a_{n-1}).\end{aligned}$$

If $i = 3$ in (7.4) we have

$$a_n(\mu_3 - b_0\mu_2)e_2(a_0+\cdots+a_{n-1}) = b_0 a_n \lambda_1(a_0+\cdots+a_{n-1})(a_0+\cdots+a_{n-2}). \tag{7.5}$$

Since $\mu_3 = b_0^3 + 2b_0\lambda_1 + b_1\lambda_1$, $\mu_2 = b_0^2 + \lambda_1$, $a_n \neq 0$, $\lambda_1 \neq 0$, (7.5) implies

$$(b_0+b_1)e_2(a_0+\cdots+a_{n-1}) - b_0(a_0+\cdots+a_{n-1})(a_0+\cdots+a_{n-2}) = 0. \tag{7.6}$$

The coefficient of a_{n-1} in (7.6) is $b_1(a_0+\cdots+a_{n-2}) \neq 0$, so a_{n-1} is uniquely determined from $b_0, b_1, a_0, \cdots, a_{n-2}$. The solution is

$$a_i = (\frac{b_0}{b_1})^i a_0.$$

The values of b_n and λ_n are determined either by applying Proposition 3, or by considering (7.4) for $i = 4$. ■

If the combinatorics of $p_n(x)$ and μ_n are known, then the combinatorics of the associated orthogonal polynomials is often easy to find. For the associated Hermite polynomials, $b_n = 0$, $\lambda_n = n-1+c$. To combinatorially interpret these polynomials and their moments, we weight one the n choices for $m(n+1)$ in the matching m by c instead of 1. An analogous technique can be applied to the associated q-Hermite polynomials (see [Ke]) and gives associated versions of Theorems 5 and 6.

Acknowledgments. We gratefully acknowledge many helpful comments by the referees, including the discussion of (1.7). This work was done while the first author was visiting Imperial College and he gratefully acknowledges the research support of both Imperial College and the University of South Florida.

References

[AC] W. A. Al-Salam and T. S. Chihara, Convolutions of orthogonal polynomials, *SIAM J. Math. Anal.* **7** (1976), 16–28.

[AI] R. Askey and M.E.H. Ismail, Recurrence relations, continued fractions, and orthogonal polynomials, *Mem. Amer. Math. Soc.* **49**, no. 300, 1984.

[Ch] T. Chihara, An Introduction to Orthogonal Polynomials, Gordon and Breach, New York, 1978.

[Fo] D. Foata, Some Hermite polynomial identities and their combinatorics, *Advances Appl. Math.* **2**, 1981, pp. 250–259.

[FS] D. Foata and V. Strehl, Combinatorics of Laguerre polynomials, in Enumeration and Design, Waterloo Jubilee Conference, Academic Press, 1984, pp. 123–140.

[GR] G. Gasper and M. Rahman, Basic Hypergeometric Series, Cambridge University Press, Cambridge, 1990.

[I] M.E.H. Ismail, The Askey–Wilson operator and summation theorems, in Mathematical Analysis, Wavelets and Signal Processing, M. E. H. Ismail, M. Z. Nashed, A. I. Zayed, and A. Ghaleb, eds, *in:* Contemporary Mathematics 190, American Mathematical Society, 1995, pp. 171–178.

[IS1] M.E.H. Ismail and D. Stanton, On the Askey–Wilson and Rogers polynomials, *Can. J. Math.* **40**, 1988, pp. 1025–1045.

[IS2] M.E.H. Ismail and D. Stanton, Classical orthogonal polynomials as moments, *Can. J. Math.* **49** (3), 1997, pp. 520–542.

[ISV] M.E.H. Ismail, D. Stanton, and G. Viennot, The combinatorics of q-Hermite polynomials, *Eur. J. Combinatorics* **8**, 1987 pp. 379–392.

[JR] S.A. Joni and G.C. Rota, Coalgebras and bialgebras in combinatorics, *Studies in Appl. Math.* **61**, 1978, pp. 93–139.

[K] I. Kaplansky, Symbolic solutions of certain problems in permutations, *Bull. Amer. Math. Soc.* **50**, 1944, pp. 906–914.

[KR] I. Kaplansky and J. Riordan, The problème des meńages, *Scripta Math.* **12**, 1946, pp. 113–124.

[K2] S. Karlin, Sign regularity of classical orthogonal polynomials, in Orthogonal Expansions and Their Continuous Analogues, ed. D. Haimo, Southern Illinois University Press, Carbondale, 1967, pp. 55–74.

[Ke] S. Kerov, Interlacing measures, preprint.

[KS] R. Koekoek and R. Swarttouw, The Askey scheme of hypergeometric orthogonal polynomials and its q-analogue, Report 94-05, Technical University Delft, 1994.

[M] N. Mendelsohn, Symbolic solutions of card matching problems, *Bull. Amer. Math. Soc.* **52**, 1946, pp. 918–924.

[RR] S.M. Roman and G.C. Rota, The umbral calculus, *Advances in Math.* **27**, 1978, pp. 95–188.

[RKO] G.C. Rota, D. Kahaner and A. Odlyzko, On the foundations of combinatorial theory, VIII. Finite operator calculus, *J. Math. Anal. Appl.* **42**, 1973, pp. 685–760.

[SS] R. Simion and D. Stanton, Octabasic Laguerre polynomials and permutation statistics, *J. Comp. Appl. Math.* **68**, 1996, pp. 297–329.

[S] S. Suslov, personal communication.

[V] G. Viennot, Une Théorie Combinatoire des Polynômes Orthogonaux Généraux, Lecture Notes, University of Quebec at Montreal, 1983.

M.E.H. Ismail
Department of Mathematics, University of South Florida
Tampa, FL 33620

D. Stanton
School of Mathematics, University of Minnesota
Minneapolis, MN 55455

Difference Equations via the Classical Umbral Calculus

Brian D. Taylor

1. Introduction

The classical umbral calculus, formalized in [7] and [8] following the classical examples of Blissard, Bell, Riordan, Touchard, etc. (see for example [4] or the papers listed in the bibliography of [8]), has two primary advantages over its more conventional modern forebears (see for example [5], [6], etc.). It allows classical results to be easily read, verified and extended while maintaining the suggestive notation which was the strength of the classical umbral calculus.

This paper is primarily concerned with the solution of difference equations in terms of Bernoulli umbrae. Just as the solution to a first order differential equation, $D_x f(x) = g(x)$, is, barring pathologies of $g(x)$, given by $f(x) = \int_0^x g(t)dt + C$ for some constant C, one might well wish for a correspondingly simple solution to the difference equation $\Delta f(x) = g(x)$, where Δ is the forward difference operator, $\Delta f(x) = f(x+1) - f(x)$. In [8] the Bernoulli umbra was developed as a method of solving such difference equations. For $g(x)$ belonging to a sufficiently well behaved class of "functions", the solutions can be written down by $f(x) \simeq \int g(t+\beta)dt$, for some suitable version of the indefinite integral. The Greek letter β in the above expression is called an *umbra*. This paper extends the theory to permit $g(x)$ to be certain power series.

The primary application is to the Loeb-Rota algebra of logarithmic type formal power series developed in [2]. The primary novelty of the current results lies in the definitions. In fact, once the appropriate definitions have been settled upon, the verification of the key results is a straightforward extension of the results in [7] and [8]. I hope that considering the muddied history of the umbral calculus and the expository aims of this paper, the reader will forbear if I occasionally belabor the obvious.

I would like to thank my advisor, Gian-Carlo Rota, for introducing me to the umbral calculus and for first suggesting that our results in [8] should extend along the lines of [2] to provide a "closed form" solution to $\sum_{i=0}^{n} \frac{1}{x+i}$. I am pleased to thank Marguerite Eisenstein Taylor for her scrutiny of this paper in several of its drafts. Finally, I am grateful to my first referee for a number of useful suggestions.

2. Basic umbral calculus

An umbral calculus (over a commutative ring F) consists of

1. A polynomial ring $F[\mathcal{A}]$ with variables chosen from a set $\mathcal{A}$ of *umbrae*.
2. An F-linear map $\mathbf{eval} : F[\mathcal{A}] \to F$ such that
   (a) $\mathbf{eval}(1) = 1$,
   (b) $\mathbf{eval}(\alpha^i\beta^j \cdots \gamma^k) = \mathbf{eval}(\alpha^i)\mathbf{eval}(\beta^j)\cdots\mathbf{eval}(\gamma^k)$, for distinct $\alpha, \beta, \ldots, \gamma \in \mathcal{A}$ and $i, j, k \in \mathbf{N}$.

The map **eval** takes the place of the "lowering operator" that, in the original symbolic method, "lowers" exponents to indices. Readers familiar with [8] will observe that we have taken the liberty of eliminating the distinguished umbra ϵ from the preceding definition. As a result we permit ourselves the convention that $0^0 = 1$. One may of course return to the language of [8] if one wants to avoid such abuses of notation.

Two polynomials, $p, q \in F[\mathcal{A}]$ (sometimes called *umbral polynomials*), are said to be *equivalent*, written $p \simeq q$, when $\mathbf{eval}(p) = \mathbf{eval}(q)$. A sequence $a_0, a_1, a_2, \ldots$ in F is *umbrally represented* by an umbra α when $\alpha^i \simeq a_i$ for all $i \geq 0$. This imposes the restriction that $a_0 = 1$. Just as an umbra can represent a sequence, we say that, in general, an umbral polynomial, p, represents the sequence $\mathbf{eval}(p^0), \mathbf{eval}(p^1), \mathbf{eval}(p^2), \ldots$. Finally, two umbral polynomials p, q are *exchangeable*, written $p \equiv q$, when $p^i \simeq q^i$ for all $i \geq 0$, or equivalently, when p and q represent the same sequence.

3. How to solve a difference equation using umbral calculus

Assume for this section that the ground ring F is the univariate polynomial ring $\mathbf{Q}[x]$ over the rationals. We call an umbra β a *Bernoulli umbra* if, for any $p(x) \in \mathbf{Q}[x]$, the solutions $f(x)$ to $\Delta f(x) = p(x)$ are given by $\int_0^x p(t+\beta)dt + C$, that is if

$$\Delta\left(\int_0^x p(t+\beta)dt\right) \simeq p(x). \tag{1}$$

Equivalently, taking the derivative of both sides, we may require that

$$\Delta p(x+\beta) \simeq p'(x).$$

To be perfectly rigorous, the above statement requires the following technical lemma. The lemma may be directly proved with little difficulty, or derived as a consequence of Corollary 4.

Lemma 1. *If T is a $\mathbf{Q}[\mathcal{A}]$-linear operator on $\mathbf{Q}[\mathcal{A}][x]$ such that $T(\mathbf{Q}[x]) \subseteq \mathbf{Q}[x]$, then $p \simeq q$ implies $Tp \simeq Tq$ for $p, q \in \mathbf{Q}[\mathcal{A}][x]$.*

The characterization of Bernoulli umbrae can be further simplified by setting $x = 0$, i.e. for $p(t) \in \mathbf{Q}[t]$,

$$\Delta p(\beta) \simeq p'(0). \tag{2}$$

This is precisely the characterization that we began with in [8].

The machinery is now in place to interpret the preceding equations. In particular, the Bernoulli umbra may be pinned down. First, by setting $p(x) = x^n$ in (2) we establish the umbral version of the recursion for the Bernoulli numbers. For all $n > 1$,

$$(\beta + 1)^n - \beta^n \simeq 0. \tag{3}$$

Expanding yields $\sum_{i=0}^{n-2} \binom{n}{i} \beta^i \simeq -n\beta^{n-1}$. Applying **eval** demonstrates that the sequence $\mathbf{eval}(\beta^i)$ satisfies the recurrence for the Bernoulli numbers and hence that Bernoulli umbrae represent the Bernoulli numbers.

Strictly speaking, it has not yet been shown that Bernoulli umbrae exist. A proof of their existence (i.e. that umbrae representing the Bernoulli numbers actually satisfy (2)) will be given in the more general setting of Section 4.

Example 1. We establish the first several values of the Bernoulli numbers using the preceding recursion. Obviously, $B_0 = 1$. Since $(\beta + 1)^2 \simeq \beta^2$, we have $2\beta + 1 \simeq 0$ i.e., $B_1 \simeq \beta \simeq -1/2$. Similarly, $(\beta + 1)^3 \simeq \beta^3$, so $3\beta^2 + 3\beta + 1 \simeq 0$, or $3B_2 = -3B_1 - 1$. Thus $B_2 = 1/6$.

Example 2. We use the above techniques to find (modulo the results of the sequel or of [8]) a closed form solution to $f(n) = \sum_{i=0}^{n-1} i^2$. We know that $\Delta f(x) = x^2$ and that $f(0) = 0$. Hence by (1),

$$f(x) \simeq \int_0^x (t + \beta)^2 dt.$$

Thus

$$f(x) \simeq \frac{x^3 + 3\beta x^2 + 3\beta^2 x}{3}.$$

Example 1. gives $\beta \simeq B_1 = -1/2$, $\beta^2 \simeq B_2 = 1/6$, hence

$$f(x) = \frac{1}{3}x^3 - \frac{1}{2}x^2 + \frac{1}{6}x.$$

4. A general setting for the solution of difference equations

For the remainder of this paper, fix a ring $\mathbf{k}$ containing $\mathbf{Q}$. This ring will play the same role as $\mathbf{Q}$ did in Section 3. As usual, the assumption that $\mathbf{k} \supseteq \mathbf{Q}$ can, in some cases, be circumvented via judicious use of divided powers.

The motivating example for this section is $\mathbf{Q}[y]\{\{x\}\}$, the formal Laurent series in x^{-1} with coefficients in $\mathbf{Q}[y]$. Since the stated aim of this section is to describe the solution of difference equations using a Bernoulli umbra, β, it will also be necessary to work with $\mathbf{Q}[y][\beta]\{\{x\}\}$. This algebra enjoys the property of being a graded module over $\mathbf{Q}[D]$ where $D = \frac{\partial}{\partial x}$ is taken to have degree -1. Further, the series $\sum_{i\geq 0} a^i \frac{D^i}{i!}(f)$ converges for all $f \in \mathbf{Q}[y][\beta]\{\{x\}\}$ and all $a \in \mathbf{Q}[y][\beta]$.

The operator $E^a = \sum_{i\geq 0} a^i \frac{D^i}{i!}(f)$ on the subring of polynomials in x is the shift homomorphism defined by $x \mapsto x + a$. Since a shift operator is defined, the expression $\frac{1}{x+y+1+\beta}$ makes sense as a formal Laurent series; it is interpreted to be $E^{y+1+\beta}(1/x)$. The operator $\frac{\partial}{\partial y}$ extends coefficientwise to these series. Similarly, the operator **eval** acts coefficientwise on Laurent series. Two series are umbrally equivalent if **eval** takes them to the same Laurent series. With this in hand, an equation like

$$\Delta \frac{1}{x+\beta+y^2} = \frac{1}{x+y^2+1+\beta} - \frac{1}{x+y^2+\beta} \simeq -\frac{1}{(x+y^2)^2} \tag{4}$$

makes perfect sense, and turns out to be true. However, the superficially similar equation

$$\frac{1}{x+y^2+1+2\beta} - \frac{1}{x+y^2+2\beta} \simeq -\frac{1}{(x+y^2+\beta)^2} \tag{5}$$

is false.

In order to examine these equations, start by considering E^β as an element of a formal power series ring, $\mathbf{Q}[\beta][[D]]$. Defining $\Delta = E^1 - I$, the key result,

$$\Delta E^\beta \simeq D, \tag{6}$$

is now easily proved. Equation (6) holds precisely when

$$\sum_{i\geq 0} \left((\beta+1)^i - \beta^i\right) \frac{D^i}{i!} \simeq D,$$

i.e. when $(\beta+1)^i - \beta^i \simeq \delta_{i,1}$ where δ is the Kronecker delta. But this is the umbral recurrence (3) used to define a Bernoulli umbra.

Most of this section is devoted to abstracting the constructions used above and developing the machinery that allows (4) to be proved while separating out the spurious case of (5). The basic technical tool is the notion of the *support* of a series, the set of umbrae contained in it. In (5), $f(x) = 1/(x + y^2 + \beta)$ fails to satisfy the technical condition that β not be contained in the support of $f(x)$. This accounts for the failure of (5) to hold. (Of course, equation (5) can be fixed using the notation of Definition 3. in the sequel; replace each occurrence of 2β by $2 \,.\, \beta$.)

In the motivating example, D was a derivation. However, for the formalisms of this section we will not impose this restriction.

In the motivating example, we worked over a completion of the ring $S = \mathbf{k}[x, x^{-1}]$ which admits an action of $\mathbf{Q}[[D]]$. In general, let $S = \bigoplus_{i \in \mathbf{Z}} S_i$ be a graded ring (with $\mathbf{k} \subset S_0$) that is also a graded $\mathbf{k}[D]$-module. Here $\mathbf{k}[D]$ is the polynomial ring in the variable D, graded by taking D to have degree -1. For $\sum_i r_i \in S$, with $r_i \in S_i$, define $\deg(r) = \max_{r_i \neq 0} i$. (This setup can be generalized to the case that S_0 does not contain $\mathbf{k}$, but this seems unnecessary for the desired applications.)

Similarly, in the original example, the algebra of coefficients could be taken to be variously, $\mathbf{Q}$, $\mathbf{Q}[y]$, or $\mathbf{Q}[y, \beta]$. In general this algebra of coefficients will take the form $F[\mathcal{A}]$ for some commutative ring F and some set (possibly empty) of umbrae $\mathcal{A}$. If $F[\mathcal{A}]$ is a $\mathbf{k}$-algebra, then $F[\mathcal{A}] \otimes_{\mathbf{k}} S$ is a graded algebra. It has a graded $\mathbf{k}[D]$-action in which D acts as $I \otimes D$. The notions of degree and homogeneity generalize in the obvious fashion. We topologize S and $F[\mathcal{A}] \otimes S$ by the metric $d(r, s) = 2^{\deg(r-s)}$. Completing S and $F[\mathcal{A}] \otimes S$ to $\widehat{S}$ and $\widehat{F[\mathcal{A}] \otimes S}$ yields the algebras of *formal power series of type* S and *formal power series of type* S *with coefficients in* $F[\mathcal{A}]$, respectively. These are topological modules over $\mathbf{k}[[D]]$ and $F[\mathcal{A}][[D]]$ respectively.

Any $\mathbf{k}$-linear operator on $F[\mathcal{A}]$ extends to a continuous $\mathbf{k}$-linear operator on $\widehat{F[\mathcal{A}] \otimes S}$. In particular, $\mathbf{eval}$ extends to $\widehat{F[\mathcal{A}] \otimes S}$. Given two formal power series r, s of type S with coefficients in $F[\mathcal{A}]$, define $r \simeq s$ when $\mathbf{eval}(r) = \mathbf{eval}(s)$.

The present section continues with several technical results used in the remainder of the paper. The point to all of these is that operators on $\widehat{F[\mathcal{A}] \otimes S}$ respect umbral equivalence. This requires a notion of umbral equivalence for operators.

Definition 1. *Let* T *be a continuous* $\mathbf{k}[\mathcal{A}]$*-linear operator on* $\widehat{F[\mathcal{A}] \otimes S}$. *Define* $\mathbf{eval}(T)$ *to be the unique continuous* $\mathbf{k}[\mathcal{A}]$*-linear operator such that* $\mathbf{eval}(T)s = \mathbf{eval}(Ts)$ *when* $\mathbf{supp}(s) = \emptyset$, *i.e. when* $s \in \widehat{F \otimes S}$ *rather than* $\widehat{F[\mathcal{A}] \otimes S}$. *Write* $T_1 \simeq T_2$ *when* $\mathbf{eval}(T_1) = \mathbf{eval}(T_2)$.

In particular, if $S = \mathbf{k}[x]$ then $T \in \mathbf{k}[\mathcal{A}][[D_x]]$ is a $\mathbf{k}[\mathcal{A}]$-linear operator on $\mathbf{k}[\mathcal{A}][x]$. If $T = \sum_{j=0}^{d} p_j D_x^j$, then $\mathbf{eval}(T) = \sum_j \mathbf{eval}(p_j) D_x^j$. As long as the umbrae involved in T_1, T_2 are distinct from the umbrae involved in $s_1, s_2 \in F\widehat{[\mathcal{A}] \otimes} S$, then $T_1 \simeq T_2$ and $s_1 \simeq s_2$ implies $T_1(s_1) \simeq T_2(s_2)$. This fact will be used to show that (6) behaves well with respect to umbral equivalence and hence may be utilized in the solution of difference equations.

Definition 2. *Let $s \in F[\mathcal{A}] \otimes_{\mathbf{k}} S$. Thus s has a unique expansion $\sum_m m s_m$ where $s_m \in F \otimes S$ and m ranges over all monomials in $\mathcal{A}$. Define* $\mathbf{supp}(s)$*, the support of s, to be the set of all umbrae in $\mathcal{A}$ appearing in some m such that $s_m \neq 0$.*

Any $r \in F\widehat{[\mathcal{A}] \otimes} S$ expands uniquely as a possibly infinite sum $r = \sum_{i \leq d} r_i$ where r_i is a degree i homogeneous element of $F[\mathcal{A}] \otimes S$. Define $\mathbf{supp}(r) = \bigcup_i \mathbf{supp}(r_i)$.

If T is a (continuous) $\mathbf{k}[\mathcal{A}]$-linear operator on $F[\mathcal{A}] \otimes S$ ($F\widehat{[\mathcal{A}] \otimes} S$), then $\mathbf{supp}(T) = \{\alpha \in \mathcal{A} : \alpha \in \mathbf{supp}(Ts)$ *for some* $s \in F \otimes S\}$

Example 3. For any polynomial, $p \in \mathbf{k}[\mathcal{A}]$, the support of p is the set of umbrae which appear in p.

Example 4. Suppose that S is a free $\mathbf{k}$-module. Fix a $\mathbf{k}$-basis $\{v_i\}$ for S. Any $s \in \mathbf{k}[\mathcal{A}] \otimes_{\mathbf{k}} S$ expands uniquely as $\sum_i p_i v_i$ with $p_i \in \mathbf{k}[\mathcal{A}]$. The support of s is then $\bigcup_i \mathbf{supp}(p_i)$.

Example 5. If $S = \mathbf{k}[D]$, then $F\widehat{[\mathcal{A}] \otimes} S = F[\mathcal{A}][[D]]$, and the support of $r \in F[\mathcal{A}][[D]]$ is the set of umbrae which appears in some monomial appearing in the coefficient of some D^i. Of course $F[\mathcal{A}][[D]]$ acts on $\widehat{S'}$ for any S' that is a graded $\mathbf{k}[D]$-module. So r may be considered as a $F[\mathcal{A}]$-linear operator on $F\widehat{[\mathcal{A}] \otimes} S'$. As such, $\mathbf{supp}(r)$ agrees with the support of r treated as a formal power series. Similarly, if p_i are $\mathbf{k}[\mathcal{A}]$-linear operators on $F[\mathcal{A}]$, then as an operator on $F\widehat{[\mathcal{A}] \otimes} S'$, $\mathbf{supp}\left(\sum_i p_i D^i\right) = \bigcup_i \mathbf{supp}(p_i)$.

The following propositions serve to transfer the umbral equivalence (6) to the $F[\mathcal{A}][[D]]$-module of formal power series of type S with coefficients in $F[\mathcal{A}]$. They generalize (and build upon) Proposition 1 of [7] which says that for $p, q \in F[\mathcal{A}]$, if $\mathbf{supp}(p)$ and $\mathbf{supp}(q)$ are disjoint, then $\mathbf{eval}(pq) = \mathbf{eval}(p)\mathbf{eval}(q)$.

Lemma 2. *Let T be a $\mathbf{k}[\mathcal{A}]$-linear operator on $\mathbf{k}[\mathcal{A}] \otimes S$ and let $s \in \mathbf{k}[\mathcal{A}] \otimes S$. If* $\mathbf{supp}(T) \cap \mathbf{supp}(s) = \emptyset$ *then* $\mathbf{eval}(Ts) = \mathbf{eval}(T)\mathbf{eval}(s)$.

Proof. (Sketch) Write $s = \sum_m m s_m$ where m ranges over all monomials in $\mathcal{A}$ and $\mathbf{supp}(s_m) = \emptyset$. Apply the third part of Definition 2. and Proposition 1 of [7]. ∎

The next result is a direct application of Lemma 2. and the $F[\mathcal{A}]$-linearity of D.

Proposition 3. *As in the discussion preceding Definition 1., let S be a graded $\mathbf{k}[D]$-module. Let F be a commutative $\mathbf{k}$-algebra. Let $T_1, T_2, \ldots$ be $\mathbf{k}[\mathcal{A}]$-linear operators on $F[\mathcal{A}]$. Suppose $f \in \mathbf{k}[T_1, T_2, \ldots][[D]]$ and $r \in \widehat{F[\mathcal{A}] \otimes S}$. If $\mathbf{supp}(f) \cap \mathbf{supp}(r) = \emptyset$ then $f(r) \simeq \mathbf{eval}(f)\big(\mathbf{eval}(r)\big)$.*

Corollary 4. *If $\mathbf{supp}(f) \cap \mathbf{supp}(r) = \emptyset$ and $\mathbf{supp}(g) \cap \mathbf{supp}(r) = \emptyset$, then $f \simeq g$ implies $f(r) \simeq g(r)$.*

The following is an immediate consequence of (6) and Corollary 4..

Corollary 5. *If β is a Bernoulli umbra then for $v \in \widehat{F \otimes S}$,*

$$\Delta E^{\beta}(v) \simeq D(v).$$

Corollary 5. provides a method for solving difference equations. Suppose D_r^{-1} is a continuous operator on S which is a right-inverse for D.

Proposition 6. *Let β be a Bernoulli umbra and suppose $g \in \widehat{F \otimes S}$. The equation $\Delta f = g$ has a solution $f \in \widehat{F \otimes S}$ where $f \simeq E^{\beta} D_r^{-1}(g)$.*

Proof. Corollary 5. implies that $\Delta(E^{\beta} D_r^{-1} g) \simeq D D_r^{-1} g = g$. Since **eval** and Δ commute (as operators on $F[\mathcal{A}] \otimes S$ and hence on $\widehat{F[\mathcal{A}] \otimes S}$), the result follows. ∎

Proposition 7. *Let β be a Bernoulli umbra. Let $F = \mathbf{k}[y]$. Let $D_y = \frac{\partial}{\partial y}$ be the derivative with respect to y. Let Δ_y be the forward difference operator with respect to y. Let E_y^{β} be the shift operator on $F[\mathcal{A}]$ that sends $y \mapsto y + \beta$. If $v \in \widehat{F[\mathcal{A}] \otimes S}$ and $\beta \notin \mathbf{supp}(v)$, then*

$$\Delta_y E_y^{\beta}(v) \simeq D_y(v).$$

Proof. It suffices to show that $\Delta_y E_y^{\beta} \simeq D_y$ on $F[\mathcal{A}]$. This is basically part 1 of Corollary 4.4 in [8]. It may be seen directly by mimicking the proof of (6) and recalling, essentially from Example 5., that equivalence as operators is the same as coefficientwise equivalence of the corresponding series in D_y. ∎

This too has direct application to solving difference equations.

Proposition 8. *Let $g(y) \in \widehat{F[y] \otimes S}$. Suppose that $\Delta_y f(y) = g(y)$ and further suppose that $f(0) = 0$. Then $f(y) = \int_0^y g(t+\beta)dt$.*

Example 6. [Polynomials] Let D be the usual derivative, $\frac{\partial}{\partial x}$ and let $S = \mathbf{k}[x]$. In this case, $E^1 - I = \Delta$ is the usual forward difference operator, $\widehat{S} = S$, and we find that $\Delta p(x+\beta) \simeq p'(x)$ for all $p(x) \in S$. As usual, regard $\mathbf{k}[\mathcal{A}] \otimes \mathbf{k}[x]$ as $\mathbf{k}[\mathcal{A}][x]$.

Example 7. Although it is most useful to assume that D is a derivation and hence that the formal shift operators, E^a, are algebra homomorphisms, this is not strictly necessary. In fact, if one takes S to be $\mathbf{k}[x]$ with x^i taken to have degree $-i$ and if D is taken to be the operator, $\mathbf{x}$, which multiplies by x, then $\widehat{\mathbf{k}[x]} = \mathbf{k}[[x]]$. In this case, taking $v = e^{yx}$ in Proposition 7. implies that

$$\Delta_y e^{(y+\beta)x} \simeq D_y e^{yx}.$$

Hence $e^{(\beta+1)x} - e^{\beta x} \simeq xe^0$. Simplifying, $e^{\beta x}(e^x - 1) \simeq x$. Applying Lemma 1. or Corollary 4., we find that $e^{\beta x} \simeq \frac{x}{e^x-1}$. To further interpret this, apply **eval** to

$$\sum_{k \geq 0} \frac{\beta^k}{k!} x^k \simeq \frac{x}{e^x - 1}.$$

But this just says that

$$\sum_{k \geq 0} \mathbf{eval}(\beta^k) \frac{x^k}{k!} = \frac{x}{e^x - 1}.$$

This deduces the well-known fact that the exponential generating function for the Bernoulli numbers is $\frac{x}{e^x-1}$.

5. Formal power series of logarithmic type

Fix a $\mathbf{k}$-algebra F and a set of umbrae $\mathcal{A}$. Let S be the ring $\mathbf{k}[x, x^{-1}, \log(x)]/(xx^{-1} - 1)$, where $x, x^{-1}, \log(x)$ are algebraically independent generators of $\mathbf{k}[x, x^{-1}, \log(x)]$. Grade this ring by setting the degree of each such generator to $1, -1, 0$ respectively. The homogeneous components S_d of S are free $\mathbf{k}$-modules. Let D be the derivation on S such that $Dx = 1$, and $D\log(x) = x^{-1}$. Define $\mathcal{L} = \widehat{F[\mathcal{A}] \otimes S}$.

This algebra is in fact a generalization of the Loeb-Rota algebra of formal power series of logarithmic type over $F[\mathcal{A}]$ defined in [2]. Technically, Loeb and Rota assume that $F[\mathcal{A}]$ is a field, but their construction remains valid more generally. Both the Loeb-Rota algebra and the algebra $\mathcal{L}$ described above are closures of S under the metric topology in which $d(p,q) = 2^{\deg(p-q)}$, $\deg(0)$ is defined to be $-\infty$.

The algebra $\mathcal{L}$ allows for convergence of some series which diverge in the Loeb-Rota algebra. In particular, convergent series in the Loeb-Rota algebra are sums of the form $\sum_{i\geq 0} p_i$ with $p_i \in S$, satisfying two criteria. First, for all $d \in \mathbf{Z}$, there exists an i such that for all $j \geq i$, $\deg(p_j) < d$. For the second criterion, define $\mathrm{Ldeg}(p)$ to be the largest power to which $\log(x)$ appears in p. In the Loeb-Rota algebra, convergence requires that $\mathrm{Ldeg}(p_i)$ is bounded above. In $\mathcal{L}$, this second criterion is unnecessarily restrictive.

As previously, notation is freely abused in as much as the tensor signs between coefficients in $F[\mathcal{A}]$ and elements of S are omitted.

The operators E^a (for $a \in \mathbf{k}$), Δ, and D on $\mathcal{L}$ all agree with their counterparts in the Loeb-Rota algebra. In particular, it is easy to see that, for all $a \in F[\mathcal{A}]$, E^a defines an algebra homomorphism. Since E^a is defined by exponentiating a derivation, for $f, g \in F[\mathcal{A}] \otimes S$, $E^a(fg) = E^a(f)E^a(g)$. Since $\mathcal{L}$ is the completion of $F[\mathcal{A}] \otimes S$, E^a is also a homomorphism in $\mathcal{L}$. Generalizing the notation of [2], we will abbreviate $E^a f(x)$ by $f(x+a)$ for all $a \in F[\mathcal{A}]$ and all $f(x) \in \mathcal{L}$.

Example 8. If β is a Bernoulli umbra, then

$$\Delta e^{\log(x+\beta)/(x+\beta)} \simeq \frac{1-\log(x)}{x^2} e^{\log(x)/x}.$$

This follows directly from Corollary 5. and the chain rule in $\mathcal{L}$ once the convergence of $\sum_{i\geq 0} \frac{1}{i!}\left(\frac{\log(x+\beta)}{x+\beta}\right)^i$ is ascertained. But since $\log(x+\beta)/(x+\beta)$ has degree -1 in S, this is immediate.

To get a better feeling for what is going on, expand the left-hand side as

$$\Delta \sum_{i\geq 0} \frac{1}{i!} \frac{\log(x+\beta)^i}{(x+\beta)^i} = \sum_{i\geq 1} \frac{1}{i!} \frac{\log(x+1+\beta)^i}{(x+1+\beta)^i} - \frac{\log(x+\beta)^i}{(x+\beta)^i}. \tag{7}$$

In order to apply **eval**, we need to expand each summand on the right-hand side in terms of β. This means we need to expand the product of the logarithmic type formal power series,

$$\log(x+a)^i = \left(\log(x) + a\frac{1}{x} - a^2/2\frac{1}{x^2} + a^3/3\frac{1}{x^3} - \cdots\right)^i$$

and

$$\left(\frac{1}{x+a}\right)^i = \left(\frac{1}{x} - a\frac{1}{x^2} + a^2\frac{1}{x^2} - \cdots\right)^i$$

in terms of a. This product can be written as

$$\frac{\log(x)^i}{x^i} a^0 + \left(\frac{i\log(x)^{i-1}}{x^{i+1}} - \frac{i\log(x)^i}{x^{i+1}}\right) a^1 + \sum_{j>1} a^j f_j$$

where $f_j \in \mathbf{Q}[\frac{1}{x}, \log(x)]$ and $\deg(f_j) < -j, -i$. The bound on degrees of the f_j guarantees convergence when the the fractions in 7 are replaced by the preceding series with a set to $\beta+1$ and β respectively. Thus,

$$\Delta e^{\frac{\log(x+\beta)}{x+\beta}} = \sum_{i\geq 1} \frac{1}{(i-1)!}\left(\frac{\log(x)^{i-1}}{x^{i+1}} + \frac{-\log(x)^i}{x^{i+1}}\right) + \\ + \sum_{j>1}\left((\beta+1)^j - \beta^j\right) f_j.$$

Applying **eval** and (3) to the above equation verifies that indeed

$$\mathbf{eval}\left(\Delta e^{\frac{\log(x+\beta)}{x+\beta}}\right) = \sum_{i\geq 0} \frac{1}{i!}\left(\frac{\log(x)^i - \log(x)^{i+1}}{x^{i+2}}\right).$$

Next we apply Proposition 6. to the solution of difference equations in $\mathcal{L}$. First observe that in $\mathcal{L}$, $\ker(D) = F[\mathcal{A}]$. The reader may obtain an elementary proof using the usual monomial basis for S. The right proof is the one contained in [2]. Simply look at the behavior under D of the Loeb-Rota basis for S, namely $\{\lambda_n^{(t)}\}_{(n,t)}$ where $(n,t) \in \mathbf{Z}\times\mathbf{N}\setminus\mathbf{Z}^-\times 0$. Recall from [2] that

$$\lambda_n^{(0)} = \begin{cases} x^n & n \geq 0 \\ 0 & n < 0 \end{cases}.$$

In particular, apply the formulae $D\lambda_n^{(t)} = n\lambda_{n-1}^{(t)}$ for $n \neq 0$ and $D\lambda_0^{(t)} = \lambda_{-1}^{(t)}$.

Choose a homogeneous basis $\mathcal{B}$ containing 1 for S. If $t > 1$, $\lambda_n^{(t)}$ is homogeneous of degree n. Let $\sigma_{\mathcal{B}}$ be the projection of S onto $\mathrm{span}(\mathcal{B}\setminus\{1\})$. This projection extends to $\mathcal{L}$. The observations of the preceding paragraph guarantee that there exists a unique map $\int_{\mathcal{B}} : \mathcal{L} \to \mathrm{im}(\sigma_{\mathcal{B}})$ such that

$$D\int_{\mathcal{B}} = I \quad \text{and} \quad \int_{\mathcal{B}} \mathrm{D} = \sigma_{\mathcal{B}}. \tag{8}$$

Hence, Proposition 6. implies that the solutions to $\Delta f(x) = g(x)$ for $f(x) \in \mathcal{L}$ are given by $f(x) \simeq E^\beta \int_{\mathcal{B}} g(x)dx + C$. The equations in (8) guarantee that $\int_{\mathcal{B}}$ is shift invariant (i.e. commutes with all shift operators) modulo $F[\mathcal{A}]$. Hence the solutions to $\Delta f(x) = g(x)$ are given by

$$f(x) \simeq \int_{\mathcal{B}} g(x+\beta)dx + C. \tag{9}$$

Example 9. We apply the preceding results to find a simple umbral expression for

$$\frac{1}{x} + \frac{1}{x+1} + \frac{1}{x+2} + \cdots + \frac{1}{x+n-1}. \tag{10}$$

This is a formal power series of logarithmic type; we are looking to find a nice expression for it.

Denote the sum in (10) by $f(x, n)$. We can approach finding an umbral description of $f(x, n)$ through either Corollary 5. or Proposition 8.. In the first method, Corollary 5. implies that $\Delta \log(x + \beta) \simeq \frac{1}{x}$. Summing the preceding result as x ranges from x to $x + n - 1$ (more strictly, applying $I + E^1 + \cdots + E^{n-1}$) yields

$$\sum_{i=0}^{n-1} \frac{1}{x+i} \simeq \log(x + \beta + n) - \log(x + \beta). \tag{11}$$

Expanding the right-hand side and applying **eval** yields equation 31 of [2].

Alternately, consider that $f(x, y)$ satisfies the equations $\Delta_y f(x, y) = \frac{1}{x+y}$ and $f(x, 0) = 0$. Thus Proposition 8. says that

$$\sum_{i=0}^{n-1} \frac{1}{x+i} \simeq \int_0^n \frac{1}{x + \beta + t} dt. \tag{12}$$

That these two solutions are indeed the same essentially amounts to verifying that the chain rule holds as expected both for D and for any **k**-linear derivation on $F[\mathcal{A}]$. Both $\int_0^y \frac{1}{x+\beta+t} dt$ and $\log(x+\beta+y) - \log(x+\beta)$ become 0 when $y = 0$. So to check equality it is enough to know that they are equal after D_y is applied. But checking whether

$$\frac{1}{x + \beta + y} = D_y \log(x + \beta + y)$$

comes down to checking that the chain rule applies for D_y.

The proper algebraic setting for the preceding argument is the iterated logarithmic algebra of [1]. In fact, when $F[\mathcal{A}]$ is a field (implying $\mathcal{A} = \emptyset$), $\mathcal{L}$ is easily seen to be a subalgebra of the iterated logarithmic algebra. Polynomials in $\log(t), t, t^{-1}$ can be regarded as functions on the iterated logarithmic algebra and a general version of the chain rule holds. The technical lemmas associated with a classical umbral calculus for the iterated logarithmic algebra are not quite so straightforward as those in Section 4 and will be dealt with elsewhere.

6. Nörlund series

The Bernoulli umbrae were defined so as to provide solutions to linear difference equations. The process can of course be iterated. Namely, if

$\Delta^2 f(x) = g''(x)$, then $\Delta f(x) \simeq g'(x+\beta) + C$. Thus a solution is given by $f(x) \simeq g(x+\beta+\beta')$ where $\beta \equiv \beta'$ are Bernoulli umbrae. The fact that β does not equal β' is necessitated by the disjoint support requirements in the results of Section 4.

Definition 3. *Recall from [8] that the symbol $n\,.\,\alpha$ for $n \in \mathbf{Z}$ is defined to be a new umbra such that*

$$e^{(n.\alpha)x} \simeq \left(\mathbf{eval}(e^{\alpha x})\right)^n.$$

These are what Ray calls negative umbral integers in [3]. In particular, if α is an umbra representing the sequence $1, a_1, a_2, \ldots$, then $m\,.\,\alpha$ represents the sequence starting with

$$1, \quad m^2a_1^2 + m(a_2 - a_1^2), \quad m^3a_1^3 + 3m^2(a_1a_2 - a_1^3) + m(a_3 - 3a_1a_2 + 2a_1^3),$$

which, when a_1 is set to 1 and a_i is set to ϕ_{i-1} for $i > 1$, is precisely the sequence 2.5 of [3].

Recall that if β is a Bernoulli umbra, then the kth *Bernoulli polynomial*, $B_k(x) \in \mathbf{Q}[x]$, is defined by $B_k(x) \simeq (x+\beta)^k$ for all $k \in \mathbf{N}$. For all $n, k \in \mathbf{Z}$, define the kth *Nörlund series* of order n to be $B_k^{(n)}(x) \in \mathbf{Q}\{\{x\}\}$ where

$$B_k^{(n)}(x) \simeq (x + n\,.\,\beta)^k.$$

For $n = 1$ these define the Bernoulli series (denoted $B_k(x)$) and for $k \geq 0$ the Bernoulli series are the Bernoulli polynomials. The Bernoulli series are defined (see [2]) in the more general context of the Loeb-Rota algebra. The present notation (carried over from [8]) overlaps unfortunately with the notation of [2]; care should be taken to distinguish the two.

In general, for $f(x) \in \mathcal{L}$, the analog of Theorem 5.2 of [8] holds, namely if β is a Bernoulli umbra, then

$$f(x + (n+j)\,.\,\beta) \simeq \left(\frac{\Delta}{D}\right)^{-n} f(x + j\,.\,\beta).$$

The proof is by induction using the equation $f(x+\beta) \simeq (\Delta/D)^{-1} f(x)$ which follows directly from (6) and Corollary 4.

It is possible to generalize other results of [8] in this fashion. Setting $p(t) = q(y+t)$ in Theorem 8.7 of [8] and the argument in Example 5. shows that,

$$(n\,.\,\beta)E_y^{n.\beta+n.\gamma} D_y \simeq -n\left[E_y^{\beta} + D_y - I\right],$$

where β is a Bernoulli umbra and $\beta + \gamma \equiv 0$. Applying Corollary 4. shows that for $g(x) \in F[\mathcal{A}]\{\{x\}\}$

$$\begin{aligned}(n\,.\,\beta)D_y g(x + y + n\,.\,\beta + n\,.\,\gamma) &\simeq \\ \simeq -n\left[g(x+y+\beta) + D_y g(x+y) - g(x+y)\right],\end{aligned} \tag{13}$$

as long as $\beta \notin \mathbf{supp}(g(x))$.

Using this result we generalize Corollary 8.8 of [8] to derive a doubly indexed recurrence for the Nörlund series. The following computations will take place in $\mathbf{Q}[y][\mathcal{A}]\{\{x\}\}$ where $\mathcal{A}$ contains at least the umbrae β, γ, $n \,.\, \beta$, $n \,.\, \beta'$, $(n+1) \,.\, \beta$, $(n+1) \,.\, \beta'$, and $n \,.\, \gamma$. Let β be a Bernoulli umbra and let $\gamma + \beta \equiv 0$.

We define a polynomial $f(x,y) = (x+y+n.\beta)^k$ which umbrally represents the Nörlund series shifted by E^y. Now $f(x,y)$ may be split into

$$(n \,.\, \beta)(x+y+n \,.\, \beta)^{k-1} + (x+y)(x+y+n \,.\, \beta)^{k-1}.$$

Observe that the left summand is $(n.\beta)D_y \frac{(x+y+n.\beta)^k}{k}$. Now, since $n.\gamma + n.\beta' \equiv 0$ and $n \,.\, \beta', n \,.\, \gamma \notin \mathbf{supp}(f(x,y))$, Proposition 1.1 from [8] says that the left summand is equivalent to

$$(n \,.\, \beta)D_y \frac{(x+y+n \,.\, \beta + n \,.\, \beta' + n \,.\, \gamma)^k}{k}.$$

Setting $g(x) = (x + n \,.\, \beta')^k/k$ in (13) shows that $f(x,y)$ is equivalent to

$$\begin{aligned} -\frac{n}{k}\Big[(x+y+(n+1) \,.\, \beta')^k + k(x+y+n \,.\, \beta')^{k-1} - (x+y+n \,.\, \beta')^k\Big] \\ + (x+y)(x+y+n \,.\, \beta)^{k-1} \\ \simeq -\frac{n}{k}\Big[(x+y+(n+1) \,.\, \beta)^k + \frac{k}{n}(n-x-y)(x+y+n \,.\, \beta)^{k-1} \\ - (x+y+n \,.\, \beta)^k\Big]. \end{aligned}$$

Multiply through by k, set $y=0$, and combine terms to find

$$(k-n)(x+n \,.\, \beta)^k \simeq -n(x+(n+1) \,.\, \beta)^k + k(x-n)(x+n \,.\, \beta)^{k-1}.$$

The $k=0$ case holds trivially. This equation can be expressed purely as an identity in Bernoulli series.

Proposition 9. *For $n, k \in \mathbf{Z}$*

$$nB_k^{(n+1)}(x) = k(x-n)B_{k-1}^{(n)}(x) + (n-k)B_k^{(n)}(x).$$

Setting $k = n$ (and a trivial check at $n = 0$) implies that $B_{n-1}^{(n)}(x) = \frac{1}{x-n}B_n^{(n+1)}(x)$. Hence for $n \geq 0$,

$$B_{-n}^{(-n+1)}(x) = \left(\frac{1}{x}\right)\left(\frac{1}{x-1}\right)\cdots\left(\frac{1}{x-n+1}\right).$$

7. Universal difference equations

In [3], Nigel Ray surveys the theory associated to the universal difference operator Δ^ϕ. Not surprisingly, difference equations involving Δ^ϕ are approachable via the techniques outlined here. In fact, the connections run much deeper and I am indebted to Nigel Ray for bringing them to my attention.

In the notation of the present paper, he considers the case where $\mathbf{k} = \mathbf{Z}$ (a departure from my assumption that $\mathbf{Q} \subset \mathbf{k}$) and where the ring F is the polynomial ring $\mathbf{Z}[\phi_1, \phi_2, \ldots]$. In Ray's treatment, the ring S is $\mathbf{Z}[x]$ and D is $\frac{\partial}{\partial x}$, so of course $\hat{S} = S$. Let ϕ be an umbra such that $\phi^k \simeq \phi_{k-1}$. For compatibility with the notation of [3], this is an exception to the convention that only umbrae are written using Greek letters. Since in [3] all functions are polynomials, the shift operator E^a, for $a \in F[\mathcal{A}]$, may be defined as $\sum_{i \geq 0} a^i D^{(i)}$ where $D^{(i)}$ is the ith divided power of $\frac{\partial}{\partial x}$ and thus the need for rational numbers is averted.

Following [3], a universal Bernoulli umbra is an umbra, $\boldsymbol{\beta}_\phi$, satisfying the defining equation

$$\boldsymbol{\beta}_\phi^n \simeq (\boldsymbol{\beta}_\phi + \phi)^n$$

for all $n > 1$. The universal Bernoulli numbers $B_n^\phi \in \mathbf{Z}[\phi_1, \phi_2, \ldots]$ are defined by $B_n^\phi \simeq \boldsymbol{\beta}_\phi^n$. This immediately implies, copying the proof of (6), that $e^{\boldsymbol{\beta}_\phi D}(e^{\phi D} - I) \simeq D$. With this equation in hand, analogs to the results of Section 4 are easily deduced. If, at variance with the philosophy of [3], $\mathbf{k}$ is taken to be $\mathbf{Q}$ rather than $\mathbf{Z}$, these results will extend to formal power series of logarithmic type and

$$\Delta^\phi f(x + \boldsymbol{\beta}_\phi) \simeq f'(x)$$

for all $f(x)$ in the Loeb-Rota algebra.

References

[1] D. Loeb, The iterated logarithmic algebra. *Adv. Math.* **86** (1991), 155–234.

[2] D. Loeb and G.-C. Rota, Formal power series of logarithmic type. *Adv. Math.* **75** (1989), 1–118.

[3] N. Ray, Universal constructions in umbral calculus. *in this volume.*

[4] J. Riordan, *An Introduction to Combinatorial Analysis*, John Wiley, New York, 1958.

[5] S. Roman, *The Umbral Calculus*, Academic Press, San Diego, CA, 1984.

[6] S. ROMAN AND G.-C. ROTA, The umbral calculus. *Adv. Math.* **27** (1978), 95–188.

[7] G.-C. ROTA AND B. D. TAYLOR, An introduction to the umbral calculus. in *Analysis, Geometry and Groups: A Riemann Legacy Volume* Hadronic Press, Palm Harbor, FL, 1993. 513–525.

[8] G.-C. ROTA AND B. D. TAYLOR, The classical umbral calculus. *SIAM J. Math. Anal.* **25** (1994), 694–711.

Department of Mathematics
MIT
Cambridge, MA 02139

An Analogy in Geometric Homology: Rigidity and Cofactors on Geometric Graphs

Walter Whiteley

Dedicated to Gian-Carlo Rota on the occasion of his 64th birthday

Abstract

From recent work in two areas of discrete applied geometry, we abstract a common pattern of families of geometric homologies for graphs realized in projective d-space (for static rigidity) or in the projective plane (for bivariate splines). Using distinct algebraic constructions for the local coefficients of the chain complexes (exterior algebra for statics and symmetric algebra for splines), we explore the underlying analogy between the two theories. The analogy starts with isomorphisms for the lowest two members of the families, moves to a conjectured isomorphism for generic realizations in the third family members (static rigidity in 3-space and C_2^1-cofactors for bivariate splines) and a proven difference with a conjectured injection for all higher family members.

The conjectured isomorphism for static rigidity in 3-space and C_2^1-cofactors for bivariate splines illustrates fundamental problems in structural rigidity, in polynomial-time algorithms for graph properties which are random polynomial, and in defining 'freest' matroids for submodular functions.

1. Introduction

Recent work in two areas of discrete applied geometry has generated two families of matroids on geometric graphs [24] (Figure 1):

(1) the rigidity matroids for graphs in projective space are abstracted from the statics and kinematics of bar frameworks in d-space;

(2) the C_s^{s-1}-cofactor matroids for graphs in the plane are abstracted from the theory of bivariate functions which are piecewise degree s polynomials and globally of continuity C^{s-1}: the bivariate C_s^{s-1}-splines.

1991 *Mathematics Subject Classification.* Primary 05B35, 52C25; 55U15 Secondary 05C75, 65D07.

Key words and phrases. geometric homology, first-order rigidity, static rigidity, multivariate splines, cofactors, exterior algebra, symmetric algebra, matroids on graphs.

Work supported, in part, by grants from NSERC (Canada).

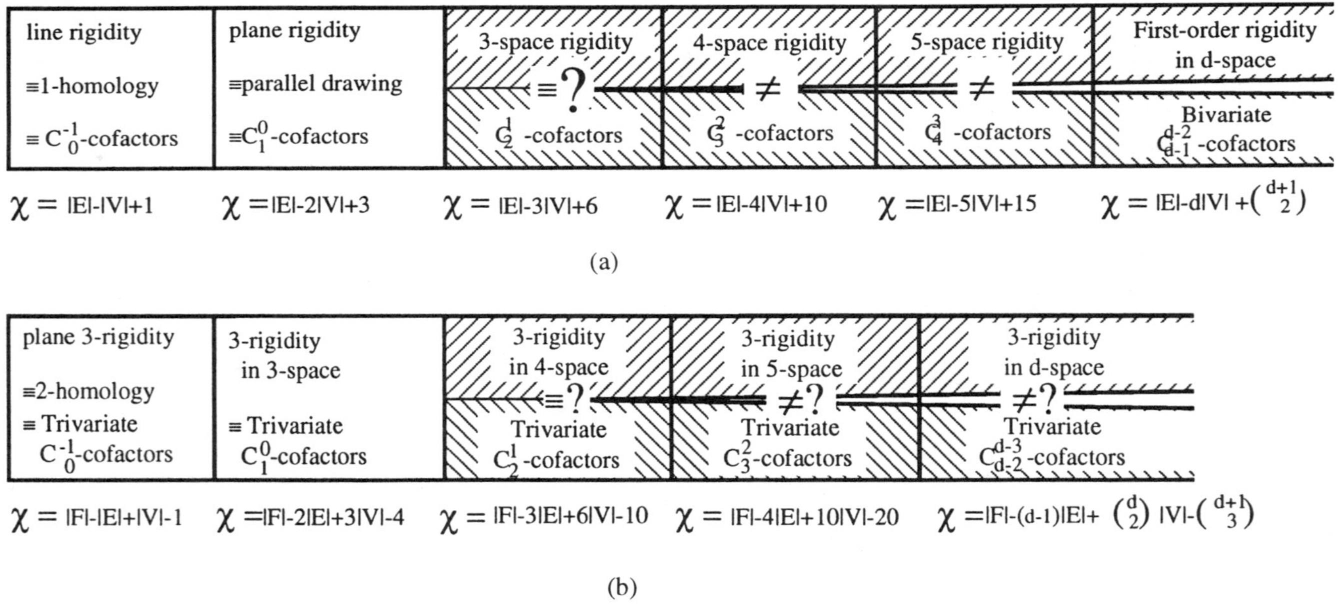

line rigidity ≡1-homology ≡ C^{-1}_0-cofactors	plane rigidity ≡parallel drawing ≡C^0_1-cofactors	3-space rigidity ≡? C^1_2 -cofactors	4-space rigidity ≠ C^2_3 -cofactors	5-space rigidity ≠ C^3_4 -cofactors	First-order rigidity in d-space Bivariate C^{d-2}_{d-1} -cofactors
$\chi = \|E\|-\|V\|+1$	$\chi = \|E\|-2\|V\|+3$	$\chi = \|E\|-3\|V\|+6$	$\chi = \|E\|-4\|V\|+10$	$\chi = \|E\|-5\|V\|+15$	$\chi = \|E\|-d\|V\| + \binom{d+1}{2}$

(a)

plane 3-rigidity ≡2-homology ≡ Trivariate C^{-1}_0-cofactors	3-rigidity in 3-space ≡ Trivariate C^0_1-cofactors	3-rigidity in 4-space ≡? Trivariate C^1_2-cofactors	3-rigidity in 5-space ≠? Trivariate C^2_3-cofactors	3-rigidity in d-space ≠? Trivariate C^{d-3}_{d-2}-cofactors
$\chi = \|F\|-\|E\|+\|V\|-1$	$\chi = \|F\|-2\|E\|+3\|V\|-4$	$\chi = \|F\|-3\|E\|+6\|V\|-10$	$\chi = \|F\|-4\|E\|+10\|V\|-20$	$\chi = \|F\|-(d-1)\|E\| + \binom{d}{2}\|V\| - \binom{d+1}{3}$

(b)

Figure 1. (a) Two families of geometric chain complexes on geometric graphs; (b) two families of geometric chain complexes on geometric 2-simplicial complexes.

The structure of these matroids (and extensions to higher dimensional geometric complexes) becomes clearer when expressed as a geometric homology: the homology of a chain complex defined for an abstract simplicial complex with local coefficients based on a geometric configuration for the vertices. We present the basic structure of these (and other) families of geometric homologies on simplicial complexes (§3) with graphs as core examples (§2).

For our two families, the local coefficients of the chains are drawn from the two classical geometric algebras: exterior algebra (rigidity §4) and symmetric algebra (cofactors §5). These two families of chain complexes have striking similarities, including identical Euler characteristics (Figure 1). For the well-understood basic case (rigidity on the plane $\equiv C_1^0$-cofactors) the chain complexes are isomorphic (§7), reflecting an old correspondence of James Clerk Maxwell [8,13,20]. Even lower, for rigidity on the line and C_0^{-1}-cofactors, we get alternate interpretations of the underlying homology of the graph. For the third level, rigidity in 3-space and C_2^1-cofactors (§8), there has been an extensive transfer of theorems and techniques supporting our conjecture that the homologies for generic configurations are isomorphic. For higher cases, such as rigidity in 4-space and C_3^2-cofactors (§9), there is both an extensive transfer of results and a known divergence which suggests that the cofactor matroids, for generic plane configurations, have more independent sets than the rigidity matroids for generic configurations in 4-space.

For the plane, the underlying matroid on edges defined by the 1-cycles at generic configurations can be defined by the submodular function $f(E) = 2|V(E)|-3$, using a theorem of Edmonds and Rota for submodular functions which are positive on singletons [10]. The matroids for statics in 3-space and plane C_2^1-cofactors are intimately related to the submodular function $f(E) = 3|V(E)| - 6$, for sets larger than singletons. Since this function is 0 for singletons, the submodular definition does not apply. It is conjectured that these matroids are, in a specific sense, the 'freest' matroids with this rank function (roughly, they give a maximal collection of independent sets) [12,24].

The complexity of combinatorial characterizations for the bases of one (or both) of these matroids, as defined by generic configurations, is also an interesting example for the theory of algorithms on graphs. There is a polynomial-time random algorithm to test for bases (pick 'random integer points' in the space and compute the rank of a matrix) but no known polynomial-time deterministic algorithm (or even exponential-time deterministic algorithm).

In exploring this comparison, we are also presenting a common pattern which is related to geometric homologies and matroids arising in a broader class of geometric studies (§10). The analogy, including the isomorphisms and conjectures, extends to geometric homologies of larger simplicial (and polyhedral) complexes related to 'skeletal rigidity' [16,17,24] and multivariate splines [1,2,23]. We emphasize the examples on graphs to focus on the surprising analogy of the exterior algebra and the symmetric algebra is this setting. This analogy has become a strong support for the development of both theories, as we transfer results and conjectures between the two fields [23,24].

Our simplicial homological approach is intimately related to, but not identical to, H. Crapo's approach to geometric cohomology for liftings of polyhedral configurations and the rigidity of general structures of points, lines, bodies etc. [4,5,6]. Crapo studies the cohomology of these structures using the more general tools of nerves and schemes of functions, rather than the elementary approach of extending simplicial homology with geometrically defined local coefficients. There are a number of important problems which can only be investigated with his more general approach.

Acknowledgments. My approach to these topics has been shaped by Gian-Carlo Rota in many ways. Gian-Carlo's culture of conjectures and his broad appreciation for the power of analogies have been a strong inspiration for this paper. The specific content of this paper was shaped by joint work with fellow students of Gian-Carlo Rota: Henry Crapo, Neil White and a second generation member of the Rota family, Tiong-Seng Tay.

2. Graph theory as homology

Consider a graph $G = (V, E)$ as an oriented simplicial complex, with the set $\overline{E}$ of oriented edges (both orientations are present), with vertices $V = \{1, \ldots, n\}$. The oriented edges are written as $[i, j] \in \overline{E}$. The 1-chains of the graph are functions from the oriented edges to the reals [3,14]: $c : \overline{E} \to \mathbb{R}$ with $c([j, i]) = -c([i, j])$. (We work over the reals $\mathbb{R}$ to match the underlying applications to frameworks and splines.) Equivalently, we pick an arbitrary orientation for each edge (say lexicographic) to form the set E and the 1-chains are vectors with coordinates indexed by E. We write $[e]$ to emphasize that this is an abstract oriented edge.

$$C_1 = \mathbb{R}^{|E|} = \oplus_{e \in E} \mathbb{R} = \{\sum_{e \in E} c_{[e]}[e] \}.$$

The 0-chains on the vertices are the vector space

$$C_0 = \mathbb{R}^{|V|} = \oplus_{i \in V} \mathbb{R} = \{\sum_{i \in V} c_i[i] \},$$

and the (-1)-chains are $C_{-1} = \mathbb{R} = c[\emptyset]$. With this notation, the augmented simplicial chain complex for the graph is

$$\mathcal{G} : \mathbf{0} \to \oplus_{e \in E} \mathbb{R} \xrightarrow{\partial_1} \oplus_{i \in V} \mathbb{R} \xrightarrow{\partial_0} \mathbb{R} \xrightarrow{\partial_{-1}} \mathbf{0},$$

with the standard simplicial boundary operators $\partial_1 c_{[i,j]}[i,j] = c_{[i,j]}[i] - c_{[i,j]}[j]$, $\partial_0 c_{[i]}[i] = c_{[i]}[\emptyset]$, and $\partial_{-1} c[\emptyset] = 0$, each extended linearly to corresponding chains. Since $\partial_0(\partial_1[i,j]) = \partial_0([j] - [i]) = (1-1)[\emptyset] = \mathbf{0}$ for all oriented edges, it is easy to check that $\partial_{i-1}\partial_i = 0$ for $i = 1, 0$.

With such a chain complex, the *i-cycles* are the i-chains $\mathbf{c}$ satisfying the linear equation: $\partial_i(\mathbf{c}) = 0$, forming the subspace $Z_i(\mathcal{G})$ [14]. The *i-boundaries* are the i-chains $\mathbf{c}$ which are $\partial_{i+1}(\mathbf{d}) = \mathbf{c}$ for some $(i+1)$-chain $\mathbf{d}$, forming the subspace $B_i(\mathcal{G})$. Since $\partial_{i-1}\partial_i = 0$, all i-boundaries are i-cycles, and $B_i(\mathcal{G}) \subseteq Z_i(\mathcal{G})$. The quotient $H_i(\mathcal{G}) = Z_i(\mathcal{G})/B_i(\mathcal{G})$ is the (reduced) *homology space.* The dimension of the i-homology is the *Betti number* $\beta_i(\mathcal{G})$. [Since we have (-1)-chains, $\mathcal{G}$ is an augmented chain complex and the homology is called reduced [14]. We will omit the word 'reduced' throughout the rest of the paper.]

By standard results of homology theory, the *Euler characteristic* of $\chi(\mathcal{G})$ has two equivalent definitions as alternating sums

$$-\beta_1(\mathcal{G}) + \beta_0(\mathcal{G}) + \beta_{-1}(\mathcal{G}) = \chi(\mathcal{G}) = -|C_1| + |C_0| - |C_{-1}| = -|E| + |V| - 1.$$

To assist the tracking of signs, sums and associated matrices, we define

$$\operatorname{Sign}\delta(i,[e]) = \begin{cases} 1 & \text{if } [e] = [i,k] \text{ for some } k > i; \\ -1 & \text{if } e = [k,i] \text{ for some } k < i; \\ 0 & \text{if } i \notin e. \end{cases}$$

With this notation, the matrix for ∂_1 as a linear transformation on the 1-cycles, with the columns as the boundaries of the oriented edges, is

$$B(G) \quad = \quad \begin{array}{c} \\ 1 \\ \vdots \\ i \\ \vdots \\ j \\ \vdots \end{array} \begin{array}{c} \begin{array}{ccccc} \cdots & [i,j] & \cdots & [e] & \cdots \end{array} \\ \left(\begin{array}{ccccc} \cdots & 0 & \cdots & \operatorname{Sign}\delta(1,[e]) & \cdots \\ \ddots & \vdots & \ddots & \vdots & \ddots \\ \cdots & 1 & \cdots & \operatorname{Sign}\delta(i,[e]) & \cdots \\ \ddots & \vdots & \ddots & \vdots & \ddots \\ \cdots & -1 & \cdots & \operatorname{Sign}\delta(j,[e]) & \cdots \\ \ddots & \vdots & \ddots & \vdots & \ddots \end{array} \right) \end{array}$$

This is the standard *incidence matrix for the oriented graph* over the reals, and the dependence and independence of these columns of the matrix define the 1-*cycle matroid*, $\mathcal{M}(G)$, on the edges of the graph [3]. (The matroid does not depend on the orientation of the graph.) The kernel of $B(G)$ is the 1-cycle space $Z_1(\mathcal{G})$, or equivalently, the 1-homology of the graph $H_1(\mathcal{G})$.

Example 2.1. The objects and terms of this homology correspond to standard terms in graph theory [3].

(1) Any oriented polygon P in the graph leads to a 1-cycle $\sum_{e\in P} \pm[e]$ where the sign is $+$ for edges oriented with the polygon and $-$ for edges oriented against the polygon. Conversely, any 1-cycle is the algebraic sum of real multiples of such polygonal cycles. This 1-cycle matroid is the standard graphic matroid with polygons as 1-cycles. The dimension of $Z_1(\mathcal{G})$ ($\beta_1(\mathcal{G})$) is also called the cyclomatic number of the graph [3].

(2) For any graph, all (-1)-chains are (-1)-cycles. For the empty graph, with no 0-chains, $\beta_{-1}(\mathcal{G}) = 1$. For any non-empty graph with a vertex v, $\partial_0(c[v]) = c[\emptyset]$, so all (-1)-chains are 0-boundaries and $\beta_{-1}(\mathcal{G}) = 0$.

(3) Since $\beta_1(\mathcal{G})$ measures cycles, $\beta_1(\mathcal{G}) = 0$ if and only if the graph is a forest. A nonempty forest F has $\beta_1(\mathcal{F}) = 0, \beta_{-1}(\mathcal{F}) = 0$, so $\beta_0(\mathcal{F}) = \chi(\mathcal{F}) = -|E| + |V| - 1$. For a non-empty forest, $\beta_0(\mathcal{F})$ is the number of trees in the forest minus one (i.e. the number of components minus one).

(4) For a general graph, $\beta_0(\mathcal{G}) + 1$ is also the number of maximal connected components (see the next example).

(5) A nonempty spanning tree T (a connected graph with no polygons) has $\beta_1(\mathcal{T}) = \beta_0(\mathcal{T}) = \beta_{-1}(\mathcal{T}) = 0$. Such a chain complex is *acyclic*, with $-\chi(\mathcal{G}) = 0$. For a tree, we see that $|E| - |V| + 1 = 0$, or $|E| = |V| - 1$.

The incidence matrix $B(G)$ has a cokernel which corresponds to the 0-cohomology of the graph. In general, i-cochains form the dual vector space to the i-chains [14]. Specifically, we write the cochains as

$$C^{-1} = \mathbb{R}^* = \{c[\emptyset]\}, \quad C^0 = (\mathbb{R}^*)^{|V|} = \{\sum_{i\in V} c^i[i]\ \}, \quad C^1 = \{\sum_{e\in E} c^{[e]}[e]\ \}.$$

The coboundary operators are

$$\delta^1 c^{[i,j]}[i,j] = 0, \qquad \delta^0 c^{[i]}[i] = \sum_{e\in E} \operatorname{Sign}\delta(i,e) c^{[i]}[e], \qquad \delta^{-1}c[\emptyset] = \sum_{j\in V} c[j],$$

extended linearly to all cochains. With this notation, the (augmented) cochain complex of the graph G is

$$\mathcal{G}_* : \mathbf{0} \xleftarrow{\delta^1} \oplus_{e\in E}\mathbb{R}^* \xleftarrow{\delta^0} \oplus_{i\in V}\mathbb{R}^* \xleftarrow{\delta^{-1}} \mathbb{R}^* \leftarrow \mathbf{0}.$$

The *i-cocycles*, $Z^i(\mathcal{G}_*)$, are the kernel of the i-coboundary; the *i-coboundaries*, $B^i(\mathcal{G}_*)$, are the image of the $(i-1)$-coboundary; and the i-cohomology is the quotient space $H^i(\mathcal{G}_*) = Z^i(\mathcal{G}_*)/B^i(\mathcal{G}_*)$.

Notice that the rows of the incidence matrix $B(G)$ are the 1-coboundaries of the vertices and the 1-cycles are orthogonal to the 1-coboundaries. Using the inner product of i-chains $\mathbf{c}$ and i-cochains $\mathbf{d}_*$, $\langle \mathbf{c}, \mathbf{d}_* \rangle = \sum_\sigma c_{[\sigma]} d_*^{[\sigma]}$, any 1-chain $\mathbf{c}$ and 0-cochain $\mathbf{d}_*$ satisfy $\langle \partial \mathbf{c}, \mathbf{d}_* \rangle = \langle \mathbf{c}, \delta \mathbf{d}_* \rangle$. This also shows that 0-cocycles are orthogonal to the 0-boundaries. The general observation that i-cycles are the orthogonal compliment to the i-coboundaries, and the dual result for i-cocycles, gives the basic theorem that the Betti numbers of the chain and cochain complex over a field are the same ([14]): $\beta_i(\mathcal{G}) = \beta^i(\mathcal{G}_*)$.

Example 2.2. The 0-cocycles are the cokernel of the incidence matrix and are generated by the independent 'component sums' (i.e. characteristic functions for the vertices of a maximal connected component). The 0-coboundaries are scalar multiples of the vertex sum $\sum_{i \in V} [v]$. This verifies that $\beta^0(\mathcal{G}_*) (= \beta_0(\mathcal{G}))$ is the number of components minus one.

The 1-coboundaries are also called the *cut space* of the graph [3] and the orthogonality of these 1-coboundaries and the 1-cycles is a basic theorem [3]. Note that these are some, but not all, of the 1-cocycles, although the cut space is sometimes called the 'cocycle space of the graph' [3].

Example 2.3. For comparison with later sections, we recall a standard technique of homology: the Mayer-Vietoris sequence for computing the homology of the union of two sets from the homologies of the original sets and of their intersection [14]. For two graphs G_1, G_2, the Mayer-Vietoris sequence is

$$\begin{aligned} \mathbf{0} \to H_1(G_1 \cap G_2) &\to H_1(G_1) \oplus H_1(G_2) \to H_1(G_1 \cup G_2) \\ &\to H_0(G_1 \cap G_2) \to H_0(G_0) \oplus H_0(G_2) \to H_0(G_1 \cup G_2) \\ &\to H_{-1}(G_1 \cap G_2) \to H_{-1}(G_1) \oplus H_{-1}(G_2) \to H_{-1}(G_1 \cup G_2) \to \mathbf{0}. \end{aligned}$$

If G_1, G_2 are non-empty trees ($\beta_1 = \beta_0 = \beta_{-1} = 0$), then:

$$\mathbf{0} \to H_1(G_1 \cup G_2) \to H_0(G_1 \cap G_2) \to \mathbf{0} \to H_0(G_1 \cup G_2) \to H_{-1}(G_1 \cap G_2) \to \mathbf{0}.$$

We conclude that

(1) $\beta_1(G_1 \cup G_2) = \beta_0(G_1 \cap G_2)$ and the number of cycles in $G_1 \cup G_2$ is the number of components in $G_1 \cap G_2$, minus one.
(2) $\beta_0(G_1 \cup G_2) = \beta_{-1}(G_1 \cap G_2)$ and we have an additional component only if the intersection is empty.

If we join two trees on an intersection $G_1 \cap G_2$, then the Mayer-Vietoris sequence shows that the union is a single tree if and only if $G_1 \cap G_2$ is a tree.

3. A general pattern for geometric homologies

Our geometric homologies will be defined for a simplicial complex Δ and a configuration $\tilde{\mathbf{p}} \in \mathbb{R}^{(d+1)|V|}$ realizing the vertices as weighted points $\tilde{\mathbf{p}}(i) = \tilde{\mathbf{p}}_i$ in projective d-space $\mathbb{P}_d$. (Geometric homologies could use affine or Euclidean points for the geometric data but we have selected projective examples.)

Recall the ordinary simplicial chain complex for Δ, with i-faces $\Delta^{(i)}$ [14]

$$\mathcal{C}(\Delta^r) : \mathbf{0} \rightarrow \oplus_{\sigma\in\Delta^{(r)}}\mathbb{R} \xrightarrow{\partial_r} \oplus_{\rho\in\Delta^{(r-1)}}\mathbb{R} \ldots \xrightarrow{\partial_1} \oplus_{v\in\Delta^{(0)}}\mathbb{R} \xrightarrow{\partial_0} \mathbb{R} \xrightarrow{\partial_{-1}} \mathbf{0},$$

with $\partial_i[\rho] = \sum_{\pi\in\Delta^{(i-1)}} \mathrm{Sign}\delta(\pi,\rho)[\pi]$.

By some construction (which will vary with the example), the combinatorial and geometric data for each simplex ρ: $\{\tilde{\mathbf{p}}_i \mid i \in \rho\}$ is converted into a vector space of *local coefficients* $V_{\rho,\tilde{\mathbf{p}}}$, with the property that $\pi \subseteq \rho \Rightarrow V_{\rho,\tilde{\mathbf{p}}} \subseteq V_{\pi,\tilde{\mathbf{p}}}$. With these coefficients, the geometric chain complex is

$$\mathcal{C}(\Delta^r;\tilde{\mathbf{p}}) : \mathbf{0} \rightarrow \oplus_{\sigma\in\Delta^{(r)}} V_{\sigma,\tilde{\mathbf{p}}} \xrightarrow{\partial_r} \oplus_{\rho\in\Delta^{(r-1)}} V_{\rho,\tilde{\mathbf{p}}} \xrightarrow{\partial_{r-1}} \cdots$$

$$\cdots \xrightarrow{\partial_1} \oplus_{v\in\Delta^{(0)}} V_{v,\tilde{\mathbf{p}}} \xrightarrow{\partial_0} V_\emptyset \xrightarrow{\partial_{-1}} \mathbf{0}$$

with the boundary operator

$$\partial_i c_{[\rho]}[\rho] = c_{[\rho]}(\partial_i[\rho]) = \sum_{\pi\in\Delta^{(i-1)}} \mathrm{Sign}\delta(\pi,\rho) c_{[\rho]}[\pi]$$

extended linearly to all i-chains. With our assumptions, it is clear that $\partial_{i-1}\partial_i = 0$, since $\partial_{i-1}\partial_i(c_{[\pi]}[\pi]) = c_{[\pi]}(\partial_{i-1}\partial_i[\pi]) = 0$ by definition.

Remark 3.1. Consider distinct vector spaces $V_{\rho,\tilde{\mathbf{p}}}$ and linear transformations $T_{\pi,\rho} : V_{\rho,\tilde{\mathbf{p}}} \rightarrow V_{\pi;\tilde{\mathbf{p}}}$ for $\pi \subset \rho$, with $T_{\psi,\pi}T_{\pi,\rho} = T_{\rho,\psi}$ and $\partial P[\rho] = \sum_\pi \mathrm{Sign}\delta(\pi,\rho)T_{\pi,\rho}P[\pi]$. These define *local coefficients* for a geometric chain complex.

The desired cohomology generalizes the usual simplicial cohomology [14]

$$\mathcal{C}^*(\Delta^r) : \mathbf{0} \leftarrow \oplus_{\sigma\in\Delta^{(r)}}\mathbb{R}^* \xleftarrow{\delta_{r-1}} \oplus_{\rho\in\Delta^{(r-1)}}\mathbb{R}^* \xleftarrow{\delta_{r-2}} \ldots \oplus_{v\in\Delta^{(0)}} \mathbb{R}^* \xleftarrow{\delta_{-1}} \mathbb{R}^* \leftarrow \mathbf{0}.$$

The geometric coefficients for the cohomology will be based on the dual vector space V^*. The space $V^*\backslash_{\sigma,\mathbf{p}}$ is a quotient space with the equivalence relation:

$$S^* \stackrel{\sigma,\tilde{\mathbf{p}}}{=} T^* \text{ iff } \left(\forall P \in V_{\sigma,\tilde{\mathbf{p}}}\right)\ (S^* - T^*)(P) = 0.$$

One point of this definition is to ensure the inner product $\langle P, S^*\rangle = S^*(P)$ is well-defined: If $S^* \overset{\sigma,\tilde{\mathbf{p}}}{=} T^*$, then $\langle P, S^*\rangle = \langle P, T^*\rangle$. The second point is that, if $\pi \subset \rho$, then (up to a simple isomorphism) $V^*\backslash_{\pi,\tilde{\mathbf{p}}} \subset V^*\backslash_{\rho,\tilde{\mathbf{p}}}$. This follows from $S^* \overset{\pi,\tilde{\mathbf{p}}}{=} T^* \Rightarrow S^* \overset{\rho,\tilde{\mathbf{p}}}{=} T^*$, since $V_{\rho,\tilde{\mathbf{p}}} \subset V_{\pi,\tilde{\mathbf{p}}}$. With these coefficients, the geometric cochain complex is

$$\mathcal{C}^*(\Delta^1; \tilde{\mathbf{p}}) : \mathbf{0} \leftarrow \oplus_{\sigma\in\Delta^{(r)}} V^*\backslash_{\sigma,\tilde{\mathbf{p}}} \xleftarrow{\delta^{r-1}} \oplus_{\rho\in\Delta^{(r-1)}} V^*\backslash_{\rho,\tilde{\mathbf{p}}} \xleftarrow{\delta^{r-2}} \dots$$
$$\dots \xleftarrow{\delta^0} \oplus_{v\in\Delta^{(0)}} V^*\backslash_{v,\tilde{\mathbf{p}}} \xleftarrow{\delta^{-1}} V^*\backslash_{\emptyset,\tilde{\mathbf{p}}} \leftarrow \mathbf{0}.$$

The orthogonality of i-cycles and i-coboundaries extends: for any $(i+1)$-chain $\mathbf{c}$ and i-cochain $\mathbf{d}_*$: $\langle \partial_{i+1}\mathbf{c}, \mathbf{d}_*\rangle = \sum_{\pi\in\Delta^{(i)}} \text{Sign}\delta(\pi,\rho) d_*^{[\pi]}(c_{[\rho]}) = \langle \mathbf{c}, \delta^i \mathbf{d}_*\rangle$.

Remark 3.2. Crapo defines a related geometric cohomology (and homology) by means of the 'nerve' of a general complex and a sheaf of functions with local geometric properties (such as locally linear functions) [4,5,6]. This gives another description of the examples we will present in terms of geometric homology. In this approach certain properties are much clearer and easier to express and others are more difficult to see.

4. Exterior algebra and rigidity of frameworks

A *geometric graph* $G; \tilde{\mathbf{p}}$ is a graph (a 1-simplicial complex) $G = (V, E)$ and a *d-configuration* $\tilde{\mathbf{p}} \in R^{(d+1)|V|}$, with $\tilde{\mathbf{p}}(i) = \tilde{\mathbf{p}}_i$ as a weighted point for each $i \in V$. In this section, the local coefficients are defined using the exterior algebra or the Grassman-Cayley Algebra Λ_{d+1} of $\mathbb{R}^{d+1}$, with the 'exterior product' written as a *join* $\tilde{\mathbf{a}} \vee \tilde{\mathbf{b}}$ [19]. For our examples on graphs, we focus on the space of 2-extensors $\Lambda_{d+1}^{(2)}$.

For an edge $e = \{i, j\}$, the space of local coefficients is the 1-dimensional subspace (the same space for both directed edges):

$$V_{e,\tilde{\mathbf{p}}} = \{c\tilde{\mathbf{p}}_i \vee \tilde{\mathbf{p}}_j \mid c \in \mathbb{R}\} = \Lambda_{d+1}^{(2)}|_{e,\tilde{\mathbf{p}}}.$$

For a vertex i, the d-dimensional space of local coefficients is:

$$V_{i,\tilde{\mathbf{p}}} = \{c\tilde{\mathbf{q}} \vee \tilde{\mathbf{p}}_i \mid c \in \mathbb{R}, \tilde{\mathbf{q}} \in \mathbb{R}^{d+1}\} = \Lambda_{d+1}^{(2)}|_{i,\tilde{\mathbf{p}}}.$$

For the empty set $\emptyset$, the coefficients are the $\binom{d+1}{2}$-dimensional space $V_\emptyset = \Lambda_{d+1}^{(2)}$. An alternate definition of these coefficients, which transfers more easily to all simplicial complexes, is $V_{\pi,\tilde{\mathbf{p}}} = \{P \in \Lambda^{(2)}_{d+1} | P \vee \tilde{\mathbf{p}}_i = \mathbf{0} \text{ for all } i \in \pi\}$ (§10).

The *d-rigidity chain complex* on a geometric graph $G; \tilde{\mathbf{p}}$ is

$$\mathcal{R}_d(G;\tilde{\mathbf{p}})\colon \mathbf{0} \to \oplus_{e\in E}\Lambda^{(2)}_{d+1}|_{e,\tilde{\mathbf{p}}} \xrightarrow{\partial_1} \oplus_{i\in V}\Lambda^{(2)}_{d+1}|_{i,\tilde{\mathbf{p}}} \xrightarrow{\partial_0} \Lambda^{(2)}_{d+1} \xrightarrow{\partial_{-1}} \mathbf{0}.$$

With the boundary maps defined following the previous section, for any i-face ρ $(-1 \le i \le 1)$: $\partial_i c_{[\rho]}[\rho] = c_{[\rho]}\partial_i[\rho]$ extended linearly to all i-chains. (Note that in cases such as $i \in e$, $\Lambda^{(2)}_{d+1}|_{e,\tilde{\mathbf{p}}} \subset \Lambda^{(2)}_{d+1}|_{i,\tilde{\mathbf{p}}}$ as required for this operation.) Using the dimensions given above, the Euler characteristic is

$$\chi(\mathcal{R}_d(G;\tilde{\mathbf{p}})) = -|E| + d|V| - \binom{d+1}{2}.$$

(a) The 1-chains $C_1(\mathcal{R}_d(G;\tilde{\mathbf{p}}))$ have coefficients $c_{[i,j]} \in \Lambda^{(2)}_{d+1}|_{[i,j],\tilde{\mathbf{p}}}$. These are multiples $\omega_{i,j}$ of the 2-extensor $\tilde{\mathbf{p}}_i \vee \tilde{\mathbf{p}}_j$ for the geometric line through $\tilde{\mathbf{p}}_i, \tilde{\mathbf{p}}_j$. Since $\omega_{i,j}\tilde{\mathbf{p}}_i \vee \tilde{\mathbf{p}}_j = -\omega_{i,j}\tilde{\mathbf{p}}_j \vee \tilde{\mathbf{p}}_i$, the underlying condition $c(-[i,j]) = -c[i,j]$ ensures that each edge has a unique induced scalar $\omega_{i,j} = \omega_{j,i}$. These 1-chains are called *static stresses*, where the line vectors $c_{[i,j]} = \omega_{i,j}\tilde{\mathbf{p}}_i \vee \tilde{\mathbf{p}}_j$ represent internal tension $(\omega_{i,j} < 0)$ or compression $(\omega_{i,j} > 0)$ as forces on the ends of the *bar* $\{i,j\}$ (Figure 2C).

(b) The 0-chains $C_0(\mathcal{R}_d(G;\tilde{\mathbf{p}}))$ have coefficients c_i which are 2-extensors of the form $\tilde{\mathbf{q}} \vee \tilde{\mathbf{p}}_i$, representing forces applied to each vertex. These are the *static loads* on the vertices (*joints*) of the framework (Figure 2E).

(c) The 1-boundary operator $\partial_1\mathbf{c}$ computes the net loads exerted by this stress $\mathbf{c}$ on each vertex of the framework.

(d) The 1-cycles $Z_1(\mathcal{R}_d(G;\tilde{\mathbf{p}}))$ are *self-stresses*, which exert the net force $\mathbf{0}$ on each vertex (Figure 2D). If $\beta_1(\mathcal{R}_d(G;\tilde{\mathbf{p}})) = 0$ and the only self-stress is the zero stress, the framework is *statically independent* (Figure 2A).

(e) The matrix for the 1-boundary operator ∂_1 is the transpose of a projective form of the *rigidity matrix* of the framework [7,12,16,18], with the 1-cycles as the kernel.

(f) The 0-boundaries $B_0(\mathcal{R}_d(G;\tilde{\mathbf{p}}))$ are the *resolved loads*: the net boundary at each vertex matches the force applied by the static load at that vertex.

(g) The (-1)-chains C_{-1} are the *wrenches* (sums of 2-extensors or line-bound vectors) which can be applied to a body in d-space.

(h) The 0-boundary operator ∂_0 computes the global wrench exerted by the loads applied to all of the vertices.

(i) The 0-cycles $Z_0(\mathcal{R}_d(G;\tilde{\mathbf{p}}))$ are the *equilibrium loads*: static loads which would exert no net translational force or rotational torque on a rigid body connecting the vertices (Figure 2E).

(j) A framework $G; \tilde{\mathbf{p}}$ is *statically rigid* if all equilibrium loads are resolved, that is, if all 0-cycles are 0-boundaries and $\beta_0(\mathcal{R}_d(G;\tilde{\mathbf{p}})) = 0$.

(k) The (-1)-chains are net wrenches on a rigid body and all are (-1)-cycles.

(l) The (-1)-boundaries are the net wrenches exerted by the loads over all the vertices of $\tilde{\mathbf{p}}$. All (-1)-cycles are (-1)-boundaries and $\beta_{-1}(\mathcal{R}_d(G;\mathbf{p})) = 0$ if, and only if, the vertices span a hyperplane in d-space (see below).

(m) If the complex $\mathcal{R}_d(G;\tilde{\mathbf{p}})$ is *homologically trivial* ($\beta_0 = \beta_{-1} = 0$) then it is *statically rigid* and the joints span at least a hyperplane in d-space. It will have a space of self-stresses of dimension $\beta_1(\mathcal{R}_d(G;\tilde{\mathbf{p}})) = |E| - d|V| + \binom{d+1}{2}$.

(n) An *acyclic complex* ($\beta_1 = \beta_0 = \beta_{-1} = 0$) will be statically rigid with $|E| = d|V| - \binom{d+1}{2}$ statically independent bars on at least d vertices. Structural engineers call such a minimal statically rigid framework on its vertices *isostatic*. In general, provided the joints are either in general position or span projective d-space, a framework is isostatic if and only if it is also a maximal statically independent framework on the vertices.

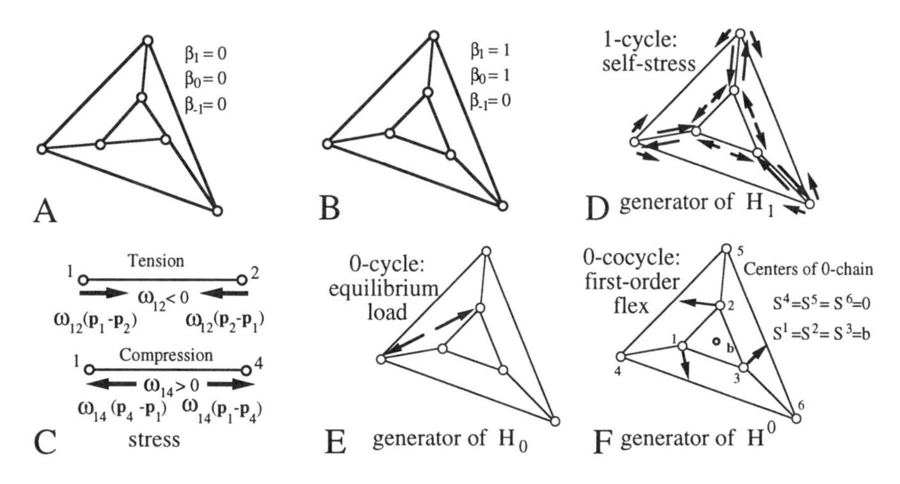

Figure 2. Two plane configurations (A,B) produce distinct homologies for the same graph.

We can also permit points at infinity which represent 'slide joints' [7]. Within traditional statics, there are interpretations for forces applied 'at infinity', including forces along the line at infinity which are *static couples* applying a net rotational wrench. It is no accident that Möbius developed barycentric coordinates and wrote a text on statics during the same period of work!

Lemma 4.1. *If the points of a geometric graph* $\{\tilde{\mathbf{p}}_i \mid i \in V\}$ *span a projective subspace of dimension k then* $\beta_{-1}(\mathcal{R}_d(G;\tilde{\mathbf{p}})) = \binom{d-k}{2}$

Proof. Assume $\{\tilde{\mathbf{p}}_i \mid i \in V\}$ spans a projective k-space. Take a basis B for $\mathbb{R}^{d+1}$ beginning with a set K of $k+1$ vertices from this subspace, and $d-k$ other points. $C_{-1}(\mathcal{R}_d(G;\tilde{\mathbf{p}}))$ is generated by the independent 2-extensors formed by the distinct pairs in B. The space $B_{-1}(\mathcal{R}_d(G;\tilde{\mathbf{p}})) = \partial(C_0(\mathcal{R}_d(G;\tilde{\mathbf{p}})))$ is generated by pairs with one or both points in K. Therefore, the quotient $H_{-1}(\mathcal{R}_d(G;\tilde{\mathbf{p}}))$ is generated by pairs both in $B-K$, a set of size $\binom{d-k}{2}$. Therefore $\beta_{-1}(\mathcal{R}_d(G;\tilde{\mathbf{p}})) = \binom{d-k}{2}$. ■

The cohomology for this complex also has a mechanical interpretation. Recall that the local coefficients for an i-face ρ are

$$V^*\backslash_{\rho,\tilde{\mathbf{p}}} = \Lambda^{*(2)}_{d+1}/\{\overset{\rho,\tilde{\mathbf{p}}}{=}\} = \Lambda^{*(2)}_{d+1}\backslash_{\rho,\tilde{\mathbf{p}}},$$

where the equivalence relation $\{\overset{\rho,\tilde{\mathbf{p}}}{=}\}$ is defined by $S^* \overset{\rho,\tilde{\mathbf{p}}}{=} T^*$ iff $(S^*-T^*)\vee \tilde{\mathbf{p}}_i \doteq 0$ for all $i \in \rho$. The cochain complex is

$$\mathcal{R}^d(G;\tilde{\mathbf{p}})\colon \mathbf{0} \leftarrow \oplus_{e\in E}\Lambda^{*(2)}_{d+1}\backslash_{e,\tilde{\mathbf{p}}} \xleftarrow{\delta^0} \oplus_{i\in V}\Lambda^{*(2)}_{d+1}\backslash_{i,\tilde{\mathbf{p}}} \xleftarrow{\delta^{-1}} \Lambda^{*(2)}_{d+1} \leftarrow \mathbf{0}.$$

Example 4.2. We offer a kinematic interpretation of the cohomology. For easier interpretation, the dual hyperplane 2-extensors $\Lambda^{*(2)}_{d+1}\backslash_{\rho,\tilde{\mathbf{p}}}$ are replaced by the equivalent point $(d-1)$-extensors $\Lambda^{(d-1)}_{d+1}\backslash_{\rho,\tilde{\mathbf{p}}}$, with $S \overset{\rho,\tilde{\mathbf{p}}}{=} T$ iff $S\vee\tilde{\mathbf{p}}_i = T\vee\tilde{\mathbf{p}}_i$ for all $i \in \rho$. These $(d-1)$-tensors as coefficients of chains are called *screw centers of motion* for the corresponding geometric object $\tilde{\rho}$ [7]. Again, for convenience, we assume the vertices are affine points $\tilde{\mathbf{p}}_i$ with weight $\omega_i = 1$ and Euclidean coordinates $\mathbf{p}_i$.

(a) The 0-chains $\sum_{v\in V} S^i[i]$ give a $(d-1)$-tensor *center* S^i for each vertex. For the point $\tilde{\mathbf{p}}_i$, $\tilde{\mathbf{M}}_i = S^i \vee \tilde{\mathbf{p}}_i$ is a hyperplane with dual coordinates $(A_i,\dots,C_i,D_i)$. For an affine point $\tilde{\mathbf{p}}_i = (x_i,\dots,z_i,1)$, the d-vector normal, $\mathbf{u} = (A_i,\dots,C_i)$ is the *velocity* or displacement of the point. The velocity is well-defined since, if $S \overset{i,\tilde{\mathbf{p}}}{=} T$ then $\tilde{\mathbf{M}}_i = S\vee\tilde{\mathbf{p}}_i = T\vee\tilde{\mathbf{p}}_i$ [7].

(b) The 0-coboundaries have a single center S assigned to all vertices. This *trivial first-order motion* gives the velocities of a *first-order congruence* which is the derivative of an analytic congruence of the configuration [7].

(c) The 1-cochains assign a scalar $s^{i,j} = S^{i,j}\vee\tilde{\mathbf{p}}_i\vee\tilde{\mathbf{p}}_j$ to each edge. This is the *strain* of the edge (the first-order change in the length of the bar).

(d) The 0-coboundary operator assigns the edge $[i,j]$ the strain

$$S^i \vee \tilde{\mathbf{p}}_i\tilde{\mathbf{p}}_j - S^j \vee \tilde{\mathbf{p}}_i\tilde{\mathbf{p}}_j = [\mathbf{u}_i\cdot\mathbf{p}_j - \mathbf{u}_i\cdot\mathbf{p}_i] - [\mathbf{u}_j\cdot\mathbf{p}_i - \mathbf{u}_j\cdot\mathbf{p}_j] = (\mathbf{p}_i - \mathbf{p}_j)\cdot(\mathbf{u}_i - \mathbf{u}_j).$$

The 0-cocycles are the velocities which induce the strain 0 on each edge (these velocities cause no first-order deformation of the lengths of the bars). These are the *first-order flexes* of the framework (Figure 2F).

(e) A framework in d-space $G;\tilde{\mathbf{p}}$ is *first-order rigid* if all first-order flexes (0-cocycles) are trivial (0-coboundaries) that is, if $H^0(\mathcal{R}_d(G;\tilde{\mathbf{p}})) = 0$. Since $\beta^0(\mathcal{R}^d(G;\tilde{\mathbf{p}})) = \beta_0(\mathcal{R}_d(G;\tilde{\mathbf{p}}))$, we observe that the framework $G;\tilde{\mathbf{p}}$ is first-order rigid if and only if $G;\tilde{\mathbf{p}}$ is statically rigid.

Remark 4.3. If we use weighted finite points $\tilde{\mathbf{p}}_i$, then the normal to $S^i \vee \tilde{\mathbf{p}}_i$ will be this weight times the velocity: the *momentum* $\mathbf{m}_i$. The induced strain $S^i \vee \tilde{\mathbf{p}}_i\tilde{\mathbf{p}}_j - S^j \vee \tilde{\mathbf{p}}_i\tilde{\mathbf{p}}_j = (\mathbf{p}_i - \mathbf{p}_j)\cdot(\mathbf{m}_i - \mathbf{m}_j)$ is a weighted strain, but the 0-cocycles (first-order flexes) are unchanged. The infinite points continue to behave kinematically, as well as statically, as 'slide joints' [7].

For a graph $G = (V,E)$ and a new vertex u, the *cone graph* $G * u$ is the graph with vertices $V \cup \{u\}$ and edges $E \cup \{\ [i,u] \mid i \in V\ \}$. The *projection* of a configuration $\tilde{\mathbf{p}}$ from a point $\tilde{\mathbf{u}}$ to a hyperplane $\tilde{\mathbf{H}}$, $\Pi_{\tilde{\mathbf{u}},\tilde{\mathbf{H}}}(\tilde{\mathbf{p}})$, replaces each point $\tilde{\mathbf{p}}_i$ by the weighted point $\Pi_{\tilde{\mathbf{u}}}(\tilde{\mathbf{p}}_i) = (\tilde{\mathbf{p}}_i \vee \tilde{\mathbf{u}}) \wedge \tilde{\mathbf{H}}$.

Theorem 4.4. *For a geometric cone graph $G * u;\tilde{\mathbf{p}},\tilde{\mathbf{u}}$, in $\mathbb{P}_{d+1}$ with $\tilde{\mathbf{u}}$ not collinear with any other edge, and the projection $\Pi_{\tilde{\mathbf{u}},\tilde{\mathbf{H}}}(\tilde{\mathbf{p}})$ into a hyperplane $\tilde{\mathbf{H}}$ not containing $\tilde{\mathbf{u}}$, the map $\Pi_{\tilde{\mathbf{p}},\tilde{\mathbf{H}}}$ induces a chain map from $\mathcal{R}_{d+1}(G * u;\tilde{\mathbf{p}},\tilde{\mathbf{u}})$ to $\mathcal{R}_d(G;\Pi_{\tilde{\mathbf{u}},\tilde{\mathbf{H}}}\tilde{\mathbf{p}})$ which gives an isomorphism of the homologies (and cohomologies) for each i.*

Proof. The chain map is implicit in the results on coning [21] and is surjective on the i-chains of each i. The Euler characteristic is unchanged in this mapping:

$$\chi(\mathcal{R}_d(G;\Pi_{\tilde{\mathbf{u}},\tilde{\mathbf{H}}}\tilde{\mathbf{p}})) = -|E| + d|V| - \binom{d+1}{2}$$
$$= -(|E| + |V|) + (d+1)(|V|+1) - \binom{d+2}{2} = \chi(\mathcal{R}_{d+1}(G * u;\tilde{\mathbf{p}},\tilde{\mathbf{u}})).$$

The isomorphisms will follow from an equality of the corresponding Betti numbers. If $\{\tilde{\mathbf{p}}_i \mid i \in V\} \cup \{\tilde{\mathbf{u}}\}$ spans a space of dimension $k+1$, then $\{\tilde{\mathbf{p}}_i \mid i \in V\}$ spans a space of dimension k. Since $\binom{d-k}{2} = \binom{d+1-(k+1)}{2}$, the equality of β_{-1} follows.

A standard result for frameworks gives an isomorphism for the 1-cycles (and therefore H_1 and β_1) [16,21]. Another version, using cohomology or kinematics ([21]), gives the isomorphism for β_0 but this equality also follows from the equality of the Euler characteristics. ■

Lemma 4.5. *For $d = 1$, and a line configuration $\tilde{\mathbf{p}}$ with distinct vertices on each edge, the complex $\mathcal{R}_1(G;\tilde{\mathbf{p}})$ is the simplicial complex $\mathcal{G}$ of the graph G.*

Proof. For $d = 1$, the 2-extensors are pseudoscalars (isomorphic to $\mathbb{R}$) and $\Lambda_2^{(2)}|_{\pi,\tilde{\mathbf{p}}}$ is isomorphic to $\mathbb{R}$ for all $\tilde{\mathbf{p}}$ with distinct vertices and all faces π. With the same coefficients and the same boundary operators, we have the simplicial complex $\mathcal{G}$. We note that this isomorphism gives a static and kinematic interpretation for the cycles and cocycles of the standard graphic matroid. ■

A critical task is to characterize the graphs which produce acyclic complexes $\mathcal{R}_d(G;\tilde{\mathbf{p}})$ for some (almost all) d-configurations $\tilde{\mathbf{p}}$ (see §6).

Lemma 4.6. *For the complete graph K_d and a general position d-configuration $\tilde{\mathbf{p}}$ on d vertices:*

(1) *the complex $\mathcal{R}_d(K_d;\tilde{\mathbf{p}})$ is acyclic (all $\beta_i(\mathcal{R}_d(K_d;\tilde{\mathbf{p}})) = 0$);*
(2) *$K_k(\tilde{\mathbf{p}})$ has $\beta_1(\mathcal{R}_d(K_k;\tilde{\mathbf{p}})) = \beta_0(\mathcal{R}_d(K_k;\tilde{\mathbf{p}})) = 0$ for all $k \le d$.*

Proof. This follows by a coning induction in d. For $k = d$, we induct from a single point which is acyclic in the simplicial homology or $\mathcal{R}_1(G;\Pi(\tilde{\mathbf{p}}_1))$. For $k < d$, we start with $\mathcal{R}_{d-k}(\emptyset;\Pi(\tilde{\mathbf{p}}))$, which has $\beta_1 = \beta_0 = 0$. ■

Lemma 4.7. *If $G';\tilde{\mathbf{p}},\tilde{\mathbf{u}}$ comes from $G;\tilde{\mathbf{p}}$ by adding a d-valent vertex u at $\tilde{\mathbf{u}}$ and the d attached vertices are in general position in d-space, then for $-1 \le i \le 1$, $\beta_i(\mathcal{R}_d(G;\tilde{\mathbf{p}})) = \beta_i(\mathcal{R}_d(G';\tilde{\mathbf{p}},\tilde{\mathbf{u}}))$ (Figure 3A).*

Proof. It is a simple check that the Euler characteristic has not been changed: $-|E| + d|V| - \binom{d+1}{2} = -(|E| + d) + d(|V| + 1) - \binom{d+1}{2}$. By the assumption that the d attached points are in general position, $\beta_{-1} = 0$ for both complexes. If we consider the boundary of any 1-cycle $\mathbf{c}$ at the vertex $[u]$, we have $\left(\sum_{i\,|\,[i,u]\in E} f_{\{i,u\}}\tilde{\mathbf{p}}_i\tilde{\mathbf{u}}\right)[u] = \mathbf{0}$. Since the points are in general position, the d 2-extensors $\tilde{\mathbf{p}}_i\tilde{\mathbf{u}}$ are linearly independent and the scalars $f_{\{i,u\}}$ are zero. Therefore the 1-cycles are isomorphic and β_1 is unchanged. Using the Euler characteristic, we conclude all Betti numbers are unchanged. ■

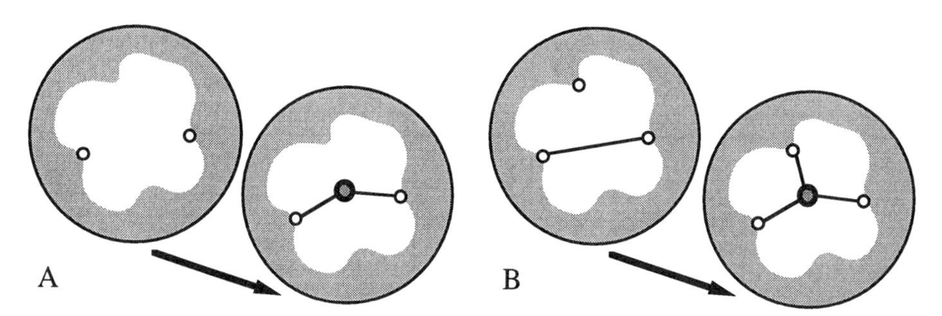

Figure 3. Two constructions: adding a 2-valent vertex (A) and an edge 2-split (B); which preserve acyclic complexes from $\mathcal{R}_2(n)$ to $\mathcal{R}_2(n+1)$ and from $\mathcal{K}_1^0(n)$ to $\mathcal{K}_1^0(n+1)$.

A graph $G = (V, E)$, with at least d vertices, is *d-simple* if there is an ordering of the vertices $\sigma(1), \sigma(2), \dots, \sigma(|V|)$ such that

(i) $G_d = K_{\{\sigma(1),\dots,\sigma(d)\}}$;
(ii) for $i \geq d$, G_{i+1} adds $\sigma(i+1)$ connected to d vertices in G_i;
(iii) $G_{|V|}$ is G.

Corollary 4.8. *If G is a d-simple graph and $\tilde{\mathbf{p}}$ is in general position in projective d-space, then $\mathcal{R}_d(G;\tilde{\mathbf{p}})$ is acyclic: $\beta_1(\mathcal{R}_d(G;\tilde{\mathbf{p}})) = \beta_0(\mathcal{R}_d(G;\tilde{\mathbf{p}})) = \beta_{-1}(\mathcal{R}_d(G;\tilde{\mathbf{p}})) = 0$, and $|E| = d|V| - \binom{d+1}{2}$.*

5. Cofactors and the symmetric algebra

Our second family of chain complexes over a graph is defined from the symmetric algebra S_3 over $\mathbb{R}^3$, interpreted as homogeneous coordinates for the projective plane $\mathbb{P}_2$. Specifically, we work with the vector space S_3^s of homogeneous forms (polynomials) of degree s in the three variables (x, y, w). This space of dimension $\binom{s+2}{2}$ will be the local coefficients for the (-1)-chains.

For the local coefficients of an i-face ρ, we define

$$V_{\rho,\tilde{\mathbf{q}}} = \langle\, f^s \mid f \in S_3^1, f(\tilde{\mathbf{q}}_i) = 0 \text{ for all } i \in \rho \,\rangle = S_3^s|_{\rho,\tilde{\mathbf{q}}}.$$

For each edge $\{i,j\}$, $S_3^s|_{[i,j],\tilde{\mathbf{q}}} = \{cL_{i,j}^s \mid c \in \mathbb{R}\}$, where $L_{i,j}$ is a linear form for the line through $\tilde{\mathbf{q}}_i, \tilde{\mathbf{q}}_j$. This has dimension 1 for all s.

For each vertex i, $S_3^s|_{i,\tilde{\mathbf{q}}}$ is a space of dimension $s+1$. [For example, if $\tilde{\mathbf{q}}_i$ is the origin $(0,0,1)$, this is the space of polynomials of degree s in the first two variables, with a basis $\{x^k y^{s-k}\}$, $0 \leq k \leq s$.]

The C_s^{s-1}-*cofactor chain complex* on a plane graph $G;\tilde{\mathbf{q}}$ is

$$\mathcal{K}_s^{s-1}(G;\tilde{\mathbf{q}}) : \mathbf{0} \to \oplus_{\rho \in E} S_{\rho,\tilde{\mathbf{q}}}^s \xrightarrow{\partial_1} \oplus_{v \in V} S_{v,\tilde{\mathbf{q}}}^s \xrightarrow{\partial_0} S_3^s \xrightarrow{\partial_{-1}} \mathbf{0}.$$

Using the dimensions given above, the Euler characteristic is

$$\chi(\mathcal{K}_s^{s-1}(G;\tilde{\mathbf{q}})) = -|E| + (s+1)|V| - \binom{s+2}{2}.$$

Remark 5.1. For connected planar graphs, this homology was defined by Billera to interpret the algebra of bivariate splines [2]. For a planar drawing of a planar graph with affine coordinates, a bivariate C_s^{s-1}-spline is an assignment of a degree s polynomial P^k to each region, meeting with continuity C^{s-1} over the common edges. For such a C_s^{s-1}-spline, the differences $f^k - f^l = c_{[i,j]} L_{i,j}^s$ induce a 1-cycle, where edge $[i,j]$ separates faces k, l. Moreover, all 1-cycles come from such bivariate splines and the coefficients are called *smoothing cofactors* [2,23]. For use in approximation theory or Computer Aided Geometric Design, the dimension of this space of bivariate C_s^{s-1}-splines can be measured as $\beta_1(\mathcal{K}_s^{s-1}(G;\tilde{\mathbf{q}})) + \binom{s+2}{2}$. We have generalized this homology to arbitrary graphs to aid comparisons with rigidity [23,24].

Lemma 5.2. *If the largest set of general position points in the configuration* $\{\tilde{\mathbf{q}}_i \mid i \in V\}$ *has size* k *then* $\beta_{-1}(\mathcal{K}_s^{s-1}(G;\tilde{\mathbf{q}})) = \binom{s+2-k}{2}$.

Proof. Assume the set $\{\tilde{\mathbf{q}}_i \mid i \in V\}$ contains k points in general position and extend this with $s+2-k$ points to a set B of $s+2$ points in general position in the plane. S_3^s is generated by the independent powers $L_{i,j}^s$ formed by the distinct pairs in B. [While I have found no reference for this, it is implicit in the inductions for coning in [24].] The rest follows as in Lemma 4.3. ∎

The cohomology will use the dual coefficients $S_3^s(\mathbb{R}^*)\backslash_{\sigma,\tilde{\mathbf{q}}}$, which can be identified with the 'power of points' $(x,y,z)^s = (x^s, x^{s-1}y, \ldots, y^s, \ldots, z^s)$ and the equivalence relation $S^* \overset{\sigma,\tilde{\mathbf{q}}}{=} T^*$ iff $S^* - T^* \in < \{(\tilde{\mathbf{p}}_i)^s \mid i \in \sigma >$ or equivalently, $\left(\forall f \in I_{\sigma,\tilde{\mathbf{p}})}^s\right) f(S^* - T^*) = 0$. The cochain complex is then defined in the obvious fashion [15,24]. There has been very little study of this cohomology, although Ripmeester [15] has recently used the 0-cocycles as 'dual splines' for some interesting results.

Again, a critical task is to characterize the graphs which produce acyclic complexes $\mathcal{K}_s^{s-1}(G;\tilde{\mathbf{q}})$ for some (almost all) plane configurations $\tilde{\mathbf{q}}$ (see §6).

Theorem 5.3. *For a geometric cone graph $G*u;\tilde{\mathbf{q}},\tilde{\mathbf{q}}_u$, in the plane $\mathbb{P}_2$ with $\tilde{\mathbf{q}}_u$ not colinear with any other edge, the homology $H_i(\mathcal{K}_{s+1}^s(G*u;\tilde{\mathbf{q}},\tilde{\mathbf{q}}_u))$ is isomorphic to $H_i(\mathcal{K}_s^{s-1}(G;\tilde{\mathbf{q}}))$ for each i.*

Proof. A recent result shows that the space of 1-cycles of $\mathcal{K}_{s+1}^s(G*u;\tilde{\mathbf{q}},\tilde{\mathbf{q}}_u)$ is isomorphic to the 1-cycles of $\mathcal{K}_s^{s-1}(G;\tilde{\mathbf{q}})$ [24]. The rest of the proof follows the proof of Theorem 4.6. ∎

Lemma 5.4. *For $s=0$ and a configuration $\tilde{\mathbf{q}}$ in general position in the plane, the complex $\mathcal{K}_0^{-1}(G;\tilde{\mathbf{q}})$ is isomorphic to the simplicial chain complex $\mathcal{G}$.*

Proof. For $s=0$, the symmetric space $S_3^0|_{\rho,\tilde{\mathbf{q}}}$ is isomorphic to $\mathbb{R}$ for all general position configurations $\tilde{\mathbf{q}}$. Again, we have the simplicial complex $\mathcal{G}$. ∎

Lemma 5.5. *For the complete graph K_{s+1} and a general position plane configuration $\tilde{\mathbf{q}}$ on $s+1$ vertices:*

(1) *the complex $\mathcal{K}_s^{s-1}(K_{s+1};\tilde{\mathbf{q}})$ is acyclic (all $\beta_i(\mathcal{K}_s^{s-1}(K_{s+1};\tilde{\mathbf{q}}))=0$);*
(2) *$K_k(\tilde{\mathbf{q}})$ has $\beta_1(\mathcal{K}_s^{s-1}(K_k;\tilde{\mathbf{q}}))=\beta_0(\mathcal{K}_s^{s-1}(K_k;\tilde{\mathbf{q}}))=0$ for all $k\le s+1$.*

Proof. This follows by a coning induction in s. For $k=s+1$, we induct from a single point which is acyclic in the simplicial homology or $\mathcal{K}_0^{-1}(G;\tilde{\mathbf{q}}_1)$. For $k<s+1$, we induct from $\mathcal{K}_{s+1-k}^{s-k}(\emptyset;\tilde{\mathbf{q}})$, which has $\beta_1=\beta_0=0$. ∎

Lemma 5.6. *If $G';\tilde{\mathbf{q}},\tilde{\mathbf{q}}_u$ comes from $G;\tilde{\mathbf{q}}$ by adding a d-valent vertex u at $\tilde{\mathbf{q}}_u$ and these attached vertices are in general position in the projective plane, then for $-1\le i\le 1$, $\beta_i(\mathcal{K}_s^{s-1}(G;\tilde{\mathbf{q}}))=\beta_i(\mathcal{K}_s^{s-1}(G';\tilde{\mathbf{q}},\tilde{\mathbf{q}}_u))$.*

Proof. It is a simple check that the Euler characteristic is unchanged

$$-|E|+(s+1)|V|-\binom{s+2}{2}=-(|E|+s+1)+(s+1)(|V|+1)-\binom{s+2}{2}.$$

By the assumption that $\tilde{\mathbf{q}}_u$ is in general position relative to its attachments the $s+1$ lines of attachment have distinct slopes. If we consider the first $s+1$ coefficients of $L^s_{i,u}$, these are essentially the rows of an $(s+1)$-by-$(s+1)$ Vandermonde matrix at distinct points (the distinct slopes [24]) which has independent rows. The rest of the argument follows Lemma 4.9. ■

Corollary 5.7. *If G is an $(s+1)$-simple graph and $\tilde{\mathbf{q}}$ is in general position in the projective plane, then $\mathcal{K}^{s-1}_s(G;\tilde{\mathbf{q}})$ is acyclic, with $\beta_1(\mathcal{K}^{s-1}_s(G;\tilde{\mathbf{q}}))=\beta_0(\mathcal{K}^{s-1}_s(G;\tilde{\mathbf{q}}))=\beta_{-1}(\mathcal{K}^{s-1}_s(G;\tilde{\mathbf{q}}))=0$ and $|E|=(s+1)|V|-\binom{s+2}{2}$.*

6. Common generic results

For each of these geometric homologies, we have direct combinatorial information: The graph G and properties such as the Euler characteristic defined by counts on the vertices and edges. However, for a fixed graph G, the Betti numbers $\beta_i(\mathcal{R}_d(G;\tilde{\mathbf{p}}))$ and $\beta_i(\mathcal{K}^{s-1}_s(G;\tilde{\mathbf{q}}))$ vary with the choice of configurations $\tilde{\mathbf{p}}\in\mathbb{R}^{(d+1)|V|}$ and $\tilde{\mathbf{q}}\in\mathbb{R}^{3|V|}$. Figure 2 A,B showed two plane configurations for a fixed graph which have different Betti numbers for the (equivalent, see below) complexes $\mathcal{R}_2(G;\tilde{\mathbf{p}})$ and $\mathcal{K}^0_1(G;\tilde{\mathbf{q}})$.

Each boundary operator in these complexes can be presented as a matrix, with entries which are polynomials in the coordinates of the configurations. If we replace these coordinates of the points by variables, the ranks of each of these matrices (and the resulting Betti numbers) are determined by maximal non-zero determinants – non-zero polynomials in the coordinate variables. The *special positions* which reduce the rank of one of these matrices form an algebraic variety. The *generic configurations* which avoid all three algebraic varieties (for δ_1, δ_0, and δ_{-1}) are an open dense subset of all configurations. In particular, if we use algebraically independent numbers for the coordinates, we have such a generic configuration for the geometric complex.

These generic configurations, which give a maximal rank for each boundary operator, must give the minimal values for the Betti numbers. These minima are combinatorial values, determined by the graph. We write $\mathcal{R}_d(G)$ and $\mathcal{K}^{s-1}_s(G)$ for the *generic complexes* at corresponding generic configurations. For the complete graph K_n, we write $\mathcal{R}_d(n)$ and $\mathcal{K}^{s-1}_s(n)$.

A basic combinatorial problem is to compute the Betti numbers for each of these generic complexes directly from the graph. Equivalently, we want to characterize the edge sets E which are *d-rigid* (resp. C_s^{s-1}*-rigid*), that is $\mathcal{R}_d\big((V(E),E)\big)$ (resp. $\mathcal{K}_s^{s-1}\big((V(E),E)\big)$) has $\beta_0 = \beta_{-1} = 0$. We note that a direct computation using 0-cycles shows that a non-empty set E is d-rigid if and only if every other edge $f \in K_{V(E)}$ generates a 1-cycle $\mathbf{c}$ supported on $E \cup \{f\}$ with $c_f \neq 0$.

The proofs of the following result use generic geometric complexes in which the added point is assigned to position $\tilde{\mathbf{u}} = \frac{1}{2}(\tilde{\mathbf{p}}_1 + \tilde{\mathbf{p}}_2)$ [18,23,24]. Since the generic values are minima, one configuration at which the Betti numbers are zero guarantees these are the generic values!

Theorem 6.1. *If G is acyclic in $\mathcal{R}_d(n)$ (resp. $\mathcal{K}_{d+1}^d(n)$) with an edge $[1,2]$, then an edge d-split which removes $[1,2]$ and inserts a new vertex 0 with $d+1$ edges $[0,1]$, $[0,2]$, $\ldots$, $[0,i_k], \ldots$ is also acyclic in $\mathcal{R}_d(n+1)$ (resp. $\mathcal{K}_{d+1}^d(n+1)$)* (Figure 3B).

Corollary 6.2. *If $G = (V,E)$ is has a Henneberg d-construction: an ordering of the vertices, $\sigma(1), \sigma(2), \ldots, \sigma(|V|)$, and a sequence of graphs, $G_d, \ldots, G_{|V|}$, such that:*

(i) $G_d = K_{\{\sigma(1)\ldots\sigma(d)\}}$;
(ii) *for $d \leq i < |V|$, G_{i+1} is G_i with an added d-valent vertex $\sigma(i+1)$ or G_{i+1} is an edge d-split on G_i which adds a vertex $\sigma(i+1)$;*
(iii) *$G_{|V|}$ is G;*

then G is acyclic in $\mathcal{R}_d(n)$ and in $\mathcal{K}_{d-1}^{d-2}(n)$.

Example 6.3. The Henneberg 2-construction for $K_{3,3}$ in Figure 4 generalizes to a Henneberg d-construction of $K_{d+1,\binom{d+1}{2}}$, by splitting all edges of K_{d+1}. This demonstrates that $K_{d+1,\binom{d+1}{2}}$ is acyclic for $\mathcal{R}_d(n)$ and $\mathcal{K}_{d-1}^{d-2}(n)$.

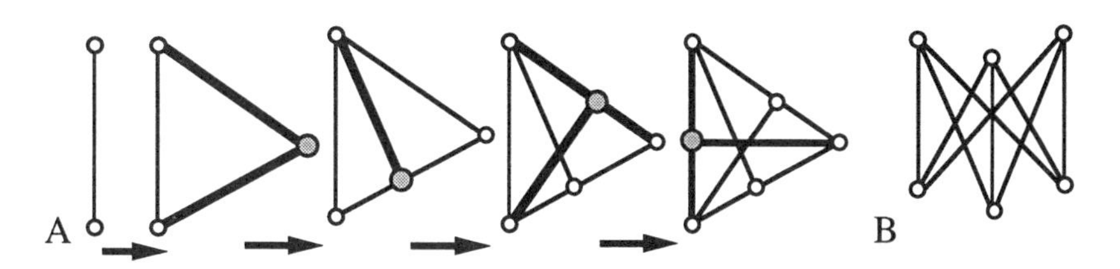

Figure 4. A Henneberg 2-construction (A) shows that the complete bipartite graph $K_{3,3}$(B) is acyclic in $\mathcal{R}_2(6)$ and $\mathcal{K}_1^0(6)$.

Theorem 6.4. *If G_1, G_2 are acyclic in $\mathcal{R}_d(n)$ (resp. in $\mathcal{K}_s^{s-1}(n)$) then*

(1) $\beta_0(\mathcal{R}_d(G_1 \cup G_2)) = 0$ *if and only if* $|V(G_1) \cap V(G_2)| \geq d$;
(2) $\beta_0(\mathcal{K}_s^{s-1}(G_1 \cup G_2)) = 0$ *if and only if* $|V(G_1) \cap V(G_2)| \geq s+1$;
(3) $\beta_1(\mathcal{R}_d(G_1 \cup G_2)) = 0$ *if and only if* $\beta_0(\mathcal{R}_d(G_1 \cap G_2)) = 0$.

Proof. (1), (3). These follow from the Mayer-Vietoris sequence for $G_1 \cup G_2$:

$$\mathbf{0} \to H_1(\mathcal{R}_d(G_1 \cup G_2)) \to H_0(\mathcal{R}_d(G_1 \cap G_2)) \to \mathbf{0}$$
$$\mathbf{0} \to H_0(\mathcal{R}_d(G_1 \cup G_2)) \to H_{-1}(\mathcal{R}_d(G_1 \cap G_2)) \to \mathbf{0}.$$

noting that $\beta_{-1}(\mathcal{R}_d(G_1 \cap G_2)) = 0$ if and only if $|V(G_1 \cap G_2)| = |V(G_1) \cap V(G_2)| \geq d$. (2) follows in an identical fashion for $\mathcal{K}_s^{s-1}$. ■

Figure 5 shows a 'vertex splitting' construction for d=3 ($s = 2$) which takes a triangulated 2-surface to a triangulated 2-surface. This construction, and its analogues for other d and s, take an acyclic complex to an acyclic complex (or rigid set to a rigid set) [1,23,24]. This construction, plus the 'gluings' above, are the essential tools in a general theory, due to Fogelsanger, that proves inductively results such as [11,24]:

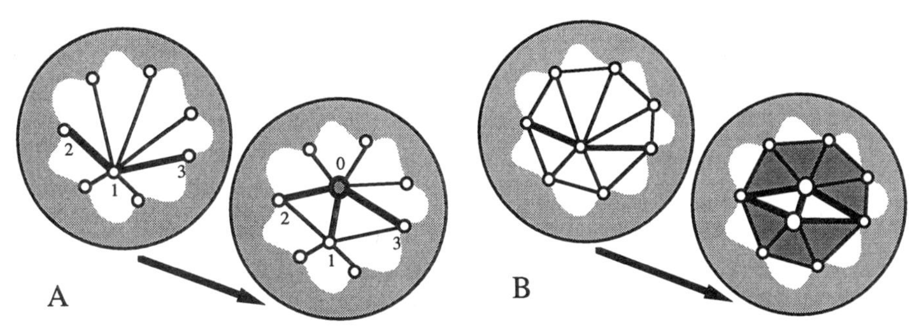

Figure 5. Vertex splitting (A) takes acyclic complexes, for $\mathcal{R}(G, \tilde{\mathbf{p}})$ and $\mathcal{K}_2^1(G, projq)$, to acyclic complexes and takes triangulated surfaces to triangulated surfaces (B).

Theorem 6.5. *For $d \geq 3$, the graph G of an abstract triangulated $(d-1)$-surface are rigid in $\mathcal{R}(G)$ and $\mathcal{K}_2^1(G)$. For a triangulated 2-sphere G, $\mathcal{R}_3(G)$ and $\mathcal{K}_2^1(G)$ are acyclic.*

7. The plane correspondence

In the previous three sections, we have seen that $\mathcal{R}_d(G; \tilde{\mathbf{p}})$ and $\mathcal{K}_s^{s-1}(G; \tilde{\mathbf{q}})$ have the same Euler characteristic for $d = s+1$. We have also seen a number of pairs of analogous results (and proofs). Finally, we have observed that $\mathcal{R}_1(G;\tilde{\mathbf{p}})$ and $\mathcal{K}_0^{-1}(G;\tilde{\mathbf{q}})$ are the same underlying simplicial complex $\mathcal{G}$.

For $\mathcal{R}_2(G;\tilde{\mathbf{p}})$ and $\mathcal{K}_1^0(G;\tilde{\mathbf{q}})$, $\tilde{\mathbf{p}}$ and $\tilde{\mathbf{q}}$ are both plane configurations. When $\tilde{\mathbf{p}} = \tilde{\mathbf{q}}$ there is a direct chain-map isomorphism of the complexes. We observe that the spaces of local coefficients are isomorphic in a natural way. $S_3^1(\mathbb{R})$ is the space of homogeneous 1-forms $\{Ax+By+Cw \mid A,B,C \in \mathbb{R}\}$, which is the same as the space of Plücker coordinates for 2-extensors in $\Lambda_3^{(2)}(\mathbb{R})$. For distinct points $\tilde{\mathbf{q}}_i, \tilde{\mathbf{q}}_j$ on the line $Ax+By+Cz=0$, A,B,C is, up to a scalar multiple, the Plücker coordinates for $\tilde{\mathbf{q}}_i \vee \tilde{\mathbf{q}}_j$. This induces an isomorphism: $S_3^1|_{[i,j],\tilde{\mathbf{q}}} = \{a_{\{i,j\}}L_{i,j}\} \equiv \{b_{\{i,j\}}\tilde{\mathbf{q}}_i \vee \tilde{\mathbf{q}}_j\} = \Lambda_3^{(2)}|_{[i,j],\tilde{\mathbf{q}}}$. A similar isomorphism holds for the coefficients of 0-chains. It is a simple task to verify that these isomorphisms induce a chain-map isomorphism, taking 1-cycles of $\mathcal{K}_1^0(G;\tilde{\mathbf{q}})$ (the C_1^0-cofactors) to 1-cycles of $\mathcal{R}_2(G;\tilde{\mathbf{q}})$ (the self-stresses) ([20,24]).

This isomorphism is a modern generalization of a classical theorem of James Clerk Maxwell for self-stresses in planar graphs and projected polyhedra [8,13,20].

Theorem 7.1. *Given a connected planar graph G, with the faces of a spherical polyhedron identified by some planar drawing of the graph, and a Euclidean plane configuration $\mathbf{p}$, the framework $G;\mathbf{p}$ supports a self-stress non-zero on all edges if and only is there is a spatial polyhedron, with planes for the faces, with points $\mathbf{p}'_i$ for the vertices such that $\mathbf{p}_i$ is the vertical projection of $\mathbf{p}'_i$ for each vertex i, and with distinct planes for each pair of faces meeting at an edge of G.*

There are a number of equivalent combinatorial characterizations of the sets E which generate acyclic graphs on their vertices [12,18,24]. We offer one short list:

Theorem 7.2. *A non-empty edge set E is acyclic in $\mathcal{R}_2(V(E))$ and $\mathcal{K}_1^0(V(E))$ if and only if:*

(1) $|E| = 2|V|-3$ *and for every nonempty subset* E', $|E'| \leq 2|V(E')|-3$ *(Laman's Theorem);*

(2) *there is a Henneberg 2-construction of* $G = (V(E), E)$ *(Henneberg's Theorem);*

(3) *for each* $\{i,j\} \in E$, *the multigraph obtained by doubling the edge* $\{i,j\}$ *is the union of two spanning trees (Recski's Theorem).*

The characterization of Laman actually shows that the matroid generated by the 1-cycles of these equivalent chain complexes is the matroid on edge sets defined by standard submodular techniques [10] from the submodular function: $f(E) = 2|V(E)| - 3$ [24]. A corollary of these results is that the edge set of any minimal 1-cycle is, itself, rigid in these equivalent chain complexes (removing any one edge leaves a graph acyclic in the complexes). These characterizations also lead to polynomial time ($O(|V|^2)$) algorithms for acyclic sets E.

8. Comparing $\mathcal{R}_3(n)$ and $\mathcal{K}_2^1(n)$

For $\mathcal{R}_3(G)$ and $\mathcal{K}_2^1(G)$, we have a number of common results but no algebraic chain maps. We also lack any nice combinatorial characterization for the acyclic graphs, comparable to Theorem 7.2.

The absence of a combinatorial characterization (and corresponding polynomial-time algorithms for acyclic sets) reflects a number of differences between these chain complexes and $\mathcal{R}_2(G) \equiv \mathcal{K}_1^0(G)$. Figure 6A shows a non-rigid minimal 1-cycle. Alternately, the submodular function $f(E) = 3|V| - 6$ is 0 on single edges and does not define a useful matroid on the edges [24]. Finally, we can build an acyclic graph graph G by Henneberg 3-constructions and vertex splits, but not all acyclic graphs are created this way. See [12,18,24] for longer discussions of these difficulties.

After a number of years of study, our current understanding can be restated in the following conjecture and summary table [24]:

Conjecture 8.1. *For every graph G, $\mathcal{R}_3(G)$ and $\mathcal{K}_2^1(G)$ have equal Betti numbers.*

	$\mathcal{R}_3(G)$	$\mathcal{K}_2^1(G)$
rank K_n, $n \geq 3$	$3n-6$	$3n-6$
vertex 3-addition	Yes	Yes
edge 3-split	Yes	Yes
3-constructions	Yes	Yes
vertex 3-split	Yes	Yes
3-coning	Yes	Yes
$K_{4,6}$	Acyclic	Acyclic
$K_{5,5}$	$\beta_1 = 1, \beta_0 = 0$	$\beta_1 = 1, \beta_0 = 0$
Δ-ted 2-spheres	Acyclic	Acyclic
Δ-ted 2-surfaces	$\beta_0 = 0$	$\beta_0 = 0$

9. Comparing $\mathcal{R}_4(G)$ and $\mathcal{K}_3^2(G)$

The next two chain complexes in our families are known to be different. Specifically, $\beta_1(\mathcal{R}_4(K_{6,6})) = 1$ but $\beta_1(\mathcal{K}_3^2(K_{6,6})) = 0$ [24]. By coning, this difference will transfer up to all larger d and $s = d - 1$.

However, we have a number of similarities and a general conjecture [24]:

Conjecture 9.1. *For every graph G and every $d \geq 4$, $\beta_1(\mathcal{R}_d(G)) \geq \beta_1(\mathcal{K}_{d-1}^{d-2}(G))$ or equivalently, $\beta_0(\mathcal{R}_d(G)) \geq \beta_0(\mathcal{K}_{d-1}^{d-2}(G))$.*

We summarize the current situation for $d = 4$ in the following table.

	$\mathcal{R}_4(G)$	$\mathcal{K}_3^2(G)$
rank K_n, $n \geq 4$	$4n - 10$	$4n - 10$
vertex 4-addition	Yes	Yes
edge 4-split	Yes	Yes
4-constructions	Yes	Yes
vertex 4-split	Yes	Yes
simplicial 4-polytopes	$\beta_0 = 0$	$\beta_0 = 0$
4-coning	Yes	Yes
$K_{5,10}$	Acyclic	Acyclic
$K_{6,6}$	$\beta_1 = 1$	$\beta_1 = 0$
$K_{6,7}$	$\beta_1 > 0$	Acyclic

10. Larger geometric simplices

This pattern of geometric chain complexes with isomorphisms and strong analogies generalizes to larger simplicial complexes (and polyhedral complexes) realized in appropriate projective spaces [24]. For example:

(1) A geometric chain complex on a triangular complex with vertices in projective 3-space, using coefficients in the symmetric algebra of $\mathbb{R}^4$ arises from cofactors of trivariate splines [1,2,24]. The corresponding local coefficients are:

$$V_{\rho,\tilde{\mathbf{q}}} = \langle f^s \mid f \in S_4^1, f(\tilde{\mathbf{q}}_i) = 0 \text{ for all } i \in \rho \rangle = S_4^s|_{\rho,\tilde{\mathbf{q}}}.$$

(2) a geometric chain complex on a triangular complex in d-space ($d \geq 2$), using coefficients in the exterior algebra of projective d-space, arises as 'skeletal rigidity' [9,16,17,24]. The required local coefficients can be defined in the obvious way:

$$V_{\rho,\tilde{\mathbf{p}}} = \{P \in \Lambda_{d+1}^{(3)} \mid P \vee \tilde{\mathbf{p}}_i = \mathbf{0} \text{ for all } i \in \rho\} = \Lambda_{d+1}^{(3)}|_{\rho,\tilde{\mathbf{p}}}.$$

The pattern of matching Euler characteristics and the known or conjectured isomorphisms for this family is displayed in Figure 1(b). Our larger paper [24] includes a summary of these higher geometric homologies for r-simplicial complexes in d-space.

Finally we note that other 'geometric matroids' and 'geometric hypermatroids' ([24]) also generate appropriate geometric chain complexes.

Example 10.1. Consider a geometric graph $G; \tilde{\mathbf{p}}$ in projective d-space, with the local coefficients:

$$V_{\rho,\tilde{\mathbf{p}}} = \{P \in \Lambda_{d+1}^{(d)} | P \vee \tilde{\mathbf{p}}_i = \mathbf{0} \text{ for all } i \in \rho\} = \Lambda_{d+1}^{(d)}|_{\rho,\tilde{\mathbf{p}}}.$$

These produce the *parallel drawing complex* $\mathcal{P}_d(G;\tilde{\mathbf{p}})$ on the geometric graph with the Euler characteristic $\chi(\mathcal{P}_d(G;\tilde{\mathbf{p}})) = -(d-1)|E| + d|V| - (d+1)$. For $d = 2$, this is the same as the rigidity chain complex $\mathcal{R}_2(G;\tilde{\mathbf{p}})$.

The simplest way to interpret this complex is via the 0-cohomology. For each finite point $\mathbf{p}_i$, the coefficient in a 0-cochain is (in point coordinates) a 1-extensor (point) $\tilde{\mathbf{q}}_i$ under the equivalence relation $\tilde{\mathbf{q}}_i \stackrel{i,\tilde{\mathbf{p}}}{=} \tilde{\mathbf{q}}_i'$ if and only if $\tilde{\mathbf{q}}_i \vee \tilde{\mathbf{p}}_i = \tilde{\mathbf{q}}_i' \vee \tilde{\mathbf{p}}_i$. This 2-extensor is directly interpreted as a Euclidean 2-vector $\mathbf{q}_i - \mathbf{p}_i$, giving an alternate configuration $\tilde{\mathbf{q}}$. For a 0-cocycle, the vector $\mathbf{q}_i - \mathbf{q}_j$ is parallel to $\mathbf{p}_i - \mathbf{p}_j$ for each edge of the graph [24]. These *parallel drawings*, and their generalizations to larger polyhedral complexes, are directly related to geometric properties such as Minkowski decomposability of polytopes and (via duality) polyhedral scene analysis [22].

In our experience, moving from a geometric matroid to a geometric chain complex clarifies and simplifies a number of mathematical patterns. One of these patterns is the analogy between exterior algebra and symmetric algebra which is the central theme of this paper.

References

[1] P. Alfeld, L. Schumaker and W. Whiteley, *The generic dimension of the space of C^1 splines of degree $d \geq 8$ on tetrahedral decompositions*, SIAM J. Numer. Anal. **30** (1993), 889–920.

[2] L. Billera, *Homology of smooth splines: generic triangulations and a conjecture of Strang*, Trans. Amer. Math. Soc. **310** (1988), 325–340.

[3] B. Bollobas, *Graph Theory: An Introductory Course*, Springer-Verlag, New York, 1979.

[4] H. Crapo, *Applications of geometric homology*, Geometry and Robotics: proceedings of the workshop, Toulouse , May 26-28, 1988, Lecture Notes in Computer Science #391, Springer-Verlag, 1990, pp. 213–224.

[5] ———, *Invariant theoretic methods in scene analysis and structural mechanics*, J. Symbolic Computation **11** (1991), 523–548.

[6] ———H. Crapo and G-C. Rota, *The resolving bracket*, Invariant Methods in Discrete and Computational Geometry, Neil White (ed.), Kluwer Academic Publisher, 1995, pp. 197–222.

[7] H. Crapo and W. Whiteley, *Statics of frameworks and motions of panel structures: a projective geometric introduction*, Structural Topology **6** (1982), 42–82.

[8] ———, *Spaces of stresses, projections and parallel drawings for spherical polyhedra*, Contributions to Algebra and Geometry **35** (1994), 259–281.

[9] ———, *3-stresses in 3-space and projections of 4-polytopes: reciprocals, liftings and parallel configurations*, Preprint, Department of Mathematics and Statistics, York University, North York, Ontario (1994).

[10] J. Edmonds and G-C. Rota, *Submodular set functions*, Abstract of the Waterloo Combinatorics Conference, U. Waterloo, Waterloo Ont., 1966.

[11] A. Fogelsanger, *The generic rigidity of minimal cycles*, Ph.D. Thesis, Department of Mathematics, Cornell University (1988).

[12] J. Graver, B. Servatius and H. Servatius, *Combinatorial Rigidity*, vol. 2, AMS Monograph, 1993.

[13] J.C. Maxwell, *On reciprocal figures and diagrams of forces*, Phil. Mag. Ser. 4 **27** (1864), 250–261.

[14] J. E. Munkres, *Elements of Algebraic Topology*, Addison-Wesley, Reading, Mass., 1984.

[15] D-J. Ripmeester, *Dimension of Spline Spaces*, Ph.D. Thesis, Universiteit van Amsterdam, (1995).

[16] T-S. Tay, N. White and W. Whiteley, *Skeletal rigidity of simplicial complexes, I, II*, Eur. J. Combin. **16** (1995), 381–403, 503–523.

[17] ———, *Homology of skeletal rigidity*, preprint (1996).

[18] T-S. Tay and W. Whiteley, *Generating isostatic frameworks*, Structural Topology **11** (1985), 21–69.

[19] N. White, *Tutorial on Grassmann-Cayley Algebra*, Invariant Methods in Discrete and Computational Geometry, Kluwer, 1995.

[20] W. Whiteley, *Motions and stresses of projected polyhedra*, Structural Topology **7** (1982), 13–38.

[21] ———, *Cones, infinity and one-story buildings*, Structural Topology **8** (1983), 53–70.

[22] ———, *A matroid on hypergraphs with applications to scene analysis and geometry*, Discrete Comput. Geometry **4** (1989), 75–95.

[23] ———, *Combinatorics of bivariate splines*, Applied Geometry and Discrete Mathematics – the Victor Klee Festschrift, DIMACS, vol. 4, AMS, 1991, pp. 587–608.

[24] ———, *Some Matroids from Discrete Applied Geometry*, Matroid Theory, J. Bonin, J. Oxley and B. Servatius (eds.), Contemporary Mathematics **197**, AMS, 1996, pp. 171–311.

Department of Mathematics and Statistics
York University, 4700 Keele Street
North York, Ontario M3J1P3, Canada
email: whiteley@mathstat.yorku.ca

The Umbral Calculus and Identities for Hypergeometric Functions with Special Arguments

Jet Wimp

Abstract

We review a little of what is known regarding closed-form expressions for generalized hypergeometric functions with special arguments. Mathematical software is now available, namely, the powerful algorithm "hyper," for determining whether certain sums can be expressed in closed form, and we illustrate it's application to some important problems. Next, we show how the umbral calculus can be used to derive identities for hypergeometric functions which, when specialized, also yield closed form expressions for important sums.

KEYWORDS: hypergeometric functions, hypergeometric functions of unit argument, associated polynomials, orthogonal polynomials, Legendre polynomials, drunkard's walk, random walk, umbral calculus, hypergeometric identities.

1. Introduction, hypergeometric functions

The generalized hypergeometric function with p numerator parameters and q denominator parameters is defined by

$$\begin{aligned} {}_pF_q\left(\begin{matrix} a_1, a_2, \dots, a_p \\ b_1, b_2, \dots, b_q \end{matrix}; z\right) &= \sum_{k=0}^{\infty} \frac{\prod_{j=1}^{p}(a_j)_k}{k!\prod_{j=1}^{q}(b_j)_k} z^k, \\ b_j &\neq 0, -1, -2, \dots, j = 1, 2, \dots, q, \end{aligned} \tag{1.1}$$

where

$$(a)_k = \left\{\begin{matrix} a(a+1)\dots(a+k-1), k > 0; \\ 1, k = 0 \end{matrix}\right\}, \tag{1.2}$$

It will save considerable writing if I use an abbreviated notation for a function of repeated parameters, and write the function above as follows

$${}_pF_q\left(\begin{matrix} [a_p] \\ [b_q] \end{matrix}; z\right) = \sum_{k=0}^{\infty} \frac{[(a_p)_k]}{k![(b_q)_k]} z^k. \tag{1.3}$$

Thus, wherever square brackets [.] are used, repetition over the indicated subscript is being invoked.

The designation *hypergeometric* was, apparently, first used by Wallis in 1655 in referring to a sequence $\{s_n\}$ whose term ratio s_{n+1}/s_n is non-constant, [wal]. Hypergeometric series with $p = 2, q = 1$ (though they were not called that) were studied by Wallace, Newton ($\sim$ 1664) and Stirling (1730) in connection with the rectification of certain algebraic curves, see [dut] for details. Euler in 1778 discussed the general series of this type, though again without using the term hypergeometric. Gauss thoroughly investigated this series in a large number of published and unpublished works, beginning in 1805. Today the corresponding function is called Gauss's hypergeometric function or, simply, the hypergeometric function, [whi], though it was in only relatively recent times that Kummer (1836) applied the term hypergeometric to Gauss's series. Pochhammer (1890) and Barnes (1907) developed the notation for the general function (1.1), see [wat].

There are four cases of (1.1):

(i) If one of the a_j is a negative integer or zero, the series always makes sense, since it terminates. The result is a polynomial in z.

Failing (i), then

(ii) If $q \geq p$, the series converges uniformly on compact subsets to an entire function of z (of finite order $q + 1 - p$);

(iii) If $p = q + 1$, the series converges and thus defines an analytic function of z on compact subsets of $|z| < 1$. The function may be analytically continued into the complex plane cut along $[1, \infty]$.

(iv) If $p > q + 1$, the series diverges for all $z \neq 0$. However, the series may still be computationally very useful, for instance, as a member of an umbral calculus of formal power series about 0.

Much interest attaches to the case where the number of denominator parameters is one greater than the number of numerator parameters, and where the variable z assumes special values; the value $z = 1$ is especially important.

If this is the case and, in addition,

$$\sum_{j=1}^{p+1} a_j + 1 = \sum_{j=1}^{p} b_j, \tag{1.4}$$

the series ${}_{p+1}F_p(1)$ is called *Saalschützian*.

The simplest non-trivial case is the formula that expresses Gauss' function with argument 1 in terms of gamma functions,

$$ {}_2F_1\left(\begin{matrix} a,b \\ c \end{matrix};1\right)=\frac{\Gamma(c)\Gamma(c-a-b)}{\Gamma(c-a)\Gamma(c-b)},\ \mathrm{Re}\ (c-a-b)>0, c\neq 0,-1,-2,\dots. \tag{1.5} $$

The derivation of this formula is easy if one works with Euler's representation (1748),

$$ {}_2F_1\left(\begin{matrix} a,b \\ c \end{matrix};z\right)=\frac{\Gamma(c)}{\Gamma(b)\Gamma(c-b)}\int_0^1 t^{b-1}(1-t)^{c-b-1}(1-tz)^{-a}dt, \quad \mathrm{Re}\ c>\mathrm{Re}\ b>0, |\arg(1-z)|<\pi, \tag{1.6} $$

but this is not what Gauss used. Readers should consult his extremely ingenious proof, given in [whi], p. 282, to see what elementary analysis can accomplish in the hands of a master. There are other values of z for which Gauss' function can also be evaluated in closed form, see the Bateman Manuscript volume [erd], v. 1, 2.8; this, however, is an ancient reference. New and bizarre identities for ${}_2F_1$'s are still being discovered, for instance,

$$ {}_2F_1\left(\begin{matrix} -n,-2n-2/3 \\ 4/3 \end{matrix};-8\right)=(-27)^n\frac{(\frac{5}{6})_n}{(\frac{3}{2})_n}, \tag{1.7} $$

see [ges].

Hypergeometric functions of the form ${}_3F_2$ with specialized arguments occur in the strangest of places. For instance, consider the path of a drunkard climbing around in 3-space, occupying unit lattice points. The drunkard starts out at the origin and makes consecutively one of any of the six movements $(\pm1,0,0),(0,\pm1,0),(0,0,\pm1)$ with equal probability. If a_n denotes the number of ways of going from the origin back to the origin in $2n$ steps, then

$$ a_n=\frac{4^n(1/2)_n}{n!}{}_3F_2\left(\begin{matrix} -n,-n,1/2 \\ 1,1 \end{matrix};4\right), \tag{1.8} $$

see [wim1]. (Note the above hypergeometric series makes sense, since it terminates.) It can be shown that the probability of return to the origin is $u=(m-1)/m$, where

$$ m=\sum_{n=0}^{\infty}\frac{a_n}{6^{2n}}=1.51638\dots. \tag{1.9} $$

Thus $u=.34053\dots$. It is known that if the drunkard moves in 1-space or 2-space, the probability of return to origin is 1. The apparently bland fact that the three-space result differs from the lower dimensional results is probably the basis for all life on earth.[1] The constant .34053... is called Pólya's constant. Pólya investigated the situation in 1921.

If the drunkard moves about in the wedge $x \geq y \geq z$, i.e., the drunkard dies if he comes into contact with one of the absorbing walls $x-y=-1, y-z=-1$, then

$$a_n = \frac{4^n(1/2)_n}{(n+1)!}\,{}_3F_2\left(\begin{matrix} -n, -n-1, 1/2 \\ 2, 2 \end{matrix}; 4\right), \tag{1.10}$$

again, see [wim1]. This type of motion is important in problems of wetting, melting, and physical dislocation, see [hus1], [hus2], [fis].

Although it is not known whether the above hypergeometric functions can be evaluated in closed form, say, as a ratio of Gamma functions, there are hundreds of results—scattered throughout the literature of mathematics and physics—that provide the explicit evaluation of hypergeometric series with special arguments, for example, the result of Andrews:

$${}_3F_2\left(\begin{matrix} -n, n+3a, a \\ 3a/2, (3a+1)/2 \end{matrix}; 3/4\right) = \begin{cases} 0, & n \not\equiv 0 (\text{mod } 3); \\ \dfrac{(3N)!(a+1)_N}{N!(3a+1)_{3N}}, & n = 3N. \end{cases} \tag{1.11}$$

The reference [ges] has this and many many more examples. However, the bestiary of these strange results grows daily, and no available list can make any pretense to completeness.

Recently, the search for such identities has gained a powerful ally in computer algebra techniques. The innovative new book of Petkovsek, Wilf and Zeilberger, quaintly entitled "$A = B$", furnishes a wonderful introduction to this remarkable field, [pet]. The authors, building on their previous work as well as the work of Bill Gosper, describe a powerful new menu of algorithms for determining whether sums whose terms are of hypergeometric type can be expressed in closed form.

Definition. (Petkovsek–Wilf–Zeilberger) The sequence $\{s_n\}$ is called *hypergeometric* if its term ratio s_{n+1}/s_n is a rational function of n. A function $f(n)$

[1] In living organisms, the underlying explanation for many vital biochemical reactions is that a walk representing the path of some enzyme system suddenly becomes constrained. In membrane-associated reactions in cells, [mss],[tru], it has been found that the turnover numbers of certain membrane bound enzyme systems is enhanced if their substrates undergo two-dimensional diffusion along membrane surfaces. Eigen [eig] formulated the principle that the reduction of dimensionality in a walk was nature's trick for overcoming the barrier of diffusion control and making multi-molecular reaction processes at low concentrations more efficient.

is said to be *in closed form* if it can be expressed as a linear combination of a fixed (independent of n) number of hypergeometric terms.

The algorithm "hyper" in "$A = B$" is designed to attack the following very general and very important question:

Let $F(n,k)$ be hypergeometric in both of its variables, n and k. Is the sum

$$\sum_k F(n,k) \tag{1.12}$$

expressible in closed form (as a function of n)?

This question is of significance in regard to some series of the form ${}_4F_3(1)$ that arise in the theory of orthogonal polynomials, see Section 3.

The algorithms given in [pet] provide complete information. For instance, it is known that the sum of the binomial coefficients squared can be expressed in closed form:

$$\sum_{k=0}^{n} \binom{n}{k}^2 = \binom{2n}{n}. \tag{1.13}$$

In that book, the authors show that the sum of the binomial coefficients cubed, $\sum_{k=0}^{n} \binom{n}{k}^3$, has no closed-form expression.

A recent important paper on computer algebra and hypergeometric identities is [koe]. This paper extends the algorithms given by Wilf and Zeilberger and gives applications to the sort of identities considered in this paper.

A fact that seems to not be generally known is the following. Although the series for ${}_{p+1}F_p(1)$ converges only for $e > 0$, where

$$e = \sum_{j=1}^{p} b_j - \sum_{j=1}^{p+1} a_j - 1, \tag{1.14}$$

the function can be analytically continued and is meromorphic in its parameter space $\mathbb{C}^{2p+1}$. It has poles at $b_j, e = 0, -1, -2, \ldots$, and, if none of these parameters differ by integers, the poles are simple, see [wim3].[2]

---

[2] Although the result in this reference is established only for ${}_3F_2(1)$, the argument is the same in the general case. One establishes a recursion relation for a related ${}_{p+1}F_p(1)$ and then uses the principle of preservation of functional equations.

2. Series of the type ${}_3F_2(1)$

These series occur in all branches of mathematics and physics. They occupy a place of prominence in physics and group representation theory as the so-called Clebsch-Gordon coefficients, which arise in the realization of the tensor product of two irreducible representations of $SU(2)$, the group of unimodular unitary matrices of second order, [vil], v. 1, p. 498 ff. The Clebsch-Gordon coefficient $C(\boldsymbol{l};\boldsymbol{j})$ can be expressed as a ${}_3F_2(1)$ whose parameters have integer values.

Gauss's formula (1.5) gave rise to the hope that the five parameter class of functions ${}_3F_2(1)$ could be evaluated in closed form, say, as a ratio of gamma functions. If this could be done, many of the formulas of mathematical physics would substantially simplify. Indeed, such formulas are known, but they involve special values of the parameters. There is Saalschütz's theorem,

$$ {}_3F_2\left(\begin{matrix} a,b,-n \\ c, a+b+1-c-n \end{matrix} ;1\right) = \frac{(c-a)_n(c-a)_n}{(c)_n(c-a-b)_n}, n=0,1,2,\ldots, \tag{2.15} $$

which sums a terminating Saalschützian ${}_3F_2$. Richard Askey has pointed out [ask] that this result was obtained in Pfaff's memoir of 1797, nearly ninety years before Saalschütz's work.

There is also Watson's theorem,

$$ {}_3F_2\left(\begin{matrix} a,b,c \\ \frac{a+b+1}{2}, 2c \end{matrix} ;1\right) = \frac{\sqrt{\pi}\,\Gamma(c+1/2)\Gamma\left(\frac{a+b+1}{2}\right)\Gamma\left(c-\frac{a+b-1}{2}\right)}{\Gamma\left(\frac{a+1}{2}\right)\Gamma\left(\frac{b+1}{2}\right)\Gamma\left(c+\frac{1-a}{2}\right)\Gamma\left(c+\frac{1-b}{2}\right)}, \quad b=1-a. \tag{2.16} $$

Many other formulas are given in [erd], v. 1, 4.4. All the available formulas which sum ${}_3F_2(1)$'s are of the above type, namely, ratios of gamma functions whose arguments are linear combinations with rational coefficients of the parameters of the function.

Unfortunately, the search for a truly general formula of this type is a chimera, as a fairly recent (1983) result shows.

Let $g : \mathbb{C}^m \to \mathbb{C}$ and

$$ g(\boldsymbol{x}+n\cdot\boldsymbol{e}) = g(\boldsymbol{x}) + \delta\, n,\ \boldsymbol{e}=(1,1,\ldots,1), \tag{2.17} $$

where n is a non-negative integer and $\delta = \delta(g)$ is a rational number. Such a function will be called a *unicial.* Note any linear function with rational coefficients is a unicial.

Theorem. The unrestricted ${}_{k+1}F_k(1),\ k > 0$, can be written in the form

$$K\lambda^L \frac{\prod_{j=1}^{p} \Gamma(\xi_j)}{\prod_{j=1}^{q} \Gamma(\omega_j)}, \tag{2.18}$$

where K, L, ξ_j, ω_j are unicial functions of some or all of the parameters of the hypergeometric function and $K = 0$, if and only if $k = 1$. ∎

Of course, the above doesn't mean that ${}_3F_2(1)$ can't be written in terms of some functions other than gamma functions.

One can actually make an even stronger statement.

Theorem. Watson's theorem (2.2) cannot be generalized by means of the formula (2.4), i.e.,

$${}_3F_2\left(\begin{matrix} a, b, c \\ d, 2c \end{matrix}; 1\right), \tag{2.19}$$

cannot be represented by means of such a formula. ∎

The proofs of these results are based on the irreducibility[3] of a second order difference equation with polynomial coefficients, see [wim2] for the details.

3. Series of the type ${}_4F_3(1)$

These functions too are ubiquitous in mathematics. They arise, for instance, in the realization of the triple tensor product of irreducible representations of the group $SU(2)$. In this context the ${}_4F_3(1)$'s have integer parameters and are called Racah coefficients or, sometimes, Wigner 6-j symbols, see [vil], p. 532 ff.

They also arise in analysis, in the theory of orthogonal polynomials. The Jacobi polynomials, $P_n^{(\alpha,\beta)}(x),\ n = 0, 1, 2, \ldots$, satisfy a three term recurrence relation,

$$\begin{gathered} p_{n+1}(x) = [A_n + xB_n]p_n(x) - C_n p_{n-1}(x),\ n = 0, 1, 2, \ldots, \\ p_{-1}(x) = 0,\ p_0(x) = 1. \end{gathered} \tag{3.1}$$

[3] A linear homogeneous difference equation with coefficients in a ring is said to be irreducible if it has no solutions in common with an equation of the same type of lower order.

The coefficients in this recursion are rational functions of n,

$$\begin{aligned} A_n &= \frac{(2n+\gamma)(\alpha^2-\beta^2)}{2(n+1)(n+\gamma)(2n+\gamma-1)}, \\ B_n &= \frac{(2n+\gamma)(2n+\gamma+1)}{2(n+1)(n+\gamma)}, \\ C_n &= \frac{(n+\alpha)(n+\beta)(2n+\gamma+1)}{(n+1)(n+\gamma)(2n+\gamma-1)}, \\ &\text{where } \gamma=\alpha+\beta+1,\ \alpha>-1,\ \beta>-1. \end{aligned} \tag{3.2}$$

The polynomials are orthogonal with respect to the measure,

$$\mu(f)=\int_{-1}^{1} f(x)(1-x)^\alpha(1+x)^\beta dx. \tag{3.3}$$

In certain problems in numerical analysis it is important to analyze the properties of the polynomials satisfying a recursion relationship obtained from (3.1) by replacing n everywhere by $n+c$, where $c>0$ is called the *association parameter*:

$$\begin{gathered} p_{n+1}(x)=[A_{n+c}+xB_{n+c}]p_n(x)-C_{n+c}p_{n-1}(x),\ n=0,1,2,\ldots, \\ p_{-1}(x)=0,\ p_0(x)=1. \end{gathered} \tag{3.4}$$

These polynomials are also orthogonal with respect to a measure supported on $[-1,1]$, but the measure is vastly more complicated than the measure (3.3), see [wim4] for details. Of course, the measure reduces to (3.3) when the association parameter c is zero. The polynomials are denoted by $P_n^{(\alpha,\beta)}(x;c)$. It is more convenient to work with the polynomials shifted to the interval $[0,1]$. These are written

$$R_n^{(\alpha,\beta)}(x;c)=P_n^{(\alpha,\beta)}(2x-1;c). \tag{3.5}$$

Hendriksen recently, [hen1], [hen2], using an ingenious argument, has found the following expression for these polynomials:

$$\begin{aligned} R_n^{(\alpha,\beta)}(x;c) &= \frac{(\gamma+2c)_{2n}}{(\gamma+c)_n(c+1)_n\, n!} \\ &\times \text{Polynomial Part}\left\{x^n{}_2F_1\left(\begin{matrix} c, c+\beta \\ \gamma+2c-1\end{matrix};\frac{1}{x}\right){}_2F_1\left(\begin{matrix} -n-c,-n-c-\beta \\ 1-2n-2c-\gamma\end{matrix};\frac{1}{x}\right)\right\}. \end{aligned} \tag{3.6}$$

One can obtain a power series for the polynomials either by taking a Cauchy product in (3.6) or from the reference [wim4]:

$$R_n^{(\alpha,\beta)}(x;c) = \frac{(-1)^n(\gamma+2c)_n(\beta+c+1)_n}{(\gamma+c)_n\, n!}$$
$$\times \sum_{k=0}^{n} \frac{x^k(-n)_k(n+\gamma+2c)_k}{(c+1)_k(\beta+c+1)_k} \tag{3.7}$$
$$\times\ {}_4F_3\left(\begin{matrix} k-n, n+\gamma+k+2c, \beta+c, c \\ \beta+k+c+1, k+c+1, \gamma+2c-1 \end{matrix}; 1\right).$$

The above ${}_4F_3(1)$'s are a special case of a class of polynomials called *extended Jacobi polynomials,*

$${}_{p+2}F_{p+1}\left(\begin{matrix} -n, n+\lambda, [a_p] \\ [b_{p+1}] \end{matrix}; z\right), \quad n = 0, 1, 2, \ldots. \tag{3.8}$$

It is known that these polynomials satisfy a linear homogeneous recurrence relation in n (with coefficients which are polynomials in n) of order $p+2$, see [wim5], where explicit formulas for the coefficients in the equation are given. If $z = 1$, the order of the recurrence drops by 1 to order $p+1$. If, further, the function is Saalschützian, the order drops by 1 more to order p, as the work of Lewanowicz shows, who gave closed form expressions for the coefficients of the difference equation in this case, [lew1]. Lewanowicz's work has been somewhat neglected; as a matter of fact, concealed in his work is the recurrence relation for a very general class of polynomials now called Askey-Wilson polynomials.

It would be extremely interesting to know whether the coefficients in (3.7) could be expressed in closed form (see the definition in Section 1) or, to set our sights lower, to know for what important values of the parameters these ${}_4F_3(1)$ coefficients can be expressed in closed form. Herbert Wilf, utilizing the powerful algorithm "hyper" described in the book [pet] has demonstrated a remarkable result, [wil].

Theorem. (Wilf) The functions

$${}_4F_3\left(\begin{matrix} k-n, n+\gamma+k+2c, \beta+c, c \\ \beta+k+c+1, k+c+1, \gamma+2c-1 \end{matrix}; 1\right), \tag{3.9}$$

for general c, β, γ, and integer $k, n, 0 \le k \le n$, cannot be expressed in closed form. ∎

Wilf accomplished this result by applying "hyper" to the second order difference equation satisfied by the function (3.9). (Since the function is Saalschützian, the work of Lewanowicz guarantees that a recurrence of second order exists, [lew1].) In the important special case $\alpha = 0, \beta = 0, c = 1$ the difference equation takes the form

$$y(n) + A_1 y(n+1) + A_2 y(n+2) = 0,$$
$$A_1 = \frac{(2n+5)(n-k+2)(n+k+3)}{(n+2)(n-k+1)(n+k+2)}, \tag{3.10}$$
$$A_2 = \frac{(n+3)(n-k+3)(n+k+4)}{(n+2)(n-k+1)(n+k+2)}.$$

Wilf was able to show that in this case a closed form solution of this recurrence does exist

$$y^{(1)}(n) = \frac{1}{(n-k+1)(n+k+2)}. \tag{3.11}$$

Using this solution, one can construct another linearly independent solution using the technique in [wim6], p. 269. Adjusting the general solution to accommodate the initial conditions as $k = n, k = n-1$, gives an expression for the ${}_4F_3(1)$ that, while not quite hypergeometric, is certainly surprisingly simple. Denote the nth harmonic sum by

$$h_n = \begin{cases} \sum_{r=0}^{n} \frac{1}{r+1}, & r \geq 0, \\ 0, & r < 0. \end{cases} \tag{3.12}$$

Theorem. The shifted associated Legendre polynomials have the explicit expression,

$$R_n^{(0,0)}(x,1) = (-1)^n (n+1) \sum_{k=0}^{n} \frac{(-n)_k (n+2)_k (h_n - h_{k-1}) x^k}{(k!)^2 (n+1-k)}. \tag{3.13}$$

■

These polynomials, which are important in Padé approximation and numerical quadrature, have been widely studied (see [wim4] and the references given there and especially Lewanowicz, [lew2], who studies the more general associated polynomials); yet the expression above seems to be new.

4. Identities and the umbral calculus

In some recent work on Padé approximants, Bernhard Beckermann and I discovered that umbral calculus could be used to secure many new ${}_4F_3(1)$ identities. I feel this approach has been underutilized. Although the method seems exotic (its usage in the cited reference was vigorously opposed by my co-author), it is very slick. Once an identity is established it is true that it can be verified by analytical means, but this process is simply a verification, not a derivation. One has to know an identity to verify it!

A number of evaluations of ${}_4F_3(1)$ are known, see, for example, [ges],[car], and [abi]. Virtually all of these are for Saalschützian functions, however. Using the umbral approach one can derive identities for non-Saalschützian ${}_4F_3(1)$ and from these will follow a multitude of ${}_4F_3(1)$ evaluations as special cases. This approach is interesting in that it is based on a two element, rather than a one element, umbral calculus.

In this section I shall show how umbral calculus can be used to derive one such identity. By varying the approach, many others are possible.

Denote by $\mathcal{H}$ the linear space of formal series with complex coefficients

$$\mathcal{H} = \left\{ h(z) \,\middle|\, h(z) = \sum_{j=-\infty}^{\infty} A_j z^j, A_j \in \mathbb{C} \right\}. \tag{4.1}$$

Define addition and scalar multiplication on these series the usual way.

Define the following projection operator onto a finite dimensional subspace of $\mathcal{H}$,

$$\prod_{r,s}^{z}\{f(z)\} = \begin{cases} \sum_{j=r}^{s} A_j z^j, r \le s, \\ 0, r > s. \end{cases} \tag{4.2}$$

This operator is called the (r, s) cut of h. The special cut operator $\prod_{0,k}$ will be important. This is simply the kth partial sum of a "Taylor" series.

The following properties are easily verified:

$$\prod_{r,s}^{z}\{f(z)\} \mid_{z\to 1/z} = z^{r-s} \prod_{-r,s-2r}^{z} \{z^{s-r} f(1/z)\}, \tag{4.3}$$

$$\prod_{r,s}^{z}\{z^p f(z)\} = z^p \prod_{r-p,s-p}^{z} \{f(z)\}, \tag{4.4}$$

$$\prod_{r,s}^{z}\prod_{p,q}^{z} = \prod_{k,l}^{z}, \; k = \max\{r,p\}, \; l = \min\{s,q\}. \tag{4.5}$$

In addition to the above series, I will employ the two-element series

$$h(z,w) = \sum_{i,j=-\infty}^{\infty} A_{i,j} z^i w^j. \tag{4.6}$$

For these series the operators (4.2) commute:

$$\prod_{r,s}^{z} \prod_{t,u}^{w} = \prod_{t,u}^{w} \prod_{r,s}^{z}. \tag{4.7}$$

The subspace of $\mathcal{H}$ of those series containing only a finite number of negative powers, i.e., $A_j = 0$ except for a finite number of $j < 0$, is a field, call it $\mathcal{H}^+$. Multiplication is Cauchy multiplication of series and division is defined recursively by synthetic division. If the two series are of hypergeometric type one uses the following to form the product

$$\begin{aligned} &{}_rF_s\left(\begin{matrix}[a_r]\\ [b_s]\end{matrix};z\right) \times {}_mF_n\left(\begin{matrix}[c_m]\\ [d_n]\end{matrix};z\right) = \\ &\sum_{k=0}^{\infty} \frac{[(c_m)_k]z^k}{k![(d_n)_k]}\, {}_{n+r+1}F_{s+m}\left(\begin{matrix}-k, [a_r], [1-k-d_n]\\ [b_s], [1-k-c_m]\end{matrix};(-1)^{m+n+1}\right). \end{aligned} \tag{4.8}$$

A fact useful for future reference is that terminating hypergeometric series can be turned around:

$${}_{r+1}F_s\left(\begin{matrix}-k, [a_r]\\ [b_s]\end{matrix};z\right) = (-1)^k z^k \frac{[(a_r)_k]}{[(b_s)_k]}\, {}_{s+1}F_r\left(\begin{matrix}-k, [1-b_s-k]\\ [1-a_r-k]\end{matrix};\frac{(-1)^{r+s}}{z}\right). \tag{4.9}$$

The subspace of $\mathcal{H}$ consisting of series with only a finite number of positive powers is also a field, call it $\mathcal{H}^-$. When z is replaced by $1/z$ in (4.8) one has a similar formula in $\mathcal{H}^-$.

There exists natural isomorphism between the field of rational complex functions and the subfield of rational functions in $\mathcal{H}^+$ or $\mathcal{H}^-$.

$$h(z) = \sum_{j=-\infty}^{m} A_j z^j, \quad m \in \mathcal{Z}, \tag{4.10}$$

is rational if and only if there is a linear difference operator acting on j with constant coefficients which annihilates the sequence $\{A_j\}_{-\infty}^{m}$. If that is the case, there is a complex rational function to which the above series converges when z is a complex variable, $|z| > r$. Conversely, one may identify any such complex series with a series in $\mathcal{H}^-$ which represents a rational function. Analogous statements are true for series in $\mathcal{H}^+$.

One of our basic results involves an umbral calculus analog of the ratio of gamma functions and its representation by means of Gaussian hypergeometric functions.

Definition (umbral gamma functions).

$$\left(\frac{\Gamma(z+c)\Gamma(z+c-a-b)}{\Gamma(z+c-a)\Gamma(z+c-b-k)}\right)$$

$$\begin{aligned} &= z^k \exp \sum_{m=1}^{\infty} \frac{(-1)^{m+1}}{m(m+1)z^m} \left[B_{m+1}(c) + B_{m+1}(c-a-b)\right. \\ &- \left.B_{m+1}(c-a) - B_{m+1}(c-b-k)\right], \qquad k \in \mathcal{Z}. \end{aligned} \tag{4.11}$$

The series on the right is as an element of $\mathcal{H}^-$ constructed in the obvious way. (The $B_n(x)$ are Bernoulli polynomials, see [erd], Section 1.13.)

Theorem.

$${}_2F_1\left(\begin{matrix} a,\ b \\ c+z \end{matrix};1\right) = \left(\frac{\Gamma(z+c)\Gamma(z+c-a-b)}{\Gamma(z+c-a)\Gamma(z+c-b)}\right), \tag{4.12}$$

$${}_2F_1\left(\begin{matrix} a,\ b \\ c-z \end{matrix};1\right) = \left(\frac{\Gamma(z+a+1-c)\Gamma(z+b+1-c)}{\Gamma(z+1-c)\Gamma(z+a+b+1-c)}\right), \tag{4.13}$$

where all series are elements of $\mathcal{H}^-$.

Remark. The series in $\mathcal{H}^-$ comprising the left hand side of (4.12) is to be constructed in the following way:

$$\begin{aligned} {}_2F_1\left(\begin{matrix} a,\ b \\ c+z \end{matrix};1\right) &= \sum_{k=0}^{\infty} \frac{(a)_k(b)_k}{(c+z)_k k!} = \sum_{k=0}^{\infty} \frac{(a)_k(b)_k}{z^k k!} \sum_{m=0}^{\infty} \frac{\mu_m(c)}{z^m} \\ &= \sum_{m=0}^{\infty} \frac{1}{z^m} \sum_{k=0}^{m} \frac{(a)_k(b)_k \mu_{m-k}(c)}{k!} = \sum_{m=0}^{\infty} \frac{P_m(a)}{z^m}, \end{aligned} \tag{4.14}$$

where $P_m(a)$ is a polynomial in a of degree m. (Note the second equality is simply the definition of $\mu_k(c)$.)

Proof. The proof of these formulas is not trivial. See the reference [wim6]. ∎

The series inside the large parentheses on the right of (4.11) is the asymptotic series for $\dfrac{\Gamma(z+c)\Gamma(z+c-a-b)}{\Gamma(z+c-a)\Gamma(z+c-b)}$ when z is a real variable, $z \to \infty$. When either a or b is a negative integer, each series in (4.12),(4.13) represents a rational function given by the terminating ${}_2F_1$ on the left or, equivalently, by the appropriate ratio of gamma functions rewritten in terms of Pochhammer symbols. One might conclude that

$$ {}_2F_1\left({a,\ b \atop c+z} ;1\right) = \frac{\Gamma(z+c)\Gamma(z+c-a-b)}{\Gamma(z+c-a)\Gamma(z+c-b)}, \tag{4.15} $$

$$ {}_2F_1\left({a,\ b \atop c-z} ;1\right) = \frac{\Gamma(z+a+1-c)\Gamma(z+b+1-c)}{\Gamma(z+1-c)\Gamma(z+a+b+1-c)}. \tag{4.16} $$

Both these equations hold in $\mathcal{H}^-$, i.e., when parentheses are placed around the right hand sides and the left hand sides are considered to be series in $\mathcal{H}^-$ constructed in the obvious way. The first equation is also true when $z \in \mathbb{C}$, z not a pole of the right-hand side. However, the second is not, since the right-hand side does not provide the correct analytic continuation of (4.15).

In applying umbral calculus to hypergeometric series, one often encounters ratios of "gamma" functions like the right hand sides of (4.12),(4.13). They are always to be considered members of $\mathcal{H}^-$ as defined by (4.14).

I will utilize cut operators operating on products of Gaussian hypergeometric functions. The basic operand is the product

$$ \begin{aligned} f_{k,l,n}(a,b,\delta;z) &\equiv f(z) = {}_2F_1\left({a, \delta \atop \delta+b} ;z\right) \\ &\times {}_2F_1\left({1-a-k, -\delta-n \atop 1-\delta-b-l-n} ;z\right), \quad k,l \in \mathcal{Z}, \quad n \in \mathcal{N}. \end{aligned} \tag{4.17} $$

Note that when an Euler transformation (see [erd], v. 1, p. 64, (22)) is used on the first ${}_2F_1$ above, the result is a ${}_2F_1$ of the form (4.12) (where $z=\delta$) and thus a member of $\mathcal{H}^-$. Similarly, the second ${}_2F_1$, by (4.13), is also a member of $\mathcal{H}^-$.

Define

$$ C = (-1)^k (a)_k (b)_l. \tag{4.18} $$

Theorem. For $k \ge l \ge 0$ or for $0 \ge k \ge l$,

$$ \begin{aligned} (\delta+b)_{n+l} &\prod_{k,n+l}^{z} \{f_{k,l,n}(a,b,\delta;z)\} \\ &= C(1+\delta-a)_{n-k} \prod_{k,n+l}^{z} \{z^n f_{-l,-k,n}(1-b,1-a,\delta;1/z)\}. \end{aligned} \tag{4.19} $$

Proof. First, assume $k \geq l \geq 0$. Obviously, one can assume that $n + l \geq k$, else the result is trivial. Denote the left hand side of (4.19) by G, the right hand side by H. Then for $\delta \in \mathcal{Z}, -n \leq \delta \leq 0$, $G = H$. (Just turn both series around, using (4.9).) I can write

$$G = H + V(\delta)_{n+1}. \tag{4.20}$$

Also,

$$H = C \prod_{k,n+l}^{z} \left\{ \sum_{m=0}^{\infty} z^{n-m} \sum_{j=0}^{m} C_{m,j}(\delta)_j(-\delta-n)_{m-j}(1+\delta+j-a)_{n-m-k} \right\}, \tag{4.21}$$

where $C_{m,j}$ is independent of δ and z. H is a polynomial of degree at most $n - k$ in δ if $k \leq n$, and is identically zero if $k > n$. One can similarly show G is a polynomial of degree at most $n + l$. Thus V is a polynomial of degree at most $l - 1$. The result is immediate if $l = 0$. Assume $l > 0$. Obviously

$$V = \prod_{0,l-1}^{\delta} \left\{ \prod_{k,n+l}^{z} \left\{ \frac{(\delta + b)_{n+l}}{(\delta)_{n+1}} (1-z)^{k-1} \sum_{m=0}^{\infty} A_m(\delta) \left(\frac{z}{z-1} \right)^m \right\} \right\}, \tag{4.22}$$
$$A_m(\delta) = 0(\delta^{-m}).$$

The δ series are considered members of H^-, the z series members of $\mathcal{H}^+$. In the previous step I have used Euler-type transformations on the ${}_2F_1$'s, see [erd], v. 1, p. 64, (22). It is easily shown that these transformations are valid in $\mathcal{H}^+$. Because of the order estimate on A_m one has

$$V = \prod_{0,l-1}^{\delta} \left\{ \prod_{k,n+l}^{z} \left\{ \frac{(\delta+b)_{n+l}}{(\delta)_{n+1}} \sum_{m=0}^{l-1} A_m(\delta)(z)^m (1 - z)^{k-m-1} \right\} \right\}. \tag{4.23}$$

(This step utilizes commutativity of the cut operators.) Since $k \geq 1$ the quantity in the inner brackets is a polynomial in z of degree at most $k - 1$, and consequently the above expression is zero. The first statement of the theorem is established.

The second statement follows from the first. Let $z \to 1/z$ and use property (4.3) of the cut operator, then property (4.4) with $p = l - k$. Finally, replace k by $-l, l$ by $-k, a$ by $1 - b, b$ by $1 - a$ and interchange the two sides of the equation. ■

Explicit polynomial expressions for the quantities $\prod_{r,s}^{z}$ can be obtained by using the formula (4.8) for multiplying hypergeometric series. Multiplying out the relevant hypergeometric functions gives

$$z^{-r}\prod_{r,r}^{z}\{f_{k,l,n}(a,b,\delta;z)\} = \frac{(1-a-k)_r(-\delta-n)_r}{r!(1-\delta-b-l-n)_r} \times {}_4F_3\left(\begin{matrix}-r, a, \delta, \delta+b+l+n-r \\ \delta+b, a+k-r, \delta+n+1-r\end{matrix};1\right). \tag{4.24}$$

When all cut operators in (4.19) are interpreted as ${}_4F_3(1)$'s via the above equation, the result is the ${}_4F_3(1)$ identity

$${}_4F_3\left(\begin{matrix}-r, a, \delta, \delta+b+l+n-r \\ \delta+b, a+k-r, \delta+n+1-r\end{matrix};1\right) = \frac{(-1)^{k+r}r!(\delta+1)_{n-r}(b)_l(b+l)_{n-r}(\delta+1-a)_{r-k}}{(n-r)!(\delta+1)_r(a+k-r)_{r-k}(\delta+b)_{l+n-r}} \times {}_4F_3\left(\begin{matrix}r-n, 1-b, \delta, \delta+1-a-k+r \\ \delta+1-a, 1-b-l-n+r, \delta+1+r\end{matrix};1\right), \quad 0 \le l \le k \le r \le n+l. \tag{4.25}$$

(One interprets the right hand side to be zero when $r > n$.) Note the functions involved are not Saalschützian, so this identity cannot be derived from known ${}_4F_3(1)$ identities, such as those given in [bai], p. 56.

Many evaluations of ${}_4F_3(1)$ will result when the parameters in (4.25) are assigned special values, for instance, let $b = 1$. One finds

$${}_4F_3\left(\begin{matrix}-r, a, \delta, \delta+l+n-r \\ \delta+1, a+k-r, \delta+n+1-r\end{matrix};1\right) = \frac{(-1)^{k+r}r!(\delta+1)_{n-r}(n-r+l)!(\delta+1-a)_{r-k}}{(n-r)!(\delta+1)_r(a+k-r)_{r-k}(\delta+1)_{l+n-r}}, \quad 0 \le l \le k \le r \le n+l. \tag{4.26}$$

Conclusions

Identities like (4.25) are useful in constructing what are called two-point Padé approximants, that is, rational expressions which approximate a given analytic function to one order at $z = 0$ and a related function to a different order at $z = \infty$, see [wim6] and the references given there.

The umbral calculus leads one effortlessly to hypergeometric identities, but one has to know what to start with. Probably the most promising approach is to start with hypergeometric products more general than (4.17).

A referee, to whom I am indebted for a scrupulous reading of this manuscript, has furnished me with additional information about results pertaining to cut operators. Roman has studied the space $\mathcal{H}^-$ in [rom1], [rom2]. Ueno, in [uen1], [uen2], developed powerful operator methods for obtaining hypergeometric identities. This suggests that another way of extending the results of section 4 of this paper is to consider more general operators.

References

[1] [abi] Abiodun, R.F.A., An application of Carlitz's formula, J. Math. Anal. Appl. **70**, 114–119 (1979).

[2] [ask] Askey, Richard, A note on the history of series, MRC Technical Summary Report #1532, March 1975, University of Wisconsin.

[3] [bai] Bailey, W.N., Generalized hypergeometric series, Cambridge University Press, Cambridge, UK (1935).

[4] [car] Carlitz, L., Summation of a special ${}_4F_3(1)$, Boll. Un. Math. Ital. **18**, 90–93 (1963).

[5] [dut] Dutkas, Jacques, The early history of the hypergeometric function, Arch. Hist. Exact Sci. **31**, 15–34 (1984).

[6] [eig] Eigen, M., Diffusion control in biochemical reactions, in Quantum statistical mechanics in the natural sciences, S.L. Mintz and S.N. Widmayer ed., Plenum Press, New York (1974).

[7] [erd] Erdélyi, A., et al, Higher transcendental functions, 3 v., McGraw-Hill, NY (1953).

[8] [fis] Fisher, M.E., Walks, walls, wetting and melting, J. Stat. Phys. **34**, 667–729 (1984).

[9] [ges] Gessel, Ira, and Stanton, Dennis, Strange evaluations of hypergeometric series, SIAM J. Math. Anal. **13**, 295–308 (1982).

[10] [hen1] Hendriksen, E., Associated Jacobi Laurent polynomials, J. Comp. Appl. Math. **32**, 125–141 (1990).

[11] [hen2] Hendriksen, E., A weight function for the associated Jacobi Laurent polynomials, J. Comp. Appl. Math. **33**, 171–180 (1990).

[12] [husl] Huse, D.A., Szpilka, A.M., and Fisher, M.E., Melting and wetting transitions in the three-state chiral clock model, Physica 121A, 363–398 (1983).

[13] [hus2] Huse, D.A., and Fisher, M.E., Commensurate melting, domain walls, and dislocations, Physical Review B 29, 239–270 (1984).

[14] [koe] Koepf, W., Algorithms for m-fold Hypergeometric Summation, J. Symbolic Computation **20**, 399–417 (1995).

[15] [lew1] Lewanowicz, S., Recurrence relations for hypergeometric functions of unit argument, Math. Comp. **45**, 521–535 (1985).

[16] [lew2] Lewanowicz, S., Results on the associated Jacobi and Gegenbauer polynomials, J. Comput. Appl. Math. **49**, 137–143 (1993).

[17] [mcc] McClosky, M.A., and Poo, M., Rates of membrane-associated reactions: reduction of dimensionality revisited, J. Cell Biology **102**, 88 (1986).

[18] [pet] Petkovsek, M., Wilf, H., and Zeilberger, D., "$A = B$", A.K. Peters Publishing, Wellesley, MS (1996).

[19] [rom1] Roman, S.M., The algebra of formal series, Adv. Math. **31**, 309–329 (1979) (erratum in Adv. Math. **35**, 274 (1980)).

[20] [rom2] Roman, S.M., The algebra of formal series. II. Sheffer sequences, J. Math. Anal. Appl. **74**, 120–143 (1980).

[21] [tru] Trurnit, J.J., Über monomoleculare Filme an Wassergrenzflechen und über Schichtfilme, Fortschr. Chem. Org. Naturst. **4**, 347–476 (1953).

[22] [uen1] Ueno, K., Umbral calculus and special functions, Adv. Math. **67**, 174–229 (1988).

[23] [uen2] Ueno, K., Hypergeometric series formulas through operator calculus, Funcialja Ekvaciog **33**, 493–518 (1990).

[24] [vil] Vilenkin, N. Ja., and Klimyk, A.U., "Representation of Lie groups and special functions," v. 1, Kluwer Academic Publishers, The Netherlands (1991).

[25] [wal] Wallace, John, "A treatise of algebra," London (1685).

[26] [wat] Watson, G.N., "A treatise on the theory of Bessel functions," 2nd Edition, Cambridge University Press, Cambridge, UK, p. 100, (1962).

[27] [whi] Whittaker, E.T., and Watson, G.N., "A course in modern analysis," 4th edition expanded, Cambridge University Press, Cambridge, UK (1962).

[28] [wil] Wilf, H., personal communication to the author, April 4 (1996).

[29] [wim1] Wimp, Jet, and Zeilberger, Doron, How likely is Polya's drunkard to stay in $x \geq y \geq z$?, J. Stat. Phys. **56**, 1129–1135 (1989).

[30] [wim2] Wimp, Jet, Irreducible recurrences and representation theorems for ${}_3F_2(1)$, Comp. & Maths. with Appls. **9**, 669–678 (1983).

[31] [wim3] Wimp, Jet, The computation of ${}_3F_2(1)$, Inter. J. Computer Math. **10**, 55–62 (1981).

[32] [wim4] Wimp, Jet, Explicit formulas for the associated Jacobi polynomials and some applications, Can. J. Math. **439**, 983–1000 (1987).

[33] [wim5] Wimp, Jet, Recursion formulas for hypergeometric functions, Math. Comp. **21**, 363–373 (1967).

[34] [wim6] Wimp, Jet, and Beckermann, B., Families of two-point Padé approximants and some ${}_4F_3(1)$ identities, SIAM J. Appl. Math. **26**, 761–773 (1995).

[35] [wim7] Wimp, Jet, "Computation with recurrence relations," Pitman Press, London (1984).

Department of Mathematics and Computer Science
Drexel University
Philadelphia, PA 19104

I

We are the Rota nerds
We are the math nerds
Leaning together
Headpiece filled with theorems. Alas!
Our dried voices, when
We whisper together
Speak utter nonsense
To sensible people
But to us they betray
An unknown beauty.

Abstractions without form, identities without reason,
Paralysed intellect, theory without foundation.

Those who have crossed
With direct eyes, to Gian-Carlo's office
Remember us–if at all–not as lost
Absent-minded souls, but only
As the math nerds
The Rota nerds.

II

A siren I cannot capture with formulas
In evening's dream kingdom
Never seen directly
There, diaphanous fabric flutters
Sunlight on rippled muscles
There, alabaster fingers beckon
And voices are
In the wind's singing
More delicate and more rarified
Than a fading nebula.

Let me be nearer
In evening's dream kingdom
Let me also wear
Such deliberate disguises
Editor, philosopher, professor
In a tower
Behaving as the wind behaves
Only to be nearer–

Yet not so near,
That intimacy brings madness.

III

This is the learned land
This is the academic land
In the light of the day
Are gathered supplicants, here to receive
A blessing from the esteemed man's lips
And bowing, the glamour of fame.

Is it like this
In the dark of the night
Sitting alone
At the hour when car alarms
Shatter the night
Lips that would kiss
Form conjectures to seductive muses.

IV

His proselytes are here
His fellow disciples are here
In this valley of mathematical minds
In these hallowed halls
This spawning ground of our theories.

In this meeting place of 64
We gather together
And make toasts
Gathered by the basin near the tumid river.

Nonsensical, unless
Harmony appears
And encapsulated action
As a multifaceted friend
Then enter the combinatorial kingdom
The hope only
Of math nerds.

V

Here we have a tax break,
A tax break, a tax break,
Here we have a tax break,
Overflowing with books.

Between the chocolate
And the caviar
Between the coca cola
And the champagne
Falls the Rota

Truth is not a teddy bear

Between the pencil
And the Macintosh
Between the philosophy
And the mathematics
Falls the Rota

This is Mickey Mouse

Between the generating function
And the cryptomorphism
Between the finite operator calculus
And the inventory
Between the endomorphism
And the Mobius function
Falls the Rota

Show me something beautiful

This is the way the Rotafest ends
This is the way the Rotafest ends
This is the way the Rotafest ends
Not with a book review but a poem.

Progress in Mathematics

Edited by:

Progress in Mathematics is a series of books intended for professional mathematicians and scientists, encompassing all areas of pure mathematics. This distinguished series, which began in 1979, includes authored monographs and edited collections of papers on important research developments as well as expositions of particular subject areas.

We encourage preparation of manuscripts in some form of TEX for delivery in camera-ready copy which leads to rapid publication, or in electronic form for interfacing with laser printers or typesetters.

Proposals should be sent directly to the editors or to: Birkhäuser Boston, 675 Massachusetts Avenue, Cambridge, MA 02139, U. S. A.

100 TAYLOR. Pseudodifferential Operators and Nonlinear PDE
101 BARKER/SALLY (eds). Harmonic Analysis on Reductive Groups
102 DAVID (ed). Séminaire de Théorie des Nombres, Paris 1989-90
103 ANGER /PORTENIER. Radon Integrals
104 ADAMS /BARBASCH/VOGAN. The Langlands Classification and Irreducible Characters for Real Reductive Groups
105 TIRAO/WALLACH (eds). New Developments in Lie Theory and Their Applications
106 BUSER. Geometry and Spectra of Compact Riemann Surfaces
107 BRYLINSKI. Loop Spaces, Characteristic Classes and Geometric Quantization
108 DAVID (ed). Séminaire de Théorie des Nombres, Paris 1990-91
109 EYSSETTE/GALLIGO (eds). Computational Algebraic Geometry
110 LUSZTIG. Introduction to Quantum Groups
111 SCHWARZ. Morse Homology
112 DONG/LEPOWSKY. Generalized Vertex Algebras and Relative Vertex Operators
113 MOEGLIN/WALDSPURGER. Décomposition spectrale et séries d'Eisenstein
114 BERENSTEIN/GAY/VIDRAS/YGER. Residue Currents and Bezout Identities
115 BABELON/CARTIER/KOSMANN-SCHWARZBACH (eds). Integrable Systems, The Verdier Memorial Conference: Actes du Colloque International de Luminy
116 DAVID (ed). Séminaire de Théorie des Nombres, Paris 1991-92
117 AUDIN/LAFONTAINE (eds). Holomorphic Curves in Symplectic Geometry
118 VAISMAN. Lectures on the Geometry of Poisson Manifolds
119 JOSEPH/ MEURAT/MIGNON/PRUM/ RENTSCHLER (eds). First European Congress of Mathematics, July, 1992, Vol. I
120 JOSEPH/ MEURAT/MIGNON/PRUM/ RENTSCHLER (eds). First European Congress of Mathematics, July, 1992, Vol. II
121 JOSEPH/ MEURAT/MIGNON/PRUM/ RENTSCHLER (eds). First European Congress of Mathematics, July, 1992, Vol. III (Round Tables)
122 GUILLEMIN. Moment Maps and Combinatorial Invariants of T^n-spaces

123 Brylinski/Brylinski/Guillemin/Kac (eds). Lie Theory and Geometry: In Honor of Bertram Kostant
124 Aebischer/Borer/Kalin/Leuenberger/ Reimann (eds). Symplectic Geometry
125 Lubotzky. Discrete Groups, Expanding Graphs and Invariant Measures
126 Riesel. Prime Numbers and Computer Methods for Factorization
127 Hörmander. Notions of Convexity
128 Schmidt. Dynamical Systems of Algebraic Origin
129 Dijkgraaf/Faber/Van der Geer (eds). The Moduli Space of Curves
130 Duistermaat. Fourier Integral Operators
131 Gindikin/Lepowsky/Wilson (eds). Functional Analysis on the Eve of the 21st Century. In Honor of the Eightieth Birthday of I. M. Gelfand, Vol. 1
132 Gindikin/Lepowsky/Wilson (eds.) Functional Analysis on the Eve of the 21st Century. In Honor of the Eightieth Birthday of I. M. Gelfand, Vol. 2
133 Hofer/Taubes/Weinstein/Zehnder (eds). The Floer Memorial Volume
134 Campillo Lopez/Narvaez Macarro (eds) Algebraic Geometry and Singularities
135 Amrein/Boutet de Monvel/Georgescu. C_0-Groups, Commutator Methods and Spectral Theory of N-Body Hamiltonians
136 Broto/Casacuberta/Mislin (eds). Algebraic Topology: New Trends in Localization and Periodicity
137 Vigneras. Représentations l-modulaires d'un groupe réductif p-adique avec $l \neq p$
138 Berndt/Diamond/Hildebrand (eds). Analytic Number Theory, Vol. 1 In Honor of Heini Halberstam
139 Berndt/Diamond/Hildebrand (eds). Analytic Number Theory, Vol. 2 In Honor of Heini Halberstam
140 Knapp. Lie Groups Beyond an Introduction
141 Cabanes (eds). Finite Reductive Groups: Related Structures and Representations
142 Monk. Cardinal Invariants on Boolean Algebras
143 Gonzalez-Vega/Recio (eds). Algorithms in Algebraic Geometry and Applications
144 Bellaïche/Risler (eds). Sub-Riemannian Geometry
145 Albert/Brouzet/Dufour (eds). Integrable Systems and Foliations Feuilletages et Systèmes Intégrables
146 Jardine. Generalized Etale Cohomology
147 DiBiase. Fatou TypeTheorems. Maximal Functions and Approach Regions
148 Huang. Two-Dimensional Conformal Geometry and Vertex Operator Algebras
149 Souriau. Structure of Dynamical Systems. A Symplectic View of Physics
150 Shiota. Geometry of Subanalytic and Semialgebraic Sets
151 Hummel. Gromov's Compactness Theorem For Pseudo-holomorphic Curves
152 Gromov. Metric Structures for Riemannian and Non-Riemannian Spaces
153 Buescu. Exotic Attractors: From Liapunov Stability to Riddled Basins
154 Böttcher/Karlovich. Carleson Curves, Muckenhoupt Weights, and Toeplitz Operators
155 Dragomir/Ornea. Locally Conformal Kähler Geometry
156 Guivarc'h/Ji/Taylor. Compactifications of Symmetric Spaces
157 Murty/Murty. Non-vanishing of L-functions and Applications
158 Tirao/Vogan/Wolf (eds). Geometry and Representation Theory of Real and p-adic Groups
159 Thangavelu. Harmonic Analysis on the Heisenberg Group

GETTYSBURG COLLEGE
3326800 0380400 6